Applied Science

Applied Science

Volume 3

Editor

Donald R. Franceschetti, Ph.D.

The University of Memphis

SALEM PRESS
A Division of EBSCO Publishing

Ipswich, Massachusetts   Hackensack, New Jersey

Library of Congress Cataloging-in-Publication Data

Applied science / editor, Donald R. Franceschetti.
  p. cm.
 ISBN 978-1-58765-781-8 (set) — ISBN 978-1-58765-782-5 (v. 1) — ISBN 978-1-58765-783-2 (v. 2) — ISBN 978-1-58765-784-9 (v. 3) — ISBN 978-1-58765-785-6 (v. 4) — ISBN 978-1-58765-786-3 (v. 5)
 1. Engineering. 2. Technology. I. Franceschetti, Donald R., 1947-
  600--dc23

2012002375

FIRST PRINTING
PRINTED IN THE UNITED STATES OF AMERICA

CONTENTS

COMMON UNITS OF MEASURE

Common prefixes for metric units—which may apply in more cases than shown below—include *giga-* (1 billion times the unit), *mega-* (one million times), *kilo-* (1,000 times), *hecto-* (100 times), *deka-* (10 times), *deci-* (0.1 times, or one tenth), *centi-* (0.01, or one hundredth), *milli-* (0.001, or one thousandth), and *micro-* (0.0001, or one millionth).

Unit	Quantity	Symbol	Equivalents
Acre	Area	ac	43,560 square feet 4,840 square yards 0.405 hectare
Ampere	Electric current	A *or* amp	1.00016502722949 international ampere 0.1 biot *or* abampere
Angstrom	Length	Å	0.1 nanometer 0.0000001 millimeter 0.000000004 inch
Astronomical unit	Length	AU	92,955,807 miles 149,597,871 kilometers (mean Earth-Sun distance)
Barn	Area	b	10^{-28} meters squared (approx. cross-sectional area of 1 uranium nucleus)
Barrel (dry, for most produce)	Volume/capacity	bbl	7,056 cubic inches; 105 dry quarts; 3.281 bushels, struck measure
Barrel (liquid)	Volume/capacity	bbl	31 to 42 gallons
British thermal unit	Energy	Btu	1055.05585262 joule
Bushel (U.S., heaped)	Volume/capacity	bsh *or* bu	2,747.715 cubic inches 1.278 bushels, struck measure
Bushel (U.S., struck measure)	Volume/capacity	bsh *or* bu	2,150.42 cubic inches 35.238 liters
Candela	Luminous intensity	cd	1.09 hefner candle
Celsius	Temperature	C	1° centigrade
Centigram	Mass/weight	cg	0.15 grain
Centimeter	Length	cm	0.3937 inch
Centimeter, cubic	Volume/capacity	cm³	0.061 cubic inch
Centimeter, square	Area	cm²	0.155 square inch
Coulomb	Electric charge	C	1 ampere second
Cup	Volume/capacity	C	250 milliliters 8 fluid ounces 0.5 liquid pint

Unit	Quantity	Symbol	Equivalents
Deciliter	Volume/capacity	dl	0.21 pint
Decimeter	Length	dm	3.937 inches
Decimeter, cubic	Volume/capacity	dm³	61.024 cubic inches
Decimeter, square	Area	dm²	15.5 square inches
Dekaliter	Volume/capacity	dal	2.642 gallons 1.135 pecks
Dekameter	Length	dam	32.808 feet
Dram	Mass/weight	dr *or* dr avdp	0.0625 ounce 27.344 grains 1.772 grams
Electron volt	Energy	eV	$1.5185847232839 \times 10^{-22}$ Btus $1.6021917 \times 10^{-19}$ joules
Fermi	Length	fm	1 femtometer 1.0×10^{-15} meters
Foot	Length	ft *or* '	12 inches 0.3048 meter 30.48 centimeters
Foot, square	Area	ft²	929.030 square centimeters
Foot, cubic	Volume/capacity	ft³	0.028 cubic meter 0.0370 cubic yard 1,728 cubic inches
Gallon (British Imperial)	Volume/capacity	gal	277.42 cubic inches 1.201 U.S. gallons 4.546 liters 160 British fluid ounces
Gallon (U.S.)	Volume/capacity	gal	231 cubic inches 3.785 liters 0.833 British gallon 128 U.S. fluid ounces
Giga-electron volt	Energy	GeV	$1.6021917 \times 10^{-10}$ joule
Gigahertz	Frequency	GHz	—
Gill	Volume/capacity	gi	7.219 cubic inches 4 fluid ounces 0.118 liter
Grain	Mass/weight	gr	0.037 dram 0.002083 ounce 0.0648 gram
Gram	Mass/weight	g	15.432 grains 0.035 avoirdupois ounce

Unit	Quantity	Symbol	Equivalents
Hectare	Area	ha	2.471 acres
Hectoliter	Volume/capacity	hl	26.418 gallons 2.838 bushels
Hertz	Frequency	Hz	$1.08782775707767 \times 10^{-10}$ cesium atom frequency
Hour	Time	h	60 minutes 3,600 seconds
Inch	Length	in or "	2.54 centimeters
Inch, cubic	Volume/capacity	in^3	0.554 fluid ounce 4.433 fluid drams 16.387 cubic centimeters
Inch, square	Area	in^2	6.4516 square centimeters
Joule	Energy	J	$6.2414503832469 \times 10^{18}$ electron volt
Joule per kelvin	Heat capacity	J/K	$7.24311216248908 \times 10^{22}$ Boltzmann constant
Joule per second	Power	J/s	1 watt
Kelvin	Temperature	K	-272.15° Celsius
Kilo-electron volt	Energy	keV	$1.5185847232839 \times 10^{-19}$ joule
Kilogram	Mass/weight	kg	2.205 pounds
Kilogram per cubic meter	Mass/weight density	kg/m^3	$5.78036672001339 \times 10^{-4}$ ounces per cubic inch
Kilohertz	Frequency	kHz	—
Kiloliter	Volume/capacity	kl	—
Kilometer	Length	km	0.621 mile
Kilometer, square	Area	km^2	0.386 square mile 247.105 acres
Light-year (distance traveled by light in one Earth year)	Length/distance	lt-yr	5,878,499,814,275.88 miles 9.46×10^{12} kilometers
Liter	Volume/capacity	L	1.057 liquid quarts 0.908 dry quart 61.024 cubic inches
Mega-electron volt	Energy	MeV	—
Megahertz	Frequency	MHz	—
Meter	Length	m	39.37 inches
Meter, cubic	Volume/capacity	m^3	1.308 cubic yards

Unit	Quantity	Symbol	Equivalents
Meter per second	Velocity	m/s	2.24 miles per hour 3.60 kilometers per hour
Meter per second per second	Acceleration	m/s²	12,960.00 kilometers per hour per hour 8,052.97 miles per hour per hour
Meter, square	Area	m²	1.196 square yards 10.764 square feet
Metric. *See* unit name			
Microgram	Mass/weight	mcg *or* μg	0.000001 gram
Microliter	Volume/capacity	μl	0.00027 fluid ounce
Micrometer	Length	μm	0.001 millimeter 0.00003937 inch
Mile (nautical international)	Length	mi	1.852 kilometers 1.151 statute miles 0.999 U.S. nautical miles
Mile (statute or land)	Length	mi	5,280 feet 1.609 kilometers
Mile, square	Area	mi²	258.999 hectares
Milligram	Mass/weight	mg	0.015 grain
Milliliter	Volume/capacity	ml	0.271 fluid dram 16.231 minims 0.061 cubic inch
Millimeter	Length	mm	0.03937 inch
Millimeter, square	Area	mm²	0.002 square inch
Minute	Time	m	60 seconds
Mole	Amount of substance	mol	6.02×10^{23} atoms or molecules of a given substance
Nanometer	Length	nm	1,000,000 fermis 10 angstroms 0.001 micrometer 0.00000003937 inch
Newton	Force	N	0.224808943099711 pound force 0.101971621297793 kilogram force 100,000 dynes
Newton meter	Torque	N·m	0.7375621 foot-pound
Ounce (avoirdupois)	Mass/weight	oz	28.350 grams 437.5 grains 0.911 troy or apothecaries' ounce

Unit	Quantity	Symbol	Equivalents
Ounce (troy)	Mass/weight	oz	31.103 grams 480 grains 1.097 avoirdupois ounces
Ounce (U.S., fluid or liquid)	Mass/weight	oz	1.805 cubic inch 29.574 milliliters 1.041 British fluid ounces
Parsec	Length	pc	30,856,775,876,793 kilometers 19,173,511,615,163 miles
Peck	Volume/capacity	pk	8.810 liters
Pint (dry)	Volume/capacity	pt	33.600 cubic inches 0.551 liter
Pint (liquid)	Volume/capacity	pt	28.875 cubic inches 0.473 liter
Pound (avoirdupois)	Mass/weight	lb	7,000 grains 1.215 troy or apothecaries' pounds 453.59237 grams
Pound (troy)	Mass/weight	lb	5,760 grains 0.823 avoirdupois pound 373.242 grams
Quart (British)	Volume/capacity	qt	69.354 cubic inches 1.032 U.S. dry quarts 1.201 U.S. liquid quarts
Quart (U.S., dry)	Volume/capacity	qt	67.201 cubic inches 1.101 liters 0.969 British quart
Quart (U.S., liquid)	Volume/capacity	qt	57.75 cubic inches 0.946 liter 0.833 British quart
Rod	Length	rd	5.029 meters 5.50 yards
Rod, square	Area	rd^2	25.293 square meters 30.25 square yards 0.00625 acre
Second	Time	s or sec	$1/60$ minute $1/3600$ hour
Tablespoon	Volume/capacity	T or tb	3 teaspoons 4 fluid drams
Teaspoon	Volume/capacity	t or tsp	0.33 tablespoon 1.33 fluid drams

Unit	Quantity	Symbol	Equivalents
Ton (gross or long)	Mass/weight	t	2,240 pounds 1.12 net tons 1.016 metric tons
Ton (metric)	Mass/weight	t	1,000 kilograms 2,204.62 pounds 0.984 gross ton 1.102 net tons
Ton (net or short)	Mass/weight	t	2,000 pounds 0.893 gross ton 0.907 metric ton
Volt	Electric potential	V	1 joule per coulomb
Watt	Power	W	1 joule per second 0.001 kilowatt $2.84345136093995 \times 10^{-4}$ ton of refrigeration
Yard	Length	yd	0.9144 meter
Yard, cubic	Volume/capacity	yd³	0.765 cubic meter
Yard, square	Area	yd²	0.836 square meter

COMPLETE LIST OF CONTENTS

Volume 1

Volume 2

Volume 3

Volume 4

Volume 5

Contents v
Common Units of Measure vii
Complete List of Contents xiii

Applied Science

G

GAME THEORY

FIELDS OF STUDY

Economics; finance; political science; biology; psychology; decision science; mathematics.

SUMMARY

Game theory is a tool that has come to be used to explain and predict decision making in a variety of fields. It is used to explain and predict both human and nonhuman behavior as well as the behavior of larger entities such as nation states. Game theory is often quite complex mathematically, especially in new fields such as evolutionary game theory.

KEY TERMS AND CONCEPTS

- **Cooperative Game:** Game in which rational actors find a means to cooperate so as to enhance the utility of each.
- **Dominant Strategy:** Strategy that gives a player a higher payoff than other strategies no matter what the opponent does.
- **Evolutionary Game:** Game that changes over time as players' strategies change and as payoffs change. These games have multiple equilibria.
- **Multiple Equilibria:** Many potential equilibria that have no clearly superior strategies.
- **Nash Equilibrium:** Position in which both players have selected a strategy and neither player can change his or her strategy independently without ending up in a worse position.
- **Noncooperative Game:** Game in which rational actors will not cooperate even though doing so may be to their advantage.
- **Repeated Game:** Game that occurs through iterations, either indefinitely or with a finite ending point.
- **Utility:** Benefit that a player achieves from following a strategy.

DEFINITION AND BASIC PRINCIPLES

Game theory is a means of modeling individual decision making when the decisions are interdependent and when the actors are aware of this interdependence. Individual actors (which may be defined as people, firms, and nations) are assumed to maximize their own utility—that is, to act in ways that will provide the greatest benefits to them. Games take a variety of forms such as cooperative or noncooperative; they may be played one time only or repeated either indefinitely or with a finite ending point, and they may involve two or more players. Game theory is used by strategists in a variety of settings.

The underlying mathematical assumptions found in much of game theory are often difficult to understand. However, the explanations drawn from game theory are often intuitive, and people can engage in decision making based on game theory without understanding the underlying mathematics. The formal approach to decision making that is part of game theory enables decision makers to clarify their options and enhance their ability to maximize their utility.

Classical game theory has quite rigid assumptions that govern the decision process. Some of these assumptions are so rigid that critics have argued that game theory has few applications beyond controlled settings.

Later game theorists developed approaches to strategy that have made game theory applicable in many decision-making applications. Advances such as evolutionary game theory take human learning into account, and behavioral game theory factors emotion into decision making in such a way that it can be modeled and predictions can be made.

BACKGROUND AND HISTORY

The mathematical theory of games was developed by the mathematician John von Neumann and the economist Oskar Morgenstern, who published *The*

Theory of Games and Economic Behavior in 1944. Various scholars have contributed to the development of game theory, and in the process, it has become useful for scholars and practitioners in a variety of fields, although economics and finance are the disciplines most commonly associated with game theory.

Some of the seminal work in game theory was done by the mathematician John Nash in the early 1950's, with later scholars building on his work. Nash along with John Harsanyi and Reinhard Selten received a Nobel Prize in Economic Sciences in 1994 for their work in game theory. The political scientist Thomas Schelling received a Nobel Prize in Economic Sciences in 2005 for his work in predicting the outcomes of international conflicts.

How It Works

Game theory is the application of mathematical reasoning to decision making to provide quantitative estimates of the utilities of game players. Implicit in the decision-making process is the assumption that all actors act in a self-interested fashion so as to maximize their own utilities. Rationality, rather than altruism or cooperation, is a governing principle in much of game theory. Actors are expected to cooperate only when doing so benefits their own self-interest.

Noncooperative Games. Most games are viewed as noncooperative in that the players act only in their own self-interest and will not cooperate, even when doing so might lead to a superior outcome. A well-known noncooperative game is the prisoners' dilemma. In this game, the police have arrested Moe and Joe, two small-time criminals, for burglary. The police have enough information to convict the two men for possession of burglar's tools, a crime that carries a sentence of five years in prison. They want to convict the two men for burglary, which carries a ten-year sentence, but this requires at least one man to confess and testify against the other. It would be in each man's best interest to remain silent, as this will result in only a five-year sentence for each. The police separate Moe and Joe. Officers tell Moe that if he confesses, he will receive a reduced sentence (three years), but if Joe confesses first, Moe will receive a harsher sentence (twelve years). At the same time, other officers give Joe the same options. The police rely on the self-interest of each criminal to lead him to confess, which will result in ten-year sentences for

both men. If the various payoffs are examined, cooperation (mutual silence), which means two five-year sentences, is the most rewarding overall. However, if the two criminals act rationally, each will assume that the other will confess and therefore each will confess. Because each man cannot make sure that the other will cooperate, each will act in a self-interested fashion and end up worse off than if he had cooperated with the other by remaining silent.

As some game theorists such as Robert Axelrod have demonstrated, it is often advantageous for players to cooperate so that both can achieve higher payoffs. In reality, as the prisoners' dilemma demonstrates, players who act rationally often achieve an undesirable outcome.

Sequential Games. Many games follow a sequence in which one actor takes an action and the other reacts. The first actor then reacts, and so the game proceeds. A good way to think of this process is to consider a chess game in which each move is countered by the other player, but each player is trying to think ahead so as to anticipate his or her opponent's future moves. Sequential games are often used to describe the decision process of nations, which may lead to war if wrong reactions occur.

Sequential games can be diagrammed using a tree that lists the payoffs at each step, or a computer program can be used to describe the moves. Some sequential games are multiplayer games that can become quite complicated to sort out. Work in game theory suggests that equilibrium points change at each stage of the game, creating what are called sequential equilibria.

Simultaneous Games. In some games, players make their decisions at the same time instead of reacting to the actions of the other players. In this case, they may be trying to anticipate the actions of the other player so as to achieve an advantage. The prisoners' dilemma is an example of a simultaneous game. Although rational players may sometimes cooperate in a sequential game, they will not cooperate in a simultaneous game.

Applications and Products

Game theory is most commonly used in economics, finance, business, and politics, although its applications have spread to biology and other fields.

Economics, Finance, and Business. Game theory was first developed to explain economic decision

Game Theory

Strategy for the Prisoners' Dilemma

	Prisoner B stays silent (cooperates)	Prisoner B confesses (defects)
Prisoner A stays silent (cooperates)	Each serves 1 month	Prisoner A: 1 year Prisoner B: goes free
Prisoner A confesses (defects)	Prisoner A: goes free Prisoner B: 1 year	Each serves 3 months

making, and it is widely used by economists, financial analysts, and individuals. For example, a firm may want to analyze the impact of various options for responding to the introduction of a new product by a competitor. Its strategists might devise a payoff matrix that encompasses market responses to the competitor's product and to the product that the company introduces to counter its competitor. Alternatively, a firm might prepare a payoff matrix as part of decision process concerning entry into a new market.

Game theory cannot be used to predict the stock market. However, some game theorists have devised models to explain investor response to such events as an increase in the interest rate by the Federal Reserve Board. At the international level, there are various games that can be used to explain and predict the responses of governments to the imposition of various regulations such as tariffs on imports. Because of the large number of variables involved, this sort of modeling is quite complex and still in its infancy.

Some businesses might be tempted to develop a game theoretic response (perhaps using the prisoners' dilemma) to the actions of workers. For example, a company can develop a game theoretic response to worker demands for increased wages that will enable the company to maximize its utility in the negotiating process. Even if a game does not play out exactly as modeled, a company gains by clarifying its objectives in the development of a formal payoff matrix. Companies can also develop an approach to hiring new employees or dealing with suppliers that draws on game theory to specify goals and strategies to be adopted.

Politics. Some of the most interesting work in game theory has occurred in explaining developments in international affairs, enabling countries to make better decisions in the future. Game theory is often used to explain the decision process of the administration of President John F. Kennedy during the Cuban Missile Crisis. Other game theoretic explanations have been developed to examine a country's decision to start a war, as in the Arab-Israeli conflicts. Game theorists are able to test their formal game results against what actually occurred in these cases so as to enhance their models' ability to predict.

Other game theorists have devised game theoretic models that explain legislative decision making. Much of the legislative process can be captured by game theoretic models that take into account the step-by-step process of legislation. In this process, members from one party propose legislation and the other party responds, then the first part often responds, and the responses continue until the legislation is passed or defeated. Most of these models are academic and do not seem to govern legislative decision making, at least not explicitly.

Biology. Some evolutionary biologists have used game theory to describe the evolutionary pattern of some species. One relationship that is often described in game theoretic terms is the coevolution of predators and prey in a particular area. In this case, biologists do not describe conscious responses but rather situations in which a decline in a prey species affects predators or an increase in predators leads to a decline in prey and a subsequent decline in predators. Biologists use this relationship (called the hawk-dove game by the biologist John Maynard Smith) to show how species will evolve so as to better fit an evolutionary niche, such as the development of coloration that enables prey to better conceal itself from predators.

IMPACT ON INDUSTRY

Much initial work in game theory was done by academics and was not readily adopted by practitioners. Over time, business and political leaders have come to understand that the application of game theoretic principles could possibly enhance their decision making. For example, in the 1950's, organizations such as the RAND Corporation developed game theoretic models that created different scenarios for how the United States should respond to the potential of nuclear war with the Soviet Union. These models were quite cumbersome and their predictability was questioned because of the rigidity of the models and the difficulties of running extensive computer simulations. More powerful computers and the evolution of game theory (including behavioral game theory and the recognition of sequential equilibria) have enabled game theorists to develop quite sophisticated simulation models, some of which can be run on laptop computers, although some international relations and financial models still require supercomputers.

Simple Interpersonal Models. Some businesses have applied the formal approach to decision making used in game theory to labor relations or in developing pay scales. Many of these simple models can be worked out quickly so that executives can make quick decisions. Executives trained in game theory can follow a somewhat formal approach that enables them to evaluate the impact of their decisions in a clear fashion. Of course, employees can also counter using game theoretic logic. The use of ad hoc game theory is becoming commonplace in many businesses.

Game Theoretic Simulations. Assisted by powerful computers that can combine massive amounts of data in multitudes of scenarios, decision makers use game theory in making major decisions that affect market share, profit, and the long-term success of firms. Some financial institutions use game theory to consider the impact of financial instruments such as derivatives. Analysts in such governmental agencies as the Congressional Budget Office use game theoretic simulations to study the potential impact of legislation such as quotas or tariffs or of providing assistance to various domestic industries. In these cases, game theoretic approaches are combined with econometric models to examine the effect of various possible courses of action. National security analysts use various games to predict the responses of foreign countries to moves made by the United States. In one case, analysts used theory to examine responses that U.S. policy makers might follow in responding to an attack with a biological weapon.

Improved Decision Making. Because nations and businesses do not act in isolation, they are starting to use game theory to predict the responses of other nations and businesses. These models are complex, but they still follow the same principles found in simple games such as the prisoners' dilemma. The development of evolutionary game theory has enabled game

---

Fascinating Facts About Game Theory

- Although the motion picture *A Beautiful Mind* (2001) is a fascinating account of John Nash's life, including his work in game theory and his struggles with mental illness, it is not always an accurate introduction to game theory.
- Most game theory texts have at least one example drawn from domestic relations, often concerning how a husband and wife might go about reaching a decision regarding a night out so as to maximize their utilities and reduce domestic discord.
- The common sense nature of game theory (in this case, a cooperative game) can be used to explain why soldiers in the trenches on both sides of the front in World War I refrained from firing at each other at certain times of the day such as mealtimes.
- Most games are not true zero-sum games in which one player wins and the other loses but instead have mixed payoffs for players.
- Game theory is governed by the garbage-in, garbage-out dictum in that poor initial assumptions lead to flawed predicted outcomes.

---

theorists to better predict the outcomes of games with repeated rounds that have multiple equilibria, as have the use of concepts from psychology.

CAREERS AND COURSE WORK

Game theory can be used by individuals to help them to make better decisions, but it is also used to develop business or political strategy, to predict aspects of financial markets, and to describe evolutionary processes in biology. Many people practice game theory, often without knowing it. Formal game theory, driven by extensive mathematical modeling and computer applications, is used in industry and government to help guide decision making. Only a few mathematicians are full-time game theorists; most game theorists are people who use game theory to enhance decisions or to describe decision processes.

Although game theory is a field of applied mathematics, many graduate programs in social sciences require students to familiarize themselves with at least the basics. Students planning on advanced work in game theory can benefit from courses in statistics and formal logic, as well as any computer programming and any course work that enhances their ability to deal with quantitative material. As behavioral game theory develops, gaining knowledge of psychology (and possibly neuroscience) will become important for some applications. Many game theorists, however, are self-taught.

Social Context and Future Prospects

Game theory is an evolving field that can become esoteric and divorced from the realities of practical decision making. It can also be an intuitive, essentially nonmathematical approach to enhancing decision making. Both formal and informal game theoretic approaches are likely to be used in business and government in the future. However, most observers agree that sophisticated actors who are aware of the principles of game theory are likely to prevail over those who follow an ad hoc approach to decision making.

Game theory is not a perfect guide to decision making. At times, it has led to overly simplistic approaches that are derived from a narrow view of utility. With the introduction of newer conceptual frameworks, game theory has become less rigid and better able to model human decision making. In the future, sophisticated computer simulations based on game theory are likely to be used for corporate and governmental decision making.

John M. Theilmann, Ph.D.

Further Reading

Binmore, K. G. *Game Theory: A Very Short Introduction.* New York: Oxford University Press, 2007. A renowned game theorist explains game theory in a nontechnical, entertaining manner, looking at famous figures such as John Nash and applications in many areas, including drivers negotiating heavy traffic.

Bueno de Mesquita, Bruce. *The Predictioneer's Game: Using the Logic of Brazen Self-Interest to See and Shape the Future.* New York: Random House, 2009. Describes various predictive applications of game theory without delving deeply into mathematics. Focuses on rationality and people's interests and beliefs.

Fisher, Len. *Rock, Paper, Scissors: Game Theory in Everyday Life.* New York: Basic Books, 2008. A scientist examines game theory by using its strategies in everyday life, such as a crowded supermarket and congested Indian roads. His emphasis is on cooperation.

Gintis, Herbert. *Game Theory Evolving: A Problem-Centered Introduction to Modeling Strategic Behavior.* 2d ed. Princeton, N.J.: Princeton University Press, 2009. Looks at how game theory is moving from a formal study of rational behavior to a way to study behavior (not always rational) in social situations such as teams.

Miller, James. *Game Theory at Work: How to Use Game Theory to Outthink and Outmaneuver Your Competition.* New York: McGraw-Hill, 2003. Basic work with business applications, based on a strong perspective from neoclassical economics.

Web Sites

Economic and Social Research Council
Centre for Economic Learning and Social Evolution
http://else.econ.ucl.ac.uk/newweb/index.php

Game Theory Society
http://www.gametheorysociety.org

See also: Applied Mathematics; Military Sciences and Combat Engineering; Paleontology; Risk Analysis and Management.

GASOLINE PROCESSING AND PRODUCTION

FIELDS OF STUDY

Chemistry; engineering; chemical engineering; chemical process modeling; fluid dynamics; heat transfer; distillation design; distillation processes; unit operation; corrosion engineering; environmental engineering; control engineering; process engineering; industrial engineering; mechanical engineering; electrical engineering; safety engineering; physics; thermodynamics; mathematics; materials science; computer science; business administration.

SUMMARY

About one-quarter of the crude oil extracted globally is processed to produce gasoline, the fuel for the internal combustion engine that powers most cars, light trucks, motorcycles, and piston engine aircraft. To meet the world's demand for quantity and quality of gasoline, refineries perform several processing steps, beginning with desalting and distillation of crude oil into separate fractions. Selected fractions undergo desulfurization, cracking, reforming, and other processes. The different components gained are blended with further additives to produce various grades of gasoline. Gasoline can also be produced as synthetic fuel from coal, oil sands, natural gas, and biomass.

KEY TERMS AND CONCEPTS

- **Additive:** Special chemical component added to improve the quality of gasoline.
- **Alkylation:** Catalytic combination process to gain high-octane isoparaffins to add to the gasoline pool.
- **Aviation Gasoline:** Gasoline for use by only piston engine aircraft, not by jet engines.
- **Blending:** Physical process of creating various grades of gasoline for consumption.
- **Catalytic Reforming:** Chemical process used on gasoline components to increase their octane ratings.
- **Cracking:** Breaking of long-chain hydrocarbon molecules into short-chain hydrocarbon molecules, some of which are suitable as gasoline components.

- **Crude Oil:** Liquid part of petroleum, the raw material from which gasoline, among many other products, is processed.
- **Desulfurization:** Process of cleansing gasoline components of unwanted sulfur.
- **Distillation:** Process in which a mix of components with different volatilities is heated to create physical separation of the components.
- **Isomerization:** Chemical process to rearrange the atoms in a molecule to form another molecule, done to create a higher octane rating for a gasoline blend.
- **Octane Rating:** Unit to measure quality of gasoline; the higher the octane rating, the less likely the gasoline will knock, or prematurely ignite in the engine.
- **Polymerization:** Chemical process used to increase the octane rating by forming longer molecules from propene and butane as gasoline stocks.
- **Refining:** Processing crude oil by various means to form different end products such as gasoline.

DEFINITION AND BASIC PRINCIPLES

Gasoline is the most effective fuel for the internal combustion engine that powers most automobiles and light trucks as well as piston engine aircraft. It is generally gained from processing crude oil at a refinery. The gasoline sold to consumers consists of a complex blend of hydrocarbons that have boiling ranges from about 38 to 205 degrees Celsius (100 to 400 degrees Fahrenheit).

The first step in processing at a refinery is generally the desalting of crude oil. Then the oil is heated, partially vaporized, and sent to a distillation tower operating at atmospheric pressure. There it condenses into separate fractions that are extracted. The yield of light straight-run (LSR) gasoline from crude oil after atmospheric distillation generally consists of only up to 10 percent. This is much too low to satisfy the global demand for gasoline. Therefore, additional, different fractions from distillation are processed into suitable components of gasoline.

Heavy straight-run (HSR) naphtha gained from atmospheric distillation and from vacuum distillation, followed by thermal, catalytic, or hydrocracking, is fed into a catalytic reformer to become reformate, a gasoline component. To boost the octane rating of the final gasoline blend, which is a key indicator of its quality as engine fuel, gasoline components also undergo such chemical processes as isomerization, polymerization, and alkylation. Special additives gained in other refining processes are mixed in to form high-quality gasoline.

Gasoline can also be produced by processing feedstock such as coal, oil sands and oil shale, natural gas, or even biofuels. These processes are far more expensive than producing gasoline from crude oil. They are usually undertaken only in special circumstances, such as periods of abundance of alternative feedstock and lack of crude oil, during wars, or as the result of a political commitment to alternative fuels.

Background and History

In 1856, in Poland and Romania, the first refineries were built to process crude oil into more useful products through distillation into different fractions. The first refinery opened in the United States in 1861. What would later be known as gasoline in the United States, a naphtha-based hydrogen compound, called petrol in Great Britain, was one of the different refinery products. The rise of the internal combustion engine in the late nineteenth century, particularly as the motor of the newly invented automobile, led to a tremendous increase in the demand for gasoline as its fuel.

To satisfy this demand, methods of increasing the yield of gasoline from crude oil were developed. Thermal cracking of other distillation fractions was invented separately by Vladimir Shukhov in Russia in 1891 and by William Merriam Burton in the United States in 1913. Thermal cracking doubled the yield of gasoline in the United States.

In 1930, the invention of thermal reforming boosted the octane rating of gasoline and lessened the stress on engines burning gasoline. In 1932, hydrogenation came into use to lower the undesirable sulfur content of gasoline, and coking created additional base stocks for gasoline out of heavier distillation fractions. In 1935, catalytic polymerization further boosted octane ratings.

The big breakthrough in producing more gasoline with higher octane ratings came with the invention of catalytic cracking in 1937 and fluid catalytic cracking in 1942. French engineer Eugene Jules Houdry is generally given credit for the first and a consortium of American university scientists and oil industry researchers for the second. Other important steps were introduction of visbreaking in 1939, alkylation and isomerization in 1940, catalytic reforming in 1952, and hydrotreating in 1954. Research to improve refinery processes to optimize gasoline yield and quality and to make processes and products more environmentally friendly continues into the twenty-first century.

Producing gasoline synthetically from coal was invented in 1913 by German chemist Friedrich Bergius. By 2010, SASOL, a South African company, had become the leading producer of gasoline from coal. In 1978, commercial production of synthetic crude oil from oil sands began in Canada; in 2009, this accounted for 13 percent of Canada's petroleum product consumption including gasoline.

How It Works

Gasoline is processed and produced from crude oil at a refinery. As refineries are very complex installations, processing a wide variety of crude oils, each refinery follows its own, customized process. However, the following are the most common processes.

Desalting. As crude oil arrives at a refinery, it is generally desalted to remove suspended salt and solid contaminants. Crude oil is heated and mixed with hot water. Salt and contaminants are washed out either by adding chemicals or by application of an electric field. Another method is to filter heated crude oil through diatomaceous earth.

Distillation. Desalted crude oil is heated for fractional distillation between 343 and 399 degrees Celsius (650 to 750 degrees Fahrenheit). The resulting vapor and liquid mix is sent to the first distillation tower that operates at atmospheric pressure. Because of the different boiling points of the different hydrocarbon molecules of crude oil, the hydrocarbons can be separated into different fractions (also called cuts). This occurs at the distillation tower, which can be as tall as 50 meters (164 feet). The heaviest hydrocarbons remain at the bottom, while the middle and lighter ones are extracted.

From the lightest hydrocarbons collected from distillation, light straight-run (LSR) gasoline (sometimes called light naphtha) is extracted at a gas separation plant that commonly contains a hydrodesulfurization unit. Among the lighter gases, butane is used for blending into the final gasoline products or as feedstock for the alkylation unit.

Oil refinery in Richmond, California. (Lawrence Migdale/Photo Researchers, Inc.)

Atmospheric distillation also yields lower-boiling heavy straight-run (HSR) gasoline (or HSR naphtha). Often after hydrotreating, this fraction is processed further at the catalytic reformer.

Cracking. Distillation also yields residue. Among its fractions is what is commonly called gas oils, middle distillate, or wax distillate. After hydroprocessing, these gas oils are sent to a cracker at the refinery to convert them into more valuable products including gasoline components. If a hydrocracker is used, no prior hydroprocessing occurs.

Cracking is the chemical conversion process used to break down longer-chain hydrocarbon molecules into shorter-chain ones that are more valuable. The naphtha gained from various cracking processes is used for further processing into gasoline components with high octane ratings.

The mildest form of thermal cracking is visbreaking. It breaks the longest hydrocarbon molecules to eventually yield also more gasoline. Delayed coking is severe thermal cracking, heating the gas oil feedstock to 500 degrees Celsius (930 degrees Fahrenheit). The product gained for gasoline processing is called coker naphtha.

Fluid catalytic cracking (FCC) is a key process to gain components for gasoline blending from the heavier feedstocks processed at a FCC unit. Fluid catalytic cracking converts nearly half of the heavy feedstocks into naphtha for gasoline production and accounts for 35 to 45 percent of the volume of the gasoline produced at U.S. refineries as of 2010. The naphtha gained from cracking is typically divided.

The light fraction is used directly for gasoline blending. The heavy fraction is sent to the catalytic reformer and functions as octane booster.

Hydrocracking is the most sophisticated, flexible, and expensive form of cracking. Hydrocrackers are less common at American refineries than at European and Asian ones. This is because gasoline is the most in-demand product in the United States, and a hydrocracker's maximum yield of gasoline components is only 8 percent more than that of a much cheaper fluid catalytic cracker. Hydrocracking combines catalytic cracking with the insertion of hydrogen. Light and heavy naphtha is gained for gasoline blending and processing.

Catalytic Reforming. To improve the octane rating of the blended gasoline, heavy straight-run naphtha and naphtha gained from cracking is sent to a catalytic reformer. During catalytic reforming, the feedstock naphtha has its hydrocarbon molecules restructured by light cracking in the presence of a platinum-based catalyst. Catalytic reforming creates a desirable high octane gasoline stock. Its aromatics that are responsible for its high octane rating have come under environmental scrutiny.

Isomerization, Polymerization, and Alkylation. There are some further processes to increase the octane rating of gasoline. Light straight-run gasoline often has at least some of its components isomerized, so that it consists of some molecules with a different structure but with the same number of atoms. Polymerization is a cost-effective way to boost octane ratings by combining the very light gases propene and butane into longer-chain olefin molecules. Alkylation is an effective but expensive process to increase the octane rating of light components of the gasoline blending pool. It refers to the addition of an alkyl group to isobutene at low temperatures of 5 to 38 degrees Celsius (41 to 100 degrees Fahrenheit) with a catalyst of sulfuric or hydrofluoric acid.

Blending and Additives. In the end, all gasoline components are blended together. At a modern refinery, this is done in a computer-controlled process. Additives are used primarily to boost octane rating. Lead was widely used once but was banned in 1996 in the United States, as well as in many other nations. With the increasing prohibition of methyl-tert-butyl-ether (MTBE) in many U.S. states, use of alternatives such as tertiary amyl methyl ether (TAME) or ethyl tert-butyl ether (ETBE) increased. There are many governments that mandate the blending of ethanol, which is done after gasoline leaves the refinery.

Gasoline from Synthetic Fuels. Gasoline can be produced from oil shale, coal, biomass, or natural gas. These processes are invariably more expensive than processing and producing gasoline from crude oil. They are employed in specific locations such as in South Africa from coal and in Canada and Venezuela from huge oil sand deposits.

APPLICATIONS AND PRODUCTS

Fuel Gasoline. By 2009, worldwide, 676 refineries produced 7.7 billion barrels of gasoline, or 325 billion gallons. This meant that globally, about one-quarter of the 30.8 billion barrels of crude oil refined was converted to gasoline. In the United States, because of a higher demand for gasoline, 141 operating refineries produced 3.2 billion barrels (135 billion gallons) of gasoline, amounting to 46 percent of all refinery output. This U.S. percentage for gasoline production has been basically stable since the 1990's, with a slight dip to 44 percent in 2008 because of the recession.

Over 90 percent of gasoline produced in the United States is used to fuel automobiles and light trucks. All over the world, gasoline as fuel is essential for transportation in any industrial and industrializing society. Its desired qualities are strong resistance to premature ignition in the engine, facilitating easy start, warm-up, and acceleration of the engine. Further desirable qualities include high mileage for the fuel consumed, prevention of vapor lock and deposit build-up in the engine, and as few polluting emissions as possible.

Gasoline Components. To optimize its qualities as fuel for the internal combustion engine, gasoline is blended from different components. Out of concern for the environment and for protection from cancer-causing components, national governments often regulate the composition of gasoline. An example is the prohibition of lead as an additive for fuel gasoline in the United States and the European Union and increasingly in many other nations.

In the United States, gasoline sold to consumers consists primarily of naphtha gained from catalytic cracking at 38 percent of the total, reformate from catalytic reforming at 27 percent, alkylate at 12 percent, and light straight-run gasoline and its isomeric form at 7 percent. Smaller contributors are the light component normal butane at 3 percent, light naphtha from hydrocracking at 2.4 percent, light coker naphtha at 0.7 percent, and polymers at 0.4 percent. Other additives, especially ethanol, account for the remainder.

Out of environmental concerns in the United States, the concept of reformulated gasoline (RFG) has been developed. This refers to a blend of gasoline that burns at least as cleanly as high methanol-content alternative fuels. As a result of federal and state regulations, in 2009, about one-third, or 1.1 billion gallons, of gasoline produced by U.S. refineries was blended with ethanol to become reformulated gasoline.

Measuring Gasoline Quality. For the consumer, the most important indicator of gasoline quality is its octane rating. The higher the octane rating, the less likely is the gasoline blend to ignite prematurely in the engine, damaging it in an event commonly called knocking. There are many different ways to calculate the octane rating. The research octane number (RON) is derived from testing the gasoline blend in a laboratory engine. The gasoline's ability to burn by controlled ignition and not ignite prematurely is related to the respective quality of a mix of iso-octane and heptanes. A RON rating of 95, for example, indicates that this gasoline burns as well as a mix of 95 percent iso-octane and 5 percent heptanes would. The motor octane number (MON) uses the same comparison but places the test engine under more stress to simulate actual driving situations. For this reason, the MON octane rating is between 8 to 10 points lower than the RON. Different nations use different octane ratings. In the United States, an average of RON and MON is used and posted at gas station pumps. It can be called PON (posted octane number), or (R+M)/2, and is also called the anti-knocking index (AKI).

In the United States, gasoline is typically available as regular unleaded gasoline, with a PON of generally 85, and premium gasoline ranges from a PON of 89 to 93. California typically offers the three grades of 87, 89, and 91 PON. Premiums are branded under different names. Because lower atmospheric pressure at high altitudes reduces pressure in the engine's combustion chamber and lessens the danger of premature ignition, gasoline sold at high elevations in the Rocky Mountains states typically have a PON of 85 up to 91, signifying that gasoline has to be blended to account for the environment where it is burned as fuel. Special gasoline for automobile racing in the United States can have a PON of 100 or higher.

Two other indicators of gasoline quality are as important but less visible to consumers than the octane rating. They are Reid vapor pressure (RVP) and boiling range of the blended gasoline. Low RVP, or low volatility, and higher boiling ranges mean the gasoline is less likely to evaporate too quickly, causing vapor lock in the engine. They also prevent higher evaporation losses that lower mileage gained from gasoline. A high RVP, or high volatility, and lower boiling ranges make the engine start easier and warm up more quickly. Outside temperature influences gasoline behavior. As a result, gasoline is blended differently for use during the summertime, with a RVP of about 7.2 pounds per square inch (psi), or 49.6 kilo Pascal (kPA), and during the wintertime with a RVP of about 13.5 psi (93.1 kPa).

E 85. Ethanol, or grain alcohol, is blended into gasoline before sale to consumers because it is a very clean-burning fuel from renewable resources. However, it gets only 70 percent of the mileage of undiluted gasoline. Several states have mandated a blend of at least 5.9 percent of ethanol into gasoline, and many gas pumps state that the gasoline sold may contain up to 10 percent ethanol. An alternate fuel, marketed as E 85, contains 85 percent ethanol to only 15 percent gasoline. Its lack of fuel efficiency is made up by a federal tax subsidy trying to encourage its consumption.

Avgas. For use as fuel for aircraft with piston engines, refineries produce relatively small quantities of aviation gasoline, called avgas. This is very different from jet fuel. In 2009, U.S. refineries produced about 211 million gallons of avgas, about 0.1 percent of their products, and down from 261 million gallons in 2004. This decline occurred because avgas, most commonly of the 10011 variety, contains a small amount of lead in the form of tetraethyl lead (TEL). As public pressure increases to completely phase out leaded gasoline, scientists have looked for an alternative to lead in aviation gasoline. Avgas must have a high octane rating and cannot lead to engine vapor lock at the low pressure encountered during flight. By 2010, there had been some promising experiments with a nonleaded aviation gasoline called G100UL.

IMPACT ON INDUSTRY

Because gasoline is the key fuel for the internal combustion engine that powers most of the world's automobiles and a significant number of small buses and light trucks, its impact on global industry is extremely significant. This begins with the quest for the raw materials from which the majority of gasoline is produced, in particular crude oil from petroleum. Exploration, extraction, and transport to processing of crude oil is a major industrial enterprise. However, the oil industry is significant not only in itself but also as a customer of highly specialized operating equipment and customized plants. Transporting gasoline from the refinery to the customer requires a large logistics effort involving pipelines, tanker trucks, and a network of gas stations. The huge costs of producing gasoline are justified only by its high value.

Government Agencies. Governments of nations with exploitable stock of petroleum benefit hugely, primarily from royalties for extraction concessions on national land or territorial waters. Gasoline taxes also make up a major revenue stream for many industrialized countries. Government agencies such as the U.S. Environmental Protection Agency (EPA) or those linked to ministries of transportation or the environment all over the industrial and industrializing world also affect processing and production of gasoline, particularly in regard to gasoline standards. A key example was the long, drawn-out fight to end the addition of lead to gasoline in the United States, in which EPA scientists became involved. This was mirrored in other nations, where governments set similarly strict standards for gasoline components.

Government agencies also set standards for transportation, storage, and distribution of gasoline, as well as for specifications of the internal combustion engines and their exhaust mechanisms, all to lessen the negative environmental impact of gasoline combustion. The requirement for catalytic converters to treat gasoline engine exhaust has been set by the United States, the European Union, Japan, and many other nations.

National governments have been concerned with safeguarding the supply of gasoline for their industries and citizens. This has led to conflicts triggered in part by disputes and fears over petroleum extraction, such as arguably the United Nation's successful termination of Iraq's occupation of Kuwait (1990-1991).

Universities and Research Institutes. Universities educate the chemists, engineers, and other scientists and professionals involved in all steps of gasoline processing and production or in petroleum extraction and production as gasoline's prime feedstock. Universities also conduct research into fuel alternatives to gasoline that are more environmentally friendly. Research institutes employ scientists and researchers to engage in the continuing project to improve the quality, cost efficiency, and the environmental impact of gasoline at all steps of production and consumption. Trade associations such as the American Petroleum Institute also conduct public relations and see themselves as an interface between the oil-gasoline industry and the general public.

Industry and Business Sectors. As part of the chemical industry, gasoline production belongs to the oil and gas industry sector. There, it is part of the midstream and downstream sector that engages in the processing, blending, transport, and marketing of gasoline to the consumer and oversees its final distribution at a gas station.

Gasoline production also affects the financial industry as its key raw material, crude oil, is one of the world's most widely traded commodities. Shock waves from swings in the oil price can affect the global economy. Manufacturers, particularly automakers and their suppliers, are also affected by gasoline production. The petrochemical industry is affected as its feedstocks such as ethane compete with the production of gasoline at refineries.

The transportation industry is involved in bringing gasoline to its individual points of sale at a network of gas stations. Gas stations, like most of the 162,000 operating in the United States in 2010, often also serve as small retail outlets. The advertising industry earns revenue from creating major gasoline brands and public relation campaigns for oil companies.

Major Corporations. Overall, gasoline processing and production tends to be highly integrated into the business of large private or national oil and gas companies. Because of significant economies of scale, the number of refineries has declined. Smaller refineries independent of big oil companies have become the exception. Although huge new world-scale refineries have been built, particularly in the Middle East, the source of much crude oil, or Asia, source of increasing demand for gasoline, the number of operable U.S. refineries has shrunk by half. By 2010,

the number had dropped to 148 from 301 in 1982. However, U.S. refinery capacity was virtually the same in 2010 as in 1982, at about 17.5 million barrels per day, which indicates consolidation of refining operations

In 2007, the world's largest oil and gas company was Saudi Aramco in Saudi Arabia. Its four sole-owned refineries in Saudi Arabia processed almost 1 billion barrels of crude oil, and a new refinery opened in 2008.

Fascinating Facts About Gasoline Processing and Production

- Americans are the world's top consumers of gasoline. In 2009, the American population of about 305 million used 377 million gallons of gasoline per day, more than one gallon for every person.

- Japan is the only industrialized nation with virtually no domestic raw materials to produce gasoline and must import all of its crude oil for processing at its thirty-one refineries.

- Gasoline production is the largest of all U.S. basic industries, exceeding the output of steel or lumber.

- South Africa is the world's leading producer of gasoline from coal, an abundant natural resource in that nation, following a process invented in Germany in 1913.

- Antarctica is the world's only continent without a refinery.

- Because of heavy taxation, gasoline is more expensive in the European Union than in the United States. On June 30, 2010, a gallon of unleaded gasoline cost on average $2.78 in the United States but cost $7.19 in the Netherlands.

- Both Norway and Venezuela produce more gasoline than they consume. However, while Norwegian taxes raise the price of a gallon of unleaded gasoline to $7.65, Venezuelan subsidies lower it to the world's cheapest price of $0.09 (as of June, 2010).

- During World War II, Japanese occupying forces requisitioned much-needed rice from Vietnamese farmers to convert into synthetic gasoline for army trucks and aircraft, exacerbating Vietnam's severe famine of 1944 to 1945.

- Aviation gasoline is artificially colored so that it can be quickly distinguished from automobile gasoline by sight. The most common 10011 variety shines a bright blue.

In 2010, ExxonMobil was the world's private oil company and operated the two largest refineries in the United States. Its Baytown, Texas, refinery had a capacity of 205 million barrels per year, and its Baton Rouge, Louisiana, refinery was a close second at 184 million barrels.

Royal Dutch Shell and BP were other leading international private oil companies heavily engaged in the refining business as part of their integrated oil and gas operations. In 2009, Shell held interests in forty refineries around the world processing about 1.5 billion barrels of crude. Similarly, worldwide, in 2009, BP held an interest in sixteen refineries that processed 1.35 billion barrels of crude oil, with BP's share being 973 million barrels.

CAREERS AND COURSE WORK

Students interested in a career in the gasoline processing and production industry should take courses in science, mathematics, and economics in high school. Refineries employ many skilled technicians for their operations and need laboratory analysts, occupations for which a two-year associate degree is helpful. A refinery also has positions for firefighters and employs members of the medical profession, from paramedics to physicians.

A bachelor's degree in chemistry, physics, computer science, mathematics, or environmental science is very useful for a career in the actual processing and production field. The same is true for an engineering degree, whether at the undergraduate or postgraduate level, especially in chemical engineering, or electrical, mechanical or computer engineering. Such degrees create good employment prospects. For a career in the purchasing department of a refinery, selecting the different crude oils for processing, for instance, a business major or master of business administration is helpful and can also lead to a higher management position.

Any master of engineering or doctoral degree in chemistry or chemical engineering serves as good preparation for a top-level career. A doctorate in a science field can lead to an advanced research position either with a company or a government agency. A research career linked to the subject of gasoline production, whether at a university, institute, or corporation, also benefits from postdoctoral work in chemistry or materials science.

Particularly for a career in private industry, the cyclical nature of the oil industry does not guarantee employment during industry downturns. As many new refineries are built outside the United States and many operate in different places of the world, global mobility is of significant advantage when pursuing an advanced career.

SOCIAL CONTEXT AND FUTURE PROSPECTS

As the most efficient fuel for the internal combustion engine, gasoline has vastly increased private and public mobility of the world's people, especially in industrialized societies. The nineteenth century increased humanity's mobility through the steam engines of railroad trains, and the twentieth century saw the rise of gasoline to enable private automobile transportation. As such, gasoline significantly affects the lives of almost every person in a developed country.

However, the emissions caused by both the production of gasoline and its use as fuel have had a negative impact on the environment that has been fiercely debated and publicly discussed. Of particular concern are carbon dioxide emissions and the side effects of many gasoline additives. The wide availability of gasoline also promotes the manufacture of automobiles and other vehicles, which generates more greenhouse gases and sometimes hazardous wastes, especially during the vehicle disposal process. Oil, gasoline, and automobiles have been attacked by some activists as the main causes of human-made environmental degradation, although they have granted freedom of mobility to a vast number of people on a scale unthinkable in the nineteenth century.

Because of environmental concerns, research seeks to minimize the aromatics content of gasoline, as additives such as benzene are particularly harmful if handled carelessly. Research is ongoing to make gasoline burn cleaner and increase the cost efficiency of its production. To improve the economics of gasoline production, very large world-class capacity refineries have been built to use economics of scale and take advantage of proximity to either raw materials or markets.

Research is under way into alternative fuels to gasoline that would offer similar fuel efficiency at a lesser environmental cost. The rise of the petrochemical industry as a competitor for gasoline's raw material, crude oil, led to more refineries increasing their output of ethane, a very light gaseous hydrocarbon, at the expense of gasoline production, as ethane is a prime and valuable petrochemical feedstock.

R. C. Lutz, B.A., M.A., Ph.D.

FURTHER READING

Duffield, John. *Over a Barrel: The Costs of U.S. Foreign Oil Dependence.* Stanford, Calif.: Stanford University Press, 2008. Notes that as demand for gasoline far outstrips domestic resources, U.S. economic and foreign policy is forced to address this problem in specific ways, including military options. Concludes with a proposal to lower the cost of dependence. Tables, figures, maps.

Gary, James, et al. *Petroleum Refining: Technology and Economics.* 5th ed. New York: CRC Press, 2007. A well-written textbook that details all the refining steps of gasoline processing and production. Five appendixes, index, photographs.

Leffler, William. *Petroleum Refining in Nontechnical Language.* Tulsa, Okla.: Penn Well, 2008. Covers all major aspects of gasoline processing and production at a refinery, from the chemical foundations to the processes themselves.

Meyers, Robert, ed. *Handbook of Petroleum Refining Processes.* 3d ed. New York: McGraw-Hill, 2004. Advanced-level technical compendium covers various industry methods for gasoline production. In-depth information for those considering a career in this field.

Raymond, Martin, and William Leffler. *Oil and Gas Production in Nontechnical Language.* Tulsa, Okla.: Penn Well, 2005. Accessible introduction to the process of producing gasoline from crude oil. Covers all major aspects. Light-hearted but very informative style.

Yeomans, Matthew. *Oil: Anatomy of an Industry.* New York: The New Press, 2004. Critical look at the oil industry, particularly in the United States. Good historical overview; addresses issues of America's dependency on oil caused by its gasoline demand, world conflicts caused by oil, and the question of alternatives like hydrogen fuel.

WEB SITES

Global Petroleum Research Institute
http://www.pe.tamu.edu/gpri-new/home

Independent Petroleum Association of America
http://www.ipaa.org

National Petrochemical and Refiners Association
http://www.npra.org

National Petroleum Council
http://www.npc.org

Society of Petroleum Engineers
http://www.spe.org/index.php

U.S. Department of Energy
Energy Information Administration, Petroleum
http://www.eia.doe.gov/oil_gas/petroleum/info_glance/petroleum.html

U.S. Environmental Protection Agency
Gasoline Fuels
http://www.epa.gov/otaq/gasoline.htm

See also: Biofuels and Synthetic Fuels; Chemical Engineering; Coal Gasification; Coal Liquefaction; Fossil Fuel Power Plants; Petroleum Extraction and Processing.

GASTROENTEROLOGY

FIELDS OF STUDY

Biology; chemistry; mathematics; physiology; health; nutrition; anatomy; nephrology; neurology; hematology; internal medicine; endocrinology and metabolism; oncology; nuclear medicine; infectious disease; geriatric medicine; pediatric medicine.

SUMMARY

The medical specialty known as gastroenterology focuses on conditions and diseases involving the digestive tract and related organs. The esophagus, stomach, duodenum, and large and small intestines are part of the gastrointestinal system. Related organs include the liver, gallbladder, and pancreas. Conditions addressed by the gastroenterologist include gastric ulcers, heartburn, abdominal discomfort, nausea and vomiting, constipation, diarrhea, inflammatory bowel diseases, hepatitis, nutritional deficiencies, and gastric cancers. Endoscopy, a diagnostic tool that allows visualization of the digestive tract, is an integral component of patient care. Gastroenterologists must be skilled in all aspects of the digestive process, and they must be caring, compassionate, and excellent team leaders.

KEY TERMS AND CONCEPTS

- **Colon:** Large intestine.
- **Colonoscopy:** Procedure using a thin tube to project images of the interior of the large intestine and rectum to a monitor.
- **Endoscopy:** Procedures using a thin, flexible fiberoptic tube with a camera to project images of the gastrointestinal tract back to a monitor.
- **Gastrointestinal Tract:** Organs associated with the digestion and elimination of food, including the esophagus, stomach, duodenum, small intestine, large intestine (colon), rectum, and anus.
- **Hepatitis:** Inflammation of the liver caused by several different types of viruses.
- **Motility:** Movement of food through the intestines.
- **Physiology:** Study of the function of organs.
- **Sigmoidoscopy:** Procedure to inspect the lining of the last third of the colon, or descending colon.

DEFINITION AND BASIC PRINCIPLES

Gastroenterology is a medical specialty focused on the diagnosis and treatment of conditions involving the digestive tract and associated organs. Gastroenterologists have a thorough understanding of the normal function of the gastrointestinal tract, which encompasses the esophagus, stomach, small and large intestines, and the rectum. This specialty also includes other organs associated with the gastrointestinal tract, such as the pancreas, gallbladder, bile ducts, and liver. Training includes knowledge of the normal movement of food through the digestive tract (motility), the absorption of nutrients as foods move through the intestines, and the removal of waste products. A detailed knowledge of normal digestive processes allows gastroenterologists to evaluate abnormal conditions and plan a course of treatment.

Gastroenterological training stresses diagnostic procedures, such as endoscopies, combined with thorough patient interviews regarding signs and symptoms. Colorectal cancer screenings are an important part of the practice. Gastroenterologists may specialize further and narrow their practice to children or the elderly, or they may become board certified in surgery. Not all gastroenterologists are surgeons, but most gastroenterologists perform minimally invasive procedures, such as removal of colon polyps. Gastroenterologists specializing in bariatric surgery generally perform procedures such as gastric bypass or lapband surgery for weight loss.

BACKGROUND AND HISTORY

Gastroenterology is a relatively new medical specialty. Until the mid-1980's, gastroenterology was considered a subspecialty of internal medicine and required only an additional one or two years of training. The introduction of endoscopy, which uses a flexible fiberoptic tube to explore the digestive tract, transformed gastroenterology and allowed the field to emerge as a separate subspecialty. The addition of hepatology, or the study of liver conditions, to gastroenterology and the rise in hepatitis further defined the field. Although gastroenterologists are first trained as internists, several more years of training in gastroenterology are required. Gastroenterologists may specialize further in gastric conditions particular to children.

The field of gastroenterology has developed unevenly across developed and developing countries. Standardized training is lacking, with nations requiring different curricula. Whereas gastroenterology has emerged as a separate subspecialty in North America, the field is still considered part of internal medicine in Europe. Fellowships are required, but they last only one or two years and consist of on-the-job training. Some developing countries have no formal training guidelines or firm requirements regarding entrance and exit examinations.

HOW IT WORKS

Patient Care. First and foremost, gastroenterologists are concerned with patient care. Most gastrointestinal conditions are chronic, allowing gastroenterologists to develop long-term relationships with their patients. Detailed medical histories are often the key to diagnosis, so gastroenterologists must be compassionate and able to listen closely to the patient's complaints.

The physical examination is an important aspect of diagnosis and treatment. Procedures such as colonoscopies and sigmoidoscopies are uncomfortable and may be embarrassing for patients, so the gastroenterologist must be skilled in relaxing patients and providing appropriate sedation.

Setting. Gastroenterologists may work in private practice or hospital settings. They also may work in nursing homes or long-term care facilities, hospices, or outpatient surgery centers.

Conditions Treated. Although they focus on the gastrointestinal system, gastroenterologists diagnose and treat a wide range of symptoms and conditions. Gastric and peptic ulcers and gastroesophageal reflux disease (GERD) are diagnosed and treated by gastroenterologists, as are gallstones. Gastroenterologists are called on to diagnose and manage abdominal pain and discomfort, hemorrhoids and bloody stools, constipation, diarrhea, nausea, and vomiting. They treat intestinal conditions, such as inflammatory bowel disease, ulcerative colitis, Crohn's disease, and diverticulitis. Gastroenterologists screen for colon and rectal cancer and remove rectal polyps. Gastroenterologists detect reasons for unexplained weight loss and poor absorption of nutrients. They also treat liver diseases such as hepatitis and jaundice.

Management Duties. Regardless of the setting, gastroenterologists are expected to handle a number of management duties in addition to their clinical caseload. The gastroenterologist in private practice manages staff and ensures that the office is run efficiently. All gastroenterologists consult with patients and other medical professionals, who offer their medical expertise. Experienced gastroenterologists train students.

Imaging Techniques. Imaging techniques are the cornerstone of the field. Endoscopy is the use of a thin, flexible, lighted tube that sends real-time video to a monitor. For upper endoscopy, also called an upper GI, the patient is sedated and the endoscopy tube is fed down through the throat into the stomach and duodenum. The gastroenterologist examines the lining of the esophagus and stomach for ulcers or other abnormalities. Tissue samples may be taken for biopsy, or the gastroenterologist can remove a foreign object that has been swallowed.

Colonoscopy, or a lower GI, uses endoscopy to examine the colon and rectum for abnormalities. Patients must prepare for the procedure by cleansing their bowels of waste materials, usually by drinking water mixed with a substance that results in rapid elimination of the contents of the bowels. Preparation is usually done at home. For the colonoscopy itself, the patient is sedated. The imaging instrument, called a colonoscope, is inserted into the rectum through the anus. As the gastroenterologist guides the tube up through the intestine, carbon dioxide gas is used to inflate the colon to allow for better

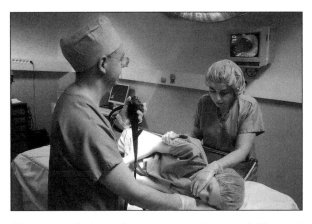

Gastroenterologist performing a gastrointestinal fiber optic endoscopy. A flexible fiber-optic endoscope is inserted through the mouth and navigates inside the digestive tract. (Garo/Phanie/Photo Researchers, Inc.)

imaging. As in an upper endoscopy, the gastroenterologist examines the intestinal wall for ulcers, polyps, or inflammation. If polyps are present, the gastroenterologist removes the growths by inserting miniature cutting tools through the colonoscope.

APPLICATIONS AND PRODUCTS

Endoscopy. Endoscopy has become the foundation of gastroenterology. Endoscopy is an imaging technique that uses a thin, flexible fiberoptic tube. As the tube is fed through the patient's gastrointestinal tract, a small camera transmits the images to a screen for viewing. The gastroenterologist manipulates the tube and camera to visually inspect the esophagus, colon, or intestines for abnormalities, lesions, or ulcers. Gastroenterologists receive extensive training on endoscopy procedures, including proper sedation of patients, and on interpretation of the images. Basic endoscopy training includes upper endoscopy, sigmoidoscopy, and colonoscopy.

Upper endoscopy examines the esophagus, stomach, and duodenum for ulcers, precancerous growths, foreign bodies, and other conditions causing pain, nausea or vomiting, bleeding, unexplained weight loss, or anemia. Gastroenterologists also receive training on using endoscopy to diagnose and dilate a narrow esophagus and to stop bleeding along the gastrointestinal tract.

Colonoscopy and sigmoidoscopy allow the gastroenterologist to visualize different portions of the colon. Colonoscopy encompasses the entire colon whereas sigmoidoscopy is limited to the last section of the colon and the rectum. Although sigmoidoscopies are quicker to perform, colonoscopies are preferred for cancer screening because they examine the entire colon. Both procedures require the patient to use a colon-cleansing product beforehand. During the procedure, the gastroenterologist examines the lining of the colon and removes polyps, or precancerous lesions.

Colorectal Cancer Screening. Colorectal cancer is the second leading cause of cancer death in the United States, according to the National Cancer Institute. Early detection through screening is an important tool in successfully treating the disease. Colonoscopy and sigmoidoscopy, along with tests measuring the presence of blood in the stool, are the tools used to screen for colon and rectal cancers. By examining the intestinal walls, the gastroenterologist detects and removes polyps, or precancerous lesions, and identifies abnormalities.

Pediatric Gastroenterology. Pediatric gastroenterology is an established subspecialty of pediatrics, whereas gastroenterology branches off from internal medicine. As such, pediatric gastroenterologists have a residency in pediatrics, and they take a separate subboard certification exam. Gastroenterologists in this subspecialty focus on gastrointestinal and nutrition-based conditions in infants, children, and adolescents. With additional training, pediatric gastroenterologists may narrow their specialization to liver transplantation, motility disorders, pancreatic diseases, endoscopic techniques for children, or nutrition.

All gastroenterologists form long-term relationships with their patients, but this is especially important for those specializing in pediatrics. Many pediatric gastric conditions are chronic and will follow the child throughout life. After the patient reaches adolescence, the pediatric gastroenterologist transfers care of the patient to a colleague; therefore, the pediatric specialist must have excellent communication skills. Pediatric gastroenterologists must consider the effect of a patient's condition on normal growth and development. Pediatric conditions often involve specialists from several disciplines, which means that the pediatric gastroenterologist must function well in teams.

Gastroesophageal Reflux Disease. The regurgitation of stomach acids into the esophagus, which causes a burning sensation in the chest known as heartburn, can lead to gastroesophageal reflux disease (GERD). Normally, a round muscle closes off the opening at the base of the esophagus leading to the stomach. Occasionally, the muscle fails to close completely, and the stomach contents wash up into the esophagus, burning the lining and causing the characteristic burning sensation. GERD may occur in adults or children.

GERD may be treated with over-the-counter antacids or other agents, but severe cases with persistent symptoms are best treated by a gastroenterologist. Upper endoscopy is useful for diagnosing GERD and evaluating treatment options. Prolonged GERD may damage the lining of the esophagus and cause bleeding ulcers.

Gastric and Peptic Ulcers. An erosion of the lining of the stomach is called a gastric ulcer; an erosion

Fascinating Facts About Gastroenterology

- As many as 70 million people in the United States are affected by a digestive disease, and about 9 percent of these people require hospitalization. An estimated 234,000 Americans die annually from digestive diseases, including gastric cancers.
- The United States has 11,704 certified gastroenterologists.
- Peptic ulcers affect 14.5 million Americans, resulting in more than 875,000 physician visits and 2 million prescriptions. Up to 20 percent of the population suffers from heartburn, or gastroesophageal reflux, at least once weekly.
- Colon polyps are small growths on the interior wall of the colon, some of which may become cancerous. Adenomas, or precancerous polyps, take about ten years to change from a polyp to a cancerous tumor.
- Women tend to have a higher incidence of constipation than men, perhaps because the colon empties slower in women than in men. Women also have less pressure in the anal sphincter compared with men, which enables men to withstand the urge to defecate longer than women.
- Colon cancer has few symptoms in the early stages, making screening important. Warning signs that may indicate colon cancer are blood in the stool, change in bowel habits (including formation of the stool), or abdominal pain.

on the duodenum is called a peptic ulcer. They are caused by infection with the bacterium *Helicobacter pylori* or the use of certain medications, such as nonsteroidal anti-inflammatory drugs (NSAIDs) but are not believed to be caused by stress. Smoking may worsen ulcers. Gastroenterologists played an important role in discovering *H. pylori* as a cause of ulcers.

Gastroenterologists diagnose ulcers by their symptoms, which include a dull or burning pain, weight loss, or vomiting, and a blood test for *H. pylori*. Upper endoscopy may be used if the patient has bleeding or if the ulcer blocks food from leaving the stomach. The gastroenterologist prescribes antibiotics and antacids to treat ulcers.

Parenteral Nutrition. Patients with serious medical conditions may be unable to eat or tolerate feedings introduced into the digestive tract. Parenteral nutrition—the injection of nutrients intramuscularly, intravenously, or subcutaneously—helps avoid complications of malnutrition. However, parenteral nutrition may be harmful in some instances, and it is up to the gastroenterologist to determine when the technique should be used. For example, parenteral nutrition has little effect in most patients following surgery, although it reduces postoperative complications in patients following surgery for esophageal or stomach cancer. Infants who are unable to eat and adults with prolonged malabsorption conditions benefit from parenteral nutrition.

Long-Term and End-of-Life Care. Patients in long-term care facilities and those with terminal illnesses receiving end-of-life care often suffer from digestive difficulties. Patients may be unable to eat properly and become malnourished. Patients receiving opioid medications for pain relief may develop constipation as a side effect of the drugs. Opioid-induced constipation causes additional pain and discomfort to patients already suffering from end-stage illnesses. Gastroenterologists are skilled at finding alternatives to taking patients off the analgesic medications, which would relieve abdominal discomfort but would leave the patients in severe pain.

Food Allergies. Gastroenterologists may be helpful in determining specific food allergies or intolerances (for example, lactose intolerance) that lead to gastric disorders. Patients with food allergies or intolerances are often seen by primary care physicians, who consult with gastroenterologists and allergists to determine the cause of the symptoms. Gastrointestinal signs of food allergy include constipation, colic, or gastroesophageal reflux in infants, or a severe reaction immediately after eating the food.

Obesity. Gastroenterologists are uniquely positioned to assist overweight patients because many conditions that require gastroenterological care result from excess weight. The gastroenterologist faces the challenge of recommending a realistic diet and exercise plan that the patient will follow. After obtaining a detailed patient history that includes a psychiatric evaluation, the gastroenterologist must determine the amount of weight the patient must lose to achieve a healthy body mass index, balanced with the patient's level of commitment to changing lifestyle habits and losing weight.

Several treatment options are available to the gastroenterologist depending on the patient's present weight, the desired amount of weight loss, and the patient's outlook. Gastroenterologists can assist patients with dietary plans using portion-controlled servings of low-fat foods and in developing realistic exercise plans to increase physical activity and energy use. The gastroenterologist helps patients develop weight-loss goals and provides support and guidance. The doctor may prescribe medications such as sibutramine (Meridia) or orlistat (Xenical). If the patient has been unable to lose weight through diet and exercise, has no unusual surgical risks, and is at risk for obesity-related complications, such as heart disease or diabetes, the doctor may perform bariatric surgery.

Research. Gastroenterology is a fertile field for research. Gastroenterologists with an interest in research perform their own studies on the cause and diagnosis of gastrointestinal disorders. They also work with pharmaceutical companies developing medications for GERD and ulcers, Crohn's disease, ulcerative colitis, inflammatory bowel disease, and other disorders, including gastrointestinal cancers.

IMPACT ON INDUSTRY

Global Standards. Gastroenterology has become a major medical specialty all over the world, but gastroenterologists are called on to perform different roles depending on the country. In developed nations, gastroenterologists follow structured educational paths to achieve certification. Advanced skills and specialized care are the primary focus. The field is complex, with emphasis on diagnostic tools such as endoscopy and medical management of gastric conditions, which include gastroenterological cancers and immunology. The addition of hepatology, which includes all types of hepatitis and cirrhosis, has greatly expanded the field, added to the complexity, and necessitated more intense education.

In developing countries, the field of gastroenterology is less well defined and the training requirements vary widely. Provision of tightly structured and focused curricula is less important than the need to quickly educate potential gastroenterologists to enable them to treat growing numbers of people with chronic liver and gastric conditions. In many developing countries, postgraduate education

is limited to two additional years of training in endoscopic procedures, compared with the five years of gastroenterological training required in the United States. In an effort to establish a global standard for gastroenterological training, the World Gastroenterology Organization (WGO) outlined educational standards in developed and developing nations, highlighting shortcomings and areas for improvement.

The WGO recommends that all gastroenterologists have a basic foundation in specific areas, including physiology, pharmacology, epidemiology, and the anatomy and development of the gastrointestinal tract. The organization also recommends that basic training cover nutrition and metabolism, diagnosis, prevention, treatment, bioethics, and cost containment.

The WGO outlines specific skills in which gastroenterologists should be proficient, regardless of the nation in which they practice or their specific patient population. These skills include endoscopy techniques and the ability to interpret laboratory data, as well as general skills such as leadership and team management, communication, and professionalism.

Research. Researchers in gastroenterology may work in academic settings conducting basic research or for pharmaceutical companies focusing on specific conditions or drug side effects.

Gastroenterological conditions are an active area of pharmaceutical development. Medications for GERD, gastric and peptic ulcers, and inflammatory bowel diseases are in demand, and active research into treatments for gastroenterological cancers is expanding as new therapies are developed.

Gastroenterologists are valuable as consultants for medications that treat conditions outside the gastroenterological tract but have side effects such as nausea, vomiting, or abdominal pain. Furthermore, they can often discern the appropriate treatment for various conditions having gastrointestinal distress as a symptom without compromising the action of medications designed to address the patient's primary condition.

Education. The gastroenterological training process in the United States requires students to train with seasoned professionals. Experienced gastroenterologists demonstrate proper interaction and communication with patients and other specialists, along with proper diagnostic techniques. They observe the students and critique their approach. The practical, personalized training provided by experienced practitioners is an important part of the educational process.

CAREERS AND COURSE WORK

Students planning a career in gastroenterology often study biology or chemistry as an undergraduate, making sure to fulfill any medical school requirements. In addition to aptitude in the life sciences, potential gastroenterologists need to demonstrate excellent listening and communication skills and compassion for their patients. After completing medical school and earning their license to practice, students must become certified in internal medicine. Internal medicine certification requires a three-year postgraduate residency, demonstrated clinical competency, and successful completion of a comprehensive final exam. To become certified in gastroenterology, students must complete three years of fellowship training exclusively in gastroenterology, complete eighteen months in clinical practice, and successfully pass a comprehensive subspecialty exam. By the time gastroenterologists become certified, they have had an additional five years of specialized training beyond medical school.

Students interested in pediatric gastroenterology must complete their three-year postgraduate residency and take the board exam in pediatrics rather than internal medicine. Students interested in specializing in treating liver problems or gastric cancers undergo additional training with specialists in those areas.

The specialized gastroenterology training is supervised by four professional societies: the American College of Gastroenterology, the American Gastroenterological Association, the American Society of Gastrointestinal Endoscopy, and the American Board of Internal Medicine. The American Board of Internal Medicine administers the final comprehensive competency exam.

During their fellowships, students become proficient in procedures essential to the gastroenterological field, including endoscopy, colonoscopy, sigmoidoscopy, and esophageal dilation. Students learn to remove colon polyps as well as foreign bodies lodged in the esophagus. Students are observed as they interact with patients and evaluated on their ability to examine patients and take complete health histories.

SOCIAL CONTEXT AND FUTURE PROSPECTS

Gastroenterology is an active field that continues to expand. As the population ages, more people will require cancer screenings and polyp removal. Conditions such as ulcers, GERD, and hepatitis are diagnosed with increasing frequency. The general population is not only growing older but also heavier. Obesity is a growing health concern. Gastroenterologists are integral in managing complications of obesity such as gallstones and GERD, and they serve as important partners in teaching overweight patients proper nutrition and weight-loss strategies, including gastric banding procedures.

In developing countries and regions with unsanitary conditions, gastroenterologists are needed to combat health concerns such as diarrhea and malnutrition. Gastroenterologists can provide basic nutritional counseling and assist patients in fighting parasitic or infectious diseases that compromise the absorption of nutrients.

Cheryl Pokalo Jones, B.A.

FURTHER READING

Butcher, Graham. *Gastroenterology: An Illustrated Colour Text.* Philadelphia: Elsevier Health Sciences, 2003. Full-color clinical photographs and detailed line drawings illustrate gastroenterological and liver diseases.

Collins, Paul. *Gastroenterology: Crash Course.* Philadelphia: Elsevier Health Sciences, 2008. Offers basic definitions and explanations of all aspects of gastroenterology in an easily understandable manner.

Grendell, James H., Scott L. Friedman, and Kenneth R. McQuaid. *Current Diagnosis and Treatment in Gastroenterology.* 2d ed. New York: Lange Medical Books, 2003. This comprehensive reference discusses all gastroenterological conditions, including hepatic, pancreatic, and biliary conditions.

Travis, Simon P. L., et al. *Gastroenterology.* 3d ed. Malden, Mass.: Blackwell, 2005. A concise, informative manual on gastroenterology with a global perspective.

WEB SITES

American College of Gastroenterology
http://www.acg.gi.org

American Gastroenterological Association
http://www.gastro.org

American Society of Gastrointestinal Endoscopy
http://www.asge.org

National Institute of Diabetes and Digestive and Kidney Diseases
National Digestive Diseases Information Clearinghouse
http://digestive.niddk.nih.gov

See also: Epidemiology; Geriatrics and Gerontology; Nutrition and Dietetics; Pediatric Medicine and Surgery.

GAS TURBINE TECHNOLOGY

FIELDS OF STUDY

Mechanical engineering; thermodynamics; aerodynamics; fluid dynamics; heat transfer; environmental engineering; control engineering; industrial engineering; electrical engineering; safety engineering; physics; mathematics; materials science; metallurgy; ceramics; process engineering; engineering; electronics; aeronautics; business administration.

SUMMARY

Gas turbine technology covers design, manufacture, operation, and maintenance of rotary conversion engines that generate power from the energy of the hot, pressurized gas they create. Key components are air and fuel intake systems, compressor, combustor, the gas turbine itself, and an output shaft in those gas turbines that are not designed to provide thrust alone as jet engines. Primary applications of gas turbines are jet engines and generators for industrial power and electric utilities. Sea and land vehicles can also be propelled by gas turbines.

KEY TERMS AND CONCEPTS

- **Airfoil:** Wing-shaped blades used in compressors and turbines to interact with gas stream.
- **Brayton Cycle:** Term for the scientific principle behind the operation of the gas turbine.
- **Cogeneration:** Purposeful use of the remaining heat from a gas turbine exhaust stream.
- **Combined Cycle:** Using a gas turbine in combination with a steam turbine to increase its fuel efficiency.
- **Combustor:** Unit where fuel mixes with pressurized air and is ignited and burned.
- **Compressor:** Unit that increases the pressure of intake air by compressing its volume.
- **Free Power:** Portion of power generated by the turbine not needed to drive the compressor.
- **Jet Engine:** Gas turbine designed to release its free power as thrust to lift, propel, and stop an aircraft.
- **Nozzle:** Apparatus through which the gas stream is injected into the turbine; also used to inject fuel into combustor.

- **Spool:** Industry name for the output shaft of the turbine conveying its power.
- **Turbine:** Rotary engine that can be fueled by a gas stream.
- **Turbofan:** Jet engine designed to provide stronger thrust by driving a fan instead of relying on exhaust.
- **Turboprop:** Jet engine designed to provide stronger thrust by driving a propeller instead of relying on exhaust.

DEFINITION AND BASIC PRINCIPLES

Gas turbine technology concerns itself with all aspects of designing, building, running, and servicing gas turbines, which provide jet engines for airplanes and represent the heart of many contemporary power plants, among other uses. Strictly speaking, the gas turbine itself is only one part of a complex engine assembly commonly given this name.

A gas turbine employs the physical fact that thermal energy can be converted into mechanical energy. Its basic principle is often called the Brayton cycle, named after George Brayton, the American engineer who developed it in 1872. The Brayton cycle involves compression of air, its heating by fuel combustion, and the release of the hot gas stream to expand and drive a turbine. The turbine creates both power for air compression and free power. The free power gained can be used either as thrust in traditional jet engines or as mechanical power driving another unit such as an electric generator, pump, or compressor.

The gas turbine is closely related to the steam turbine. The gas turbine got its name from the fact that it operates with air in gaseous form. A wide variety of fuel can be used to heat the gaseous air.

BACKGROUND AND HISTORY

The oldest known reference to an apparatus utilizing the physical principles of a gas turbine is a design by Leonardo da Vinci from 1550 for a hot-air-powered roasting spit. In 1791, British inventor John Barber obtained a patent for the design of a combined gas and steam turbine. However, the lack of suitable materials to withstand the heat and pressure needed for a working gas turbine impeded any practical applications for a long time.

In 1872, German engineer Franz Stolze designed a gas turbine, but driving its compressor used more energy than the turbine generated. In 1903, the first gas turbine with a surplus of power was built by Norwegian inventor Aegidius Elling. The idea to use gas turbines to build jet engines was pioneered by British aviation engineer and pilot Sir Frank Whittle by 1930. At the same time, German physicist Hans von Ohain developed a jet engine on his own. On August 27, 1939, the German Heinkel He 178 became the first flying jet airplane. That year, the first power plant using a gas turbine became operational in Switzerland. Since that time, the twin use of gas turbine technology either to propel jet airplanes or serve as a source for generating power on the ground, particularly electricity, has been subject to many technological advances.

HOW IT WORKS

Brayton Cycle. A gas turbine engine is a device to convert fuel energy via the compression and subsequent expansion of air. Its working process is commonly called a Brayton cycle. The Brayton cycle begins with the compression of air, in gaseous form, for which energy is expended. For this reason, a gas turbine assembly needs an air intake and a compressor where air is pressurized. Next, more energy is added to the compressed air through the heat from the combustion of fuel. Fuel is burned, most commonly, internally in the engine's combustor. However, in a variation called the Ericsson cycle, the fuel is burned externally and the generated heat is relayed to the compressed air. The third step of the cycle comes with the release of the heated gas stream through nozzles into the gas turbine proper. The gas expands as it loses some of its pressure and cools off. The energy released by gas expansion is captured by the gas turbine, which is driven in a rotary fashion as its blades are turned by the exiting hot gas stream. In the last step, the expanding air releases its leftover heat into the atmosphere.

Power Generation. As the hot gas stream flows along stator vanes to hit the airfoil-shaped rotor blades arranged on a disk inside the gas turbine, it drives the blades in rotary fashion, creating mechanical power. This power is captured by one or more output shafts, called spools. There are two uses for this power. The first is to drive the compressor of the gas turbine assembly, feeding power back to the first

Engineers assembling a jet aircraft gas turbine engine. (RIA Novosti/Photo Researchers, Inc.)

step of operations. Any free power remaining can be used to drive external loads or to provide thrust for jet engines, either directly through exhaust or by driving a fan or propeller.

The engineering challenge in gas turbine technology is to gain as much free power as possible. Attention has been focused both on materials used inside the gas turbine, looking for those that can withstand the most heat and pressure, and in arranging the individual components of the gas turbine to optimize its output. The metal of single-crystal cast alloy turbine blades can withstand temperatures of up to 1,940 degrees Fahrenheit. Ingenious air-cooling systems enable these blades to deal with gas as hot as 2,912 degrees Fahrenheit. Top compressors can achieve a 40:1 ratio. At the same time, designing gas turbines with multiple shafts to drive low- and high-pressure compressors or adding a power turbine behind the first turbine used only to gather sufficient power for the compressor has also improved efficiency. While early gas turbines used between 66 and 75 percent of the power they generated from fuel to drive their own compressor, leaving only 25 to 34 percent of free power, contemporary gas turbines for industrial power can achieve up to 65 percent of free power.

Fuel. Fuel for gas turbines is variable and ranges from the hydrocarbon product kerosene for jet engines to coal or natural gas for industrial gas turbines. The engineering challenges have been to optimize fuel efficiency and to lower emissions, particularly of nitrogen oxides.

Operation and Maintenance. Gas turbines require careful operation as they are very responsive, and malfunctions in either the compressor or turbine can happen in fractions of a second. The primary control systems are handled by computer and are hydromechanical and electrical. From start and stop and loading and unloading the gas turbine, operating controls cover speed, temperature, load, surges, and output. The control regimen ranges from sequencing to routine operation and protection control.

Gas turbines have some accessories to facilitate their operation. These include starting and ignition systems, lubrication and bearings, air-inlet cooling and injection systems for water, and steam or technical gases such as ammonia to control nitrogen oxide emissions.

To facilitate maintenance, gas turbines used for aircraft and those modeled after these have a modular design so that the individual components such as compressor, combustor, and turbine can be taken out of the assembly individually. Those gas turbines allow also for a borescopic inspection of their insides by an optical tube with lens and eyepiece. Heavy industrial gas turbines are not designed for borescopic inspection and must be dismantled for inspection and maintenance. Preventive maintenance is very important for gas turbines.

APPLICATIONS AND PRODUCTS

Jet Engines. The first jet engines were designed to use the free power of gas turbines exclusively for thrust. This gave them a speed advantage over piston- engine aircraft but at the price of very low fuel economy. As a result, gas turbine technology developed more economical alternatives. For helicopters and smaller commercial airplanes, the turboprop system was developed. Here, the gas turbine uses its free power to drive the propeller of the aircraft.

The turbofan jet engine is used in 90 percent of contemporary medium to large commercial aircraft. Efficiency is increased by the addition of a fan that acts like a ducted propeller and that is driven by the free power of the gas turbine. In contemporary high-bypass turbofan jet engines in use since the 1970's and continuously improved since, much of the air taken in bypasses the compressor and is directly propelled into the engine by the turbine-driven fan, up to a bypass ratio of 5:1. Only a smaller part of air is taken into a low- and then a high-pressure compressor, joined with fuel burned in the combustor and driving both a high- and low-pressure turbine. The net resulting thrust, primarily from the fan, is achieved with high fuel economy and relatively little noise.

Military aircraft, the only ones flying at supersonic speed, also use afterburners with their gas turbines. This is done by adding another combustor behind the turbine blades and before the exhaust nozzle, creating extra thrust at the expense of much fuel.

Aero-Derivative Turbines. Because of their relatively low weight, gas turbines based on jet engine design have been used on the ground for power generation and propulsion. Especially with the contemporary trend toward turbofan jet engines, an aero-derivative turbine that uses one or more spools, or output shafts, to provide a mechanical drive needs very little adaptation from air to ground use. There are also hybrid gas turbines that use an aero-derivative design but replace the jet engine's lighter roller and antifriction ball bearings with hydrodynamic bearings typical of the heavy industrial gas turbine.

With higher shaft speed but lower airflow through the turbine, aero-derivative turbines require less complex and shorter maintenance than other ground gas turbines. They are often used in remote areas, where they are employed to drive pumps and compressors for pipelines, for example. Because of their quicker start, stop, and loading times, aero-derivative gas turbines are also used for flexible peak load power generation and for ground propulsion.

Heavy Industrial Gas Turbines. Gas turbines for use on the ground can be built more sturdily and larger than jet engines. These heavy units have been generally used for power generation or to drive heavy industrial pumps or compressors, with power generation increasingly important. Gas turbine technology has experimented with a variety of designs for these turbines. One decision is whether to place the output shaft (spool) at the "hot" end where the gas stream exits the turbine, or at the "cold" end in front of the air intake. "Cold" end drives are easier to access for service and do not have to withstand the hot environment at the turbine end, but their position has to be carefully designed so as not to disturb the air intake. If the output shaft would cause a turbulence or vortex in the air flowing into the compressor this could lead to a surge potentially destroying the whole engine.

There are also design differences regarding the numbers of shafts (spools) and turbines within a contemporary heavy industrial gas turbine. The basic form has one output shaft rotating at the speed of the compressor and turbine. At the output, this speed can be geared up or down depending on the speed desired for the application the gas turbine is driving. This design is almost exclusively used for power generation. An alternative to minimize gear losses is to put a second, free-power turbine behind the first gas turbine driving the compressor. This means that the speed of the free-power turbine can be regulated independently of the turbine speed needed to drive the compressor, which makes it an attractive design when pumps or compressors are driven by the gas turbine. This design is only possible with a "hot" end configuration. Finally, there are gas turbine designs that use more than one shaft (spool). A dual spool split output shaft gas turbine, for example, employs three output shafts to operate independently with a high-pressure and low-pressure turbine-compressor assembly as well as a free- power turbine.

Because a single-cycle, stand-alone gas turbine has a fuel efficiency of as little as 17 percent, meaning 83 percent of the energy created is used for the compressor, engineers have combined gas turbines for power generation in cogeneration or combined-cycle power plants. In a cogeneration plant, the remaining heat that exits the gas turbine is used for industrial purposes, such as heating steam for a refinery. In a combined-cycle power plant, the heat from typically two gas turbines fuels a steam turbine. This can create fuel efficiencies ranging from 55 to 65 percent or more. These gas turbines typically create about 250 to 350 megawatts of electrical power each.

Marine and Tank Propulsion. Aero-derivative gas turbines are also used to propel ships, particularly military vessels. Military requirements of high speed outweigh the fuel and construction cost disadvantages that make gas turbines too expensive for commercial ships. Gas turbines can also be used as tank engines, for example in the American M1A1 Abrams tank or the Russian T-80. However, their high fuel use provides an engineering challenge, particularly at idle speed. The M1A1 tanks have been retrofitted with batteries to use for idling, and the Russian T-80 was replaced by the diesel-engine-powered T-90.

Turbochargers. Their low fuel efficiency makes gas turbines unsuitable for car propulsion. However, small gas turbines working as turbochargers are commonly added to increase the power of diesel car engines. The power from the turbine is used to compress the air taken in by the diesel engine, increasing its performance.

IMPACT ON INDUSTRY

Gas turbine technology has a vast global impact in the fields of electricity, industrial power generation, and civil and military aviation. Because of the capital and research-intensive nature of the field, gas turbines for most applications are designed and built by a few large global companies with headquarters in North America, Western Europe, Russia, Japan, India, and China. However, gas turbines are used worldwide, both on the ground and powering most of the world's commercial airliners and military aircraft. The key engineering challenges that drive research across public and private institutions are to make gas turbines more efficient and operate with fewer emissions and noise.

Government Agencies. Traditionally, governments of industrialized countries at war or in prolonged political antagonism toward another have been prime sponsors of gas turbine technology, especially in the field of the jet engine. World War II and the Cold War saw key government support for jet engine and gas turbine design to gain a military advantage in the air and on water. In the West, gas turbine technology was one of the key products of the military-industrial complex, as Western governments sponsored research by private industrial corporations, and these military products were later adapted for civilian use. Governmental focus in industrial nations has generally shifted away from military applications and

Fascinating Facts About Gas Turbine Technology

- Gas turbines are small mechanical wonders, with up to 4,000 individual parts often handled and fitted by hand. Maintaining them requires the same great technical skills as building them.
- With the Messerschmitt Me 262, Germany developed the world's first jet fighter, but by the end of World War II, the British had their own with the Gloster Meteor. The two jets never fought each other, though.
- Industry analysts expect that by 2018, about 7,000 state-of-the-art jet engines will be built and commissioned every year, the vast majority of the turbofan design.
- The AGT1500 gas turbine power plant can propel the American M1A1 Abrams battle tank up to 41.5 miles per hour across level terrain. However, even the U.S. Army balked at the fuel costs and retrofitted the engine with a battery pack to power the tank while idling.
- Since the late eighteenth century folk craftsmen in Germany's Erz Mountains have used the principle of the gas turbine to move small wooden Christmas ornaments: They revolve on disks powered by rotor blades on a shaft driven by hot air rising from candles.
- Once engineers designed ways to utilize the remaining heat from the exhaust of a gas turbine employed for power generation, overall system efficiency almost doubled to as much as 65 percent.
- A contemporary gas turbine airfoil blade is grown as a single crystal from a molten super alloy to achieve great heat resistance in the turbine.

toward sponsoring research in cleaner, "greener," and more efficient gas turbines for electric utilities and power generation. These governments funnel considerable sums into public and commercial research and set policy guidelines for cleaner power generation that drives respective innovations.

Universities and Research Institutes. Public research in almost all industrially advanced nations is focused on improving efficiency, emissions, and innovation in gas turbine technology. One new research avenue concerned micro gas turbines which, if they could be constructed at affordable cost and with reasonable efficiency, may become a source of decentralized and customized power generation. There was primary research to discover a technologically feasible way for gas turbines to run on hydrogen, which would constitute a technological quantum leap. There was also great research in material sciences, particularly ceramics, to create new components for gas turbines that could cope with even higher temperatures and pressure. Research also centered on integrating gas turbines in ever-more efficient power plant assemblies. Gas turbine technology was also pursued in aeronautics for improved jet engine designs.

Industry and Business. The power industry has represented a huge market for gas turbines. From 2000 to 2009, the total global electric output of gas turbines rose from 2.5 million megawatts to 3.7 million megawatts per year and is expected to rise to 5 million megawatts by 2020. This created a huge demand for the design and manufacture of new gas turbines as well as their operation and maintenance around the world. Even though the annual market value for new gas turbines was reduced by almost half during the 2007-2009 recession (from about $20 billion to just above $10 billion), economic recovery increased this value again.

In the aviation industry, the market for jet engines has been very robust. For the period from 2007 to 2016, market analysts have predicted an annual global value of new engine sales ranging from a conservative $18 billion to an enthusiastic $30 billion. Of this, 85 percent are for civilian and 15 percent for military aircraft. As all operating jet engines have to be maintained regularly, there was an annual global market value of about $25 billion for engine maintenance, repair, and operations (MRO) in 2005, which analysts expected to grow as the world added more operating aircraft.

Major Corporations. America's General Electric has been the market leader both in manufacturing of gas turbines for power generation (at more than 40 percent market share throughout the 2000's) and in jet engine manufacturing (holding about 30 percent market share during the same time). In the power industry, Germany's Siemens company, Japan's Mitsubishi Heavy Industries, France's Alstom, and India's state-owned Bharat Heavy Electricals have been major players.

England's Rolls-Royce Group has been a close contender for market-share leader in jet engines, achieving this feat with a 32 percent share in 2001. Another key jet engine manufacturing company is America's Pratt & Whitney. Other niche players include Germany's Daimler, Russia's NK Engines Company, Japan's Kawasaki Heavy Industries, or mainland China's Shenyang Aeroengine Research Institute. Cooperation and partnership among companies in the gas turbine manufacturing industry have been very common.

CAREERS AND COURSE WORK

Gas turbine technology has been a key to two of the world's leading industries, power and aviation, so job demand in the field should remain very strong. Students interested in the field should take science courses in high school, particularly physics, as well as mathematics and computer science. An associate's degree in an engineering or science field (engineering or industrial technology) will provide a good entry.

A bachelor of science in an engineering discipline, particularly mechanical, electrical, or computer engineering, is excellent preparation for an advanced job. A B. S. in physics or mathematics would point to a more theoretically informed career, perhaps in design. A bachelor of arts in environmental studies is also useful, as emission control is becoming a major part of gas turbine technology. A minor in any science is always beneficial.

If one's career focus is on advanced work, a master of science in mechanical, electrical, and computer engineering, or in environmental science and management, could be chosen. For top scientific positions, a Ph.D. in these disciplines, together with some postdoctoral work, is advisable.

Because of the global nature of the field, students should maintain a general openness to work abroad or in somewhat remote locations, with the exception of those purely interested in design. The field can also be attractive to students with expertise in support functions, including those who have earned a B.A. in English, communications, economics, or biology. A master of business administration would serve as preparation for the business end of the field.

SOCIAL CONTEXT AND FUTURE PROSPECTS

As more nations industrialize and global development continues, the demand for power and mobility, including air travel, is expected to increase. Especially with the key applications of jet engines and gas turbines for power generation, the field of gas turbine technology is likely to keep its great relevancy. The quest for more efficient gas turbines that combust their fuel with as little emissions as possible will continue to motivate major developments in the field. If gas turbines can become an ever-more efficient and low-emission source of generating power they have the potential, like fuel cells, to become part of the next generation of power sources. There is also much promise linked to micro gas turbines as a source of efficient, affordable, clean, and decentralized power.

For jet engines, the design challenge is to reduce fuel consumption and noise and increase power. There is ongoing research to employ gas turbines in combination with electric hybrid car engines to lower overall carbon dioxide emissions from personal transport. This will grow in importance as more people in developing nations acquire cars.

R. C. Lutz, B.A., M.A., Ph.D.

FURTHER READING

Boyce, Meherwan P. *Gas Turbine Engineering Handbook.* 3d ed. Burlington, Mass.: Gulf Professional Publishing, 2006. Focuses on design, components, operation, and maintenance issues with emphasis on gas turbines used in combined cycle power plants. Each chapter has own bibliography.

Giampaolo, Tony. *Gas Turbine Handbook: Principles and Practices.* 4th ed. Boca Raton, Fla.: CRC Press, 2009. Comprehensive coverage of technology and all applications. Clear organization, figures, bibliography.

Kehlhofer, Rolf, et al. *Combined-Cycle Gas and Steam Turbine Power Plants.* 3d ed. Tulsa, Okla.: PennWell, 2009. Application of gas turbine as part of cutting-edge power-generation technology. Presents concepts, components, applications, operations of these plants. Each chapter has illustrative figures.

Peng, William W. *Fundamentals of Turbomachinery.* Hoboken, N.J.: John Wiley & Sons, 2008. Chapter 8 discusses gas turbines and covers thermodynamics, design, efficiency, and performance, as well as applications; illustrated.

Rangwala, A. S. *Turbo-Machinery Dynamics: Design and Operation.* New York: McGraw-Hill, 2005. Comprehensive textbook, very useful presentation of the field. Covers all applications, component design, materials, and manufacture for gas turbines. Excellent illustrations.

Soares, Claire. *Gas Turbines: A Handbook of Air, Land and Sea Applications.* Burlington, Mass.: Butterworth-Heinemann, 2008. Good, comprehensive overview of all aspects of the field from a practical viewpoint.

WEB SITES

Gas Turbine Association
http://www.gasturbine.org

Gas Turbine Builders Association
http://www.gtba.co.uk

International Gas Turbine Institute
http://igti.asme.org

See also: Ceramics; Electrical Engineering; Engineering; Fluid Dynamics; Mechanical Engineering; Metallurgy.

GEMOLOGY AND CHRYSOLOGY

FIELDS OF STUDY

Geology; mineralogy; chemistry; economic geology; crystallography.

SUMMARY

Gemology is the study of minerals that are attractive enough to be worn as jewelry after being cut and polished. Gemstones often are transparent and internally reflect light, are colored, and have a brilliant luster. Sometimes the term "gemstone" is used to refer to attractive aggregates of minerals (such as turquoise), organic materials (such as amber), rocks (such as marble), or synthetically made gemstones. Only about seventy of the many thousands of known minerals have varieties that are of gem quality. Diamond (carbon) is the most important gemstone. Chrysology refers to the study of precious metals such as gold, silver, and platinum. Precious metals are opaque and often have colors (such as the yellow of gold) that make them attractive in jewelry.

KEY TERMS AND CONCEPTS

- **Carat:** Measure of weight used for gemstones, equal to 0.2 gram.
- **Cleavage:** Tendency of a mineral or gemstone to break along smooth, plane surfaces.
- **Density:** Mass per volume of a substance.
- **Hardness:** Resistance of a mineral surface to scratching by another mineral. The hardness scale ranges from 1 to 10, with diamond having a hardness of 10 and the common mineral quartz having a hardness of 7.
- **Luster:** Appearance of a gem's surface in reflected light; some gemstones appear vitreous (reflect like glass), pearly, silky, adamantine (refracts light strongly like diamond), and metallic.
- **Mineral:** Naturally occurring element or compound that has a definite composition and arrangement of atoms that is reflected by its physical characteristics.
- **Refractive Index:** Ratio of the speed of light in air (300,000 kilometers/second) to the speed of light in a mineral (much less speed in the mineral).
- **Silicate:** Mineral that consists of silicon-oxygen negative ions combined with other positively charged metals.
- **Transparent:** Describes a gemstone that transmits light so that an object may be seen through it.

DEFINITION AND BASIC PRINCIPLES

Gemstones must be attractive to viewers. Their attractiveness depends on the color, transparency, and luster of the gem and the way it is cut. The most important gemstones are diamonds, sapphires, rubies, and emeralds. Diamonds are often yellowish and transparent, with a brilliant adamantine luster. Some of the most valuable diamonds, however, contain more intense green, red, blue, or black colors; their rarity often makes them more valuable.

Gemstones ideally should be harder than the common mineral quartz and should not fracture easily so that the gemstone is not easily scratched or broken when worn as jewelry. About ten minerals have these characteristics. Diamond is the hardest known mineral; however, diamonds can be fractured along cleavage planes if excessive force is applied to them. Gemstones softer than quartz will retain their luster and beauty as long as they are not scratched. Valuable gemstones should not contain many small impurities such as minerals or fractures that might degrade the beauty of the gem.

Gemstones are often cut in certain ways to enhance their beauty. The choice of cutting method depends in part on the characteristics of the stone. Diamond, for instance, has four major planes of cleavage (octahedral cleavage) along which the diamond may be broken. Diamonds are cut along these octahedral cleavage planes so that light is internally reflected within the diamond, producing many flashes of light as the gem is viewed.

In contrast to gemstones, precious metals such as gold and silver are opaque and have a metallic luster. Their beauty comes from their color in reflected light.

BACKGROUND AND HISTORY

Gemstones and precious metals are mentioned in ancient writings from cultures around the world, and they have been found during many archaeological digs, suggesting that they have been used for thousands of years. For instance, ancient Babylonians had gemstones of jasper (red, fine-grained quartz) and lapis lazuli (blue, the mineral lazulite, a sodium calcium aluminum silicate sulfate, mixed with other minor minerals).

A number of gemstones were found in Egyptian tombs dating from 5000 to 3000 B.C.E. These gems include lapis lazuli, agates (silica-rich mineral), jasper, and emeralds (dark green, transparent beryl, a beryllium aluminum silicate). Some of these gems may be traced to specific geologic sources, suggesting that trading occurred over large areas. An ancient Egyptian papyrus contains a map showing the location of several gold mines and the miners' quarters. Gold was likened to the skin of the gods, was used in funerary art, and worn only by royalty.

Many gemstones were worn not only because of their beauty but also because many people believed that the gems could protect them from evil forces and diseases and make them wise. Astrologers claimed that certain gems gave the wearer special powers in certain months. The Bible mentions a variety of gemstones, such as emeralds, sapphires (blue, transparent corundum, and aluminum oxide mineral), diamonds, agates, and amethysts (purple, transparent quartz), and gold and silver, which were sources of wealth and used for decoration.

How It Works

Traditional Identification Methods. Most gemstones and precious metals are minerals or are found in minerals, so the classic methods of identifying minerals using physical properties may be used. These physical properties are crystal forms (such as cubes and octahedons), color, hardness, luster (metallic or nonmetallic), cleavage, density, and microscopic characteristics. Experienced gemologists may be able to identify a gemstone if they can determine its luster, color, and other features by visual inspection, using no more than a magnifying glass. Some properties such as hardness, however, may be of limited use to identify a valuable gemstone because checking for hardness could damage the specimen.

Some specialized techniques may be used to identify a gemstone. Determining the gemstone's density can be very useful if the gemstone is not mixed with other minerals and is not mounted in jewelry. Gemologists might place the gem in a series of liquids of known densities. If the gemstone is denser than the liquid, it will sink; if it is less dense, it will float. Most minerals have very different densities, so this process can confirm the identity of the material.

Identification Using Instruments. A binocular microscope with a ten- to sixty-power magnification may be used to observe a gemstone using light reflected from the specimen or light transmitted through the specimen. Imperfections within a gemstone, for instance, can be seen using transmitted light.

The refractive index of a gemstone may be obtained using a refractometer. Most minerals have a unique refractive index, so this determination can also help confirm the identity of the gemstone. Refractive indices commonly range from a low of 1.4 to a high of 3.0. A gemstone typically should have one flat face that is placed on the stage of the refractometer in immersion oil with an index of refraction of 1.81. The mineral and light source are moved relative to each other until the light is internally reflected, and the refractive index is read directly on a scale. Some minerals have only one refractive index, so the readings are straightforward; others have more than one refractive index in different directions, so determining these values is more challenging.

Another approach to determining the refractive index is to place a gemstone in a series of liquids of known and different refractive indices and observe them under a microscope. If the refractive index of the gem is the same as the refractive index of the liquid, then the mineral nearly disappears; if the refractive index of the gem is substantially different from the refractive index of the oil, then the mineral edges stand out in bold relief.

If these methods fail to identify a gemstone, then other more destructive techniques such as X-ray diffraction, electron microprobe analysis, and internally coupled plasma mass spectrometry might be used for identification. Often these techniques use material that has been scraped off parts of the gemstone that will not be exposed to a viewer. X-ray diffraction techniques can be especially useful in confirming the identity of a mineral because each mineral presents a unique pattern.

Applications and Products

The most important gemstones are varieties of diamond, corundum (aluminum oxide; sapphires and rubies), beryl (beryllium aluminum silicate;

especially emeralds), tourmaline (sodium calcium, lithium silicate borate hydroxide mineral), spinel (magnesium aluminum oxide), topaz (aluminum silicate hydroxide), and quartz (silica, such as amethyst).

Diamond, because of its hardness, high refractive index, numerous colors, rarity, and ease of cutting, is the most important gemstone. Diamonds are the most common gemstones in jewelry stores. The majority of diamonds, however, are not of gem quality, but they can be used for industrial purposes such as in cutting materials. Most diamonds have a yellow tinge, but some rare diamonds with few imperfections may have intense green, blue, gold, red, and black colors that make them very valuable. The

Beth Leinonen, a gemologist, examines a diamond for imperfections at a store run by Brilliant Earth, which uses conflict-free diamonds from Canadian mines, in San Francisco, California, on June 5, 2009. (Bloomberg via Getty Images)

largest diamond ever found was larger than 3,100 carats. It was cut into smaller diamonds, the most famous of which is the colorless 530-carat Star of Africa diamond.

Rubies are the red gemstone of corundum; sapphires can be blue, yellow, green, or violet. Only diamond is harder than rubies or sapphires. Rubies and sapphires are difficult to cut because they have no cleavage and are brittle. Some of the most famous rubies and sapphires are much more expensive per carat than diamonds. One of the most famous sapphires is the Star of India (53 carats), and one of the most famous rubies is the Edward ruby (167 carats).

Beryl has many colors, but the dark green, transparent emeralds are the most expensive. Blue beryls of gem quality are called aquamarines. Some of the biggest and most beautiful specimens of emeralds are larger than 2,200 carats.

Tourmaline, spinel, and topaz occur in many colors. Pink and green varieties of tourmaline are the most popular, and the red varieties of spinels are the most valued. Good specimens of pink topaz are the most expensive.

Quartz is usually transparent and colorless, but it may be violet, black, blue, yellow, pale green, and pink. Amethyst is the name for violet quartz, which is generally more valuable than the other varieties of quartz.

Lesser Gemstones. Many other gemstones of lesser importance are commonly used. For example, spodumene (lithium aluminum silicate) can occur as transparent yellow, green, and colorless specimens. Turquoise (calcium aluminum silicate) is a blue to green, opaque mineral that is usually mixed with other dark impurities or some copper minerals; it has a hardness less than quartz and a good cleavage, so care must be taken not to damage it.

Artificial Gemstones. Artificial gemstones have been produced since the early part of the twentieth century. They are often difficult to distinguish from natural gemstones, thereby somewhat diminishing the demand for some natural gemstones. Artificial gemstones can be made in several ways. Some are produced by precipitation of the gems from a hot, steam solution; others are created by melting, at the appropriate temperature and pressure, a mixture of components in a ratio that approximates that of the natural gem they will resemble.

Since the 1950's, artificial diamonds have been created by heating carbon to a very high temperature and pressure. Most of these diamonds are not of gem quality but can be used as abrasives or for other industrial purposes. Artificial diamonds can be distinguished from natural diamonds by the presence of certain impurities, such as nitrogen or boron used in their manufacture.

Precious Metals. Gold and silver have been used for thousands of years as both a medium of exchange and for ornamentation. Gold, a malleable metal, and silver were used to make plates, cups, drinking vessels, ornamental objects, religious icons, and jewelry. Gold and silver have also been used in applications such as dentistry, electronics, and engineering. The refining of platinum was not perfected until the nineteenth century, so fewer applications have been developed. Platinum began to be used more commonly in jewelry in the 1960's, and it is also used as a catalyst in some chemical reactions.

IMPACT ON INDUSTRY

The cost of gemstones is driven by supply and demand. The most desirable gemstones—diamonds, rubies, emeralds, and sapphires—have the highest cost per carat. The price of a particular gemstone depends on the rarity of its color, its luster, the number of imperfections, and the desirability of its cut. The revenue from diamonds accounts for about 90 percent of that from all gemstones.

Similarly, the prices of precious metals such as gold, silver, and platinum are driven by supply and demand. In the twentieth century, the cost of gold ranged from as little as about $35 per ounce from 1935 to 1967 to as high as $615 per ounce in 1980. In 2010, the price of gold rose to more than $1,300 per ounce, while silver sold for around $23 per ounce, and platinum for more than $1,600 per ounce.

Industry and Business. The discovery and distribution of gemstones is complicated and depends on the source and type of the gemstone. The De Beers Group has dominated the diamond market since the early part of the twentieth century, controlling most aspects of supply, distribution, and price. However, the discovery of diamonds in Russia, Australia, and Canada and the creation of artificial diamonds for the industrial market has somewhat altered the market.

Fascinating Facts About Gemology and Chrysology

- Platinum is used in drugs such as cisplastin to fight cancer.
- The addition of silver halide crystals to eyeglass lenses allows them to rapidly change from clear to dark. Ultraviolet light hits the crystals, causing a chemical change that blocks light and darkens the lenses.
- Near Barstow, California, 1,926 silver-coated mirrors direct sunlight to nitrate-filled tubes that are used to run generators and provide electric power for 10,000 homes.
- One treatment for prostrate cancer involves inserting three grains of gold into the prostrate, using ultrasound. This allows physicians to accurately determine the position of the prostrate during radiation treatments.
- In Japan, sake, tea, and other foods containing thin gold flakes are believed to be beneficial to one's health.
- Ordinary household switches uses silver. Silver does not corrode, which would result in overheating and pose a fire hazard.
- Several companies produce artificial diamonds from the cremated remains of people or pets.
- Because the origin of amethyst was believed to be related to Bacchus, the Greek god of wine, it was once believed that wearing amethyst would prevent a person from becoming drunk.
- Ancient Greeks believed that topaz made its wearer stronger and invisible in case of an emergency.

Diamonds are sorted and valued by De Beers and sent to London, where the diamonds are sold in groups to a few select individuals at a certain price. These groups of diamonds are then sold to diamond exchanges in Belgium, the Netherlands, New York City, and Israel. These diamond exchanges can then sell diamonds to other places such as wholesale jewelry stores. As the diamonds all come from De Beers, prices tend to remain steady. However, the diamonds emerging from Russia, Canada, and Australia do not go through these distribution channels and their prices are governed by market forces.

Although the markets for other gemstones are not dominated by a single producer, those gemstones that come from limited sources may become scarce, and their prices may rise rapidly. For instance, tanzanite (calcium aluminum silicate) can be mined only at one site in Tanzania, and few new specimens are being found. This scarcity has driven up the price of tanzanite.

Precious metals are refined from ores from mines or produced from recycled materials. The firms involved in gold production generally are involved in exploration, development, or production, and are not vertically integrated. Silver is recovered from silver mines and as a by-product of gold mining. Firms are generally involved in mining, refining, fabrication, and manufacture. Most of the world's platinum comes from Russia and South Africa. Platinum production is capital intensive and dominated by large, vertically integrated firms. Gold is used primarily for jewelry but also in electronic products. Silver is used for jewelry, silverware, photography, and industrial uses, such as in batteries and electronics. Platinum is commonly used as an autocatalyst and also as jewelry and in electronic products.

Governments and Universities. Governments are primarily involved in the regulation of the gemstone and precious metals industry. Regulation can include measures to ensure the safety of workers in mines or refineries, to deal with environmental issues arising from mining or refining, or to control how gemstones and precious metals are bought and sold. Smuggling of gemstones, often involving rebel forces, organized crime, or drug cartels, remains a problem in many nations. Unstable governments have had negative effects on the supply of gemstones. For example, in 1969, the government of Burma, the main supplier of rubies, closed all ruby mines within its borders.

Universities conduct research on the composition and uses of gemstones, mineral identification techniques, and all aspects of mining operations. They also are involved in research regarding the possible uses of precious metals in medicine, engineering, and manufacturing. They also provide training in mineralogy, geology, and mining operations.

CAREERS AND COURSE WORK

A career in gemology or chrysology begins with a bachelor's degree in geology or mining engineering, with course work in mineralogy and gemology. A doctorate in one of these fields is required for conducting research on gemstones or precious metals. Additional course work in chemistry and physics may be helpful. Research involving applications that use gemstones or precious metals requires course work and degrees in those areas, such as medicine or electronics.

Those desiring to work as a gemstone identifier and grader in a jewelry store or as a jewelry maker or designer can obtain specialized training in these areas. For example, the Gemological Institute of America offers a graduate gemologist degree or a jewelry manufacturing arts degree, both six-month programs.

SOCIAL CONTEXT AND FUTURE PROSPECTS

The demand for gemstones is likely to continue, rising and falling with upturns and downturns in the economy. Gemstones, like many minerals, are found in only a limited number of locations worldwide; therefore, the supply of a given gemstone can vary considerably. A major mine may become depleted, or a new source of the gemstone may be discovered. Diamond prices are the most stable because the market is dominated by De Beers; however, as other diamond producers increase their presence in the market, prices may begin to fluctuate. The total worldwide diamond production exceeded $19 billion in 2009.

The best rubies and sapphires are so rare that the price per carat for individual specimens may exceed that for diamonds. Burma is the source for the best rubies and sapphires, and Colombia produces the finest emeralds. In 2006, total world emerald production was 5,400 kilograms, total ruby production was 10,000 kilograms, and total sapphire production was 32,500 kilograms. These levels of production are expected to be maintained.

According to the U.S. Geological Survey, in 2009, worldwide production of gold was 2,350 metric tons; silver, 21,400 metric tons; and platinum metals, 178,000 kilograms. The demand for gold and its price was increasing, while the demand for platinum had fallen because of the decline in the automotive industry. However, precious metals as a whole were predicted to increase in demand, especially as sources become depleted.

Robert L. Cullers, B.S., M.S., Ph.D.

FURTHER READING

Babcok, Loren E. *Gemstones and Precious Metals.* Rev. ed. Hoboken, N.J.: Wiley Custom Services, 2009. Examines both gemstones and precious metals, looking at history, applications, and their value.

Crowe, Judith. *The Jeweler's Directory of Gemstones.* Buffalo, N.Y.: Firefly Books, 2006. Examines how to appraise and use gemstones in jewelry making.

Desautels, Paul E. *The Gem Kingdom.* New York: Random House, 2000. Discusses gemstone quality, history, cutting, and the making of jewelry.

Gasparrini, Claudia. *Gold and Other Precious Metals: Occurrence, Extraction, Applications.* 3d ed. Toronto: Space Eagle, 2000. Examines where precious metals are found and how they are mined and used.

Klein, Cornelis, and Barbara Dutrow. *Manual of Mineral Science.* Hoboken, N.J.: John Wiley & Sons, 2008. Describes how to identify minerals using physical properties and modern analytical techniques. It has a chapter on gemstones.

Rutland, E. H. *Gemstones.* New York: Hamlyn Publishing, 1974. Describes the origin, identification, jewelry, history, and collection of gemstones.

Schumann, Walter. *Gemstones of the World.* New York: Sterling, 2009. Contains many photographs of cut and uncut gemstones. Discusses the origin, properties, deposits, cutting, polishing, and identification of gemstones.

WEB SITES

International Colored Gemstone Association
http://www.gemstone.org

International Gem Society
http://www.gemsociety.org

International Platinum Group Metals Association
http://www.ipa-news.com/en

The Silver Institute
http://www.silverinstitute.org

World Gold Council
http://www.gold.org

See also: Mineralogy.

GENETICALLY MODIFIED FOOD PRODUCTION

FIELDS OF STUDY

Genetic engineering; genetic manipulation; gene technology; recombinant DNA technology; biotechnology; food production; agriculture.

SUMMARY

Genetically modified food production is a subset of biotechnology and genetic engineering. This developing field offers both hope and concern for global food production. Genetically modified food production is the direct result of the development of genetically modified organisms. Conventional plant breeding is a slow process, and it take several years to develop plants with desirable traits. Advances in genetic engineering have allowed scientists to speed up the process of developing plants with the most desirable traits and with greater predictability. These plants help farmers increase production and obtain higher yields. However, the use of genetically modified organisms in food production is controversial in some countries because the effect of these organisms on humans has not been assessed.

KEY TERMS AND CONCEPTS

- **Biotechnology:** Technology that uses biological systems, living organisms, or derivatives thereof to create or modify organisms, existing products, or processes.
- **DNA (Deoxyribonucleic Acid):** Nucleic acid found in all cells that contains the genetic material for the growth and functioning of an organism.
- **Gene:** Piece of the DNA code that regulates biological processes in living organisms and contains the information needed to make a specific protein in that organism; it is the functional and physical unit of heredity passed from a parent to its offspring.
- **Genetically Modified Organism (GMO):** Organism that has been modified or altered using genetic engineering techniques for a specific purpose; also known as a genetically engineered organism.
- **Genetic Engineering:** Technique used to remove, modify, or add genes to a DNA molecule of a living organism to change its genetic contents and en-

able it to make different proteins or perform different functions.

- **Protein:** Large molecule containing one or more chains of amino acids arranged in a specific order; formed according to information coded in a gene.
- **Recombinant DNA:** Form of DNA that has been created by combining two or more DNA sequences that do not usually occur together.
- **Recombinant Protein:** Protein derived from recombinant DNA.
- **Transgenic Plant Or Animal:** Plant or animal that has been altered by the introduction of a gene from another species.

DEFINITION AND BASIC PRINCIPLES

Genetically modified food production is the creation of food products using genetically modified organisms. Some of the food production problems that can be addressed using genetically modified organisms are limiting or eliminating the damage caused by pests, weeds, and diseases, and providing tolerance to specific herbicides, extreme temperatures, and drought, as well as other production-related issues.

One of the initial goals of developing genetically modified plants was to achieve higher crop yields by creating versions of plants such as corn and soybeans that offered greater resistance to diseases caused by pests and viruses. Genetically modified foods, also known as genetically engineered food, are developed, produced, and marketed mainly because they present an advantage over traditionally produced food to either the farmer/producer or the consumer. The benefit to the producers comes from increased productivity and the reduction of lost crops. The consumer benefits from genetically engineered food by having access to food with better nutritional value and, in some cases, lower prices.

BACKGROUND AND HISTORY

People began harvesting plants and domesticating animals around 10,000 B.C.E., and they soon were selecting the best seed and breeding the best animals through a process of trial and error. In the latter part of the nineteenth century, the monk Gregor Mendel used peas to demonstrate heredity in plants, thereby laying the foundation for modern plant breeding and genetics. The early work of Mendel and others led to the development of commercial crops by the 1930's. The field of molecular biology, which emerged in the twentieth century, has led to a better understanding of the cells and molecular processes of living organisms. This understanding has allowed researchers to develop genetically engineered plants and animals to address and solve many of the problems faced by farmers in crop and livestock production.

The first field trials of crops developed using genetically modified organisms took place in 1990, and by 1992, the first genetically modified corn was approved for use by the U.S. Food and Drug Administration (FDA). In the 1990's, additional advances led to genetically engineered vaccines and hormones as well as to the cloning of animals. Since their introduction in the United States, genetically modified crop plants have been widely used by American farmers, and the percentage of acreage planted with genetically modified crops has been steadily increasing, reaching more than 90 percent for some crops. Soybeans and cotton genetically modified to tolerate herbicides are the two most widely adopted genetically modified crops. Cotton and corn with insect resistance are the next most common crops grown by American farmers.

How It Works

Genetic engineering allows scientists to insert a gene from one organism into another, resulting in a genetically modified organism. Therefore, genetic engineering begins with the identification and isolation of a gene that expresses a desirable trait. This gene can be found in a relative of the target species or in a completely unrelated species. A recipient plant or animal is selected, and the gene is inserted and incorporated into the genome of the recipient. The desired gene is inserted by various techniques such as using *Agrobacterium* as a vector or using a gene gun. The newly inserted gene becomes part of the genome of the recipient and is regulated in the same way as its other genes. Genetic engineering confers a new ability on the organism that has received the new gene. One advantage of genetic engineering is that genes can be introduced in a plant even if they do not occur in the genome of the target plant.

The use of genetically modified food production, a new and valuable tool in agriculture, fisheries, and forestry production, has allowed significant improvements in food production to meet the needs of the ever-expanding world population. With conventional plant breeding techniques, researchers crossbred plants, taking five to seven years to generate a plant with the desired traits. In conventional breeding, half of an individual's genes come from each parent, but in genetically modified organisms, one or more genetically desirable traits has been added to the genetic material of the desired plant. One of the main differences between conventional breeding and genetic engineering is that in the conventional process, crosses are possible only between close relatives, whereas with genetic engineering, scientists can transfer genes between plants that are not related and might not be able to crossbreed in nature.

For example, in the case of Bt-corn, a gene from a naturally occurring soil bacterium, *Bacillus thuringiensis*, was inserted into corn to provide resistance to the corn borer. The gene from the bacterium produces a protein, Bt delta endotoxin, which kills the European and southwestern corn borer larvae. Bt-corn eliminates the need to spray insecticides to control corn borers. Although planting these crops reduces the amount of pesticides released into the environment, the long-term effects of Bt-corn on human health and the environment are not known.

Companies involved in the research and development of plant and animals derived from genetic engineering patent these new products and processes. The patent allows them to protect their investment; however, it costs farmers who use the seed. A contract between the farmers and these companies prohibits the farmers from saving seed for use the following year, reselling seed to a third party, or exchanging seed with other farmers.

Scientists envisage numerous future applications of genetically modified food. For example, food could be used to produce drugs to address human health problems including infectious diseases. One of the crops that is being considered for such an application is bananas, which could be used to produce a human vaccine.

Applications and Products

Over the years, many biotechnology firms and well-established companies have become involved in

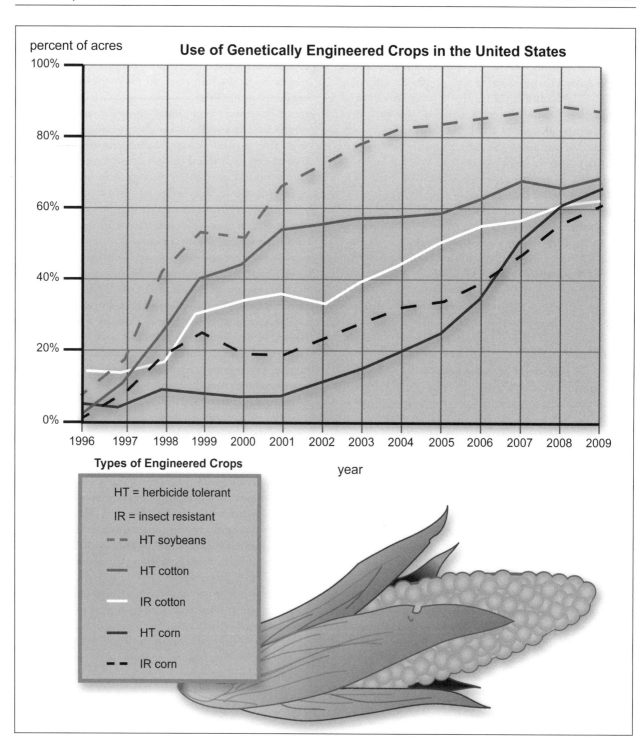

percent of acres

Use of Genetically Engineered Crops in the United States

year

Types of Engineered Crops

HT = herbicide tolerant

IR = insect resistant

- - - HT soybeans

—— HT cotton

—— IR cotton

—— HT corn

- - - IR corn

research and development using genetically modified organisms. The genetic engineering technique and biotechnology are likely to have many potential commercial uses. Genetically modified organisms have many applications in food production and livestock. Many products, including plants and animals, have been developed using genetic engineering techniques.

Insect Resistance. The use of biotechnology to achieve insect resistance in food plants destined for human or animal consumption is accomplished by incorporating a gene into a particular food plant such as corn. In the case of Bt-corn, the gene for toxin production from the bacterium *B. thuringiensis* was incorporated into the corn plant. This toxin is an insecticide commonly used in agriculture and is safe for human consumption. Genetically modified plants that permanently produce this toxin have been shown to require lower quantities of insecticides in specific situations, for example, where pest pressure is high.

Virus Resistance. Virus resistance is achieved through the introduction of a gene from certain viruses that cause disease in plants. Virus resistance makes plants less susceptible to diseases caused by such viruses, resulting in higher crop yields.

Herbicide Tolerance. Herbicide tolerance is achieved through the introduction of a gene from a bacterium conveying resistance to some herbicides. In situations where weed pressure is high, the use of such crops has resulted in a reduction in the quantity of herbicides used. For example, corn resistant to the popular Monsanto weedkiller Roundup (glyphosate) has been developed so that farmers can treat their crops with Roundup without damaging their corn crop. This means that farmers can reduce the number and amount of herbicides used in any given year and situation.

Genetically Modified Animals. Genetic engineering of animals has taken place for different purposes and using a number of different species such as cattle, sheep, goats, rabbits, pigs, chickens, and fish. For example, well-performing bulls have been cloned to create better breeding stock, and animals have been used to produce useful human proteins. Research using genetic engineering in cattle production is trying to produce cows that are resistant to mad cow disease or that have the capacity to produce milk with higher levels of protein.

Scientists have been able to genetically engineer a variety of salmon that grows at twice the rate of Atlantic salmon. The fish is no bigger than the Atlantic salmon, but it reaches that size in half the time. Scientists inserted a Chinook salmon growth hormone gene into a fertilized egg of an Atlantic salmon. To ensure that the gene remains active all year round, scientists added a "switch" from the ocean pout to the Atlantic salmon. This genetically engineered fish reduces the production cost of salmon. AquaBounty Technologies, which developed the AquAdvantage salmon applied to the FDA for permission to market the fish. In September, 2010, the FDA concluded that the fish was safe to eat but felt that more scientific research was needed, particularly on the possible environmental impact of the modified salmon.

Possible Risks. Most scientists and regulators agree that the use of genetically modified organisms in food production may pose some risks. These potential risks usually fall into two basic categories: the effects on human and animal health and the impact on the environment where the genetically modified organisms are grown. Scientists and regulators advise that care must be exercised to reduce these risks, especially the possibility of transferring toxins or allergenic compounds used in genetically modified organisms to ordinary plants or animals. This cross-contamination could result in unexpected allergic reactions in humans and animals. One of the major risks to natural resources and the environment is the possibility of outcrossing, the transfer of genes from genetically modified organisms to regular crops or related wild species. For example, the use of herbicide-resistant corn and soybeans raises the possibility of outcrossing, which could lead to the development of more aggressive weeds or herbicide-resistant wild relatives of these cultivated plants. This outcrossing could upset the balance of the natural ecosystem. The introduction of genetically modified plants could also lead to a loss of biodiversity as traditional varieties of plants are displaced by a smaller number of genetically modified varieties.

Genetically modified organisms are not always advantageous: The cost of research and development can be prohibitively high, much cheaper ways to control the undesired pests or diseases may exist, and the still unknown effects on humans and the environment may potentially result in lawsuits against the developers of the plant or animal.

IMPACT ON INDUSTRY

The private sector is responsible for the major investments in research and development in the field of genetically modified food production. These investments have targeted mostly agriculture plants and livestock to address the quality of these plants and animals and their vulnerability to disease.

Government and University Research. The U.S. government was quick to create a research and regulatory framework under which modified organisms could be used in food production. Under this framework, corporations have funded research at American universities that has played a major role in the development of genetically modified food. However, some academics are concerned about the power of these corporations. The U.S. Department of Agriculture (USDA), the agency trusted with regulation of genetically modified food production, does not conduct any tests of its own to assess the risks posed by these crops but instead relies on data provided by the industry. Some scientists have pointed out that as long as the USDA relies on these companies to provide data on the risks and benefits of genetically modified crops, the agency cannot be sure that the reports it is receiving are not biased. In addition, university researchers have complained of company intervention in their research. In at least one case, several companies did not grant a researcher permission to conduct research using their seeds. In another case, a researcher was asked to sign an agreement that he would not test the seed if he wanted access to it. Companies have control over the seed of crops derived from genetic engineering because of the patent rights they hold.

Industry and Business Sector. The industry and business sector has been working very hard to develop new products and techniques, as well as solutions to some unsolved problems. For example, the seed industry is developing new techniques and products to prevent disease from destroying crops, to make crops tolerant to specific herbicides, and to improve the nutritional quality of certain foods using genetic engineering. These new crops are helping solve hunger problems in various countries around the world.

Major Corporations. U.S. and European corporations have played a major role in the research and development of genetically modified crops. These corporations are primarily seed companies and producers of agricultural chemicals.

Syngenta, a global seed and agricultural chemicals company based in Switzerland, has created corn resistant to the European corn borer by inserting the *cry1Ab* gene from *B. thuringiensis*, corn resistant to rootworm by transforming corn with a modified *cry3A* gene, and insect-resistant and herbicide-tolerant maize by conventionally crossbreeding genetically modified parents. Aventis CropScience, a France-based crop production and technologies company, developed maize that is both resistant to insects and tolerant of glufosinate ammonium herbicide by inserting genes encoding Cry9C protein from *B. thuringiensis* subspecies tolworthi and phosphinothricin acetyltransferase (PAT) from *Streptomyces hygroscopicus*. It also created an insect-resistant and herbicide-tolerant maize by conventional crossbreeding of genetically modified parental lines.

Pioneer Hi-Bred International, an Iowa-based developer and supplier of plant genetics, created maize

---

Fascinating Facts About Genetically Modified Food Production

- As of 2010, pest-resistant Bt-cotton was produced on about 93 percent of the acreage devoted to cotton production in the United States.
- Genetically engineered soybeans resistant to the herbicide glyphosate (Roundup) occupied about 93 percent of the soybean acreage in the United States in 2010.
- Vitamin-enriched corn has been derived from a South African white corn hybrid (M37W). The new corn has bright orange kernels that have 169 times more beta carotene, 6 times more vitamin C, and twice the folate content of the regular corn.
- Rice, a major staple in Asian countries, has been modified by the introduction of three new genes, two from daffodils and one from a bacterium, to create Golden Rice, which is high in beta carotene, a precursor to vitamin A.
- It is estimated that genetic engineering has increased U.S. farmers' income by $1.5 billion a year.
- The Grocery Manufacturers Association estimates that about 80 percent of processed foods in the United States contain genetically modified organisms.
- In New Zealand, cows were genetically modified to produce milk with a higher protein content, which allows for the production of more cheese from the same amount of milk.

that is male-sterile and tolerant of glufosinate ammonium herbicide by inserting genes encoding DNA adenine methylase and PAT from *Escherichia coli* and *Streptomyces viridochromogenes*, respectively. In collaboration with Pioneer, Dow Agrosciences, an Indiana-based subsidiary of Dow Chemical, produced a rootworm-resistant maize by inserting the *cry34Ab1* and *cry35Ab1* genes from *B. thuringiensis*.

Monsanto, a U.S.-based agricultural biotechnology company, produced maize resistant to the European corn borer by inserting the *cry1Ab* gene from *B. thuringiensis* subspecies kurstaki and a rootworm-resistant maize by inserting the *cry3Bb1* gene from *B. thuringiensis* subspecies kumamotoensis. It used conventional crossbreeding of genetically modified parent plants to create an insect-resistant and enhanced lysine content maize and an insect-resistant and glyphosate-tolerant maize. In 1998, Monsanto bought Dekalb Genetics and acquired rights to an herbicide-tolerant maize and a insect-resistant and herbicide-tolerant maize that Dekalb Genetics had developed.

International Organizations. Several international organizations address issues related to genetically modified food production. Among them are the World Trade Organization, the World Health Organization, the Food and Agriculture Organization of the United Nations, the Codex Alimentarius Commission, the World Organisation for Animal Health, and the International Plant Protection Convention. These organizations address genetically modified food production from different angles: its environmental impact, the effect on human health, and the regulatory aspects.

Over the years and with the involvement of the World Trade Organization, the opportunities for global agricultural trade have increased dramatically. The organization has mainly concerned itself with reducing tariffs and subsidies, but in 1994, it adopted the Agreement on the Application of Sanitary and Phytosanitary Measures, which establishes that countries retain their right to ensure that the food, animal, and plant products that they import are safe. The agreement also states that countries should use internationally agreed standards developed by the Codex Alimentarius Commission for food safety, the World Organisation for Animal Health for animal health, and the International Plant Protection Convention for plant health.

CAREERS AND COURSE WORK

Career pathways in agricultural production, biotechnology, food production, gene technology, genetic engineering, genetic manipulation, and recombinant DNA technology offer great opportunities for students interested in issues dealing with environmental challenges, human health, and reducing world hunger. These fields are all rapidly changing with new applications, techniques, and innovations being developed all the time. These fields of study can be loosely termed "biotechnology." Students can prepare themselves for a career in biotechnology in many ways. The opportunities in the field include positions as government research scientists, corporate scientists, laboratory technicians, engineers, process technicians, and maintenance and instrumental technicians. Students can chose among a great many private and public sector employers and a wide range of work environments and types of jobs.

A biotechnology degree can lead to employment in the private sector, with drug and chemical companies, seed and agricultural companies, or environmental remediation companies; in government agencies such as the USDA's Agricultural Research Services or the Food and Drug Administration; or in educational institutions, teaching and doing research.

SOCIAL CONTEXT AND FUTURE PROSPECTS

The social debate about the use of genetically modified food centers on the level of risk to health and the environment that they present and whether these types of food are necessary. The possible risks to human health presented by genetically modified food fall into two categories, direct and indirect. Direct effects include toxicity, an allergic reaction (also called allergenicity), and negative nutritional effects, and indirect effects include the stability of the inserted gene, outcrossing, and unintended effects as a result of the gene insertion. Therefore, genetically modified foods must be tested very carefully and thoroughly to ensure that the benefits of such plants outweigh their risks and the hidden costs of developing them.

The debate about whether genetically modified food needs to be created to deal with hunger among people in the developing world is taking place in many venues, both political and scientific. Some scientists opposed to genetically engineered food argue that there is more than enough food in the world and that the hunger crisis in some countries is the result of problems with food distribution and the politics in those countries rather than production levels and systems. They also argue that offering food with unknown levels of risk to those in need is unethical. This argument assumes that genetically modified foods have risks not present in traditional foods; however, the proponents of genetically modified food argue that traditional foods are not devoid of risk. However, as production of genetically modified foods has increased and no major adverse effects have emerged, some of the earlier criticism has disappeared. However, a consensus does not exist among U.S. scientists on the risks and the safety of genetically modified plants and animals.

Lakhdar Boukerrou, B.Sc., M.S., Ph.D.

FURTHER READING

Fedoroff, Nina, and Nancy Marie Brown. *Mendel in the Kitchen: A Scientist's View of Genetically Modified Food.* 2004. Reprint. Washington, D.C.: Joseph Henry Press, 2006. Fedoroff, an expert in plant molecular biology and genetics, teams with a science writer to dispel many misconceptions about genetically modified foods and describes them as an "environmentally conservative" way to increase the food supply.

Henningfeld, Diane Andrews, ed. *Genetically Modified Food.* Detroit: Greenhaven Press, 2009. Contains a collection of essays that looks at the issues surrounding genetically modified food from both sides, pro and con.

King, Robert C., William D. Stansfield, and Pamela Khipple Mulligan. *A Dictionary of Genetics.* 7th ed. New York: Oxford University Press, 2006. Provides information on all aspects of genetics in alphabetical order by topic.

Lurquin, Paul F. *High Tech Harvest: Understanding Genetically Modified Food Plants.* Boulder, Colo.: Westview Press, 2002. A research biologist examines genetically modified plants, explaining the science behind them. He is generally supportive of biotechnology.

Weasel, Lisa. *Food Fray: Inside the Controversy over Genetically Modified Food.* New York: American Management Association, 2009. Provides a historical perspective on the debate over genetically modified food. Has some very useful information on the use of genetically modified food in places such as India and Africa as well as on the future of genetically modified food.

WEB SITES

Food and Agriculture Organization of the United Nations
Biotechnology
http://www.fao.org/biotech/stat.asp

GMO Compass
http://www.gmo-compass.org/eng/home

U.S. Department of Agriculture, Economic Research Service
Adoption of Genetically Engineered Crops in the United States
http://www.ers.usda.gov/Data/BiotechCrops

U.S. Food and Drug Administration
Genetically Engineered Foods
http://www.fda.gov/NewsEvents/Testimony/ucm115032.htm

World Health Organization
Twenty Questions on Genetically Modified (GM) Foods
http://www.who.int/foodsafety/publications/biotech/en/20questions_en.pdf

See also: Animal Breeding and Husbandry; Food Preservation; Food Science; Genetically Modified Organisms; Genetic Engineering; Plant Breeding and Propagation.

GENETICALLY MODIFIED ORGANISMS

FIELDS OF STUDY

Biology; chemistry; genetics; biotechnology; recombinant DNA technology; reproductive science; genetic engineering; molecular biology; botany; entomology; plant pathology; agricultural science; environmental science; medical science.

SUMMARY

Genetically modified organisms are produced through genetic engineering and biotechnology and basically involve genetic modifications in which genetic material is added or removed to alter the genetic structure of the organism. Many organisms have undergone genetic modification, including bacteria and viruses, plants and animals, and even human beings. The majority of genetically modified organisms are created for therapeutic reasons, such as medicine and food for human consumption. Such organisms have the potential to affect all members of human society and their surrounding environment and have therefore become one of the most controversial ethical and ecological issues of the early twenty-first century.

KEY TERMS AND CONCEPTS

- **Crop Yield:** Amount of plant crop harvested, as opposed to grown, in a given area for a given time.
- **DNA (Deoxyribonucleic Acid):** Nucleic acid found in a cell that contains all the genetic material and instructions used for growth and development of living organisms.
- **Gene:** Basic unit of heredity; occupies a precise position on a chromosome within an individual organism's DNA.
- **Genetically Modified Organisms (GMOs):** Organisms whose genetic makeup has been manipulated by genetic engineering techniques (gene technology), through either the addition or removal of genetic material; also known as genetically engineered organisms (GEO) and bioengineered organisms.
- **Natural Selection:** Process whereby a species evolves over time as individual organisms possessing the most practical and useful characteristics survive and reproduce; their offspring will, in turn, possess those same positive characteristics so as to survive and continue breeding.
- **Outcrossing:** Transfer of genes from genetically modified plants into crops or related plant species in the wild.
- **Plasmid Technology:** Technology related to plasmids, which are circular, double-stranded units of DNA. Plasmids are able to replicate inside a cell independently of the cell's chromosomal DNA and are thus useful in recombinant DNA technology and research in the transfer of genes between different cells.
- **Recombinant DNA Technology:** Techniques in genetic biology that alter an organism's genes by removing a specific gene from the cell of one organism and inserting it into the cell of another. This splicing together of gene fragments from different species produces a new organism that would not occur naturally.

DEFINITION AND BASIC PRINCIPLES

Humans have selectively bred and crossbred plants and animals for desired traits since almost the dawn of agriculture, but advances in genetic technology have given people novel ways in which to manipulate plants and animals. These advances are motivated by the desires to develop new medical treatments for genetic diseases and disorders and to increase food production to satisfy the world's growing population. Most advances involve recombinant DNA technology, in which an organism's genes are altered by removing a specific gene from the cell of one organism and inserting it into the cell of another. This splicing together of gene fragments from different species produces a new organism that would not be produced through natural reproduction processes or would not be feasible because of the impossibility of interspecies breeding. This new organism is defined as a genetically modified organism (GMO).

The advancement of genetic technology and the introduction of GMOs into the human food chain has prompted controversy over the ethics of manipulating nature and the potential for GMOs in worldwide agricultural production and medicine. Although many experts state that GMOs are safe for human consumption and offer myriad benefits to humankind, others claim that the production and consumption of GMOs is unethical and untested, which means that GMOs involve unknown consequences, which potentially could be dangerous.

BACKGROUND AND HISTORY

The process of natural selection, first described by Charles Darwin in 1859 in his seminal *On the Origin of Species by Means of Natural Selection*, states that species evolve over time. Individual organisms that possess the most desired and useful characteristics survive, reproduce, and give birth to offspring; these offspring, in turn, possess the same positive characteristics. For many hundreds of years, people have manipulated the process of natural selection though traditional agricultural selection and crossbreeding to create or eliminate specific characteristics in plant and animal species, producing a wide variety of cereal crops, livestock animals, and pets.

Human interference has altered many plants and animals through crossbreeding or selection, but the desirable traits initially appeared through naturally occurring genetic variation. Because the desired traits were already in existence, human interference in the breeding process was often viewed as relatively benign and within natural bounds. Although humans have manipulated the breeding of plants and animals based on phenotypic characteristics for a long time, the ability to directly manipulate the genotype developed much later. Specifically, to feed a growing and hungry world population and develop medical treatments, medical and agricultural scientists have researched and advanced genetic modification technology.

Although genetic engineering is a phenomenon of the late twentieth century, the building blocks for such technology began with the first isolation of DNA in 1869 and the subsequent awareness of its relevance to heredity in 1928. The first accurate double-helix model of DNA was developed in 1953 by James D. Watson and Francis Crick, and the first gene sequence and recombinant DNA was created in 1972 by researchers from Stanford University. The latter discovery truly heralded the beginning of the biotechnological industry and the development of GMOs.

Genetic engineering research continued during the 1970's, and the first publicly and commercially available GMO, a form of human insulin produced by bacteria, was developed in the United States in 1982. However, for the most part, the majority of commercial GMOs sold and used in the twenty-first century are found in agriculture and food production. GMO research scientists believe that genetic engineering is the only method that will guarantee global food production, particularly as predications regarding global climates have indicated that traditional agriculture practices will fail to meet demand.

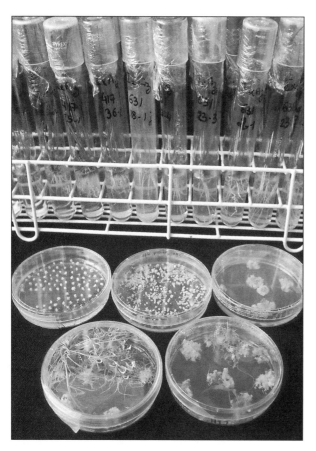

Petri dishes and test tubes showing different stages in the research and cultivation of second-generation transgenic rice. Transgenic plants are ones that have been genetically modified by adding genes from different species. (Pascal Goetgheluck/Photo Researchers, Inc.)

HOW IT WORKS

Fundamentally, the development and manufacture of GMOs is the replacement of natural selection processes with artificial genetic manipulation. At its most basic, GMO technology relies on a sound understanding of DNA and involves the subtraction of specific genetic material or substitution of material from one species with that from another. Genetic engineering, a complex endeavor, deals with the most fundamental building blocks of an organism. Within a cell are tiny strandlike structures called chromosomes, which contain a nucleic acid called DNA. This molecule contains all the genetic material required for inheritance and thus is the basis for genetic manipulation technology.

Initially, the term genetic manipulation referred to a vast array of techniques for the modification of organisms through reproduction and gene inheritance. Later, however, the definition became more restricted and refers specifically to recombinant DNA technology, a form of genetic engineering in which the genome of a cell or organism is artificially modified. The fundamental concept of this technology is that genetic material from different species is combined to create a new species or organism. That is, molecules of DNA from more than one source are united together inside a cell, which is then inserted into a new organism or host, where it is able to reproduce. Because the genome is passed on to offspring, the modification is considered to be self-perpetuating. An organism's biological activities and physical characteristics are controlled by its genome, so modification of the genome can significantly influence the organism's biological functions and traits. The objective behind such technology is to advance the fields of medicine and agriculture to develop more effective medical treatments and to improve crop yield and disease resistance.

APPLICATIONS AND PRODUCTS

The possible applications and products of genetic engineering are vast, perhaps limited only by the imagination. For the most part, however, the major function of genetic engineering, and hence the development of GMOs, is related to their potential in agricultural, medical, and environmental applications.

Agricultural Applications. The continuing rapid expansion of the human population is necessitating an increase in the supply of food. Providing adequate food for a hungry world has become a significant issue for science. The need for food has been instrumental in promoting and advancing genetic

---

Fascinating Facts About Genetically Modified Organisms

- Genetically modified seeds are used in the agricultural industry for the production of cereal crops, such as soybeans and canola.
- The first commercial GMO, a synthetic human insulin, was introduced in 1982.
- GMOs are being heralded as the solution to the world's continuing increase in population and food demand.
- GMOs offer great potential in the fight against the 3,000 genetic diseases and disorders that adversely affect people.
- In 2007, an estimated 12 million farmers in more than twenty countries were growing genetically modified crops. Roughly 90 percent of these farmers were categorized as resource-poor and were mostly in twelve developing countries.
- Bt-corn, a genetically modified version of corn, incorporates a gene from the bacterium *Bacillus thuringiensis* that acts as an insecticide.
- Cheeses and canola oil are just two of the genetically modified foods commercially available.

---

modification techniques to produce new and improved organisms, particularly those that, for example, have higher yields or are drought and disease resistant. Through recombinant gene technology, it has become possible to create plant species that are capable of surviving in extreme temperatures and with low rainfall, that can convert atmospheric nitrogen into a useable form (thereby eliminating the need for nitrogen fertilizer), and that have the ability to produce their own resistance to pests and pathogens (thereby eliminating the need for chemical pesticides). Versions of soybeans, canola, corn, potatoes, sugar beets, and cotton that have been genetically modified to increase herbicide tolerance and resist insects are all available for purchase.

With some specific exceptions, research and the development of genetically engineered animals has proven to be less straightforward than the genetic modification of plants and certainly more ethically problematic. In addition, although the public definitely shows some resistance to the idea of introducing genetically modified plants into the human food chain, most people express much greater resistance to the idea of directly consuming genetically modified animals. Therefore, research on genetically modified animals for use in agriculture has stayed in a relatively early stage of development. However, there has been some research into and experimentation with the genetic manipulation of animals to increase production and meat yield; such experimentation is most promising in fish species rather than in hoofed farm animals.

Medical Applications. Although agricultural applications are very important, the potential medical applications of GMOs are perhaps even more significant. The world's first commercial applications of genetic engineering were, in fact, medically oriented and included synthetic human insulin, approved for public sale in the United States by the Food and Drug Administration (FDA) in 1982, and a human hepatitis-B vaccine, approved by the FDA in 1987. Before the 1980's, synthetic human insulin (produced from animals) was available only in relatively limited quantities. Since the 1980's, research into medical applications for GMOs has rapidly advanced. Of particular benefit is the ability of genetic engineering to produce GMOs on a previously unavailable scale.

Perhaps the most significant potential application of genetic engineering and GMOs is in the treatment and possible cure of genetic diseases. Human society is plagued with both serious diseases and mild disorders, more than 3,000 of which are genetic in origin and therefore difficult to cure using conventional medicine. Although this technology is still in its infancy, the potential of gene therapy to assist people with genetic disease is perhaps limitless.

Environmental Applications. Increasing human populations are important not only in relation to agricultural food production but also in terms of their impact on the environment. Genetic engineering and GMOs could potentially solve some of the world's most serious ecological problems. Research has produced genetically modified viruses that can be used to create ecologically friendly lithium batteries, modified bacteria that can produce biodegradable plastic, and genetically manipulated bacteria that have been encoded for use in bioremediation. Genetic modification technology may even be of use in the fight for survival of some of the world's most vulnerable and endangered species.

IMPACT ON INDUSTRY

Major Organizations. Many agricultural organizations, the largest and most influential being the U.S.-based agricultural biotechnology corporation Monsanto, are directly involved in the promotion and advancement of GMOs for food—specifically cereal crops and meat production. Monsanto is the world's largest supplier of modified seeds (some 90 percent of the market). Genetically modified seeds for soybeans, canola, corn, potatoes, sugar beets, and cotton are commercially available.

Many organizations are involved in both the production of GMOs and their regulation. There are many environmental and ethical concerns regarding the creation of GMOs and the technologies involved in the genetic modification of organisms, particularly animals, for food and medicine. According to the World Health Organization, there are three basic concerns regarding GMOs and their possible impact on human health: increased antibiotic resistance because of the use of antibiotics in GMO manufacture; increased human allergic reactions to foods and medicines; and gene transfer.

To regulate these possible effects, the United Nations established the Convention on Biological Diversity, an international treaty that has been signed by some 190 nations. The United Nations was also responsible for the establishment of the Protocol on Biosafety, which aims to conserve biodiversity through a reduction in the potentially harmful consequences of biotechnology, specifically living GMOs, and to monitor and regulate GMO movement and international trade across borders and between countries.

Government Regulation. A number of countries, such as Australia and New Zealand, have distinguished between in-country production of GMO crops and the sale of imported GMOs. Both countries allow the sale of genetically modified soybean, canola, corn, potato, sugar beets, and cotton seeds or plants, which have been altered mostly to increase herbicide tolerance and minimize damage from insects. Australia and New Zealand have been quite measured in their approval of growing genetically modified crops in comparison with countries such as Canada and the United States. In Canada, for example, genetically modified foods are nutritionally assessed and treated the same as unmodified foods. Additionally, both the Canadian and the United

States governments do not require genetically modified foods to be labeled as such.

Numerous governments around the world, however, have banned the domestic growing of GMOs for fear of contamination with native plant species and because of the possible unknown side effects of consumption, particularly by livestock. Many environmental organizations have lobbied the world's governments and suggested a total moratorium on GMOs, claiming they will have an adverse effect on the environment and cause loss of biodiversity through unintended crossbreeding with native plants and conventional crops.

CAREERS AND COURSE WORK

Students who wish to pursue a career in genetic engineering usually obtain an undergraduate degree in science or medicine. Typical majors include molecular biology, biomedical engineering, and genetics, although some universities offer specific undergraduate courses in genetic engineering. Graduate studies are essential for those wishing to pursue a career in genetic engineering. Following graduation, students studying genetically modified organisms will understand methods and processes involved in recombinant DNA technology and techniques, including DNA cloning, recombining genes, nucleic acid hybridization, gel transfers, and DNA sequencing.

Genetic engineering, while controversial, certainly offers potential as a significant tool in solving problems in agriculture, medicine, environmental science, and basic biology. Genetic technology is likely to play a central role in almost all areas of innovative biological sciences. Rapid advances in the field correspond to significant career potential and multiple avenues. Students involved in genetic engineering research and the application of GMOs can pursue careers in medical diagnosis, treatment, and gene therapy, agricultural food production, environmental bioremediation, and resource management within the private sector, nongovernmental organizations, specialized government organizations and agencies, and universities undertaking teaching and research.

SOCIAL CONTEXT AND FUTURE PROSPECTS

Genetically modified organisms potentially could be very advantageous to people. Specifically, scientists claim that GMOs will be vital to the future of food production and therapeutic medicine. Given the possibility of climate change due to global warming, crops that can produce higher yields, resist pests and pathogens, and better tolerate drought are very attractive. The use of GMOs in the treatment of genetic disorders makes them potentially life-saving. Supporters of GMO technology argue that genetic engineering has become an economic and environmental necessity in regard to agriculture, environmental bioremediation, and medicine.

Despite their obvious benefits, GMOs also hold many potential dangers. In addition, GMOs are not well received by the public. Surveys in some countries have revealed that the majority of people are actually against the creation and production of genetically modified foods, animals in particular. This opinion is shared by many environmental organizations, which claim that the undeniable benefits of GMOs are far outweighed by their possible effects on ecosystems, native flora and fauna, and human health. Of particular concern is that many of the potential risks of GMOs are as yet unknown. Opponents of GMO technology have stated that imposing GMOs onto the public without long-term rigorous testing is irresponsible and that more research is required.

Christine Watts, Ph.D., B.App.Sc., B.Sc.

FURTHER READING

Howe, Christopher. *Gene Cloning and Manipulation.* 2d ed. New York: Cambridge University Press, 2007. A comprehensive look at advances in recombinant DNA techniques, with both a broad and a concise examination of the concepts and principles involved.

Nelson, Gerald C., ed. *Genetically Modified Organisms in Agriculture: Economics and Politics.* San Diego, Calif.: Academic Press, 2001. Examines and analyzes the economic, ecological, and social factors involved in the production of GMOs for agriculture.

Nicholl, Desmond S. T. *An Introduction to Genetic Engineering.* 3d ed. New York: Cambridge University Press, 2008. Introduces basic molecular biology, the methods used to manipulate genes, and the technology's applications.

Primrose, Sandy B., Richard M. Twyman, and Robert W. Old. *Principles of Gene Manipulation.* 6th ed. Oxford, England: Blackwell Scientific, 2003. Discusses the genetic engineering of plants, animals, and microbes; the use of nucleic acids as diagnostic tools; and modern plant breeding.

Watson, James D., et al. *Recombinant DNA: Genes and Genomes—A Short Course*. 3d ed. New York: W. H. Freeman, 2007. One of the landmark texts of recombinant DNA technology, this work presents the fundamental concepts of genetics and genomics, the Human Genome Project, bioinformatic and experimental techniques, and a survey of epigenetics and RNA interference.

Young, Tomme R. *Genetically Modified Organisms and Biosafety: A Background Paper for Decision-Makers and Others to Assist in Consideration of GMO Issues*. Gland, Switzerland: International Union for Conservation of Nature, 2004. A detailed look at biosafety and genetically modified organisms, from species conservation, to sustainable livelihoods, to sociocultural policy.

WEB SITES

Union of Concerned Scientists, Food and Agriculture
Engineered Foods Allowed on the Market
http://www.ucsusa.org/food_and_agriculture/science_and_impacts/science/engineered-foods-allowed-on.html

U.S. Food and Drug Administration
Genetically Engineered Foods
http://www.fda.gov/NewsEvents/Testimony/ucm115032.htm

World Health Organization
Twenty Questions on Genetically Modified (GM) Foods
http://www.who.int/foodsafety/publications/biotech/en/20questions_en.pdf

See also: Agricultural Science; Animal Breeding and Husbandry; Bioengineering; Genetically Modified Food Production; Genetic Engineering; Plant Breeding and Propagation.

GENETIC ENGINEERING

FIELDS OF STUDY

Genetics; biology; molecular genetics; pharmacology; botany; cell biology; ethnobotany; ecology; developmental biology; evolutionary biology; microbiology; molecular biology; xenobiology; soil science; geology; chemistry; parasitology; zoology; biophysics; agroecology; agronomy; agricultural engineering; biological systems engineering; food engineering; food science; animal husbandry; agrology; plant science; bioengineering; environmental engineering; experimental evolution; biotechnology.

SUMMARY

Genetic engineering, also known as genetic modification, is an interdisciplinary scientific technique using molecular techniques to directly alter the basic genetic blueprint (DNA) of bacteria, plants, animals, humans, and other living organisms to achieve or enhance a specific trait or useful characteristic. Genetic engineering is used in diverse areas, including medicine and agriculture, to diagnose and treat diseases, produce industrial products, neutralize pollutants, create hardier crops, and perform scientific research. The genetic engineering process uses the tools of molecular genetics to explore and change living systems on a fundamental level and has revolutionized scientists' ability to understand, modify, and enhance the natural world.

KEY TERMS AND CONCEPTS

- **Biotechnology:** Use of modified living organisms in industrial and production processes. This technology has applications in industrial, agricultural, medical, and other production arenas.
- **Classic Selection:** Selective breeding of plants or animals with a desired feature so that the feature becomes more common or dominant in the subsequent generations.
- **Cloning:** Creating an identical copy of a gene, cell, or organism.
- **DNA (Deoxyribonucleic Acid):** Material within a cell that contains the genetic instructions used in the growth, development, replication, and functioning of living organisms. DNA forms the basic building blocks of genes.
- **Gene:** Basic unit of inheritance that contains information and instructions for the creation and maintenance of a living organism; consists of a segment of DNA.
- **Genetically Modified Food (GMF):** Food directly produced by genetic engineering or containing material from a genetically engineered plant or animal.
- **Genetically Modified Organism (GMO):** Organism whose genes have been purposefully modified through genetic technologies for a particular purpose.
- **Genome:** Collection of all genes and DNA present in a particular organism.
- **Insertion:** Addition of one or more pieces of genetic material into a particular DNA sequence or gene.
- **Isolation:** Process of removing a desired DNA segment or gene and placing it into a carrier organism (vector) for later insertion into particular gene or DNA sequence.
- **Ligase:** Type of enzyme that can be used to link or glue together DNA segments.
- **Plasmid:** Circular form of DNA used frequently as a vector to carry a desired gene segment into the organism being modified.
- **Recombinant DNA Technology:** Procedure used to move specific genetic information from one organism into another.
- **Restriction Enzyme:** Bacterial chemical or enzyme that cuts DNA at a specific site; used to remove a particular gene or gene segment from an organism.
- **Transformation:** In genetic engineering, the genetic alteration of a cell as a result of inserting new genetic material into an organism's standard genetic code.
- **Transgene:** Gene or genetic material that has been transferred from one organism into another organism of a different species.
- **Vector:** Virus or other chemical carrier that is used to deliver genetic material to a cell.

DEFINITION AND BASIC PRINCIPLES

Genetic engineering is the direct and purposeful alteration of an organism's DNA, the basic genetic blueprints of a bacterium, plant, animal, human, or other living organism to add or enhance a specific characteristic or trait. Although genetic engineering is most often discussed in the controversial arenas of crop production or theoretical human genetic manipulation, genetic engineering is used in diverse areas such as medicine, industry, and agriculture to treat disease, diagnose problems, produce industrial products, convert industrial waste, create hardier crops, and perform better scientific research.

The focus of genetic engineering is the gene. Genes are the basic units of inheritance that contain information and instructions for the creation, maintenance, and reproduction of living organisms. Genes are composed of DNA, a highly organized molecule located in almost every cell of an organism's body. In genetic engineering, scientists add very specific pieces of useful genetic material to another organism's genes to change an organism's natural characteristics.

Genetic engineering was made possible by the development of new molecular genetic procedures, often called recombinant DNA technology, that can identify the particular DNA sequence of a gene or an entire genome, allow scientists to find the genetic material that codes for useful or desired features, and then insert the new material into the correct place in another organism's genetic code.

Organisms that have had new genetic material inserted into their code are referred to as genetically modified organisms (GMOs) or genetically engineered organisms (GEOs). Examples of GMOs range from corn that has been engineered to produce an innate insecticide to cows that produce milk containing human insulin.

BACKGROUND AND HISTORY

Before modern genetic engineering was possible, farmers had long selected for desired traits by breeding plants and animals with the desirable traits. Brewers and bakers also changed grains and flour into preferred products such as beer and bread through the use of small organisms called yeast and the process of fermentation.

By the early twentieth century, plant scientists had begun to use the work done by Gregor Mendel in the nineteenth century on the inheritance patterns of specific plant features to more formally introduce improvements in a plant species in a process called classic selection. However, the features of the basic unit of inheritance were not known until James D. Watson and Francis Crick identified the structure of DNA in 1953.

The nature of DNA and the technology to manipulate and modify the genetics of an organism was not available until twenty years later, when the first successful recombinant organism was created by Herbert Boyer and Stanley Cohen. Boyer and his laboratory had isolated an enzyme that could precisely cut segments of DNA in an organism, and Cohen found a way to introduce antibiotic-carrying plasmids into bacteria and a way to isolate and clone the genes in the plasmids. They combined their knowledge to create a way to clone genetically engineered molecules in foreign cells. Their discoveries led to the creation of a quick and easy way to make chemicals such as human growth hormone and synthetic insulin.

After Boyer and Cohen, many other scientists worked with recombinant DNA techniques to improve the procedures and develop a variety of genetically modified organisms (GMOs) designed to meet specific scientific, agricultural, industrial, and medical needs. Over time, these techniques and applications in genetic engineering spawned the multibillion-dollar biotechnology industry.

As the biotechnology industry grew and genetically modified organisms became more widespread, it became important to define which organisms were genetically modified organisms and which were products of classic selection. It also became necessary to determine if living organisms produced through genetic engineering could be patented by the companies and universities designing them. In 1980, the U.S. Supreme Court ruled that genetically altered life-forms can be patented.

In 1982, the U.S. Food and Drug Administration (FDA) approved the first consumer product developed through modern genetic engineering: a biosynthetic human insulin, sold under the trademark Humulin. The bacterially produced insulin created by Genentech and marketed by Eli Lilly revolutionized the treatment of diabetes, as it produced fewer immune reactions and its supply no longer depended on the availability of animals.

In 1996, Genzyme Transgenics (which in 2002 became GTC Biotherapeutics) created a transgenic goat that produced milk containing a cancer-fighting protein. It soon created additional transgenic animals that could produce specific human proteins to treat human disease. The ability to produce human hormones, enzymes, and other therapeutic products has decreased the risk of disease transmittal from donors to recipients of human products, increased supply, decreased immune reactions, and decreased the variability between medication batches that had been seen in the past.

In the 1990's, scientists sought to develop genetically modified plants and crops. By 1992, the first plant designed for human consumption (the Flavr Savr tomato) was approved for commercial production by the U.S. Department of Agriculture. In 1994, the European Union approved genetically modified tobacco in France. After these genetically modified crops gained approval, genetically engineered plants and other organisms became more widespread in the United States. It has been estimated that 60 to 70 percent of food products on store shelves may contain at least a small quantity of genetically engineered crops.

The Human Genome Project (1990-2003), a collaborative international scientific research initiative spearheaded by the National Institutes of Health, advanced genetic engineering by its publication of human DNA sequencing data. These data allowed scientists to learn more about the physical and functional aspects of genes and DNA. By its completion, the Human Genome Project had provided a basic genetic road map for scientists to find human DNA segments of interest.

During the 1900's, scientists also used genetic engineering technology to develop numerous varieties of investigational organisms with very specific characteristics for use in research. Some genetically modified organisms were used to learn more about the natural progression of particular diseases. Others were created to test experimental therapies before moving to humans. These genetically modified organisms have helped scientists learn more about genetic disease, cancer, aging, and other chronic diseases.

In academic and industry laboratories, modern genetic engineering continues to solve problems related to health, disease, industry, and agriculture. Additional applications in humans and human

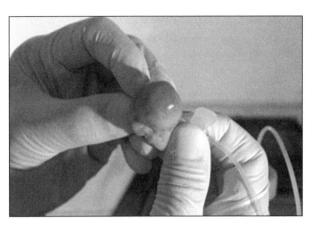

Researcher using laboratory equipment to harvest eggs (female reproductive cells) from the ovary of a cow. The egg cells will be used to create a clone. (Philippe Psaila/Photo Researchers, Inc.)

disease have been assisted by government-funded initiatives such as the Human Genome Project. Although controversial at times, genetic engineering is a modern tool to be used in addressing a wide range of problems.

HOW IT WORKS

Although the types of organisms modified vary substantially in genome size and structure, all genetic engineering involves several general steps: identifying the desired feature or end application, isolating the gene segment that codes for the feature, inserting the gene segment into a vector, and adding it to the target organism, a process called transformation.

Identification. To create genetically modified organisms that will meet a specific need or solve a particular issue, the best way to engineer a solution must be determined. For example, if a large oil spill required cleanup, the first step would be to determine the type of organism and the desired features that would be most effective at removing the spilled oil. Issues to consider in solving a problem through genetic engineering include the desired size and type of organism to be modified, the availability of desired characteristics or features with a known DNA segment, and the possible positive and negative environmental impact resulting from a modified organism.

Isolation of the Proper Gene Segment. To modify an organism through insertion of a gene or DNA segment, the specific DNA segment in the donor organism must be known and be able to be effectively removed from its host genome. In some cases, the DNA segment coding for the desired characteristic is known and available because of previous scientific work. In other cases, this step can be very labor intensive and require long-term research.

After the DNA segment is identified, a particular recombinant technique, often a restriction enzyme, is used to cut the desired gene or DNA segment out of the donor organism and move it into a vector. Depending on the genetic engineering requirements, other techniques such as polymerase chain reaction or agarose gel electrophoresis can be used to isolate a gene or gene segment.

Insertion. Insertion is the genetic engineering step during which the desired gene or DNA segment is integrated into the vector. In this step, restriction enzymes are used to cut the vector open in a particular place so that the desired gene or DNA segment can attach itself to the vector. Then a special enzyme glue called ligase is used to attach the DNA segment to the vector. The most commonly used vectors in genetic engineering are circular form of DNA called bacterial plasmids; however, the type of vector used for a particular application depends on the size of the gene or segment being moved and the organism being modified.

Transformation. The next step of genetic engineering is called transformation. During this step, the desired gene or DNA segment is introduced and successfully added to the organism being modified. A variety of methods can be used to send the vector containing the desired DNA segment into the organism being modified in such a way that the new genetic information is added to the organism's standard genes. These methods include microinjection, use of a gene gun, electroporation, or use of viruses. In each of these methods, the vector carrying the desired genetic segment is forced into the new cells of the organism being modified. Completion of the transformation step relies on testing that determines whether the inserted DNA segment is producing the desired effect or trait in the newly modified organism.

APPLICATIONS AND PRODUCTS

Genetic engineering has far-reaching applications in food production, industry, medicine, and research.

Crops. One of the most widespread but also most controversial uses of genetic engineering is in the creation of genetically modified crops and food. The goal of genetic modification varies from crop to crop. Soybeans have been modified with a DNA segment conveying resistance to herbicides sprayed over fields to kill weeds growing amid the soybeans. The Flvr Savr tomato was engineered to decrease ripening time and increase shelf life. Varieties of rice and corn have been engineered using DNA segments for other plant genomes to have increased levels of vitamins.

From the first commercially grown genetically engineered product for human consumption (the Flavr Savr tomato), adoption of genetically engineered crops in the United States has increased quickly. According to data from the U.S. Department of Agriculture's Economic Research Service, fields planted with genetically engineered cotton (herbicide-tolerant and insect-resistant cotton) reached 88 percent of the total cotton acreage in 2009. That same year, herbicide-tolerant soybeans and biotech corn accounted for 91 percent and 85 percent, respectively, of their crop populations.

Supporters of genetically modified crops feel that the plants can increase food production to meet the world's needs using lower amounts of pesticides and increasing farmer profits. Opponents of genetically engineered crops are concerned about perceived safety issues regarding food produced from these crops, ecological issues around increased use of herbicides, contamination between genetically modified and naturally grown crops because of cross-pollination of fields, and economic difficulties dealing with patents on genetically modified crops.

Livestock. Although farmers have bred particular varieties of livestock such as cows, goats, chickens, and sheep for thousands of years to maximize desirable qualities, genetic engineering allows a more rapid introduction of specific qualities that may or not occur naturally in the animals. The benefits of genetically engineered livestock are numerous and affect the producers, environment, and consumers. Producers benefit by having disease-resistant, increasingly productive, or fast-growing animals. For example, the gene responsible for regulating milk production in cows can be modified to increase milk production. Also, if animals are engineered to have milder waste, the environment will benefit. The FDA is reviewing genetically modified pigs that are better able to digest and process phosphorus in ways that release up to 70 percent less phosphorus in their waste. Consumers benefit from more nutritious, vitamin-enriched meat, as in the case of pigs that are engineered to produce omega-3 fatty acids through the expression of a roundworm gene.

Many of the concerns that apply to crops also apply to livestock. As of 2010, the FDA had not approved any genetically engineered meat products for human consumption. In September, 2010, the FDA stated that a type of salmon called AquAdvantage, genetically modified to grow twice as fast as conventional Atlantic salmon, was safe for human consumption. The FDA has placed several genetically modified animals under review in a category called food-drug. The FDA considers the DNA segment to be like a drug and is regulating transgenic animals in the same way it oversees animals that receive growth hormones or antibiotics.

Diagnosis. Genetic engineering has allowed the development of faster, cheaper, and more accurate diagnostic tests for certain diseases to be used both in the laboratory and in the body. The tests based on genetic engineering are used to identify infectious diseases, hormonal changes, pregnancies, cancer, and other diseases and conditions. For example, a series of faster and more accurate tests for the presence of the human immunodeficiency virus (HIV) have been developed based on genetically modified HIV antigens. Other tests can diagnose diseases by detecting particular substances in specific locations in the body. These exams rely on genetically modified antibodies with markers that can be injected into the body.

Medications. The use of genetically modified organisms to produce human hormones, enzymes, vaccines, and medications has revolutionized the pharmaceutical industry. Since 1982, when Genentech's biosynthetic human insulin was introduced, the ability to manufacture new products in a controlled environment instead of collecting similar substances from the limited supply of human and animal sources has led to more readily available, effective, and reliable medications. Products include human growth hormone to treat children with insufficient growth, plasminogen activator to dissolve blood clots, and erythropoietin to treat low blood iron (anemia). In 1994, genetic engineering also led to innovative treatments for rare genetic disorders such as Gaucher

disease with the production of specific human enzymes in genetically modified Chinese hamster ovary cells. The point of this enzyme replacement therapy is to replace the enzyme that the affected individuals are missing through intravenous injections of genetically engineered human enzymes. In multiple situations, the use of genetically engineered organisms to create medications has saved lives and decreased the burden of disease in ways that could not be imagined before. Future applications of genetic engineering in medicine are likely to focus on the creation of better medications for life-threatening indications.

Disease Cures. Genetic engineering made it possible to develop gene therapy. Genetic diseases are inherited conditions that occur because of one or more genetic changes or mutations that prevent

---

Fascinating Facts About Genetic Engineering

- Biotechnology has created more than two hundred new therapies and vaccines, including products to treat genetic diseases, cancer, diabetes, acquired immunodeficiency syndrome (AIDS), and autoimmune disorders.

- Scientists successfully manipulated the genetic sequence of a rat to grow a human ear on its back.

- If all the DNA in one person were laid in a straight line, it would stretch to the Sun and back more than thirty times.

- Glow-in-the-dark zebra fish are available for purchase at some pet stores. The glowing fish was made by adding naturally occurring genes from fluorescent organisms such as jellyfish and sea coral to the zebra fish's genome.

- Genetically engineered plants are being developed to detoxify pollutants in the soil or absorb and accumulate polluting substances in the air.

- The Human Genome Project produced a complete catalog of the human genome. This catalog is the size of several hundred average-sized telephone books and contains nothing but the letters A, T, G, and C in various permutations and combinations.

- Scientists still do not know the function of more than 80 percent of human DNA.

- Cows have been genetically engineered to produce medically useful proteins in their milk.

the correct functioning of a particular gene. Most genetic diseases do not have a treatment or cure. However, with genetic engineering techniques, scientists hope that they will be able to transform an affected individual's mutated gene into a working gene by replacing it with a functional copy of the gene. Gene therapy has shown some success in helping individuals with severe combined immunodeficiency (SCID), hemophilia type B, and several other genetic diseases; however, this type of treatment is still under investigation to determine how to safely and permanently cure genetic conditions.

Research. Genetic engineering and genome sequencing have been used to improve investigative techniques through the ability to manipulate organisms on a basic, genetic level. In genetic research, genetic engineering techniques have allowed scientists to create mice and other organisms affected by a specific gene change for detailed study of a specific genetic disease. For example, a genetically modified mouse that lacks the gene to produce amyloids has been used to study Alzheimer disease. On a broader scale, genetic engineering allows detailed analysis of an organism's structure, function, and development. Through the insertion of a marker in or near a gene coding for a product of interest, scientists can track the location of that gene's product over time.

Industrial Applications. Genetically engineered organisms are used in several manufacturing arenas in production, processing, and waste removal. Most industrial applications of biotechnology are based on naturally occurring processes using modified bacteria, yeast, and other small organisms to digest, transform, and synthesize natural materials from one form into another. More specifically, genetically modified microorganisms have been used to produce industrial chemicals such as ethylene oxide (for making plastics), ethylene glycol (antifreeze), and alcohol. Bacteria have also been engineered to remove toxic wastes from the environment, for example, the varieties of genetically engineered bacteria that consume oil by chemically transforming its compounds into usable basic molecules. Future directions in industry include production of textile fibers, fuels, plastics, and other industrial chemicals out of industrial wastes or raw materials.

IMPACT ON INDUSTRY

In 2009, biotechnical companies in the United States, Europe, Canada, and Australia had an aggregate net profit of $3.7 billion. In their 2010 report, Ernst and Young found that biotechnology companies in the United States, Europe, and Canada raised $23.2 billion in capital for 2009. This means that there is plenty of capital to create and market new genetic engineering applications. The biotechnology companies also report rapid job growth, which has been fueled primarily by growth in research, testing, and medical laboratories. The biotechnology industry is an important area of economic growth in the world as major advancements and innovations in medicine, agricultural, and industry based on genetic engineering continue globally.

Government and University Research. Funding for basic genetic engineering research often begins with the government-funded National Institutes of Health (NIH). The NIH performs on-site genetic research and awards grants to researchers in academic and university centers.

Beyond direct funding, various sectors of the government also encourage and stimulate research in genetic engineering. For example, the FDA is responsible for regulating genetically engineered foods and requires research studies on the safety and efficacy of particular gene changes in plants and animals before approval for consumption. In addition, the U.S. Department of Agriculture tracks the use of genetically modified crops worldwide and routinely performs research in the United States and internationally on food, farming, and biotechnology in order to issue industry reports.

Because of the wide scope of genetic engineering, research has been conducted by multiple academic groups, including the Zinc Finger (ZF) Consortium, whose members have focused their research on one method of targeting specific regions of DNA, and the Human Genome Project, an international public consortium of research laboratories led by the United States that worked together to complete the first sequence of the human genome. Groups such as the Registry of Standard Biological Parts, a consortium database of standardized gene sequences in vectors with known characteristics ready to be used in genetic engineering, have been formed primarily to encourage data sharing.

Industry and Business. Genetic engineering has led to the creation of an entirely new industry called biotechnology. Under the umbrella of biotechnology are diverse groups of companies that apply genetic engineering principles in bacteria, plants, animals, and humans to research, manufacturing, health issues, and marketing. Because of the fast-paced nature of scientific discovery, these companies are designed to fund cutting-edge research and then quickly integrate their findings into production plans. The four main sections of the biotechnology industry are agricultural feedstock and chemicals, drugs and pharmaceuticals, medical devices and equipment, and research-testing-medical laboratories. The use of genetically engineered organisms in production, processing, and waste removal stimulates research and development internally in these companies as they develop more efficient, high-yield processes.

CAREERS AND COURSE WORK

Courses in molecular genetics, biochemistry, human genetics, developmental biology, cell biology, microbiology, biological systems engineering, and biotechnology are the foundational requirements for students interested in pursuing careers in genetic engineering. Depending on the student's desired area of study, courses in animal husbandry or medicine might be required. A bachelor's degree in biology, chemistry, applied biotechnology, or genetics is appropriate preparation for graduate work in genetic engineering. In most circumstances a master's degree, medical degree, or doctorate is necessary for the most advanced career opportunities in both academia and industry. However, technician and laboratory roles for those without advanced degrees are often available.

Other career possibilities include marketing and sales staff for pharmaceutical companies, doctors and nurses involved in prenatal genetic testing, genetic ethicists, or genetic counselors. Whether students pursue a scientific, academic, or socially oriented position, they should take a variety of courses beyond the natural sciences to be aware of cultural and societal issues surrounding genetic engineering.

SOCIAL CONTEXT AND FUTURE PROSPECTS

Genetic engineering has already altered the course of agriculture, industry, and medicine with its life-changing applications. Crops have been modified so that they are more nutritious and naturally produce pesticides. Life-saving medications made of human hormones integrated into bacteria are widely available in consistent and purified forms.

Bacteria that convert toxic chemicals into harmless basic elements have been developed. Great strides have been made in using gene therapy to cure genetic diseases. However, these significant scientific strides also come with important ethical questions and safety concerns.

Environmentalists are concerned about the impact of genetically modified crops on ecosystems, in particular whether the genes introduced into genetically modified crops will be transferred to conventional crops through cross-pollination.

Advocacy groups such as Greenpeace and the World Wildlife Fund are concerned about the safety of genetically modified food and feel that the available data do not prove that there are no risks to human health from consumption. Despite statements from the Royal Society of Medicine and the U.S. National Academy of Sciences in support of the safety of such foods, these groups have called for additional and more rigorous testing before genetically engineered foods are marketed.

Other countries have significant concerns about the safety of genetically modified foods. The European Union regulates genetically modified food imported from other nations, including the United States. Venezuela has banned the growing of genetically modified crops, and India has issued a moratorium on the cultivation of genetically modified foods pending an investigation into safety concerns. Other counties such as Japan and Zambia also have registered concerns over the safety of genetically modified foods.

There is significantly less controversy over the use of genetically modified organisms in industrial production and medicines. However, the use of genetic engineering techniques for human gene therapy and related applications has touched off a firestorm of ethical debate.

Dawn A. Laney, M.S., C.G.C., C.C.R.C.

FURTHER READING

Avise, John C. *The Hope, Hype, and Reality of Genetic Engineering: Remarkable Stories from Agriculture, Industry, Medicine, and the Environment.* New York: Oxford University Press, 2004. Contains a series of well-written essays about more than sixty genetically engineered organisms. Each short essay discusses the procedures and challenges involved in engineering each organism or trait.

Hodge, Russ. *Genetic Engineering: Manipulating the Mechanisms of Life.* New York: Facts On File, 2009. Covers genetics from the beginning to genetic engineering. Examines both the science and the social issues.

Nicholl, Desmond S. T. *An Introduction to Genetic Engineering.* New York: Cambridge University Press, 2008. Describes the basic principles of genes, molecular biology, methods used in genetic engineering, and applications of genetic engineering.

Shanks, Pete. *Human Genetic Engineering: A Guide for Activists, Skeptics, and the Very Perplexed.* New York: Nation Books, 2005. A well-written discussion of genetic engineering and biotechnology that covers cloning, stem cells, gene therapy, and the genetic engineering of food. The author, who does not support genetic engineering, provides a good overview of genetic engineering debates.

Yount, Lisa. *Biotechnology and Genetic Engineering.* 3d ed. New York: Facts On File, 2008. A compact overview of issues in the genetic engineering field. Examines law and genetic engineering and key individuals involved in genetic engineering. Contains an extensive time line and glossary.

WEB SITES

Union of Concerned Scientists
Impacts of Genetic Engineering
http://www.ucsusa.org/food_and_agriculture/science_and_impacts/impacts_genetic_engineering/impacts-of-genetic.html

U.S. Department of Agriculture
Agricultural Biotechnology Website
http://www.usda.gov

World Health Organization
Twenty Questions on Genetically Modified Foods
http://www.who.int/foodsafety/publications/biotech/20questions/en

See also: Animal Breeding and Husbandry; Bioengineering; Biosynthetics; Cloning; DNA Analysis; DNA Sequencing; Environmental Microbiology; Genetically Modified Food Production; Genetically Modified Organisms; Genomics; Human Genetic Engineering; Plant Breeding and Propagation.

GENOMICS

FIELDS OF STUDY

Genetics; bioinformatics; molecular genetics; microbial genetics; epigenetics; behavioral genetics; neurogenetics; phylogenetics; population genetics; ancient DNA; genetic anthropology; paleogenetics; evolutionary biology; evolutionary biochemistry; archaeology; forensics; biology; cell biology; microbiology; molecular biology; biophysics; synthetic biology; systems biology; biochemistry; chemistry; systematics; taxonomy; virology.

SUMMARY

Genomics is the branch of biotechnology that focuses on the genome, the entire set of genes in an organism, as well as the interaction of individual genes with one another and the organism's environment. Broad applications include comparative and functional genomics.

KEY TERMS AND CONCEPTS

- **Ancient DNA:** DNA sequences from museum specimens, archaeological finds, fossils, mummified remains, and ancient microorganisms that were embedded in ice, rock, or amber.
- **Bioinformatics:** Discipline that uses computer programs to scan sequences of genomic DNA to retrieve and analyze specific genes.
- **Chromosome:** Structural unit of genetic material that consists of a lone, double-stranded molecule, found in eukaryotes.
- **Comparative Genomics:** Field of biological research that compares the genomes of different species.
- **DNA (Deoxyribonucleic Acid):** Long linear polymer found in the nucleus of a cell, consisting of nucleotides in the shape of a double helix.
- **Gene:** Segment of DNA that serves as the functional and physical unit of heredity.
- **Genome:** Entire set of DNA found in an organism.
- **Genomic Library:** Collection of bacteria that have been genetically engineered to hold the whole genome of an organism
- **Messenger RNA (mRNA):** Single-stranded molecule of RNA that functions as a template for the production of protein on the ribosomes of the cell.
- **Mitochondrial DNA (mtDNA):** Genetic material within the mitochondria that is passed from the female parent to her children.
- **Ribosomal RNA (rRNA):** RNA found in the ribosome, which translates mRNA and synthesizes proteins.
- **RNA (Ribonucleic Acid):** Single-stranded polymer consisting of ribose nucleotides formed by transcription of DNA or viruses from another RNA strand.
- **Sequencing:** Technique that elucidates the sequence of the molecules that form an organism's genomic content.

DEFINITION AND BASIC PRINCIPLES

Genomics studies organisms' entire genomes, focusing on structure, function, and inheritance. It examines complex diseases, including cancer, heart disease, and diabetes, in which both genetic and environmental factors play a part. In contrast, genetics focuses on individual genes and their role in inheritance and heritable genetic diseases such as phenylketonuria and cystic fibrosis. Of central importance in genomics is determining the sequences of DNA that make up an organism. DNA sequencing can be used to find mutations or variations that may be involved in causing a disease. Functional genomics examines how individual genes function and interact with each other within a given genome at the functional level, and comparative genetics examines the relationships between different species. Comparative genomics may be used to study evolutionary relationships between species by comparing their chromosomes.

BACKGROUND AND HISTORY

Scientists became aware of the existence of genomes in the late nineteenth century when they first viewed chromosomes under a microscope, although the word "genome" was not coined until 1920. In the twentieth century, scientists studied the frequency at which chromosomes exchanged parts during meiosis to map the genes on chromosomes. However, this technique was useful for mapping primarily genes that had mutant phenotypes, which accounted for only a small part of the total genome.

Numerous scientists discovered the base-pair sequences of many human genes, but mapping the entire human genome remained a challenge. Determining the entire human genome was generally regarded as a worthwhile endeavor, but the process would take billions of dollars, taking funding away from more established, mainstream biomedical research. Ethicists, scientists, and economists debated the merits of the research versus its exorbitant cost. Nevertheless, the Human Genome Project was launched in 1990, supported by the U.S. Department of Energy (Human Genome Program, directed by Ari Patrinos) and the National Institutes of Health (National Human Genome Research Institute, directed by Francis S. Collins). The Human Genome Project became an international effort, with scientists from around the globe joining the team. Newly developed computer software helped facilitate the process. In 1998, Celera Genomics, a private company headed by J. Craig Venter, began an independent effort to complete the mapping of the human genome using an unorthodox shotgun, or whole-genome, method of sequencing. In 2000, Collins, Venter, and Patrinos jointly announced the completion of a rough draft of the human genome. In 2003, fifty years after Francis Crick and James D. Watson discovered the double-helix structure of DNA, the human genome had been sequenced, and the project was declared complete, although final papers were published in 2006.

HOW IT WORKS

Determining an organism's genome requires the sequencing of its DNA, which involves finding the order of the bases in a given stretch of DNA. Several sequencing methods were developed in the 1970's, but sequencing was a very slow process.

One problem was that the total amount of DNA necessary for sequencing and analyzing a genome of interest could be several times the total amount of available DNA. To increase the amount of DNA, scientists turned to cloning. The DNA fragments are replicated inside a bacterial cell, then the cloned DNA is extracted and placed in a machine for sequencing. The polymerase chain reaction, developed in 1983, allows ancient DNA or degraded DNA to be sequenced using special protocols to amplify the sample before sequencing.

Genomic sequences are usually elucidated with automatic sequencers; shotgun, (whole-genome) se-

quencing, developed by Venter, is the most popular DNA sequencing technique. The DNA is randomly cut into fragments, and then the ends of each fragment are sequenced, resulting in two reads per fragment. The reads are used to reconstruct the original DNA sequence. Newer sequencing techniques generally follow this model but may use different strategies along the way. As the fragments are sequenced, the data are set aside. When enough sequences have been gathered, they are linked through sequence overlaps. The resulting genomic sequence is then deposited into a publicly accessible database.

The genomic sequence is analyzed to discover the individual genes and how they are regulated. Bioinformatics is the discipline that relies on computer programs to search for genes in DNA sequences, using what is known about the gene of interst. After scientists find the sequences that make up a gene, the structure and function can be compared with similar gene sequences in other organisms.

APPLICATIONS AND PRODUCTS

Genomics has many applications. Among the most important applications, because of their impact on society, are those that deal with evolution and the environment. Other areas, such as synthetic biology and personal genomics, may lead to innovations that

Craig Venter, president and chief scientific officer of Celera Genomics, stands next to a map of the human genome in his office. (AP Photo)

can be of great benefit to individuals and populations and to public health.

Evolutionary Genomics. Until about 30,000 years ago, Neanderthals, the closest relatives of modern humans, inhabited Europe and Asia. Since the early twentieth century, anthropologists and paleontologists have attempted to demonstrate an evolutionary relationship between Neanderthals and modern humans, who emerged about 400,000 years ago. In 1997, Svante Pääbo, director of the Department of Genetics at the Max Planck Institute for Evolutionary Anthropology, furthered the understanding of the genetic relationship of Neanderthals and modern humans when he sequenced Neanderthal mtDNA. In 2009, scientists from the Max Planck Institute announced that they had generated a draft sequence equivalent to more than 60 percent of the complete Neanderthal genome and had begun comparing it to that of humans. In May, 2010, these researchers revealed that the human genome contains some of the same genes as were found in the Neanderthal genome. Their continued efforts may shed light on the origin of humankind as well as the evolutionary process.

Environmental Genomics. In 2009, the International Union for Conservation of Nature estimated that Earth is home to 8 million to 14 million animals and plants, of which only 1.8 million have been identified and classified. Naturalists such as Charles Darwin classified plants and animals using taxonomic systems. Later scientists compared the genetic differences among species; a species was defined as a group of organisms that could breed and produce fertile offspring. Determining where one species ends and another begins is difficult, as in the case of wolves and coyotes. Modern technology allows species to be identified and classified by comparing modern and ancient samples of genomic DNA and tracking how a species descended from an ancestor. The question of what constitutes a species is important in determining which species can be considered endangered. Global climate change has led to the extinction of many plant and animal species, and many scientists hope to be able to preserve the world's biodiversity using knowledge gleaned from genomics.

In 2004, Venter and his colleagues at the Institute for Biological Energy Alternatives (later part of the J. Craig Venter Institute), of Rockville, Maryland, applied whole-genome sequencing to microbial populations collected from the Sargasso Sea, close to Bermuda, where researchers expected a low diversity of species. They analyzed more than 1 billion base pairs and found 1.2 million new genes and 1,800 species of microbes (including 150 new species of bacteria). According to Venter and his colleagues, the number of species suggests that microbial life in the ocean is more plentiful and diverse than previously thought.

Synthetic Biology. One of the goals of synthetic biology, which creates artificial biological systems (not existing in nature) and redesigns existing biological systems, is intelligent design. Synthetic biology applies the principles of large-scale engineering to biology and is built on the premise that organisms can be divided into discrete parts. In 2010, researchers at the J. Craig Venter Institute created a self-replicating bacterial cell containing totally synthetic DNA. The cell's genome was designed in the computer. No natural DNA was used, and chemical synthesis was used to make a viable cell.

Synthetic biologists believe that someday it may be possible to program bamboo to grow into preformed chairs or to program trees to spew oil from their stems. Reprogrammed bacteria may be able to heal, not harm, humans and animals. Some experts, such as David Rejeski of the Woodrow Wilson International Center for Scholars in Washington, D.C., believe that synthetic biology may fundamentally change the way things are made within the next one hundred years, creating manufacturing shift as significant as the Industrial Revolution.

IMPACT ON INDUSTRY

Genome Decoding for Individuals. The cost of sequencing the first human genome was about $3 billion. However, advances in technology have caused the price of obtaining a complete genome to drop rapidly. In 2010, Ilumina, the market leader in DNA-sequencing machines, offered to provide people with their genome for $9,500 if they might benefit from the information (as is the case with people with rare cancers). Life Technologies, Ilumina's closest competitor, is working with a group of cancer research centers to see how genetic information might help in the treatment of cancer. It is also offering a sequencer that sells for around $100,000, in contrast to the typical $700,000. The DNA-sequencing industry saw the entrance of Pacific Biosciences, which offers sequencing in a matter of minutes rather than days, and Complete Genomic, which has created a sequencing facility for drug companies to use.

Although genome decoding had focused on healthy individuals, in March of 2010, two research teams independently decoded the entire genome of patients who had genetic diseases to find the exact genes that caused their diseases. The lower cost of DNA sequencing may revive the largely unsuccessful efforts to identify the genetic causes of major killers such as heart disease, diabetes, and Alzheimer's disease. For example, James R. Lupski, a medical geneticist who has a Charcot-Marie-Tooth, a neurological disease, used whole-genome sequencing on his own DNA in an effort to better understand his illness. He was able to find the genetic variation causing his condition, although earlier research had not been able to pinpoint it.

Genomic Ventures. Many genomics companies formed in the 1990's and 2000's, but many are still in the research and development process and have yet to market products. Most of these companies apply genomics to the creation of pharmaceuticals, although some use genomic studies for environmental purposes or in synthetic biology. Many are involved in cooperative ventures with pharmaceutical companies, and mergers and acquisitions are common.

In October, 2006, Venter (former head of Celera Genomics) brought together several affiliated organizations, The Institute for Genomic Research (TIGR), The Center for the Advancement of Genomics (TCAG), The J. Craig Venter Science Foundation, The Joint Technology Center, and the Institute for Biological Energy Alternatives to create the J. Craig Venter Institute, with offices in Rockville, Maryland, and San Diego, California. The institute, known for high-throughput DNA sequencing, has research groups working on human genomic medicine, microbial and environmental genomics, plant genomics, infectious diseases, and synthetic biology and bioenergy. Its approach toward infectious disease is to examine the genome of the microbes and viruses that cause disease and the microbial flora in various cavities of the human body. Another focus is its bioinformatic group, which works on the software, databases, and mathematics to analyze the data created by DNA sequencing.

Other smaller companies are also active in various aspects of genomics. Millenium: The Takeda On-cology Company, based in Cambridge, Massachusetts, was established in 1993 as a genomics company and has evolved into a biopharmacological company focused on cancer. Its leading product is injectable VELCADE (bortezomib) for the treatment of people with multiple myeloma and a type of lymphoma. Rockville, Maryland-based Human Genome Sciences (founded 1992) is developing drugs to treat hepatitis C and systemic lupus erythematosus. Incyte, based in Wilmington, Delaware, is developing a drugs to treat rheumatoid arthritis, psoriasis, and cancer.

Affymetrix, based in Santa Clara, California, developed the first microarray (for genome-wide analysis) in 1989 and the first commercial microarray in 1996. Its genomic analysis tools are invaluable for researchers and pharmaceutical companies. San Diego, California-based Verdezyne, switched from being a provider of synthetic genes to drug and industrial enzyme companies to focus on a fermentation process to produce renewable energy fuels and chemicals, such as ethanol and adipic acid. It is using genomics technology to create proprietary metabolic pathways in yeast for enhanced conversion of hexose and pentose sugars to ethanol.

Government Agencies. One of the main governmental agencies involved in genomics is the National Human Genome Research Institute, part of the National Institutes of Health. It began as part of the Human Genome Project and continues to conduct research into the genetic basis of human disease. It is divided into seven branches: Cancer Genetics, Genetic Disease Research, Genetics and Molecular Biology, Genome Technology, Inherited Disease Research, Medical Genetics, and Social and Behavioral Research.

The U.S. Department of Energy is also involved in genomics research. Its Genomic Science Program focuses on using genomics to deal with issues regarding the environment, energy, and the climate. The agency is involved in the genome sequencing of microbes that can help cycle carbon from the atmosphere, clean up toxic waste, and create biofuels. The DOE Joint Genome Institute, operated by the University of California, Berkeley, concentrates on providing clean energy and environmental solutions.

CAREERS AND COURSE WORK

There are many pathways to a career in genomics. However, a combination of biology and mathematics

Fascinating Facts About Genomics

- In 2005, the dog genome sequence was published in *Nature* magazine. The DNA sequenced came from Tasha, a purebred female boxer. Scientists think that the genetic contribution to disease may be easier to determine in dogs than in humans. About 5 percent of the human genome is also present in dogs.

- Humans and chimpanzees share 96 percent of their genomes.

- The average difference in the genomes of two people is 0.2 percent, or one in five hundred bases.

- About 97 percent of the DNA in the human genome consists of so-called junk DNA—sequences with no known functions.

- Scientists at the J. Craig Venter Institute are studying the genome of the SARS coronavirus, which causes severe acute respiratory syndrome, in humans and animals in an effort to determine how the virus crosses the species barrier.

- In 2010, the complete draft genome sequence of the soybean was published in *Nature* magazine. It was expected to increase understanding of the nitrogen-fixing process and to help scientists develop better soybeans, including a more digestible version.

is essential for genomics. High school students interested in genomics should take courses in the basic sciences—biology, chemistry, mathematics, and physics—and computer science. At the college level, recommended courses are genetics, physiology, biology, cell and molecular biology, microbiology, developmental biology, physics, linear algebra, probability and statistics, and computer science (in particular, programming). Suitable majors include biology, biochemistry, bioinformatics, genetics, and computer science. Laboratory experience is also necessary for the would-be researcher. A master's or doctorate in genetics or genomics is required for most research, whether in private industry, academic research centers, or governmental agencies.

SOCIAL CONTEXT AND FUTURE PROSPECTS

The U.S. Department of Energy and the National Institutes of Health have devoted between 3 and 5 percent of their annual genome project budgets toward the study of ethical, legal, and social issues. Societal concerns include privacy and confidentiality issues, possible social stigma and discrimination, and how genetic information will be used by insurance companies, the legal system, and academia. Various government agencies have taken the lead in addressing the existing and potential social and ethical issues that may arise from genome-centered biology.

The Genetic Information Nondiscrimination Act of 2008 prohibits insurance companies and employers from discriminating against people based on the results of genetic testing. In addition, under the law, employers and insurers are prohibited from requesting or demanding that individuals take genetic tests.

In an interview in 2010, National Institutes of Health director and former Human Genome Project member Francis S. Collins expressed support for genomic research because it is likely to enable people to prevent and treat disease. He also noted that the patenting of human genes, once very controversial, has become more common, with more than 20 percent of all know genes having been patented. The patents, he argued, help private biotech companies fund costly research into genetic diseases. He also made a point of stating that his religious beliefs do not interfere with his work in evolutionary genetics.

Aware of the legal and social issues and of most people's limited understanding of genetics, the J. Craig Venter Institute has created a division to promote understanding of genomics among policy makers and the general public and to foster a positive image for the biotechnology industry.

The possible benefits of genomic research are immense. Some experts feel that better understanding of the genomics may lead to radical innovations in disease treatment and prevention, and environmental and synthetic genetics may produce ways to create renewable fuels. The decreasing cost of obtaining a person's genome may mean that physicians can take an individualized approach to medicine. Therefore, although controversy over genetics remains, this field is likely to remain an active area of research, yielding numerous applications, and providing numerous work opportunities to those interested in the field.

Cynthia F. Racer, M.A., M.P.H.

FURTHER READING

Davies, Kevin. *The $1,000 Genome: The Revolution in DNA Sequencing and the New Era of Personalized Medicine.* New York: Free Press, 2010. Looks at how less expensive, faster means of obtaining a person's genome will change medicine and make it more tailored to the individual.

DeSalle, Michael, and Michael Yudell. *Welcome to the Genome: A User's Guide to the Genetic Past, Present, and Future.* Hoboken, N.J.: Wiley-Liss, 2005. Starts with a brief history of genetics and description of the science before examining how the genome was sequenced. Analyzes the likely medical and agricultural applications.

Fairbanks, Daniel J. *Relics of Eden: The Powerful Evidence of Evolution in Human DNA.* Amherst, N.Y.: Prometheus Books, 2007. An examination of the field of evolutionary genomics that asserts that there is no dichotomy between religion and science.

Gee, Henry. *Jacob's Ladder: The History of the Human Genome.* New York: W. W. Norton, 2004. Examines what human genome sequencing reveals and how this information may be used in the future.

Gibson, D. G., et al. "Reation of a Bacterial Cell Controlled by a Chemically Synthesized Genome." *Science Express* (May 20, 2010). Announces the creation of a self-replicating bacterial cell governed by a synthetic genome.

Shreeve, James. *The Genome War: How Craig Venter Tried to Capture the Code of Life and Save the World.* New York: Alfred A. Knopf, 2004. Describes the competition between Venter and Collins to decode the human genome.

WEB SITES

DOE Joint Genome Institute
http://www.jgi.doe.gov

Genomic Science Program, U.S. Department of Energy
http://genomicscience.energy.gov/index.shtml

The Human Genome Project
http://www.ornl.gov/sci/techresources/Human_Genome/home.shtml

J. Craig Venter Institute
http://www.jcvi.org

National Human Genome Research Institute
http://www.genome.gov

See also: Bioenergy Technologies; Bioengineering; Biofuels and Synthetic Fuels; Bioinformatics; Biosynthetics; Cloning; DNA Sequencing; Genetically Modified Food Production; Genetically Modified Organisms; Genetic Engineering; Human Genetic Engineering; Metabolic Engineering; Paleontology.

GEOINFORMATICS

FIELDS OF STUDY

Calculus, analytic geometry, trigonometry, computer science, earth science, statistics, geography, mapmaking, digital imaging and graphics design, database management systems.

SUMMARY

The term "geoinformatics" refers to a collection of information systems and technologies used to create, collect, organize, analyze, display, and store geographic information for specific end-user applications. The field represents a paradigm shift from traditional discipline-based systems such as cartography, geodesy, surveying, photogrammetry, and remote sensing to a data systems management protocol that includes all earlier technologies and combines them to create new models of spatial information. Computation is an essential foundation of all geoinformatics systems.

KEY TERMS AND CONCEPTS

- **Datum:** Starting point from which a system of measurements of the Earth's size is computed.
- **Geodatabase:** Essential structure of a geographic information system (GIS) including selected geographic data sets, particular object definitions, and descriptions of relationships. Information is stored in a variety of formats and includes nominal, ordinal, interval, ratio, and scalar data.
- **Geographic Data Model:** Core component of a GIS system. The data model provides a geographic template for a particular industry application.
- **Geographic Information System (GIS):** Collection of computer hardware, software, and geographic data used to reference spatial geographic information. It is an analytic system to input, edit, manipulate, study, compare, and output geographic data collected from a variety of sources.
- **Geography:** Study of the Earth's surface and climate and the interrelations between them.
- **Geoid:** Physical model of the Earth. It is recognized by its irregular shape.
- **Geoprocessing:** Refers to a set of operations that can be selected to create particular geographic data outputs. These include software applications for file conversions, geographic data overlays, intersection and union of particular features, data extraction, proximity analytics, and user-facilitated data management.
- **Map:** Visual representation of the surface of the Earth or other celestial body.
- **Map Projection:** Projection of a three-dimensional figure onto a flat two-dimensional plane.
- **Query:** Carefully structured data search format designed to collect particular information from a database set.
- **Raster Data:** Digital information stored on a grid of pixels assigned a particular value. These data sets create numeric data well suited for the storage of continuous data collected over large regions.
- **Scale:** Numerical statement of the mathematical relationship between a unit of distance on the ground and its cartographic representation.
- **Spatial Cognition:** Cognitive ability to make judgments about spatial patterns and relationships.
- **Topology:** Mathematical study of objects the shapes of which are distorted under manipulation. For example, a circle can be stretched into the shape of an ellipse.
- **Vector Model:** Geographic data with lines and shapes plotted on an x-y coordinate system.

DEFINITION AND BASIC PRINCIPLES
Geoinformatics is a complex, multidisciplinary field of knowledge specializing in the creation, collection, storage, classification, manipulation, comparison, and evaluation of spatially referenced information for use in a variety of public and private practices. Its technologies are rooted in mapping, land surveying, and communication technologies that are thousands of years old. The exponential growth of science-based knowledge and mathematical expertise during the nineteenth and twentieth centuries greatly assisted the accumulation of verifiable geographic information describing the Earth and its position among the myriad celestial entities occupying the known universe. Detailed geometric descriptions and photographic materials make it possible to translate, measure, and order the surface of the Earth into multidimensional coordinate systems

that provide a rich and detailed visual language for understanding the relationships and features of locations too vast to be easily comprehended by the human senses. Geographic communication systems are greatly enhanced by the proliferation of computation technologies including database management systems, laser-based surveys, digital satellite photo technologies, and computer-aided design (CAD).

BACKGROUND AND HISTORY

The components and design structures of maps form the fundamental language of geographic information systems (GIS). Maps illustrate a variety of environments, both real and imagined, and the structures and life forms that fill them. The geographies of the Earth's land masses and the relationships of land and water to the Moon and stars are the foundations of advanced mathematics and the physical sciences, subjects that are continually modified by new measurements of spatial-temporal coordinates. Aerial and nautical photography introduced a new and vital reality to cartographic representation in the twentieth century. Advances in sonar and radar technologies, the telephone and telegraph, radio broadcast, mass transportation, manned space flight, spectrometry, the telescope, nuclear physics, biomedical engineering, and cosmology all rely on the power of the map to convey important information about the position, structure, and movement of key variables.

Cartography. Mapping is an innate cognitive ability; it is common in a variety of animate species. In human practice, mapping represents an evolutionary process of symbolic human communication. Its grammar, syntax, and elements of style are composed of highly refined systems of notation, projections, grids, symbols, aesthetics, and scales. These time-honored features and practices of cartographic representation demand careful study and practice. Maps are the most ancient documents of human culture and civilization and, as such, many are carefully preserved by private and public institutions all over the world. The World Wide Web has made it possible to share, via the Internet, facsimiles of these precious cultural artifacts; the originals are protected for posterity. Scholars study the intellectual and technical processes used to create and document maps. These provide valuable clues about the beliefs and assumptions of the cultures, groups, and individuals that contract and prepare them.

Geodesy. The cartographer artfully translates the spatial features of conceptual landscapes into two- and three-dimensional documents using data and symbols selected to illustrate specific relationships or physical features of a particular place for the benefit or education of a group or individual enterprise. Geodesy is an earth science the practitioners of which provide timely geographical measurements used by cartographers to create accurate maps. Since earliest recorded times, geodesists have utilized the most current astronomical and geographical knowledge to measure the surface of the Earth and its geometric relationship to the Sun and Moon. The roots of geodetic measurement systems are buried in the ancient cultures and civilizations of Egypt, China, Mesopotamia, India, and the Mediterranean. Enlightened scholars and astute merchants of land and sea traveled the known world and shared manuscripts, instruments, personal observation, and practical know-how for comprehending the natural world. Hellenic scholars brilliantly advanced the study of astronomy and the Earth's geography. Their works formed the canon of cartographical and astronomical theory in the Western world from the time of the great voyages of discovery in the fifteenth and sixteenth centuries.

Alexandria was the intellectual center of the lives of the Greek polymath Eratosthenes of Cyrene and the Roman astronomer Ptolemy. Both wrote geographical and astronomical treatises that were honored and studied for centuries. Eratosthenes wrote *Peri tes avametreoeos tes ges* (On the Measurement of the Earth) and the three-volume *Geographika* (Geography), establishing mathematics and precise linear measurements as necessary prerequisites for the accurate geographical modeling of the known world. His works are considered singular among the achievements of Greek civilization. He is particularly noted for his calculations of the earth's circumference, measurements that were not disputed until the seventeenth century.

Ptolemy, following Eratosthenes, served as the librarian at Alexandria. He made astronomical observations from Alexandria between the years 127 and 141, and he was firm in his belief that accurate geographical maps were derived from the teachings of astronomy and mathematics. The *Almagest* is a treatise devoted to a scientific, methodical description of the movement of the stars and planets. The beauty and integrity of his geocentric model set a standard for inquiry that dominated astronomy for centuries. He also wrote the eight-volume *Geographia* (Geography), in which he established latitude and longitude coordinates for the major landmasses and mapped the world using a conic projection of prime and linear meridians and curved parallels. His instructions for creating a coordinate system of mapping, including sectional and regional maps to highlight individual countries, are still in practice.

At the end of the nineteenth century, geophysics became a distinct science. Its intellectual foundations provided rising petroleum corporations new technologies for identifying and classifying important land features and resources. In the United States, extensive surveys were conducted for administrative and military purposes. The U.S. Coast Survey was established by President Thomas Jefferson in 1807. The Army Corps of Engineers, the Army Corps of Topographical Engineers, and the United States Naval Observatory supported intensive geophysical research including studies of harbors and rivers, oceans and land topographies. In 1878, the U.S. Coast Survey became the U.S. Coast and Geodetic Survey. In 1965, the U.S. Coast and Geodetic Survey was reincorporated under the Environmental Sciences Services Administration, and in 1970 it was reorganized as the National Oceanic and Atmospheric Administration (NOAA).

Underwater acoustics and geomagnetic topographies were critical to the success of naval engagements during World War I and World War II. The International Union of Geodesy and Geophysics (IUGG) was founded in 1919. As of 2011, it is one of thirty scientific unions participating in the International Council for Science (ICSU). That same year the American National Committee of the International Union of Geodesy and Geophysics combined with the Committee on Geophysics of the National Research Council; in 1972 the committee was independently organized as the American Geophysical Union. Graduate geophysics programs became prominent in major universities after World War II.

Geodesy is a multidisciplinary effort to calculate and document precisely the measurements of the Earth's shape and gravitational field in order to define accurate spatial-temporal locations for points on its surface. Land surveys and geomensuration, or the measure of the Earth as a whole, are essential practices. Geodetic data is used GIS—these include materials from field surveys, satellites, and digital maps. The Earth's shape is represented as an ellipsoid in current mathematical models. Three-dimensional descriptors are applied to one quadrant of the whole in a series of calculations. All cartographic grid systems and subsequent measurements of the Earth begin with a starting point called a datum. The first datum was established in North America in 1866 and its calculations were used for the 1927 North American Datum (NAD 27). In 1983, a new datum was established (NAD 83) and it is the basis of the standard geodetic reference system (GRS 80). It was again modified and became the World Geodetic System in 1984 (WGS 84). WGS 84 utilizes constant parameters for the definition of the Earth's major and minor axes, semimajor and semiminor axes, and various ratios for calculating the flattening at the poles. The geoid is another geodetic representation of the Earth, as is the sphere.

Photogrammetry. The word photogrammetry refers to the use of photographic techniques to produce accurate three-dimensional information about the topology of a given area. Accurate measurements of spaces and structures are obtained through various applications including aerial photography, aerotriangulation, digital mapping, topographic surveys, and database and GIS management. Precise, detailed photographs make it possible to compare and analyze particular features of a given environment. Aerial photographs are particularly useful to engineers, designers, and planners who need visual information about the site of a project or habitat not easily accessible by other means. The International Society for Photogrammetry and Remote Sensing (ISPRS) advances the knowledge of these technologies in more than 100 countries worldwide.

Remote Sensing. Like photogrammetry, remote sensing is an art, science, and technology specializing in the noncontact representation of the Earth and the environment by measuring the wavelengths of different types of radiation. These include passive and active technologies, both of which collect data from natural or emitted sources of electromagnetic energy. These include cosmic rays, X rays, ultraviolet light, visible light, infrared and thermal radiation, microwaves, and radio waves. Aerial photography and digital imaging are traditional passive remote

Fascinating Facts About Geoinformatics

- The 2011 State of the Birds report is the first of its kind to document the value of public lands to the stability of migratory birds in North America. The report was prepared by combining data sets provided by the National Gap Analysis Program (GAP) at the University of Idaho, the U.S. Geological Survey, and the Cornell Lab of Ornithology. The Cornell Lab Bird Database and Species Distribution Modeling programs provided geographic data representing 600,000 bird checklists collected from 107,000 locations. This information was used as an overlay correlated to the GAP Protected Area Database of the United States. (PAD-US). Combined, these data sets allowed researchers to use statistics to evaluate the relationship of public land management to the stability of bird populations in those areas.

- Peak Freaks has been providing guided expeditions of high-altitude mountains for twenty years. This year's Everest 2011 team is collaborating with 3D Reality Maps to create highly detailed three-dimensional mappings of Mount Everest. The technology used was created by DigitalGlobe, a global provider of high-resolution imagery, and the German Aerospace Center (DLR).

- GIS technologies can be used to create simulations of crowds and their movements in particular locations during different types of community events. These simulations include geometric descriptions and sets of rules of engagement. Such descriptions are particularly useful for planners of street festivals and carnivals, shopping malls, art galleries, and coordinators of large religious pilgrimages and ceremonies.

- The Magellan space probe was launched from the space shuttle in 1989. It conducted six 243-day cycles around the planet Venus, using synthetic aperture radar (SAR) image-mapping technologies to create detailed maps of the planet's surface. The fourth, fifth, and sixth orbits created mappings of the planet's gravitational field using radio signals to Earth. These images have provided detailed visual records of the geological dynamics of the planet including studies of volcanoes, lava flows, and terrestrial tectonic movements.

- The Landsat Program is jointly administered by the National Aeronautics and Space Administration (NASA) and the U.S. Geological Survey. Its satellites are used to provide digital photographs of the Earth's surface. The first Landsat satellite was launched on July 23, 1972. The Landsat Data Continuity Mission (LDCM) is scheduled to launch in December, 2012, and will provide essential data for the study of the Earth's climate and ecosystems.

---

sensing technologies based on photographic techniques and applications. Later applications include manned space and space shuttle photography and Landsat satellite imagery. Radar interferometry and laser scanning are common examples of active remote sensing technologies. These products are used for the documentation of inaccessible or dangerous spaces. Examples include studies of particular environments at risk or in danger.

Global Navigation Satellite Systems (GNSS). Satellites and rapidly advancing computer technologies have transformed GIS products. Satellite systems of known distance from the Earth receive radio transmissions, which are translated and sent back to Earth as a signal giving coordinates for the position and elevation of a location. Navstar is an American Global Positioning System (GPS) originating from a World War II radio transmission system known as loran, which is an acronym for long-range navigation. It is the only GNSS system in operation. On April 27,

2008, the European Space Agency announced the successful launch of the second Galileo In-Orbit Validation Element (GIOVE-B), one of thirty identical satellites planned to complete a European satellite navigation system similar to the Global Positioning System in use in the United States and the Global Navigation Satellite System (GLONASS) in Russia. Students can monitor the activities of remote sensing satellite systems such as the Landsat, Seasat, and Terra satellites used to create new map profiles of terrestrial and extraterrestrial landscapes, many of which are available online for study. These satellite systems are used to collect data including measurements of the Earth's ozone shield, cloud mappings, rainfall distributions, wind patterns, studies of ocean phenomena, vegetation, and land-use patterns.

HOW IT WORKS

Geographic information systems are built on geometric coordinate systems representing particular locations and terrestrial characteristics of the Earth or other celestial bodies. The place and position of particular land features form the elementary and most regular forms of geographic data. Maps provide visual information about key places of interest and structural features that assist or impede their access. Land-survey technologies and instrumentation are ancient and are found in human artifacts and public records that are thousands of years old. They are essential documents of trade, land development, and warfare. Spatial coordinates describing the Moon and stars and essential information about particular human communities are chinked into rocks, painted on the walls of caves, and hand-printed on graphic media all over the world. Digital photographs, satellite data, old maps, field data, and measurements provide new contexts for sharing geographic information and knowledge about the natural world and human networks of exchange.

GIS devices include high-resolution Landsat satellite imagery, light detecting and ranging (lidar) profiles, computer-aided design (CAD) data, and database management systems. Cross-referenced materials and intricately detailed maps and overlays create opportunities for custom-designed geographic materials with specific applications. Data that is created and stored in a GIS device provide a usable base for building complex multi-relational visualizations of landscapes, regions, and environments.

How a GIS component is used depends on the particular applications required. Service providers will first conduct a needs assessment to understand what information will be collected, by whom, and how it will be used by an individual or organization. The careful design of relationships connecting data sets to one another is an essential process of building an effective GIS system. Many data sets use different sets of coordinates to describe a geographical location and so algorithms need to be developed to adjust values that can have significant effects on the results of a study.

Depending on the needs of an organization, the rights to use some already-established data sets can be acquired. Other data can be collected by the user and stored in appropriate data files. This includes records already accumulated by an individual or organization. Converting and storing records in GIS format is particularly useful for creating the documents needed for a time-series analysis of particular

land features and regional infrastructures. Digital records protect against the loss of valuable information and create flexible mapping models for communicating with internal and external parties.

APPLICATIONS AND PRODUCTS

Satellite and computer technologies have transformed the way spatial information is collected and communicated worldwide. With exponential increases in speed and detail, satellites provide continuous streams of information about Earth's life systems. More than fifty satellites provide free access to GPS coordinates. These are used on a daily basis by people from all walks of life, providing exceptional mapping applications for determining the location and elevation of roadways and waterways.

The Environmental Systems Research Institute (ESRI) is a pioneer in the development of GIS landscape-analysis systems. This includes the development of automated software products with Web applications. Users can choose from a menu of services including community Web-mapping tools; ArcGIS Desktop with ArcView, ArcEditor, and ArcInfo applications; and two applications for use in educational settings, ArcExplorer Java Edition for Education (AEJEE) and Digital Worlds (DW). ESRI software is an industry standard used worldwide by governments and industries. The ESRI International User Conference is attended by thousands of users and is one of the largest events of its kind.

The Global Geodetic Observing System (GGOS) is another application developed by the International Association of Geodesy. It provides continuing observations of the Earth's shape, gravity field, and rotational motion. These measurements are integrated into the Global Earth Observing System of Systems (GEOSS), an application that provides high-quality reference materials for use by groups such as the Group on Earth Observations and the United Nations.

IMPACT ON INDUSTRY

The Open Geospatial Consortium (OGC) comprises 417 corporate, government, and university entities working to establish freely available standards for use by private and commercial institutions. Electronic location resources and technologies continue to be integrated into market-driven applications in industry and government. These include three-dimensional information systems; information processing standards for architecture, engineering, and construction; building information models for life-cycle manufacturing, defense and intelligence systems, homeland security, disaster management, and emergency services; sustainable natural resource and ecosystem management; mass market Web services; and university research. The OGC Web site lists all of its members with e-mail and Web links to their resources.

CAREERS AND COURSE WORK

Course work and certification in geoinformatics at the undergraduate and graduate levels can be completed in tandem with course work in other earth science and engineering fields. Courses include selections of introductory materials, computer programming, database management and design, statistics, bioinformatics, geostatistics, remote sensing, various foundational courses in mapping techniques and spatial analysis, and computer laboratory exercises to gain familiarity with a number of GIS software programs. Some course work will emphasize applications in land development, transportation analysis, environmental science, public health, and a variety of engineering and architectural design schematics.

Careers specializing in geoinformatics technologies are directly involved in data-driven hardware and software applications. Practitioners must master the organizational and design skills necessary to produce highly detailed, error-free maps and related documents for use in private industry and government-related agencies. Familiarity with field research and land-survey technologies are also desirable.

SOCIAL CONTEXT AND FUTURE PROSPECTS

As a result of the World Wide Web, computer and telecommunications technologies continue to provide novel platforms for connecting individuals, groups, and communities worldwide. Location is an essential feature of such networks, and geographic information systems are needed to provide timely spatial information for individual and cooperative ventures. The safety and integrity of global communications systems and data systems continue to challenge political values in free and democratic societies. Nevertheless, geographic information systems will continue to be integrated into a variety of applications, contributing to the safety of individuals, the integrity of the world's natural resources, and the profitability of enterprises around the globe.

Victoria M. Breting-García, B.A., M.A.

FURTHER READING

Bender, Oliver, et al., eds. *Geoinformation Technologies for Geocultural Landscapes: European Perspectives*. Leiden, the Netherlands: CRC Press, 2009. These essays explain the many ways geoinformation technologies are used to model and simulate the changing reality of landscapes and environments.

DeMers, Michael N. *Fundamentals of Geographic Information Systems*. 4th ed. Hoboken, N.J.: John Wiley & Sons, 2009. Provides useful historical background for the technologies and systems presented. Each chapter ends with a review of key terms, review questions, and a helpful list of references for further study.

Galati, Steven R. *Geographic Information Systems Demystified*. Boston: Artech House, 2006. Introductory textbook provides graphic examples of the concepts presented and includes a very useful glossary of terms.

Harvey, Francis. *A Primer of GIS: Fundamental Geographic and Cartographic Concepts*. New York: Guildford Press. 2008. This well-organized and clearly written textbook presents a thorough review of GIS with questions for analysis and vocabulary lists at the end of every chapter.

Kolata, Gina Bari. "Geodesy: Dealing with an Enormous Computer Task." *Science* vol. 200, no. 4340 (April 28, 1978): pp. 421-466. This article explains the enormous task of computing the 1983 North American Datum.

Konecny, Gottfried. *Geoinformation: Remote Sensing, Photogrammetry and Geographic Information Systems*. London: Taylor & Francis, 2003. This detailed introduction to geoinformation includes hundreds of illustrations and tables, with a chapter-by-chapter bibliography.

Rana, Sanjay, and Jayant Sharma, eds. *Frontiers of Geographic Information Technology*. New York: Springer, 2006. This collection of essays covers rising geographic technologies. Chapter 1: "Geographic Information Technologies—An Overview" and Chapter 4: "Agent-Based Technologies and GIS:

Simulating Crowding, Panic, and Disaster Management" are of note.

Scholten, Henk J., Rob van de Velde, and Niels van Manen, eds. *Geospatial Technology and the Role of Location in Science*. New York: Springer, 2009. The essays collected in this volume represent a series of presentations made at the September, 2007, conference held by the Spatial Information Laboratory of the VU University Amsterdam. They address the use of geospatial technologies at the junction of historical research and the sciences.

Thrower, Norman J. W. *Maps and Civilization: Cartography in Culture and Society*. 3d ed. Chicago: University of Chicago Press. 2007. This book is an excellent introduction to the field of cartography.

WEB SITES

American Geophysical Union
http://www.agu.org

Geodetic Systems
Photogrammetry Information
http://www.geodetic.com/photogrammetry.htm

International Union of Geodesy and Geophysics
http://www.iugg.org

Library of Congress
Map Collections
http://memory.loc.gov/ammem/gmdhtml/gmdhome.html

University of California, Berkeley, Library
Earth Sciences & Map Library
http://www.lib.berkeley.edu/EART/browse.html

See also: Calculus; Computer-Aided Design and Manufacturing; Computer Science; Geometry; Maps and Mapping; Plane Surveying; Trigonometry.

GEOMETRY

FIELDS OF STUDY

Civil engineering; architecture; surveying; agriculture; environmental and conservation sciences; mechanical engineering; computer-aided design; molecular design; nanotechnology; physical chemistry; biology; physics; graphics; computer game programming; textile and fabric arts; fine arts; cartography; geographic information systems; Global Positioning Systems; medical imaging; astronomy; robotics.

SUMMARY

Geometry, which literally means "earth measurement," is absolutely critical to most fields of physical science and technology and especially to any application that involves surfaces and surface measurement. The term applies on all scales, from nanotechnology to deep space science, where it describes the relative physical arrangement of things in space. Although the first organized description of the principles of geometry is ascribed to the ancient Greek philosopher Euclid, those principles were known by others before him, and certainly had been used by the Egyptians and the Babylonians. Euclidean, or plane, geometry deals with lines and angles on flat surfaces (planes), while non-Euclidean geometry applies to nonplanar surfaces and relationships.

KEY TERMS AND CONCEPTS

- **Conic Section:** Geometric form that can be defined by the intersection of a plane and a cone; examples are a circle, an ellipse, a parabola, a hyperbola, a line, and a point.
- **Euclidean:** Describing geometric principles included in the range of principles described by Euclid.
- **Fibonacci Series:** Geometric sequence described by Italian mathematician Leonardo of Pisa (known as Fibonacci) in about 1200, beginning with zero and in which each subsequent number is the sum of the previous two: 0, 1, 1, 2, 3, 5, 8, 13, 21, 34, 55, . . .
- **Geometric Isomers:** Chemical term used to denote molecular structures that differ only by the relative arrangement of atoms in different molecules.
- **Golden Ratio:** Represented as φ or Φ, a seemingly ubiquitous naturally occurring ratio or proportion having the approximate value 1.618018513; also known as the golden mean, the golden section, and the divine proportion.
- **Pi:** Represented as π, an irrational number that is equivalent to the ratio of the circumference of a circle to its diameter, equal to about 3.14159265358979323846264 (the decimal part is believed to be nonrepeating and indeterminate, or unending).
- **Plane:** Flat surface that would be formed by the translation of a line in one direction.
- **Polygon:** Two-dimensional, or plane, geometric shape having a finite number of sides; designated as a regular polygon if all sides are of equal length.
- **Polyhedron:** Three-dimensional geometric shape having a finite number of planar faces; designated as a regular polyhedron if all faces are equivalent.
- **Postulate:** Rule that is accepted as true without proof.
- **Pythagorean:** Describing a relation to the theorem and principles ascribed to Pythagoras, especially the properties of right triangles.
- **Theorem:** Rule that is accepted as true but requires a rigorous proof.
- **Torus:** Structure formed by the translation of a circle through space along a path defined by a second circle whose plane is orthogonal to that of the first circle.

DEFINITION AND BASIC PRINCIPLES

Geometry is the branch of mathematics concerned with the properties and relationships of points, lines, and surfaces, and the space contained by those entities. A point translated in a single direction describes a line, while a line translated in a single direction describes a plane. The intersections and rotations of various identities describe corresponding structures that have specific mathematical relationships and properties. These include angles and numerous two- and three-dimensional forms. Geometric principles can also be extended into realms encompassing more than three dimensions, such as with the incorporation of time as a fourth dimension in Albert Einstein's space-time continuum. Higher dimensional analysis is also possible and is the subject of theoretical studies.

The basic principles of geometry are subject to various applications within different frames of reference. Plane geometry, called Euclidean geometry after the ancient Greek mathematician Euclid, deals with the properties of two-dimensional constructs such as lines, planes, and polygons. The five basic principles of plane geometry, called postulates, were described by Euclid. The first four are accepted as they are stated and require no proof, although they can be proven. The fifth postulate differs considerably from the first four in nature, and attempts to prove or disprove it have consistently failed. However, it gave rise to other branches of geometry known as non-Euclidean. The first four postulates apply equally to all branches of Euclidean and non-Euclidean geometry, while each of the non-Euclidean branches

uses its own interpretation of the fifth postulate. These are known as hyperbolic, elliptic, and spherical geometry.

The point, defined as a specific location within the frame of reference being used, is the common foundation of all branches of geometry. Any point can be uniquely and unequivocally defined by a set of coordinates relative to the central point or origin of the frame of reference. Any two points within the frame of reference can be joined with a single line segment, which can be extended indefinitely in that direction to produce a line. Alternatively, the movement of a point in a single direction within the frame of reference describes a line. Any line can be translated in any orthogonal direction to produce a plane. Rotation of a line segment in a plane about one of its end points describes a circle whose radius is equal to the length of that line segment. In any plane, two lines that intersect orthogonally produce a right angle (an angle of 90 degrees). These are the essential elements of the first four Euclidean postulates. The fifth, which states that nonparallel lines must intersect, could not be proven within Euclidean geometry, and its interpretation under specific conditions gives rise to other frames of reference.

BACKGROUND AND HISTORY

The Greek historian Herodotus soundly argued that geometry originated in ancient Egypt. During the time of the legendary pharaoh Sesostris, the farmland of the empire was apportioned equally among the people, and taxes were levied accordingly. However, the annual inundation of the Nile River tended to wash away portions of farmland, and farmers who lost land in this way complained that it was unfair for them to pay taxes equal to those whose farms were complete. Sesostris is said to have sent agents to measure the loss of land so that taxation could be made fair again. The agents' observations of the relationships that existed gave rise to an understanding of the principles of geometry.

It is well documented that the principles of geometry were known to people long before the ancient Greeks described them. The Rhind papyrus, an Egyptian document dating from 2000 B.C.E., contains a valid geometric approximation of the value of pi, the ratio of the circumference of a circle to its diameter. The ancient Babylonians were also aware of the principles of geometry, as is evidenced by the

Euclid of Alexandria was a Hellenistic Greek mathematician who lived circa 300 B.C.E. He is known as the "father of geometry," best known for his book Elements. *(Science Source)*

inscription on a clay tablet that is housed in Berlin. The inscription has been translated as an explanation of the relationship of the sides and hypotenuse of a right triangle in what is known as the Pythagorean theorem, although the tablet predates Pythagoras by several hundred years.

From these early beginnings to modern times, studies in geometry have evolved from being simply a means of describing geometric relationships toward encompassing a complete description of the workings of the universe. Such studies allow the behavior of materials, structures, and numerous processes to be predicted in a quantifiable way.

HOW IT WORKS

Geometry is concerned with the relationship between points, lines, surfaces, and the spaces enclosed or bounded by those entities.

Points. A point is any unique and particular location within a frame of reference. It exists in one dimension only, having neither length nor width nor breadth but only location. The location of any point can be uniquely and unequivocally defined by a set of coordinate values relative to the central point of the particular reference system being used. For example, in a Cartesian coordinate system—named after French mathematician René Descartes although the method was described at the same time by Pierre de Fermat—the location of any point in a two-dimensional plane is described completely by an x-coordinate and a y-coordinate, as (x, y), relative to the origin point at $(0, 0)$. Thus, a point located at $(3, 6)$ is 3 units away from the origin in the direction corresponding to positive values of x, and 6 units away from the origin in the direction corresponding to positive values of y. Similarly, a point in a three-dimensional Cartesian system is identified by three coordinates, as (x, y, z). In Cartesian coordinate systems, each axis is orthogonal to the others. Because orthogonality is a mathematical property, it can also be ascribed to other dimensions as well, allowing the identification of points in theoretical terms in n-space, where n is the number of distinct dimensions assigned to the system. For the purposes of all but the most theoretical of applications, however, three dimensions are sufficient for normal physical representations.

Points can also be identified as corresponding to specific distances and angles, also relative to a coordinate system origin. Thus, a point located in a

spherical coordinate system is defined by the radius (the straight-line distance from the origin to the point), the angle that the radius is swept through a plane, and the angle that the radius is swept through an orthogonal plane to achieve the location of the point in space.

Lines. A line can be formed by the translation of a point in a single direction within the reference system. In its simplest designation, a line is described in a two-dimensional system when one of the coordinate values remains constant. In a three-dimensional system, two of the three coordinate values must remain constant. The lines so described are parallel to the reference coordinate axis. For example, the set of points (x, y) in a two-dimensional Cartesian system that corresponds to the form $(x, 3)$—so that the value of the y-coordinate is 3 no matter what the value of x—defines a line that is parallel to the x-axis and always separated from it by 3 units in the positive y direction.

Lines can also be defined by an algebraic relationship between the coordinate axes. In a two-dimensional Cartesian system, a line has the general algebraic form $y = mx + b$. In three-dimensional systems, the relationship is more complex but can be broken down into the sum of two such algebraic equations involving only two of the three coordinate axes.

Planes and Surfaces. Planes are described by the translation of a line through the reference system, or by the designation of two of the three coordinates having constant values while the third varies. A plane can be thought of as a flat surface. A curved surface can be formed in an analogous manner by translating a curved line through the reference system, or by definition, as the result of a specific algebraic relationship between the coordinate axes.

Angles. Intersecting lines have the property of defining an angle that exists between them. The angle can be thought of as the amount by which one line must be rotated about the intersection point in order to coincide with the other line. The magnitude, or value, of angles rigidly determines the shape of structures, especially when the structures are formed by the intersection of planes.

Conic Sections. A cone is formed by the rotation of a line at an angle about a point. Conic sections are described by the intersection of a plane with the cone structure. As an example, consider a cone formed in the three-dimensional Cartesian system by rotating the line $x = y$ about the y-axis, forming both a positive and a negative cone shape that meet at the origin point. If this is intersected by a plane parallel to the x-z plane, the intersection describes a circle. If the plane is canted so that it is not parallel to the x-z plane and intersects only one of the cone ends, the result is an ellipse. If the plane is canted further and positioned so that it intersects the positive cone on one side of the y-axis and the negative cone on the other side of the y-axis, the intersection defines a hyperbola. If the plane is canted still further so that it intersects both cones on the same side of the y-axis, then a parabola is described.

APPLICATIONS AND PRODUCTS

It is impossible to describe even briefly more than a small portion of the applications of geometry because geometry is so intimately bound to the structures and properties of the physical universe. Every physical structure, no matter its scale, must and does adhere to the principles of geometry, since these are the properties of the physical universe. The application of geometry is fundamental to essentially every field, from agriculture to zymurgy.

GIS and GPS. Geographical information systems (GIS) and Global Positioning Systems (GPS) have been developed as a universal means of location identification. GPS is based on a number of satellites orbiting the planet and using the principles of geometry to define the position of each point on the Earth's surface. Electronic signals from the various satellites triangulate to define the coordinates of each point. Triangulation uses the geometry of triangles and the strict mathematical relationships that exist between the angles and sides of a triangle, particularly the sine law, the cosine law, and the Pythagorean theorem.

GIS combines GPS data with the geographic surface features of the planet to provide an accurate "living" map of the world. These two systems have revolutionized how people plan and coordinate their movements and the movement of materials all over the world. Applications range from the relatively simple GPS devices found in many modern vehicles to precise tracking of weather systems and seismic activity. An application that is familiar to many through the Internet is GoogleEarth, which presents a satellite view of essentially any place on the planet at a level of detail that once was available from only the most top secret of military reconnaissance satellites. The system also allows a user to view traditional

---

Fascinating Facts About Geometry

- The translation of an ancient Babylonian clay tablet shows that the relationship between the sides of a right triangle, commonly called the Pythagorean theorem, was known at least 2,000 years before Pythagoras lived.
- The ancient Greek mathematician Thales of Miletus is credited with devising the principles of geometry known as Euclid's postulates and is also heralded as the father of modern engineering.
- The fifth postulate of Euclidean geometry has never been proven or disproven.
- The golden ratio, or ϕ, is found almost everywhere in nature and even describes processes and trends in such nonphysical fields as economics and sociology.
- The value of pi, the ratio of the circumference of a circle to its diameter, is a finite value whose nonrepeating decimal portion is infinitely long; it has been calculated to more than a trillion (10^{12}) decimal places.

---

map images in a way that allows them to be scaled as needed and to add overlays of specific buildings, structures, and street views. Anyone with access to the Internet can quickly and easily call up an accurate map of almost any desired location on the planet.

GIS and GPS have provided a whole new level of security for travelers. They have also enabled the development of transportation security features such as General Motors' OnStar system, the European Space Agency's Satellite Based Alarm and Surveillance System (SASS), and several other satellite-based security applications. They invariably use the GPS and GIS networks to provide the almost instantaneous location of individuals and events as needed.

CAD. Computer-aided design (CAD) is a system in which computers are used to generate the design of a physical object and then to control the mechanical reproduction of the design as an actual physical object. A computer drafting application such as AutoCAD is used to produce a drawing of an object in electronic format. The data stored in the drawing file include all the dimensions and tolerances that define the object's size and shape. At this point, the object itself does not exist; only the concept of it exists as a collection of electronic data. The CAD application can calculate the movements of ancillary robotic machines that will then use the program of instructions to produce a finished object from a piece of raw material. The operations, depending on the complexity and capabilities of the machinery being directed, can include shaping, milling, lathework, boring or drilling, threading, and several other procedures. The nature of the machinery ranges from basic mechanical shaping devices to advanced tooling devices employing lasers, high-pressure jets, and plasma- and electron-beam cutting tools.

The advantages provided by CAD systems are numerous. Using the system, it is possible to design and produce single units, or one offs, quickly and precisely to test a physical design. Adjustments to production steps are made very quickly and simply by adjusting the object data in the program file rather than through repeated physical processing steps with their concomitant waste of time and materials. Once perfected, the production of multiple pieces becomes automatic, with little or no variation from piece to piece.

Metrology. Closely related to CAD is the application of geometry in metrology, particularly through the use of precision measuring devices such as the measuring machine. This is an automated device that uses the electronic drawing file of an object, as was produced in a CAD procedure, as the reference standard for objects as they are made in a production facility. Typically, this is an integral component of a statistical process control and quality assurance program. In practice, parts are selected at random from a production line and submitted to testing for the accuracy of their construction during the production process. A production piece is placed in a custom jig or fixture, and the calibrated measuring machine then goes through a series of test measurements to determine the correlation between the features of the actual piece and the features of the piece as they are designated in the drawing file. The measuring machine is precisely controlled by electronic mechanisms and is capable of highly accurate measurement.

Game Programming and Animation. Basic and integral parts of both the video game and the motion-picture industry are described by the terms "polygon," "wire frame," "motion capture," and "computer-generated imagery" (CGI). Motion capture uses a series of reference points attached to an actor's body. The reference points become data points in a computer file, and the motions of the actor are recorded as the geometric translation of one set of data points into another in a series. The data points can then be used to generate a wire-frame drawing of a figure that corresponds to the character whose actions have been imitated by the actor during the motion-capture process. The finished appearance of the character is achieved by using polygon constructions to provide an outward texture to the image. The texture can be anything from a plain, smooth, and shiny surface to a complex arrangement of individually colored hairs. Perhaps the most immediately recognizable application of the polygon process is the generation of dinosaur skin and aliens in video games and films.

The movements of the characters in both games and films are choreographed and controlled through strict geometric relationships, even to the play of light over the character's surface from a single light source. This is commonly known as ray tracing and is used to produce photorealistic images.

IMPACT ON INDUSTRY

One cannot conceive of modern agriculture without the economic assessment scales of yield, fertilizer rates, pesticide application rates, and seed rates, all on a per hectare or per acre basis. Similarly, one cannot envisage scientific applications and research programs that do not rely intimately on the principles of geometry in both theory and practice. Processes such as robotics and CAD have become fundamental aspects of modern industry, representing entirely different paradigms of efficiency, precision, and economics than existed before. The economic value of computer-generated imagery and other computer graphics applications is similarly inestimable. For example, James Cameron's film *Avatar* (2009), which featured computer-generated imagery, returned more than $728 million within two weeks of its release date. Indeed, many features simply would not exist without computer-generated imagery, which in turn would not exist without the application of geometry. Similarly, computer gaming, from which computer-aided imagery developed, is a billion-dollar industry that is relies entirely on the mathematics of geometry.

Academic Research. All universities and colleges maintain a mathematics department in which practical training and theoretical studies in the applications of geometric principles are carried out. Similarly, programs of research and study in computer science departments examine new and more effective ways to apply geometry and geometric principles in computing algorithms. Graphic arts programs in particular specialize in the development and application of computer-generated imagery and other graphics techniques relying on geometry. Training in plane surveying and mechanical design principles are integral components of essentially all programs of training in civil and mechanical engineering.

Industry and Business. A wide variety of businesses are based almost entirely on the provision of goods and services that are based on geometry. These include contract land-surveying operations that serve agricultural needs; environmental conservation and management bodies; forestry and mining companies; municipal planning bodies; transportation infrastructure and the construction industries; data analysis; graphics programming; advertising; and cinematographic adjuncts, to name but a few. The opportunity exists for essentially anyone with the requisite knowledge and some resources to establish a business that caters to a specific service need in these and other areas.

Another aspect of these endeavors is the provision of materials and devices to accommodate the services. Surveyors, for example, require surveying equipment such as transit levels and laser source-detectors in order to function. Similarly, construction contractors typically require custom-built roof trusses and other structures. Graphics and game design companies often subcontract needed programming expertise. Numerous businesses and industries exist, or can be initiated, to provide needed goods.

CAREERS AND COURSE WORK

Geometry is an essential and absolutely critical component of practically all fields of applied science. A solid grounding in mathematics and basic geometry is required during secondary school studies. In more advanced or applied studies at the post-secondary level, in any applied field, mathematical training in geometrical principles will focus more closely on the applications that are specific to that field of study. Any program of study that integrates design concepts will include subject-specific applications of geometric principles. Applications of geometric principles are used in mechanical engineering, manufacturing,

civil engineering, industrial plant operations, agricultural development, forestry management, environmental management, mining, project management and logistics, transportation, aeronautical engineering, hydraulics and fluid dynamics, physical chemistry, crystallography, graphic design, and game programming.

The principles involved in any particular field of study can often be applied to other fields as well. In economics, for example, data mining uses many of the same ideological principles that are used in the mining of mineral resources. Similarly, the generation of figures in electronic game design uses the same geometric principles as land surveying and topographical mapping. Thus, a good grasp of geometry and its applications can be considered a transferable skill usable in many different professions.

SOCIAL CONTEXT AND FUTURE PROSPECTS

Geometry is set to play a central role in many fields. Geometry is often the foundation on which decisions affecting individuals and society are made. This is perhaps most evident in the establishment and construction of the most basic infrastructure in every country in the world, and in the most high-tech advances represented by the satellite networks for GPS and GIS. It is easy to imagine the establishment of similar networks around the Moon, Mars, and other planets, providing an unprecedented geological and geographical understanding of those bodies. In between these extremes that indicate the most basic and the most advanced applications of geometry and geometrical principles are the typical everyday applications that serve to maintain practically all aspects of human endeavor. The scales at which geometry is applied cover an extremely broad range, from the ultra-small constructs of molecular structures and nano-technological devices to the construction of islands and buildings of novel design and the ultra-large expanses of interplanetary and even interstellar space.

Richard M. J. Renneboog, M.Sc.

FURTHER READING

Bar-Lev, Adi. "Big Waves: Creating Swells, Wakes and Everything In-Between." *Game Developer* 15, no. 2, (February, 2008): 14-24. Describes the application of geometry in the modeling of liquid water actions in the computer graphics of gaming and simulations.

Bonola, Roberto. *Non-Euclidean Geometry: A Critical and Historical Study of Its Development*. 1916. Reprint. Whitefish, Mont.: Kessinger, 2007. A facsimile reproduction of a classic work on the history of alternate geometries.

Boyer, Carl B. *History of Analytic Geometry*. Mineola, N.Y.: Dover Publications, 2004. Traces the history of analytic geometry from ancient Mesopotamia, Egypt, China, and India to 1850.

Darling, David. *The Universal Book of Mathematics: From Abracadabra to Zeno's Paradoxes*. 2004. Reprint. Edison, N.J.: Castle Books, 2007. An encyclopedic account of things mathematical.

Heilbron, J. L. *Geometry Civilized: History, Culture, and Technique*. Reprint. New York: Oxford University Press, 2003. A very readable presentation of the history of geometry, including geometry in cultures other than the Greek, such as the Babylonian, Indian, Chinese, and Islamic cultures.

Herz-Fischler, Roger. *A Mathematical History of the Golden Number*. Mineola, N.Y.: Dover Publications, 1998. A well-structured exploration of the golden number and its discovery and rediscovery throughout history.

Holme, Audun. *Geometry: Our Cultural Heritage*. New York: Springer, 2002. An extensive discussion of the historical development of geometry from prehistoric to modern times, focusing on many major figures in its development.

Livio, Mario. *The Golden Ratio: The Story of Phi, the World's Most Astonishing Number*. New York: Broadway Books, 2002. Describes how the occurrence of the golden ratio, also known as the divine ratio, defined by the ancient mathematician Euclid, seems to be a fundamental constant of the physical world.

Szecsei, Denise. *The Complete Idiot's Guide to Geometry*. 2d ed. Indianapolis: Alpha Books, 2007. A fun, step-by-step presentation that walks the reader through the mathematics of geometry and assumes absolutely no prior knowledge beyond basic arithmetic.

West, Nick. "Practical Fluid Dynamics: Part 1." *Game Developer* 14, no. 3 (March, 2007): 43-47. Introduces the application of geometric principles in the modeling of smoke, steam, and swirling liquids in the graphics of computer games and simulations.

WEB SITES
American Mathematical Society
http://www.ams.org

Mathematical Association of America
http://www.maa.org

See also: Applied Mathematics; Computer-Aided Design and Manufacturing; Computer Graphics; Engineering; Engineering Mathematics; Fractal Geometry; Graphics Technologies; Measurement and Units; Topology; Trigonometry; Video Game Design and Programming.

GERIATRICS AND GERONTOLOGY

FIELDS OF STUDY

Medicine; nursing; physical therapy; occupational therapy; psychology; sociology; political science; anthropology; public policy; education; statistics; dentistry; biology; pharmacy; social work.

SUMMARY

Examining life-span development, particularly the later years, has never been so important. Servicing an aging population requires diverse teams of medical doctors, biologists, psychologists, and other professionals, and training such individuals has produced dedicated schools and degrees focused exclusively on the aging process. Students interested in the study of the aged have never before had so many opportunities to expand their careers. Further, the need for teachers and specialists in geriatrics and gerontology is expanding at a rapid pace.

KEY TERMS AND CONCEPTS

- **Activities Of Daily Living:** Basic chores of everyday life, such as eating and bathing, which are often used to determine level of independence.
- **Alzheimer's Disease:** Type of dementia that shrinks the brain, destroying memory and, in later stages, functional ability.
- **Centenarian:** Person who is one hundred years old or older.
- **Chronic Conditions:** Ongoing diseases that account for the most deaths and disability in the United States, such as heart disease and cancer.
- **Compression Of Morbidity:** Belief that a lifetime of healthy activities produces a small period of disability at the end of life.
- **Executive Functions:** Higher-order cognitive abilities including attention, decision making, planning, adapting to change, and inhibition.
- **Preventable Diseases:** Leading causes of death in the United States—heart disease, stroke, cancer, rheumatoid arthritis, and diabetes—that are often the direct result of poor lifestyle choices.

DEFINITION AND BASIC PRINCIPLES

Before the first baby boomer turned sixty in 2006, geriatrics, the medical subspecialty of treating the aged, and gerontology, the comprehensive field of aging studies, both experienced a substantial growth of interest. Although old age might have been viewed as a period of disengagement, as of 2011 nothing is further from the truth. Aging is now often couched in terms of "successful" or "productive"; in fact, many seniors remain as busy after retirement with volunteering activities and the like as when they were employed. Reframing the negative language surrounding aging has changed the perception of aging from a period of consuming goods and services to a period of continued growth and productivity. With an estimated 69 million baby boomers turning sixty-five by 2029, the fields of geriatrics and gerontology have never been more in demand.

Maintaining the health and well-being of this graying section of society involves extensive education, research, and policy initiatives. Thus, the field of gerontology, by necessity, is interdisciplinary in nature. No one field encapsulates the varied systems of inquiry—especially when so many seniors now remain fit and active well into their nineties.

BACKGROUND AND HISTORY

Gerontology as an official field of inquiry first began in 1903, when Russian biologist Élie Metchnikoff coined the term "gerontology" or the "study of old men." Six years later, Austrian physician Ignatz Nascher created the field of geriatrics and in 1914 published the first book on geriatrics. The Social Security Act, enacted in 1935, helped to pull millions of seniors, and others, out of poverty. The first organizations solely dedicated to the study of the aging process were formed in the 1940's: the American Geriatrics Society, founded in 1942, and the Gerontological Society of America, founded in 1945. In 1957, the National Institutes of Health formed the Center for Aging Research. In 1963, physician Sidney Katz and colleagues published the seminal work on gauging the independence of the elderly, the index of Activities of Daily Living—a concept still widely used. Irving Rosow published *Social Integration of the Aged* in 1967, which laid the groundwork for later theories on aging. The White House Conference on Aging, begun in 1961, spawned a formal division on aging at the National Institutes of Health. In 1976, Robert Butler was appointed the first director of that new National Institute on Aging. The 1980's through the 2000's saw an explosion in interest on aging as the baby boomers began to reach retirement age. Countless university centers on aging and degrees in geriatrics and gerontology have been created, and at many universities students may now declare gerontology as their major field of study.

HOW IT WORKS

Theories of Aging. Scientific theories organize the how and why of empirical findings and provide a certain epistemology from which to examine new data. Using supported data-driven theory contextualizes known parameters, permitting a foundation of knowledge from which to build new projects and inquiries into the aging process. Previous data sets and explanations about interrelated phenomena streamline future interventions and research investigations. In geriatrics, biological theories of aging are the norm; however, the interdisciplinary nature of gerontology creates barriers to building comprehensive theories of aging. Although gerontology includes biological theories of aging, these two fields will be examined separately.

Geriatric theories on aging abound, and the more popular theories are briefly reviewed here. The free-radical theory of aging posits that self-multiplying free radicals cause damage to deoxyribonucleic acid (DNA) and healthy cells. The hypothesized antidote is consumption of antioxidants, commonly found in vegetables. The programmed theory of aging proposes that every organism has an expected life span. Similarly, the wear-and-tear theory likens the human body to a machine where constant use degrades the machine, eventually leading to failure. Lastly, the immune-system theory holds that, with age, the human body becomes less able to fight off infection. In a world where health is increasingly viewed as more than the mere absence of disease, geriatricians use biological theories of aging to dispense health-promotion advice: Exercise both mind and body daily, consume the daily requirements of fruits and vegetables, and remain socially engaged in meaningful activities.

Gerontological theories of aging, often referred to as social theories of aging, are diverse in scope. Although largely refuted, disengagement theory suggests with increased age individuals slowly remove themselves from society. Alternatively and largely supported by early twenty-first-century research literature, activity theory holds a positive relationship between activity levels and happiness

Cedar Ridge Inn nursing facility resident Mildred Secrest is overwhelmed with excitement as she gets a spare while playing the Nintendo Wii Sports bowling game on January 11, 2008. (AP Images)

and health. Similarly, continuity theory maintains that new roles should be substituted for lost roles; for example, volunteering can replace retirement from a paying job. The life-course perspective on aging holds that aging is in fact a lifelong process and adaptation and change are continuous rather than enacted on by a specific age (sixty-five). These, and other social theories on aging, guide the development of new research endeavors and assimilate such new findings into the evolving language of aging research.

Geriatrics in the Field. Geriatricians have advanced training and certifications in medicine and aging and work under the broad field of geriatrics. Such individuals actively develop health plans, treat comorbidities, promote health and wellness, focus on the prevention of disease and disability, and perform clinical evaluations. Unlike other disciplines, such as cardiology, geriatrics is not focused on a single organ or disease. Therefore other medical personnel with specialty training in aging often make up a treatment team; such a team can include osteopathic physicians, nurses, social workers, physical and occupational therapists, psychologists, and others. Such a team can create a complete wellness portfolio and assess a patient's activities of daily living.

Geriatricians have moved from a prevention of disease model to a health-promotion model that uses the best available research evidence, clinical knowledge, and patient feedback to create a holistic model of well-being. Such evidence-based approaches are now much in demand; with rising health care costs insurance companies demand proven treatments. Geriatrics as a subspecialty of medicine is especially important as older adults can often differ from younger persons in symptoms related to illness and react differently to treatment methods. The success of geriatrics is important for measures of public health as well, where reducing disability and disease would have far-reaching effects on the overburdened medical system.

Generally a person should consider seeing a geriatrician when he or she turns sixty-five, although individual health complications could necessitate an earlier visit. Often by the age of seventy-five, many older adults have multiple chronic conditions, such as sensory and cognitive impairments, that require the services of a geriatrician. Because the life span is seen as continuous, rather than demarcated by specific ages, it is important that individual decisions

made throughout one's life are well-informed, and geriatricians are an important piece of the puzzle. Because the rate of aging is determined by the interaction of genetic and environmental conditions, which differ for every individual, it is important that aging seniors see a specialist. Only a qualified medical doctor should make medical decisions.

Gerontology in the Field. Applied gerontologists examine, study, and directly train the aging population and those who work with them in a variety of ways. Such areas include: learning to operate hearing aids, using assistive devices such as canes or walkers, maintaining proper nutrition, adjusting driving habits, proper use of corrective visual aids, and any other area affected by the aging process. Research by gerontologists suggests that change is the key to successful aging; static lives produce static minds and bodies.

Gerontologists also teach and promote preventative interventions to ensure successful aging. Such interventions can retrain mental acuity, strengthen ailing muscles and skeletal structure, teach positive behavioral methods to cope with loss and grieving. Gerontologists have led the aging revolution, where seniors are living longer, healthier, and more engaged lives. Accordingly, seniors are contributing to a level of human capital never before seen. For example, in 2010 Senior Corps, a federal governmental agency, saw its largest increase in senior volunteers since 2004. Part of the reason for this increase in productivity is because technology is changing the way productivity is perceived. No longer is physical stamina required for ongoing employment; technology has permitted older workers to stay in the workforce, thereby increasing social contribution and delaying age-related functional declines.

Gerontologists are quick to point out that productive aging includes activities outside of standard market contributions, such as volunteering. Such a revolution has caused many to rethink the very concept of old age. Often, age is a mixture of chronological age, biological age, psychological age, and sociological age. Functional age is a good marker of the aging process and aids in determining between three age conditions: normal aging, pathological aging, and successful aging. Another method of categorizing the diverse array of seniors is simply via chronological age. Gerontologists see three subcategories of seniors, young old (sixty-five through seventy-four), middle old (seventy-five through eighty-four) and old-old (eighty-five and older). Whatever the method of categorization, grouping the vast growing senior population is an arduous process given the inherent wide variability in human aging. Further, such grouping permits large-scale comparisons of health and wellness.

Unfortunately, the United States lags behind other countries, especially Great Britain, and the World Health Organization (WHO) recommendations on prevention and restraint of chronic diseases. The WHO's guide to *Global Age-Friendly Cities* provides eight guidelines for communities aiming for improvement: outdoor spaces, transportation, housing, social participation, respect, civic participation, communication and community support, and health services. In London and other European cities, free exercise playgrounds for the elderly are the norm. Costa Rica; Sardinia, Italy; and Ikaria, Greece, are examples of locales that possess the right mix of cultural and social factors that permit many seniors to live healthy lives into their nineties and beyond. Generally, such countries have a culture of respect for the aged. For example, those who study and treat the aging population, in Great Britain in particular, are held in high esteem. Conversely, the United States has historically stigmatized the aging and those who work with them. A slow tide of change is occurring as baby boomers prominently age in the American society, but changing preconceived notions is a slow process and stereotypes of the aged still abound.

APPLICATIONS AND PRODUCTS

Baby boomers possess a higher level of education than any previous generation; thus, the expectation is this cohort will be savvy consumers desiring the best proven treatments. Where daily life choices are more predictive of health status than genetic composition alone, the previous niche areas of applications and products for the aging is now a rampant growth industry spanning every conceivable field.

Medicine. Medical implantation devices provide relief to ailing organs and prolong well-being. The left ventricle assist device aids the normal functioning of failing hearts while an implanted defibrillator prevents cardiac arrest by shocking the heart back into a normal rhythm. Cochlear implants are placed directly under the skin behind the ear and return the gift of sound to many hearing-impaired

individuals. Cameras encapsulated in pill casings that the patient swallows take video of the intestinal track, eliminating the need for costly and invasive scoping procedures.

The increasing need for evidence-based medicine has produced some creative solutions to gathering information from patients. Wireless home-based transfer of medical information from accelerometers, glucose monitoring, and implanted devices, such as pacemakers, allows a patient to provide real-time health data while remaining independent. Cell phones with Global Positioning Systems (GPSs) track exercise regimens and allow for intermittent queries about self-perceived health and well-being. For example, patients newly released from the hospital can transmit responses to doctor-initiated questions eliminating the need and cost of in-person follow-ups. This trend in distance-based medicine includes genetic-testing kits, available at local drug stores, that allow the user to mail in his or her sample for analysis.

Although currently emerging technologies sound like science fiction, many of these products are closer to the marketplace than one might imagine. Thought-controlled mechanical limbs that receive feedback from the environment, for instance temperature and pressure, can closely mimic an individual's lost arm or leg. In development are microscopic cleaning robots, called blood bots, that can be guided to clean plaque-filled veins and take biopsies. Noninvasive blood, saliva, and urine tests for Alzheimer's disease, cancer, and other difficult-to-detect diseases are currently in development. There is even an experimental Breathalyzer test in development that may replace expensive blood and urine analysis. Semipermanent prescription tattoos might be able to respond to glucose levels in a diabetic's bloodstream by changing color when placed under a handheld infrared light, eliminating the need for painful blood monitoring.

Common memory-storage cards, like those found in a digital camera, are being used as a portable patient medical archive. The cards fit into a wallet and facilitate communication and accuracy between the various health professionals many seniors visit. A computerized medical information system will likely soon replace inefficient paper-based records. Such an electronic system will permit comprehensive care while anywhere in the world and coordinate the spectrum of health care services seniors receive.

Pharmacology. Perhaps the field most engaged with the aging population is pharmacology. Clinical drug trials deliver numerous pharmaceuticals to the market each year—many designed to treat and extend wellness into advancing age. Drugs are being developed that may fight obesity and even change one's DNA. Drug encapsulation involves the coating of medicine either to delay activation or enter affected areas. For example, most oral medications are unable to pass the blood-brain barrier, which means they do not enter the brain. Encapsulating, or masking, the active drug compound could permit the body to pass the drug into the brain, eliminating the need for invasive surgeries.

Although, often with variable scientific evidence, herbs and supplements for the aged have expanded exponentially in the early twenty-first century. In 2011, a senior can take a pill that is purported to cure any ailment. However, geriatric researchers have found minimal scientific evidence for many of these claims. Ginkgo biloba was claimed to improve memory for many years; however, numerous large-scale clinical trials have found no such evidence. Large annual doses of vitamin D, thought to improve bone health, was found to increase fractures. Conversely, some herbs and supplements have proven effective. Omega-3 fatty acids have shown promise in improving heart health. Capsaicin, found in hot peppers, has recently been added to topical arthritis creams because of its analgesic properties. However, medicines work only when taken as directed. Medication non-adherence costs the health care industry millions of dollars per year. Patients frequently take the wrong dose or fail to fill the prescription, necessitating additional doctor visits. Accordingly, an industry has developed to correct this problem. Pill bottles with reminder alarms, automatic medication dispensers, Internet-based and cell-phone text reminders to assist with medication adherence are widely available.

Assistive Technology. Numerous home-based assistive devices are available to prolong independence: swing-down shelves in kitchens, easy-open door handles along with a bevy of structural changes to accommodate individuals with decreased strength, decreased stature, complications due to arthritis, and the like. Such assistive technologies have shown to decrease the need for outside personal assistance,

---

Fascinating Facts About Geriatrics and Gerontology

- Globally, vast disparities exist in expected life span. The United States lags behind France, Sweden, Canada, and Japan in mean life expectancy. In some African countries people have a life expectancy of forty years.
- By 2050, 21 percent of the world's population will be sixty or older. Almost 1.5 million people are expected to be one hundred years old or older by 2050.
- In 2011, the first wave of baby boomers will reach age sixty-five and by this point boomers will be twice the population of the entire country of Canada.
- The aged do not suffer more crime than younger persons, although they are more fearful of crime.
- Many older adults enjoy a healthy sex life even into old age.
- Memory loss is not an inevitable part of the aging process. Most senior citizens do not get Alzheimer's disease.
- Men over fifty years old have the highest suicide rates; those eighty-five or older have the highest overall suicide rates.
- The majority of old people do not live alone.
- Religiosity does not increase with age.
- Although physical strength does decline, strength-training exercises maintain and improve flexibility and strength even into old age.

---

further prolonging independence in the home. Motion-sensor systems eliminate wandering, which is often associated with later stages of dementia. Additional home-based applications for seniors include special bathtubs, mechanical chairs that climb stairs, motorized wheelchairs, and remote home-monitoring of health status.

Often advances in care for the aging have spillover benefits for the rest of the population. For example, the physical and occupational therapy fields have created user-friendly work environments for employees of all ages. Accordingly, ergonomics is now a household name with companies and therapists building optimal sitting and standing workstations that relieve pressure and support working and moving bodies.

The automobile industry has responded with adjustable gas and brake pedals, backup cameras, audible turn-by-turn directions, and parallel-parking assistance.

Education. Lifelong learning colleges offer continuing education to seniors through a variety of formats. Online centers of learning, interactive CD and DVD training programs, and book and workbook training manuals abound that purport to increase memory, mental speed, and generally bolster one's brain power. Many traditional university and community colleges offer vacant classroom seats to seniors at a discount. Train the Trainer is an emerging public health program that trains seniors to educate other seniors, often in classroom-like settings. Select groups of seniors undergo an extensive program to learn the latest health-improvement information and how to deliver such information effectively. They each educate a room full of seniors, thereby increasing the scope of health-promotion programs.

Physical activity is the single best way to improve one's health at any age, and there are many products and services that promote an active lifestyle. The Nintendo Wii video game system has enjoyed popularity with seniors across the country. The Wii is a low-impact, hand-eye coordination system that is believed to increase balance, strength, and cognitive performance. Many senior centers have created Wii bowling leagues, and online goal-setting and health-improvement Web sites permit seniors from across the globe to post their scores to encourage other seniors to maintain healthy lifestyles.

IMPACT ON INDUSTRY

Across the world, governments, businesses, academia, and individuals are responding to the needs of the aging population. Elder rights, rural aging, poverty, retirement, and pensions are a few of the issues facing the unprecedented number of seniors in the first half of the twenty-first century. China, in particular, will have an elder population that outnumbers that in the United States by 2050, and almost 70 percent of China's aging population is not covered by a public pension system. The cost to the United Kingdom's economy because of elderly discrimination is estimated in the tens of billions of dollars. In Latin American countries where a minimum pension exists, elderly poverty rates are low; World Bank research indicates that enacting similar programs in

the remaining South American countries would drastically reduce elderly poverty.

Government and University Research. The National Institute on Aging (NIA) is solely dedicated to funding basic and applied research projects that address the aging process. The scope of their endeavor is impressive and important. For example, basic science research at the molecular level drives animal models of investigation, which in turn drives clinical trials that result in new drugs and products. The ultimate goal of such research is the increased health and wellness of the aging population. Research at the university level is a critical link in this process. University faculty are often the main recipients of NIA research grants. The large infrastructure of the university environment permits many highly qualified individuals to conduct cutting-edge aging research. For example, New York University recently discovered that a gene linked to Alzheimer's disease is involved in removing toxic proteins—excessive amyloid protein deposits in the brain are hallmark characteristics of Alzheimer's disease.

Industry and Business. As discussed above, the opportunity for new aging-related products has produced a boom in the manufacturing and service industries. Entire departments and companies cater solely to the needs of the aged. The burgeoning cognitive-training, physical fitness, nutrition, legal, political advocate, and construction industries have responded in-kind. The various industries that cater to seniors have begun to target middle-aged individuals as well as it is believed that lifelong behaviors often predict health and well-being in one's later years. Accordingly, memory- and attention-training products exist for any age. Although once confined to installation of grab rails in bathtubs, portions of the construction industry now specialize in senior-friendly housing where lower countertops and no-slip surfaces are commonplace.

Major Corporations. Founded in 1958, AARP is the largest corporation catering to the aging population. Formerly known as the American Association of Retired Persons, AARP struggled with the outdated concept of retirement and opted to go by AARP instead. Following in their footsteps, many view their later years as a period of being redirected over the static concept of retirement. The comparable academic organization to AARP is the Gerontological Society of America (GSA). The GSA provides an organizational structure for the numerous educators, researchers, and other individuals interested in aging. In addition, GSA holds a yearly conference where academics share their current research. The Cognitive Aging Conference is a biennial gathering of researchers specifically interested in cognition and the elderly. Because Alzheimer's is one of the most feared diseases, maintaining and improving cognitive skills, thought to buffer the disease's effects, is a multimillion-dollar industry. The Cogmed company, in particular, has produced Internet-based memory- and attention-training programs that have solid scientific evidence supporting their effectiveness.

CAREERS AND COURSE WORK

The rapidly aging baby boomers have already created numerous and varied career opportunities in geriatrics and gerontology. Educators and social workers have experienced tremendous growth opportunities in the last ten years as the need for coordination of health education and social services increases. The shortage of nurses is particularly prevalent in hospitals, nursing homes, and other retirement-oriented settings. By 2030, the projected ratio is 1 geriatrician for every 4,200 seniors, so future job prospects are excellent.

Geriatrics involves the completion of a bachelor's degree, four years of medical school, a three-year internship in general medicine and aging populations or subspecialty, such as immunology or cardiology, and often a one- to two-year program in geriatrics. Typically labeled a geriatrician, the individual spends his or her days compassionately caring for the medical and quality-of-life issues of the aged. Geriatric nurses have earned a specialized bachelor's or master's degree and often play the major role in the daily care of the infirmed. Presently, geriatric psychiatrists are in short supply, leaving many seniors with severe mental health issues searching for adequate care.

Conversely, gerontology accreditation varies with the specialty. Typically, employment-specific master's-level course work is required for most therapist-oriented careers, such as physical therapy or counseling. Completion of doctoral-level course work is required for clinical careers, clinical geropsychologist, educators, university professors, and other aging-related careers, such as dentistry, pharmacy, policy, and political science. Physical or occupational therapists spend their days retraining aging muscular and

structural systems, and counselors commonly deal with depression, loss, grieving, and other aging-related mental health issues. Clinical geropsychologists commonly treat more extensive cognitive-impairment issues such as dementia and mild to moderate mental-health problems.

SOCIAL CONTEXT AND FUTURE PROSPECTS

In 2011, aging in America is viewed through the experience of the baby boomers—those individuals born between 1946 and 1964—who started turning sixty in 2006. The most educated of any senior cohort, baby boomers grew up during unprecedented economic growth and, accordingly, possess a unique view on the aging process. The boomers are not taking retirement lying down: This group is more redirected than retired. Traditional leisure activities, continuing education, volunteering, and often part-time employment now replace the full-time work-week. The baby boomers in particular expect an unprecedented retirement lifestyle. In response to this expectation, retirement communities now resemble theme parks with golf courses, activity centers, and staff solely dedicated to planning events.

Future prospects have become reality. In central Florida, an entire city was developed for the retired. Specialized golf cart highways and parking spaces connect shopping malls and doctor offices to homes. Daily activities can include concerts, speeches, exercise facilities, and college classes. If such retirement cities become the norm, a complete redefining of aging will likely take place. Active and engaged theories of aging will continue to replace outmoded theories of aging such as the concept of disengagement. Whether the stigma of aging and corresponding stereotypes will also be replaced is yet to be determined.

Biotechnology holds great promise for the future of aging research. Caloric-restrictive diets have extended longevity in mice, and such research has led investigators to explore the possibility of "turning off" mechanisms responsible for fat storage. Emerging research suggests B vitamins may decrease depression, phenolic compounds might kill certain cancer cells, a juice elixir may prevent the common cold, and capsaicin, mentioned earlier in connection with arthritis, may promote weight loss. Cortisone and prednisone, often found in anti-inflammatory medications, might have an unintended side effect of bone loss as recently reported in the journal *Cell Metabolism*. Perhaps the most visible biotechnology project involves using stem cells to regenerate aging cells.

Dana K. Bagwell, B.S.

FURTHER READING

Antonucci, Toni, and James Jackson, eds. *Annual Review of Gerontology and Geriatrics: Life-Course Perspectives on Late Life Health Inequalities.* New York: Springer, 2010. Yearly review of the vast field of aging studies. This volume's emphasis is on health disparities and aging.

Chodzko-Zajko, Wojtek, Arthur F. Kramer, and Leonard W. Poon, eds. *Enhancing Cognitive Functioning and Brain Plasticity.* Champaign, Ill.: Human Kinetics, 2009. A review of recent research supporting the notion that physical activity and brain exercises strengthen the brain. The fairly new concept of neural plasticity, that aging brains can grow and form new neural connections, is discussed extensively.

Halter, Jeffrey, et al. *Hazzard's Geriatric Medicine and Gerontology.* 6th ed. New York: McGraw-Hill, 2009. A compendium of evidence-based medicine and clinical applications for treating the aged. Includes 300 illustrations, numerous tables and figures, and additional online resources.

Palmore, Erdman B. *The Facts on Aging Quiz: A Handbook of Uses and Results.* New York: Springer, 1988. Contains several quizzes that test common misconceptions of the aging process.

Schaie, K. Warner, and Laura L. Carstensen, eds. *Social Structures, Aging, and Self-Regulation in the Elderly.* New York: Springer, 2006. Examines the evolution of personal and social roles in aging with particular emphasis on familial changes, immigration, and increased life span.

WEB SITES

AARP International
http://www.aarpinternational.org

American Geriatrics Society
http://www.americangeriatrics.org

American Psychological Association
Division 20 (Aging)
http://apadiv20.phhp.ufl.edu

Gerontological Society of America
http://www.geron.org

See also: Pharmacology; Surgery.

GLASS AND GLASSMAKING

FIELDS OF STUDY

Optics; physics; mechanical engineering; electrical engineering; materials science; ceramics; chemistry; architecture; mathematics; thermodynamics; microscopy; spectroscopy.

SUMMARY

Glassmaking is a diverse field with applications ranging from optics to art. Most commercially produced glass is used to make windows, lenses, and food containers such as bottles and jars. Glass is also a key component in products such as fiber-optic cables and medical devices. Because many types of glass are recyclable, the demand for glass is expected to increase. Most careers in glassmaking require significant hands-on experience as well as formal education.

KEY TERMS AND CONCEPTS

- **Ceramic:** Material made from nonorganic, nonmetallic compounds subjected to high levels of heat.
- **Crystal:** Solid formed from molecules in a repeating, three-dimensional pattern; also refers to a high-quality grade of lead glass.
- **Glazier:** Professional who makes, cuts, installs, and replaces glass products.
- **Pellucidity:** Rate at which light can pass through a substance; also known as transparency.
- **Silica:** Silicon dioxide, the primary chemical ingredient in common glass.
- **Tempered Glass:** Glass strengthened by intensive heat treatment during manufacturing.
- **Viscosity:** Rate at which a fluid resists flow; sometimes described as thickness.
- **Vitreous:** Being glasslike in texture or containing glass.

DEFINITION AND BASIC PRINCIPLES

Glass is one of the most widely used materials in residential and commercial buildings, vehicles, and many different devices. Glass windows provide natural light while protecting indoor environments from changes in the outside temperature and humidity. The transparency of glass also makes it a good choice for lightbulbs, light fixtures, and lenses for items such as eyeglasses. A nonporous and nonreactive substance, glass is an ideal material for food packaging and preparation. Most types of glass can be recycled with no loss in purity or quality, a factor that has increased its appeal.

Although glass resembles crystalline substances found in nature, it does not have the chemical properties of a crystal. Glass is formed through a fusion process involving inorganic chemical compounds such as silica, sodium carbonate, and calcium oxide (also known as lime). The compounds are heated to form a liquid. When the liquid is cooled rapidly, it becomes a solid but retains certain physical characteristics of a liquid, a process known as glass transformation.

BACKGROUND AND HISTORY

The earliest known glass artifacts come from Egypt and eastern Mesopotamia, where craftspeople began making objects such as beads more than five thousand years ago. Around 1500 B.C.E., people began dipping metal rods into molten sand to create bottles. In the third century B.C.E., craftspeople in Babylon discovered that blowing air into molten glass was a rapid, inexpensive way to make hollow shapes.

Glassmaking techniques quickly spread throughout Europe with the expansion of the Roman Empire in the first through fourth centuries. The region that would later be known as Italy, led by the city of Venice, dominated the glass trade in Europe and the Americas for several hundred years.

With the development of the split mold in 1821, individual glass objects no longer needed to be blown and shaped by hand. The automation of glass manufacturing allowed American tinsmith John L. Mason to introduce the Mason jar in 1858. Over the next few decades, further innovations in glassmaking led to a wide range of glass applications at increasingly lower costs.

One of the most noteworthy advancements in glass use came in the 1960's and 1970's with the design and rollout of fiber-optic cable for long-range communications.

HOW IT WORKS

Most types of glass have translucent properties, which means that certain frequencies, or colors, of light can pass through them. Transparent glass can transmit all frequencies of visible light. Other types of glass act as filters for certain colors of light so that when objects are viewed through them, the objects appear to be tinted a specific shade. Some types of glass transmit light but scatter its rays so that objects on the other side are not visible to the human eye. Glass is often smooth to the touch because surface tension, a feature similar to that of water, binds its molecules together during the cooling process. Unless combined with certain other compounds, glass is brittle in texture.

There is a widespread but incorrect belief that glass is a liquid. Many types of glass are made by heating a mixture to a liquid state, then allowing it to reach a supercooled state in which it cannot flow. The process, known in glassmaking as the glass transition, causes the molecules to organize themselves into a form that does not follow an extended pattern. In this state, known as the vitreous or amorphous state, glass behaves like a solid because of its hardness and its tendency to break under force.

Chemical Characteristics. The most common type of glass is known as soda-lime glass and is made primarily of silica (60 to 75 percent). Sodium carbonate, or soda ash, is added to the silica to lower its melting point. Because the presence of soda ash makes it possible for water to dissolve glass, a third compound such as calcium carbonate, or limestone, must be added to increase insolubility and hardness. For some types of glass, compounds such as lead oxide or boric oxide are added to enhance properties such as brilliance and resistance to heat.

Manufacturing Processes. The melting and cooling of a mixture to a liquid, then supercooled state is the oldest form of glassmaking. Most glass products, including soda-lime, are made using this type of process. As the liquid mixture reaches the supercooled state, it is often treated to remove stresses that could weaken the glass item in its final form. This process is known as annealing. The item's surface is smoothed and polished through a stream of pressurized nitrogen. Once the glass transformation is complete, the item can be coated or laminated to increase traits such as strength, electrical conductivity, or chemical durability.

Specialized glass can be made by processes such as vapor deposition or sol-gel (solution-gel). Under vapor deposition, chemicals are combined without being melted. This approach allows for the creation of thin films that can be used in industrial settings. Glass created through a sol-gel process is made by combining a chemical solution, often a metal oxide, with a compound that causes the oxide to convert to a gel. High-precision lenses and mirrors are examples of sol-gel glass.

APPLICATIONS AND PRODUCTS

Glass is one of the most widely used materials in the world. Some of the most common uses for glass include its incorporation into buildings and other structures, vehicles such as automobiles, fiberglass

A glazier at work on the new Aviva Stadium in Dublin, Ireland. (AP Photo)

packaging materials, consumer and industrial optics, and fiber optics.

Construction. The homebuilding industry relies on glass for the design and installation of windows, external doors, skylights, sunrooms, porches, mirrors, bathroom and shower doors, shelving, and display cases. Many office buildings have floor-to-ceiling windows or are paneled with glass. Glass windows allow internal temperatures and humidity levels to be controlled while still allowing natural light to enter a room. Light fixtures frequently include glass components, particularly bulbs and shades, because of the translucent quality of glass and its low manufacturing costs.

Automotive Glass. Windshields and windows in automobiles, trucks, and other vehicles are nearly always made of glass. Safety glass is used for windshields. Most safety glass consists of transparent glass layered with thin sheets of a nonglass substance such as polyvinyl butyral. The nonglass layer, or laminate, keeps the windshield from shattering if something strikes it and protects drivers and passengers from injury in an accident. A car's side and back windows generally are not made of safety glass but rather of tempered glass treated to be heat resistant and to block frequencies of light such as ultraviolet rays.

Fiberglass. Russell Games Slayter, an employee of Owens-Corning, invented a glass-fiber product in 1938 that the company named Fiberglas. The product was originally sold as thermal insulation for buildings and helped phase out the use of asbestos, a carcinogen. The tiny pockets of air created by the material's fabriclike glass filaments are the source of its insulating properties. Fiberglass is also used in the manufacturing of vehicle bodies, boat hulls, and sporting goods such as fishing poles and archery bows because of its light weight and durability. Because recycled glass makes up a significant portion of many fiberglass products, the material is considered environmentally friendly.

Packaging. Processed foods and beverages are often sold in glass bottles and jars. The clarity of glass allows consumers to see the product and judge its quality. Glass is a nonporous, nonreactive material that does not affect the flavor, aroma, or consistency of the product it contains. Consumers also prefer glass because it can be recycled. A disadvantage of glass packaging is the ease with which containers can be shattered. Glass also is heavier than competing packaging materials such as plastic, aluminum, and paper.

Consumer Optics. Glass lenses can be found in a wide range of consumer-oriented optical products ranging from eyeglasses to telescopes. The use of glass in eyeglasses is diminishing as more lenses are made from specialized plastics. Cameras, however, rarely use plastic lenses, as they tend to create lower quality images and are at greater risk of being scratched. Binoculars, microscopes, and telescopes designed for consumer use are likely to contain a series of glass lenses. These lenses are often treated with coatings that minimize glare and improve the quality of the image being viewed.

Industrial Optics. For the same reasons that glass is preferred in consumer optics, it is the material of choice for industrial applications. Many specialized lenses and mirrors used for telescopes, microscopes, and lasers are asymmetrical or aspherical in shape. The manufacture of aspherical lenses for high-precision equipment was difficult and expensive until the 1990's, when glass engineers developed new techniques based on the processing of preformed glass shapes at relatively low temperatures. These techniques left the surface of the glass smooth and made it more cost-effective to manufacture highly customized lenses.

Fiber Optics. The optical fibers in fiber-optic communication infrastructures are made from glass. The fibers carry data in the form of light pulses from one end of a glass strand to the other. The light pulses then jump through a spliced connection to the next glass fiber. Fiber-optic cable can transmit information more quickly and over a longer distance than electrical wire or cable. Glass fibers can carry more than one channel of data when multiple frequencies of light are used. They also are not subject to electromagnetic interference such as lightning strikes, an advantage when fiber-optic cable is used to wire offices in skyscrapers.

IMPACT ON INDUSTRY

The worldwide market for glass is divided by regions and product types. By its nature, glass is heavy and fragile. It is subject to breaking when shipped over long distances. For this reason, most glass is manufactured in plants located as close to customers as possible.

Flat glass is one of the largest glass markets globally. An estimated 53 million tons of flat glass, valued at $66 billion, were made in 2008. The majority of

Fascinating Facts About Glass and Glassmaking

- The oldest known glassmaking handbook was etched onto tablets and kept in the library of Ashurbanipal, king of Assyria in the seventh century B.C.E.
- In 1608, settlers in Jamestown, Virginia, opened the first glassmaking workshop in the American colonies. It produced bottles used by pharmacists for medicines and tonics.
- Nearly 36 percent of glass soft drink and beer bottles and 28 percent of all glass containers were recycled in 2008. About 80 percent of recovered glass is used for new bottles.
- Toledo, Ohio, is nicknamed the Glass City because of its industrial heritage of glassmaking. In the early twentieth century, Toledo began providing glass to automobile factories in nearby Detroit.
- China manufactures about 45 percent of all glass in the world. Most Chinese glass is installed in automobiles made locally.
- The first sunglasses were built from lenses made out of quartz stone and tinted by smoke. The glasses were worn by judges in fourteenth-century China to hide their facial expressions during court proceedings.
- The National Aeronautics and Space Administration has developed NuStar (Nuclear Spectroscopic Telescope Array), a set of two space telescopes carrying a new type of glass lens made from 133 glass layers. NuStar is about one thousand times as sensitive as space telescopes such as Hubble.

flat glass, about 70 percent of the total market, was used for windows in homes and industrial buildings. Furniture, light fixtures, and other interior products made up 20 percent of the market, while the rest (10 percent) was used primarily in the automotive industry. The manufacturing of flat glass requires significant capital investment in large plants, so the market is dominated by a few large companies. In 2009, the world's leading makers of flat glass were the NSG Group (based in Japan), Saint-Gobain (France), Asahi Glass (Japan), and Guardian Industries (United States). The market is growing at about 4 to 5 percent each year. Much of this growth is fueled by

demand from China, where the construction and automotive industries are expanding at a faster rate than in North America and Europe. New flat glass products incorporate nonglass materials such as polymers, which enhance safety and the insulating properties of windows.

The glass packaging industry worldwide is led by bottles used for beverage products. In 2009, glass container manufacturers produced more than 150 billion beer bottles and more than 60 billion bottles for spirits, wines, and carbonated drinks. Research firm Euromonitor International predicts that demand for glass packaging will grow, but growth rates across package types will vary significantly. Glass bottles face competition from polyethylene terephthalate (PET) bottles, which are much lighter and more durable. A rise in median consumer income across the Asia-Pacific region in the 2000's has increased demand for processed food products sold in jars and bottles.

Niches such as glass labware are also facing pressure from items made with nonglass materials. The global market for glass labware could be as high as $3.87 billion by 2015, according to Global Industry Analysts. New plastics developed for specialized applications such as the containment of hazardous chemicals may slow down the rate at which laboratories purchase glass containers. However, glass will continue to be a laboratory staple as long as it remains inexpensive and widely available.

Fiber-optic cable continues to fuel growth for glass manufacturing. In late 2010, many telecommunications firms were preparing to invest in expanding their fiber-optic networks. Telecom research firm CRU estimated that companies worldwide ordered more than 70 million miles of new fiber-optic cable between July, 2009, and June, 2010. The growth in demand in Europe was nearly twice that of the United States in 2009 and 2010.

CAREERS AND COURSE WORK

The skills required to work in the glassmaking industry depend on the nature of the glass products being made. Manufacturing and installing glass windows in buildings is a career within the construction industry, while the design of glass homewares involves the consumer products industry. The making of lenses for eyeglasses is a career track that differs significantly from the production of specialty lenses for precision equipment such as telescopes.

Glaziers in construction gain much of their knowledge from on-the-job training. Many glaziers enter apprenticeships of about three years with a contractor. The apprentice glazier learns skills such as glass cutting by first practicing on discarded glass. Tasks increase in difficulty and responsibility as the apprentice's skills grow. This work is supplemented by classroom instruction on topics such as mathematics and the design of construction blueprints. Similarly, makers of stemware and high-end glass products used in homes learn many of their skills through hands-on work with experienced designers.

Careers in optical glass and lens design, on the whole, require more formal education than other areas of glass production do. Several universities offer course concentrations in glass science, optics, or ceramics within undergraduate engineering programs. Graduate programs can focus on fields as narrow as electro-optics (fewer than ten American schools offer this option).

Students planning careers in optical glass take general classes such as calculus, physics, and chemistry. Course work on topics such as thermodynamics, microscopy, spectroscopy, and computer-aided design are also standard parts of an optical engineer's education.

SOCIAL CONTEXT AND FUTURE PROSPECTS

Demand for glaziers in the construction industry rises and falls with the rate of new buildings being built. The early 2000's saw a sharp increase in demand for glaziers, particularly those with experience in installing windows in new houses. This trend reversed with the end of the construction boom in the late 2000's. At the same time, the development of window glass with energy-efficient features led to the upgrading of windows in office buildings and homes, which has increased the need for skilled glaziers. The U.S. Bureau of Labor Statistics predicts that jobs in this field will grow at about 8 percent from 2008 to 2018, an average rate compared to all other occupations.

Within the field of optical glass, manufacturers have developed a wide range of glass types for specialized applications such as lasers. The need for increasingly precise lenses has grown steadily with developments in fields ranging from microsurgery to astronomy.

Julia A. Rosenthal, B.A., M.S.

FURTHER READING

Beretta, Marco. *The Alchemy of Glass: Counterfeit, Imitation, and Transmutation in Ancient Glassmaking.* Sagamore Beach, Mass.: Science History Publications/USA, 2009. Examines the history of glassmaking, from Egypt and Babylonia to modern times, looking at the effects of technology.

Hartmann, Peter, et al. "Optical Glass and Glass Ceramic Historical Aspects and Recent Developments: A Schott View." *Applied Optics* (June 1, 2010): D157-D176. A brief but informative overview of optical glass from the Italian Renaissance to the modern times.

Le Bourhis, Eric. *Glass: Mechanics and Technology.* Weinheim, Germany: Wiley-VCH, 2008. Explores the properties of glass and their physical and chemical sources.

Opie, Jennifer Hawkins. *Contemporary International Glass.* London: Victoria and Albert Museum, 2004. Photographs and artistic statements from sixty international glass artists, based on pieces in the collection of the Victoria and Albert Museum in London.

Shelby, J. E. *Introduction to Glass Science and Technology.* Cambridge, England: Royal Society of Chemistry, 2005. A college-level textbook with detailed information on the physical and chemical properties of glass.

Skrabec, Quentin, Jr. *Michael Owens and the Glass Industry.* Gretna, La.: Pelican, 2007. A biography of American industrialist Michael Owens, whose mass production techniques for glass bottles and cans led to major changes in the lives of consumers.

WEB SITES

American Scientific Glassblowers Society
http://www.asgs-glass.org

Association for the History of Glass
http://www.historyofglass.org.uk

Corning Museum of Glass
http://www.cmog.org

Society of Glass Technology
http://www.societyofglasstechnology.org.uk

See also: Mirrors and Lenses; Optics.

GRAMMATOLOGY

FIELDS OF STUDY

Linguistics; anthropology; archaeology; history; literature psychology; classics; English; foreign languages; computer science.

SUMMARY

Grammatology is the science of writing. It is closely related to the field of linguistics but concentrates on written expression whereas linguistics is primarily oriented toward oral expression. Grammatology is an interdisciplinary field that is applied to research in a variety of fields studying human interaction and communication and the various cultures and civilizations. Anthropologists, psychologists, literary critics, and language scholars are the main users of grammatology. Electracy, digital literacy, is a concept that applies the principles of grammatology to electronic media.

KEY TERMS AND CONCEPTS

- **Abjad:** Script using consonants only, no vowels.
- **Abugida:** Script using consonants that contain inherent vowels, modifications to the character to denote other vowels; also known as alphasyllabary.
- **Alphabetic:** Script using both consonants and vowels.
- **Cuneiform Writing:** System of wedge-shaped marks used for communication.
- **Logography:** System of writing using a sign to indicate a word or a combination of words; also known as word writing.
- **Script:** Set of basic symbols or characters.
- **Syllabary:** Writing system using a combination of symbols for a word.
- **Text:** Object produced by writing that disseminates information.
- **Writing:** Set of marks or characters used for communication.

DEFINITION AND BASIC PRINCIPLES

Grammatology is the scientific examination of writing. Oral communication is the natural means of communication for human beings. Oral language uses neither instruments nor objects that are not part of the human body. Writing, by contrast, is not natural language; it requires an object (paper, stone, wood, or even sand) on which marks can be made and some kind of tool with which to make the marks. Although it is possible to use one's fingers to make marks in a soft medium such as sand, such "writing" lacks an essential feature of writing as defined by grammatology: Writing lasts beyond the moment of its creation. It preserves thoughts, ideas, stories, and accounts of various human activities. In addition, writing, in contrast to pictography (drawn pictures), can be understood only with knowledge of the language that it represents, because writing uses symbols to create a communication system. The discipline defines what writing is by using a system of typologies, or types. An example of this is the classification of languages by script (the kind of basic symbols used). Another way of classifying languages is by the method used to write them; linear writing systems use lines to compose the characters.

Grammatology also investigates the importance of writing in the development of human thought and reasoning. Grammatologists address the role of writing in the development of cultures and its importance in the changes that occur in cultures. Grammatology proposes the basic principle that changes in ways of communication have profound effects on both the individual and the societal life of people. It addresses the technology of writing and its influence on the use and development of language. It deals extensively with the impact of the change from oral to handwritten communication, from handwritten texts to printed texts, and from printed texts to electronic media texts.

BACKGROUND AND HISTORY

Polish American historian Ignace Gelb, the founder of grammatology, was the first linguist to propose a science of writing independent of linguistics. In 1952, he published *A Study of Writing: The Foundations of Grammatology*, in which he classified writing systems according to their typologies and set forth general principles for the classification of writing systems. As part of his science of writing systems, Gelb traced the evolution of writing from pictures that he referred to as "no writing" through semasiography to full writing or scripts that designate sounds by characters. Gelb divided languages into three typologies: logo-syllabic systems, syllabic systems, and alphabetic systems. Peter T. Daniels, one of Gelb's students, added two additional types to this typology: abugida and abjad. Gelb also included as part of the science of writing an investigation of the interactions of writing and culture and of writing and religion as well as an analysis of the relationship of writing to speech or oral language. Although Gelb did not elaborate extensively on the concept of writing as a technology, he did address the importance of the invention of printing and the changes that it brought about in dissemination of information. This aspect of writing (by hand or device) as technology, as an apparatus of communication, has become one of the most significant areas of the application of grammatology to other disciplines.

Gelb's work laid the foundation for significant research and development in several fields involved in the study of language. Eric Havelock, a classicist working in Greek literature, elaborated on the idea of how the means of communication affects thought. He developed the theory that Greek thought underwent a major change or shift when the oral tradition was replaced by written texts. In Havelock's *Preface to Plato* (1963), he presented his argument that the transition to writing at the time of Plato altered Greek thought and consequently all Western thought by its effect on the kind of thinking available through written expression. Havelock stated that what could be expressed and how it could be expressed in written Greek was significantly different from what could be expressed orally. He devoted the rest of his scholarship to elaboration of this premise and investigation of the difference between the oral tradition and writing. Havelock's theory met with criticism from many of his colleagues in classics, but it received a favorable reception in many fields, including literature and anthropology.

Starting from Havelock's theory, Walter J. Ong, a literary scholar and philosopher, further investigated the relationship between written and spoken language and how each affects culture and thought. He proposed that writing is a technology and therefore has the potential to affect virtually all aspects of an oral culture. Addressing the issue of writing's effect on ways of thinking, Ong stated that training in

writing shifted the mental focus of an individual. Ong reasoned that oral cultures are group directed, and those using the technology of writing are directed toward the individual. His most influential work is *Orality and Literacy: The Technologizing of the Word* (1982). The British anthropologist Sir John "Jack" Rankine Goody applied the basic premise of grammatology (that the technology of writing influences and changes society) in his work in comparative anthropology. For Goody, writing was a major cause of change in a society, affecting both its social interaction and its psychological orientation.

In *De la grammatologie* (1967; *Of Grammatology*, 1976), French philosopher and literary critic Jacques Derrida examined the relationship of spoken language and writing. By using Gelb's term "grammatology" as a term of literary criticism, Derrida narrowed the significance of the term. Grammatology became associated with the literary theory of deconstruction. However, Derrida's work had more far-reaching implications, and for Derrida, as for Gelb, grammatology meant the science of writing. Derrida was attempting to free writing from being merely a representation of speech.

In *Applied Grammatology: Post(e)-pedagogy from Jacques Derrida to Joseph Beuys* (1985), Gregory Ulmer moved from deconstructing writing to a grammatology that he defined as inventive writing. Although Ulmer's work moves away from the areas of systems and classifications of writing as established by Gelb, it does elaborate on Gelb's notion of how literacy or the technology of writing changes human thought and interaction. Ulmer's work brings Gelb's theories and work into the electronic age. Gelb and his followers investigated the development of writing systems and the technologies of handwritten and printed texts and their influences on human beings, both in their social interactions and in their thought processes. Just as print influenced the use, role, and significance of writing, electronic media is significantly affecting writing, both in what it is and what it does.

HOW IT WORKS

Classification by Typologies. Written languages are classified based on particular features of the language. Grammatologists analyze written languages to determine whether the basic element of the language is a syllable, a letter, a word, or some other element. They also look for similarities among languages that they compare. From these analyses, they classify languages in existing typologies or create new ones. The established typologies are logographic, syllabic, alphabetic, abjad, and abugida.

Mechanisms and Apparatuses. Grammatology identifies the mechanisms or the apparatuses of writing. It examines how writing is produced, what type of tools are used to produce it, and the kind of texts that result from writing. It also considers the speed and breadth of dissemination of information through writing. Grammatology looks at writing systems from a practical viewpoint, considering questions of accuracy and ease of usage.

Relationship of Oral and Written Language. Grammatologists investigate the relationship of oral and written language by comparing a language in its oral and written form. They look for differences in sentence structure, vocabulary, and complexity of usage.

Writing and Social Structure. Grammatologists note the time that writing appears in a culture and analyze the social, political, economic, and family structures to ascertain any and all changes that appear after the introduction of writing. They also attempt to identify the type of thinking, ideas, and beliefs prevalent in the society before and after the introduction of writing and determine if writing brought about changes in these areas. Once they have established evidence of change, they address the ways in which writing causes these modifications. In addition, they investigate these same areas when the mechanism of writing in a culture changes, such as when print replaces handwritten texts and when electronic media replaces print.

APPLICATIONS AND PRODUCTS

Pedagogy. The science of grammatology is important in the development of pedagogies that adequately address the needs and orientations of students. The way information is presented and taught varies with the medium by which students are accustomed to receiving knowledge. In an oral culture, stories, poems, and proverbs that use repetitive signifiers are employed to enable students to memorize information. Handwritten texts are suited to cultures in which a small quantity of information is disseminated among a limited number of students. Print text cultures rely heavily on the dissemination of information through books and other printed sources.

Fascinating Facts About Grammatology

- The Roman alphabet reached its definitive form about seven centuries after the first writing was done in Latin. During this time, the alphabet underwent considerable graphic changes.

- Languages that use systems of writing or scripts that contain no vowels are classified as abjad. Two of these languages are Arabic and Hebrew.

- Chinese is a language whose basic element is the logogram. Japanese contains many of the same logograms. However, an individual capable of reading a Japanese text is not able to read a Chinese text because of the disparity in the grammar of the two languages.

- Throughout history, human beings have believed that writing possessed magical powers. The writing of poetry has been traditionally associated with divine inspiration. The poet and the poet's muse play an important role in concepts of artistic creation in writing.

- The writing and the architecture of a specific period often share certain characteristics. Handwriting from the Carolingian period resembles Romanesque architecture by its use of roundness, and handwriting from the later Gothic period of European history reflects Gothic architecture by its angularity.

- After the arrival of Europeans, especially missionaries, in North America, members of certain Indian tribes such as the Cherokee and Algonquin attempted to invent writing systems for their spoken languages.

- The Egyptian civilization used a system of writing known as hieroglyphics, which survived for more than three thousand years.

Instruction is classroom based with face-to-face contact between the teacher and student. Assignments are written out and turned in to the professor to be corrected and returned. Cultures that rely on electronic media call for different pedagogical structures. Grammatology provides a means of analyzing and addressing this need.

Anthropology. Anthropologists use grammatology as an integral part of their research techniques to ascertain the development of civilizations and cultures. The time when writing appeared in a culture (or the lack of writing in a culture), the type of writing, and the classes of society that use writing are all important to the study and understanding of a culture or civilization.

Literary Criticism. Grammatology enables literary critics to dissect or deconstruct texts and perform close analysis of both the structure and the meaning of a work. It has also been the basis for the creation of new types of literary criticism, including deconstruction and postdeconstruction.

Linguistics. Grammatology, which deals with the written form of language, plays an important role in linguistics, the study of languages. It classifies languages, addresses how graphemes (units of a writing system) and allographs (letters) are used, compares written to oral forms of languages, and identifies archaic forms and vocabularies. The study of writing forms is necessary to linguists tracing the development of a particular language. Texts written in the language during different periods permit linguists to compare vocabulary, syntax, and prosaic and poetic forms of the language throughout its development. Written texts also provide evidence of borrowings from other languages.

Archaeology. The ability to analyze and classify written language is an essential part of the work of the archaeologist. Knowledge regarding when particular written forms were used and what cultures used them plays an important role in the dating of artifacts.

Electronic Media. Grammatology is highly applicable to cyberlanguage and electronic media as these phenomena affect writing. Texting and e-mail have created what may be viewed as a subtext or alternative to the traditional print culture. The principles and techniques of analysis of grammatology provide a means for classifying the language of texting and e-mail and for evaluating the effects of these new forms of language on thought and on societal interaction.

IMPACT ON INDUSTRY

The academic community is the sector in which grammatology is primarily used; therefore, its impact on industry is indirect. However, because grammatologists describe the changes that occur in human thought processes as the means and mechanisms of written communication change, they contribute significantly to the development of strategies of teaching and presentation of material in ways appropriate to the needs of those receiving the information.

Government, University, and Private Foundation Grants. Many sources of funding are available to researchers using grammatological principles and techniques of investigation in their research projects. The United States government funds many projects through the National Endowment for the Humanities. Individual universities such as those in the University of California system provide research grants and fellowships for graduate students, postdoctoral candidates, and faculty involved in research into writing and its importance and implications. Private foundations such as the Guggenheim Foundation also provide grants and residencies to encourage research in areas involving grammatology.

Cyberlanguage and Electronic Communication. Universities and graduate schools such as the European Graduate School are developing departments addressing cyberlanguage and electronic communication studies in which grammatologists who are interested in applying their skills to electronic communication can find interesting and challenging positions.

CAREERS AND COURSE WORK

Courses in linguistics, the history of language, and the composition and grammar of the language or languages with which an individual wishes to work form the basis of preparation for a career in grammatology as a subdiscipline of linguistics. However, grammatology is also relevant to a wide variety of careers. Majors in the history of language and in various disciplines such as anthropology, pedagogy, history, literature, psychology, electronic media, and cyberlanguage can be combined with the study of grammatology. Grammatologists, for the most part, work in one of these fields and apply their knowledge of writing systems and their importance to that field. They usually have a doctorate and are employed as university professors or in scientific research centers or consulting firms.

SOCIAL CONTEXT AND FUTURE PROSPECTS

With the changes in the mechanisms and apparatus of writing that have taken place in the second half of the twentieth century and continue to take place in the twenty-first, grammatology continues to be an essential science. Research in electronic media and its relation to human thought processes will be an area in which grammatology will play an important role. Because grammatology is a science used in conjunction with a considerable number of areas of scientific research, it should continue to play a significant role in human intellectual life.

Shawncey Jay Webb, Ph.D.

FURTHER READING

Daniels, Peter T., and William Bright, eds. *The World's Writing Systems*. New York: Oxford University Press, 1996. These essays continue and refine the work done by Gelb.

Gelb, Ignace. *A Study of Writing: The Foundations of Grammatology*. 1952. Reprint. Chicago: University of Chicago Press, 1989. The foundational text for grammatology.

Olson, David R., and Michael Cole, eds. *Technology, Literacy and the Evolution of Society: Implications of the Work of Jack Goody*. Mahwah, N.J.: Lawrence Erlbaum Associates, 2006. An interdisciplinary look at Goody's work on modes of communication in societies and their influence.

Ong, Walter J. *Orality and Literacy: The Technologizing of the Word*. 2d ed. Reprint. New York: Routledge, 2009. Ong elaborates on the work of Havelock concerning the impact of writing on oral culture. Identifies writing as a learned technology.

Rogers, Henry. *Writing Systems: A Linguistic Approach*. Malden, Mass.: Wiley-Blackwell, 2005. A very detailed presentation of world languages and their writing systems that describes the classification of writing systems.

Ulmer, Gregory. *Applied Grammatology: Post(e)-pedagogy from Jacques Derrida to Joseph Beuys*. 1985. Reprint. Baltimore: The Johns Hopkins University Press, 1992. Presents Ulmer's early thoughts on how electronic media was changing the technology of writing. Also provides an excellent explanation of Derrida's views.

_____. *Electronic Monuments*. Minneapolis: University of Minnesota Press, 2005. Ulmer's view of the effects of electracy, or digital literacy, on society and human interaction.

WEB SITES

American Association for Applied Linguistics
http://www.aaal.org

International Cognitive Linguistics Association
http://www.cognitivelinguistics.org

Linguistic Society of America
http://www.lsadc.org

See also: Communications; Computer Languages, Compilers, and Tools; Human-Computer Interaction; Printing; Telecommunications; Typography; Wireless Technologies and Communication.

GRAPHICS TECHNOLOGIES

FIELDS OF STUDY

Mathematics; physics; computer programming; graphic arts and animation; anatomy and physiology; optics.

SUMMARY

Graphics technology, which includes computer-generated imagery, has become an essential technology of the motion picture and video-gaming industries, of television, and of virtual reality. The production of such images, and especially of animated images, is a complex process that demands a sound understanding of not only physics and mathematics but also anatomy and physiology.

KEY TERMS AND CONCEPTS:

- **Anti-Aliasing:** A programming technique for producing smooth curve lines in an image by designating the color of an edge pixel according to how much of the pixel extends beyond the defined curve.
- **Morphing:** The process of converting one image object into another through a series of tweened images.
- **Resolution:** The fineness of detail that can be displayed in an image, determined by the number of pixels contained within a specified area.
- **Sprite:** A small static image, like a desktop icon, that is used to animate subjects that have a constant appearance.
- **Tweening:** The process of creating images to fill in the space between key images of a sequence, used in animations.

DEFINITION AND BASIC PRINCIPLES

While graphics technologies include all of the theoretical principles and physical methods used to produce images, it more specifically refers to the principles and methods associated with digital or computer-generated images. Digital graphics are displayed as a limited array of colored picture elements (pixels). The greater the number of pixels that are used for an image, the greater the resolution of the image and the finer the detail that can be portrayed. The data that specifies the attributes of each individual pixel are stored in an electronic file using one of several specific formats. Each file format has its own characteristics with regard to how the image data can be manipulated and utilized.

Because the content of images is intended to portray real-world objects, the data for each image must be mathematically manipulated to reflect real-world structures and physics. The rendering of images, especially for photo-realistic animation, is thus a calculation-intensive process. For images that are not produced photographically, special techniques and applications are continually being developed to produce image content that looks and moves as though it were real.

BACKGROUND AND HISTORY

Imaging is as old as the human race. Static graphics have historically been the norm up to the invention of the devices that could make a series of still pictures appear to move. The invention of celluloid in the late nineteenth century provided the material for photographic film, with the invention of motion picture cameras and projectors to follow. Animated films, commonly known as cartoons, have been produced since the early twentieth century by repeatedly photographing a series of hand-drawn cels. With the development of the digital computer and color displays in the last half of the century, it became possible to generate images without the need for hand-drawn intermediaries.

Computer graphics in the twenty-first century can produce images that are indistinguishable from traditional photographs of real objects. The methodology continues to develop in step with the development of new computer technology and new programming methods that make use of the computing abilities of the technology.

HOW IT WORKS

Images are produced initially as still or static images. Human perception requires about one-thirtieth of a second to process the visual information obtained through the seeing of a still image. If a sequential series of static images is displayed at a rate that exceeds the frequency of thirty images per second, the images are perceived as continuous motion. This is the basic principle of motion pictures, which are nothing more than displays of a sequential series of still pictures. Computer-generated still images (now indistinguishable from still photographs since the advent of digital cameras) have the same relationship to computer animation.

Images are presented on a computer screen as an array of colored dots called pixels (an abbreviation of "picture elements"). The clarity, or resolution, of the image depends on the number of pixels that it contains within a defined area. The more pixels within a defined area, the smaller each pixel must be and the finer the detail that can be displayed. Modern digital cameras typically capture image data in an array of between 5 and 20 megapixels. The electronic data file of the image contains the specific color, hue, saturation, and brightness designations for each pixel in the associated image, as well as other information about the image itself.

To obtain photorealistic representation in computer-generated images, effects must be applied that correspond to the mathematical laws of physics. In still images, computational techniques such as ray tracing and reflection must be used to imitate the effect of light sources and reflective surfaces.

For the virtual reality of the image to be effective, all of the actual physical characteristics that the subject would have if it were real must be clearly defined as well so that when the particular graphics application being used renders the image to the screen, all of the various parts of the image are displayed in their proper positions.

To achieve photorealistic effects in animation, the corresponding motion of each pixel must be coordinated with the defined surfaces of the virtual object, and their positions must be calculated for each frame of the animation. Because the motions of the objects would be strictly governed by the mathematics of physics in the real world, so must the motions of the virtual objects. For example, an animated image of a round object bouncing down a street must appear to obey the laws of gravity and Newtonian mechanics. Thus, the same mathematical equations that apply to the motion and properties of the real object must also apply to the virtual object.

Other essential techniques are required to produce realistic animated images. When two virtual objects are designed to interact as though they are real, solid objects, clipping instructions identify where the virtual solid surfaces of the objects are located; the instructions then mandate the clipping of any corresponding portions of an image to prevent the objects from seeming to pass through each other. Surface textures are mapped and associated with underlying data in such a way that movement corresponds to real body movements and surface responses. Image animation to produce realistic skin and hair effects is based on a sound understanding of anatomy and physiology and represents a specialized field of graphics technology.

APPLICATIONS AND PRODUCTS

Software. The vast majority of products and applications related to graphics technology are software applications created specifically to manipulate electronic data so that it produces realistic images and animations. The software ranges from basic paint programs installed on most personal computers (PCs) to

Virtual reality game. Screen shot from the 3-D virtual reality video game "Red Planet" showing a spacecraft flying through a canal on Mars. (Peter Menzel/Photo Researchers, Inc.)

full-featured programs that produce wireframe structures, map surface textures, coordinate behaviors, movements of surfaces to underlying structures, and 360-degree, three-dimensional animated views of the resulting images.

Other types of software applications are used to design objects and processes that are to be produced as real objects. Computer-assisted design is commonly used to generate construction-specification drawings and to design printed-circuit boards, electronic circuits and integrated circuits, complex machines, and many other real-world constructs. The features and capabilities of individual applications vary.

The simplest applications produce only a static image of a schematic layout, while the most advanced are capable of modeling the behavior of the system being designed in real time. The latter are increasingly useful in designing and virtual-testing such dynamic systems as advanced jet engines and industrial processes. One significant benefit that has accrued from the use of such applications has been the ability to refine the efficiency of systems such as production lines in manufacturing facilities.

Hardware. The computational requirements of graphics can quickly exceed the capabilities of any particular computer system. This is especially true of PCs. Modern graphics technology in this area makes use of separate graphics processing units (GPUs) to handle the computational load of graphics display. This allows the PC's central processing unit (CPU) to carry out the other computational requirements of the application without having to switch back and forth between graphic and nongraphic tasks.

Many graphics boards also include dedicated memory for exclusive use in graphics processing. This eliminates the need for large sectors of a PC's random access memory (RAM) to be used for storing graphics data, a requirement that can render a computer practically unusable.

Another requirement of hardware is an instruction system to operate the various components so that they function together. For graphics applications, with the long periods of time they require to carry out the calculations needed to render a detailed image, it is essential that the computer's operating system be functionally stable. The main operating systems of PCs are Microsoft Windows, its open-source competitor Linux, and the Apple Mac OS. Some versions of other operating systems, such

Fascinating Facts About Graphic Technologies

- Graphic technologies began with simple cave drawings and has progressed to full-motion, photorealistic, CGI motion pictures.
- The brain is tricked into seeing a rapid sequential series of still images as motion by the physiological process called persistence of vision.
- Jaggies are steplike edges that result when rectangular pixels are used to portray curves in images.
- Ray-tracing applies the mathematics of optical physics to a text file to produce realistic lighting effects in a corresponding image.
- A fractal is an infinitely detailed, self-similar shape that results when an image operation is repeated many times.
- Graphics technology is rapidly progressing to the point in which printed books will be able to incorporate video images displayed as part of the printed page.

as Sun Microsystems' Solaris, have been made available but do not account for a significant share of the PC market.

The huge amounts of graphics and rendering required for large-scale projects such as motion pictures demand the services of mainframe computers. The operating systems for these units have a longer history than do PC operating systems. Mainframe computers function primarily with the UNIX operating system, although many now run under some variant of the Linux operating system. UNIX and Linux are similar operating systems, the main difference being that UNIX is a proprietary system whereas Linux is open-source.

Motion Pictures and Television. Graphics technology is a hardware- and software-intensive field. The modern motion picture industry would not be possible without the digital technology that has been developed since 1980. While live-action images are still recorded on standard photographic film in the traditional way, motion picture special effects and animation have become the exclusive realm of digital graphics technologies. The use of computer generated imagery (CGI) in motion pictures has driven the development of new technology and continually raised the standards of image quality. Amalgamating live action with CGI through digital processing and manipulation enables film-makers to produce motion pictures in which live characters interact seamlessly with virtual characters, sometimes in entirely fantastic environments. Examples of such motion pictures are numerous in the science-fiction and fantasy film genre, but the technique is finding application in all areas, especially in educational programming.

Video Gaming and Virtual Training. The most graphics-intensive application is video gaming. All video games, in all genres, exist only as the graphic representation of complex program code. The variety of video game types ranges from straightforward computer versions of simple card games to complex three-dimensional virtual worlds.

Many graphics software applications are developed for the use of game designers, but they have also made their way into many other imaging uses. The same software that is used to create a fictional virtual world can also be used to create virtual copies of the real world. This technology has been adapted for use in pilot- and driver-training programs in all aspects of transportation. Military, police, and security personnel are given extensive practical and scenario training through the use of virtual simulators. A simulator uses video and graphic displays of actual terrain to give the person being trained hands-on experience without endangering either personnel or actual machinery.

IMPACT ON INDUSTRY

Graphics technology has revolutionized the way in which industry functions by streamlining processes for speed and efficiency. Barcoding is one very simple graphics technology by which this has been achieved. By scanning a barcode with a laser barcode-reader, the pertinent information about that object is available and automatically updated in just a few seconds. Barcode technology is universally applicable and has been used for tasks ranging from pricing goods in stores and cataloging library books, to animal identification tattooing and monitoring goods in transport.

Graphics technology in product design has greatly facilitated the development of new products and made existing processes both more efficient and cost effective. Printed circuit-board layout is one area

in which graphics technology is important. Using a software application designed for the purpose, an electronics engineer or technologist can lay out the circuitry and component placement of a new circuit board in a short time and then use the same software to run virtual tests of the performance of the new design. The preparation and assembly of an actual circuit board, even a simple one, is a material- and time-consuming process when carried out for the production of a small number of test cases. Virtual testing of the graphic design bypasses that aspect of development and enables the rapid prototyping of the design in development.

Process modeling is closely related to virtual testing. With the use of process modeling, graphic software applications, engineers, and designers can create complex systems in the virtual world and then animate them to see how they work. The complex mathematics of the physics involved in the operation of the system is built into the model to make it as accurate as possible. This allows designers to make changes to the system to optimize its function before it is built rather than having to disassemble and reconstruct various components afterward.

The economic value of graphics technology is demonstrated by the motion picture industry and the video-gaming industry. The top-grossing films have historically been science-fiction and fantasy titles that use CGI graphics. The cost of production of any one of these films is measured in tens of millions of dollars, and in some cases, in hundreds of millions of dollars. The gross proceeds of these same films are often ten times as much, and in a few cases have moved into the billion-dollar range. The special effects field is a multibillion dollar industry in its own right, intimately linked to the computer hardware and software industries.

The video-gaming industry is the second area of note. A survey conducted by the industry trade magazine *Game Developer* and published in April, 2011, reported that more than 48,000 game developers were active throughout North America alone. This is a low number in actuality because it does not include developers in other nations, many of whom produce more video game titles than are produced in North America. The survey presented data for the associated professions of programming, art and animation, and design within the video game industry. Of those, programmers garnered an average annual salary of

$85,733; artists and animators earned $71,354 annually, on average; and game designers earned an annual average of $70,223. If one carries out the simple calculation using the overall average of these groups ($75,770), then game developers represent a combined annual economic value of at least $3.6 billion.

Video game development and graphics have also impacted the course of postsecondary education. *Game Developer*, for example, routinely carries advertisements from several colleges and universities for dedicated programs of study in video game design at both undergraduate and graduate levels. The magazine also carries recruitment ads for several video-game production companies.

CAREERS AND COURSEWORK

Many colleges and universities now offer undergraduate and graduate degree programs in applied graphics technology, and in many other traditional programs, graphics technology holds a central position as a tool of the trade. Students interested in pursuing studies aimed at the development of graphics technology must have an advanced understanding of real-world physics and mathematics as a minimum requirement. While it is possible to acquire computer programming skills and expertise with graphics software programs without formal training, it is unlikely that anything less than formal training in those areas will be acceptable to established game-development organizations and other graphics-technology users. Students should therefore expect to take courses in advanced programming to become proficient in assembly language and the Java and C/C++ programming languages. All of these areas are utilized in the production of graphics programming.

Depending on the desired area of specialization, students should also expect to study human and animal anatomy, physiology, kinesiology (the study of movement), and general graphic arts.

Students interested in the technical side of graphics technology should take courses of study in electronics engineering and technology and in computer science. These areas also require a basic knowledge of physics and mathematics and will include a significant component of design principles.

SOCIAL CONTEXT AND FUTURE PROSPECTS

Graphics technology is inextricably linked to the computer and digital electronics industries. Accordingly, graphics technology changes at a rate that at minimum equals the rate of change in those industries. Since 1980, graphics technology using computers has developed from the display of just sixteen colors on color television screens, yielding blocky image components and very slow animation effects, to photorealistic full-motion video, with the capacity to display more colors than the human eye can perceive and to display real-time animation in intricate detail. The rate of change in graphics technology exceeds that of computer technology because it also depends on the development of newer algorithms and coding strategies. Each of these changes produces a corresponding new set of applications and upgrades to graphic technology systems, in addition to the changes introduced to the technology itself.

Each successive generation of computer processors has introduced new architectures and capabilities that exceed those of the preceding generation, requiring that applications update both their capabilities and the manner in which those capabilities are performed. At the same time, advances in the technology of display devices require that graphics applications keep pace to display the best renderings possible. All of these factors combine to produce the unparalleled value of graphics technologies in modern society and into the future.

Richard M. Renneboog, M.Sc.

FURTHER READING

Abrash, Michael. *Michael Abrash's Graphics Programming Black Book*. Albany, N.Y.: Coriolis Group, 1997. Consisting of seventy short chapters, the book covers graphics programming up to the time of the Pentium processor.

Brown, Eric. "True Physics." *Game Developer* 17, no. 5 (May, 2010): 13-18. This article provides a clear explanation of the relationship between real-world physics and the motion of objects in animation.

Jimenez, Jorge, et al. "Destroy All Jaggies." *Game Developer* 18, no. 6 (June/July, 2011): 13-20. This article describes the method of anti-aliasing in animation graphics.

Oliver, Dick, et al. *Tricks of the Graphics Gurus*. Carmel, Ind.: Sams, 1993. Provides detailed explanations of the mathematics of several computer graphics processes.

Ryan, Dan. *History of Computer Graphics: DLR Associates Series*. Bloomington, Ind.: AuthorHouse, 2011. A comprehensive history of computer graphics and graphics technologies.

See Also: Applied Mathematics; Applied Physics; Cinematography; Communication; Computer-Aided Design and Manufacturing; Computer Engineering; Computer Graphics; Computer Languages, Compilers, and Tools; Computer Science; Internet and Web Engineering; Kinesiology; Lithography; Photography; Software Engineering; Television Technologies; Typography; Video Game Design and Programming; Virtual Reality.

HAZARDOUS-WASTE DISPOSAL

FIELDS OF STUDY

Biology; chemistry; sanitary and sewage engineering; emergency response; environmental science; geology; hydrology; law; toxicology.

SUMMARY

Hazardous-waste disposal involves transporting hazardous materials to appropriate facilities for treatment, recycling, and possible storage in a manner that will protect the environment and public health. Materials that are radioactive, toxic, corrosive, ignitable, irritating, poisonous, and infectious are examples of hazardous wastes that threaten human health. Moreover, some hazardous wastes, such as mercury, present a danger because they accumulate in the environment when they are improperly managed. Although manufacturing companies are most often thought to be the entities that generate hazard wastes, there is a diversity of generators from farms to nuclear power plants to households.

KEY TERMS AND CONCEPTS

- **Corrosive Wastes:** Acidic wastes that are able to corrode metal storage containers.
- **Generators:** Entities that produce hazardous materials from physical processes or disposal of nonrenewable resources.
- **Hazardous Wastes:** Wastes that exhibit one of four characteristics: ignitability, corrosivity, reactivity, or toxicity.
- **Household Hazardous Wastes:** Consumer products that have the characteristics of hazardous wastes.
- **Ignitable Wastes:** Wastes that can spontaneously combust and have low flash points.
- **Reactive Wastes:** Unstable wastes that can explode or release toxic gases when heated, compressed, or mixed with water.
- **Toxic Materials:** Materials harmful to humans through ingestion or absorption and to the environment through groundwater pollution.
- **Transporter:** Company that removes and transports hazardous wastes for generators.

DEFINITION AND BASIC PRINCIPLES

Hazardous-waste disposal is the cradle-to-grave management of materials that threaten human health and the environment. Hazardous wastes are ubiquitous and their proper disposal is a global issue. In fact, industrialized nations are still involved in cleanup of hazardous wastes that contaminated land and waters before disposal laws were adopted and enforced. One method of recycling hazardous waste sites is to clean up the wastes and designate the land as a brownfield site that can be reused by commercial and industrial enterprises.

Arriving at effective methods for disposal of hazardous wastes is difficult, because these wastes come in many forms, including solid, liquid, sludge, and gas. Governments regulate hazardous wastes and require their disposal in designated facilities. Hazardous wastes can be stored in sealed containers in the ground or recycled into new products such as fertilizers. The methods used and locations for disposing of hazardous wastes are not the same as those employed in solid waste disposal. Nonhazardous solid wastes or trash can be disposed of in nonsecure landfills, while hazardous wastes should be deposited in secure landfills so they do not contaminate the groundwater.

BACKGROUND AND HISTORY

Disposal of wastes has existed since ancient cultures burned, recycled, and buried their trash. After World War II, toxic materials were manufactured in greater quantities, and industrialized nations allowed uncontrolled dumping, burning, and disposal of hazardous wastes until the public demanded regulation because of human health

Barrels of hazardous chemical waste are loaded into an underground missile silo near Grandview, Idaho. (David R. Frazier/Photo Researchers, Inc.)

and environmental threats. In the United States, Congress charged the Environmental Protection Agency (EPA) with regulating hazardous-waste disposal through the Resource Conservation and Recovery Act (RCRA) of 1976, which amended the Solid Waste Disposal Act of 1965. The EPA is authorized to control hazardous wastes from cradle to grave under RCRA, which means regulating generators, transporters, and the facilities involved in treatment, recycling, storage, and disposal.

From the 1940's through the 1960's research showed a relationship between hazardous-waste disposal and contamination of groundwater. By employing hydrology studies of underground water movement, scientists were able to pinpoint the sources of hazardous waste contamination. In 1980,

the EPA established the Superfund program for the clean up of sites contaminated with hazardous wastes.

How It Works

Manifest Disposal System. In the United States, hazardous-waste disposal is governed by the RCRA manifest system. Generators must prepare a form that discloses the type and quality of wastes to be transported to an off-site facility for treatment, storage, recycling, or disposal subject to the Hazardous Materials Transportation Act (HMTA). Because the generator remains liable for proper disposal, a copy of the manifest form, signed by all handlers of the hazardous wastes, is returned to the generator to verify its delivery. However, some wastes are not transported but are treated and disposed of on-site.

Disposal methods for hazardous wastes vary depending on the dangers posed by the wastes. The goal is to remove the characteristics of the wastes that are harmful to human health and the environment. Some methods include treatment, disposal in secure landfills, and incineration.

Treatment. Treatment involves neutralization of chemical, biological, and physical characteristics that make a waste hazardous to the environment. For example, acids that cause corrosion can be neutralized with basic substances, reagents, or through pH adjustment. Cement has also been used to decrease toxicity of some hazardous wastes through stabilization, often those found in sludge. Industrial hazardous wastes may receive either chemical treatments, such as chlorination, oxidation, and chemical bonding, or the application of physical techniques not limited to distillation and filtration. Medical wastes are considered biohazards that can cause the spread of disease and present different treatment issues. Treatment techniques for these biological wastes include steam sterilization and chemical decontamination before disposal.

Landfill Disposal. Hazardous materials that are not stored on the surface can be disposed of in hazardous-waste landfills. The materials must first be sequestered from nonhazardous wastes and be treated. Moreover, because hazardous wastes can interact with each other they must be segregated and stored by type before they are disposed of in a landfill. The hazardous wastes are then placed in secure storage containers that are buried in landfills with plastic and clay liners that are thicker than the ones used in solid-waste landfills to prevent leaching into the groundwater.

Incineration. Incinerating some hazardous wastes, such as oils and solvents, at high temperatures will reduce the amount of hazardous material through destruction. Although gases released through incineration can generate energy, incinerating facilities must comply with the Clean Air Act. Starved air incineration, which controls the combustion rate of hazardous wastes, is one of the newer technologies that aids in reducing the amount of air pollution caused by incineration. Another treatment method known as pyrolysis may be used to destroy concentrated organic wastes such as polychlorinated biphenyls (PCBs) and pesticides. Pyrolysis is different from high-temperature incineration in that it employs ultrahigh temperatures in the absence of oxygen, often under pressure.

APPLICATIONS AND PRODUCTS

Energy. Hazardous wastes such as solvents or used oil are often burned directly to produce heat or electricity. This disposal application is regulated throughout the world, and in the United States, the EPA requires the combustion units employed in the burning of hazardous materials to comply with specific standards that prevent harmful air pollution. Some companies, however, have developed proprietary processes that treat the hazardous materials using plasma-enhanced waste-recovery systems that turn hazardous wastes into nontoxic, synthetic, gas-alternative energy sources. Unlike direct burning, these new applications claim to destroy all hazardous components that might pollute the air.

Storage. Some hazardous wastes are disposed of through storage. Storage facilities must be able to withstand natural disasters such as seismic events. Storage drums and containers must be secure from fire and water intrusion, and an entire industry has developed to provide appropriate containers that will not leak into the groundwater or explode on impact. Hazardous wastes must first be characterized before they are disposed of in storage containers. Containers used to store hazardous wastes must be designed to accommodate diverse characteristics such as toxicity, corrosivity, and ignitability, and the stored materials should not react with the container or be mixed with other incompatible materials. Storage containers must be sealed and appropriately labeled

with the contents and characteristics of the stored hazardous wastes, and a permit is required in the United States to allow for permanent storage. Liquid hazardous wastes, however, are often disposed of in underground injection wells. The EPA regulates the construction, operations, and closure of such wells, which are also subject to regulation under the Safe Drinking Water Act.

Remediation. Many companies specialize in hazardous-waste remediation by providing a variety of disposal services and equipment, not limited to catalytic oxidizers and carbon adsorption systems. Remediation includes reduction and cleanup of hazardous wastewater by using oil-water separators to separate oil and solids from wastewater effluents of petroleum-based industries. Other hazardous-waste remediation services include demolition and removal of hazardous materials such as asbestos, removal of leaky fuel tanks, cleanup of contaminated soils, and the construction of slurry walls to aid in the remediation of groundwater that is polluted with hazardous waste. Bioremediation is also a growing field that involves disposal of medical and laboratory wastes often by using microorganisms to break down the hazardous materials.

Recycling and Reclamation. Recycling and reclamation are important disposal applications because they help reduce hazardous wastes and the amount of raw materials that are consumed. Some materials that may be classified as hazardous wastes can be recycled and used again, such as rechargeable batteries and heating and air-conditioning thermostats. Spent solvents are an example of a product that can be made pure and reused for the same purpose, as can used oil, which can also be processed to create new oil-based products. In addition, the EPA allows some hazardous materials to be treated, recycled, and disposed of on the land in fertilizers or as an ingredient in asphalt. Components of hazardous wastes can also be reclaimed for use in new products. Examples include mercury reclaimed from thermometers, silver from photographic fixers, and scrap metal from auto bodies or manufacturing processes.

Telephones, televisions, audiovisual equipment, computers, computer components, circuit boards, and handheld devices that store music and books are all examples of products that are disposed of by businesses and individual consumers as electronic wastes (e-wastes) when they become obsolete. E-wastes,

Fascinating Facts About Hazardous-Waste Disposal

- Annual hazardous-waste generation in the United States could fill the Louisiana Superdome more than 1,500 times.

- Estimates of annual e-waste disposal include more than 100 million phones in European nations and 30 million computers in the United States. These wastes contain hazardous materials such as lead, cadmium, mercury, and chromium. Computer monitors contain about four to five pounds of lead.

- The Environmental Protection Agency (EPA) has estimated that about 1.6 million tons of household hazardous wastes are generated in America annually, and include paints, pesticides, cleaning products, and used motor oil.

- Billions of batteries are thrown away each year, resulting in a majority of the mercury and cadmium in public landfills.

- When household hazardous-waste products are poured down sinks into storm sewers or onto the ground, it takes very little of the wastes to contaminate groundwater and soil.

- A single gallon of motor oil can contaminate 250 gallons of drinking water.

- About 5 million tons of oil are dumped or spilled into the world's oceans each year.

- Golf courses generate many hazardous wastes including pesticides, herbicides, and gasoline and engine oil.

however, contain hazardous materials such as lead, cadmium, and some precious metals that can be reclaimed and reused. Recycling of e-wastes is an important hazardous-waste disposal application throughout the developed world to mitigate the growth of this type of waste.

Hazardous Material (Hazmat) Products. Hazardous-waste disposal is not always carried out effectively, so an entire industry has grown out of spills that need to be cleaned up. Hazmat team members will wear differing types of hazmat suits, often with ventilators, depending on the characteristics of the hazardous material to be cleaned up. Specially manufactured boots, gloves, and socks are also part of the necessary clothing. Those involved in hazmat-cleanup applications will use spill kits that enable first

responders to determine whether the hazardous material presents a chemical, biological, or radioactive hazard. These products are employed in many situations that include not only hazardous-material spills, but also disposal of illegal drugs and abandoned storage drums and cleanup of any dump site that contains suspicious materials. After the hazardous wastes have been cleaned up and properly disposed of, decontamination is necessary. Decontamination products include temporary shelters, showers, and cleaning solutions.

IMPACT ON INDUSTRY

Although there are global initiatives to ensure adequate hazardous-waste disposal, the United States has some of the strictest laws dealing with the subject, while developing nations such as China, are struggling with environmental damage due to lack of environmental laws and improper disposal of hazardous materials.

Government Agencies. In the United States, both the EPA and the Nuclear Regulatory Commission (NRC) play significant roles in enforcing laws pertaining to hazardous-waste disposal. The NRC is instrumental in regulating radioactive-waste disposal. Each state has its own agency to regulate hazardous-waste disposal, and municipalities throughout the country are involved not only with regulation, but also with many hazardous-waste disposal activities including collection and recycling. The Occupational Safety and Health Administration (OSHA) is involved in emergency preparedness and response to hazardous-waste accidents. In addition, the Departments of Defense and Energy play a role in ensuring legal disposal of hazardous materials.

University Research. During the 1990's, when the EPA was involved in cleanup of contaminated sites throughout the United States, universities were actively involved in developing new techniques for effective hazardous-waste disposal and aiding in remediation of polluted sites. As of 2011, universities have established hazardous-waste research facilities, but much of the research is concerned with RCRA's mission of mitigating hazardous wastes and better recycling of wastes and reusable components to reduce hazardous-waste disposal in landfills. Moreover, educational institutions are working with local governments to educate the public in the proper recycling of household hazardous wastes.

International Organizations and Treaties. The United Nations is involved in treaties concerning the international disposal of hazardous materials and certifying containers that may be used in the transportation of hazardous wastes by airplanes, ground vehicles, and ships. The United States is one of several nations that disposes of some of its hazardous waste in other countries. These countries usually welcome the additional revenue that compensates for such disposal, but they have little expertise in proper disposal techniques. Some nations have entered into treaties concerning hazardous-waste imports and exports to prevent developed nations from taking advantage of some of the underdeveloped countries. One of these agreements is known as the Bamako Convention, which bans the import of hazardous wastes into Africa and regulates trans-boundary movement of hazardous wastes within Africa. The Basel Convention, which preceded the Bamako Convention, also regulates hazardous-waste imports and exports but not radioactive wastes, which are often exported to the less developed countries.

Generators. There are many types of businesses and industries that generate hazardous wastes. These include manufacturers, oil refineries, professional offices, commercial facilities such as dry cleaners, service industries including beauty salons, automobile repair shops, and exterminators and medical facilities, hospitals, and laboratories. Based on the RCRA manifest system, these generators have cradle-to-grave responsibility for the proper disposal of the hazardous wastes in the United States.

Transporters. Hazardous-waste transportation is a growing business. In the United States, the Hazardous Materials Transportation Act (HMTA) requires strict compliance with federal laws and applies not only to transporters of hazardous materials but also to generators who engage the services of such transportation companies. Transporters must be issued an identification number and are required to use the RCRA manifest system. The Department of Transportation is involved in ensuring compliance with the HMTA.

CAREERS AND COURSE WORK

Hazardous-waste disposal involves interdisciplinary course work such as engineering courses concerning the characteristics and sources of hazardous wastes and their treatment, destruction, and recycling; biology and chemistry classes, especially related to human-health issues and the environment; and law and economics studies that consider disposal regulations and their costs. Certifications that may be necessary for a career in this field include environmental hazardous materials technology and emergency response.

Career opportunities are growing in this field including positions with agencies at all levels of government and with waste-management firms and engineering companies involved in transporting, treating, recycling, and disposing of hazardous wastes and remediating and cleaning up sites where there has been contamination. Hazardous-waste generators, such as utility companies, hospitals, and manufacturers, often hire in-house hazardous-waste experts. Educational institutions and laboratories also employ educators and researchers with doctoral degrees in engineering and hazardous-waste management.

Those interested in careers as chemical, environmental, sanitary, and sewage engineers will need at least a bachelor's degree to obtain a job as a hazardous-waste management specialist. Careers that require a minimum of an associate's degree and some additional training and certifications include workers involved in collecting, transporting, treating, and destroying hazardous wastes, in addition to emergency response and cleanup. Common job titles for hazardous-waste workers are hazardous-waste technician, field technician, and environmental technician.

SOCIAL CONTEXT AND FUTURE PROSPECTS

Nations must continue to strive for hazardous-waste disposal that proactively reduces the amount of hazardous wastes and, where possible, recycles these wastes to conserve the consumption of raw materials. Although globally some progress has been made in ensuring proper hazardous-waste disposal, the funds needed for education of the public and cleanup of improperly managed hazardous-waste sites are not always available. International cooperation is also necessary to resolve several hazardous-waste disposal issues. Among these matters are the disposal of radioactive and nuclear wastes and the exporting of hazardous materials by developed nations to underdeveloped countries that are incapable of proper disposal. As of 2011, it is estimated that less than one-quarter of global e-wastes are recycled. Appropriate laws, disposal systems, and international treaties are needed to manage these rapidly growing e-wastes, which are mostly disposed of in landfills.

Disposal of household hazardous wastes (HHWs), such as used batteries, cleaning products, and medications, continues to be problematic because they contaminate groundwater when disposed of in landfills. Although the United States federal government exempts HHWs from regulation, some municipalities require the separation of household hazardous wastes from other solid wastes for collection or disposal at a municipal facility. Better public education and enhanced laws will help mitigate the damage caused by unregulated HHW disposal.

Carol A. Rolf, B.S.L.A., M.B.A., M.Ed., J.D.

FURTHER READING

Blackman, William C., Jr. *Basic Hazardous Waste Management.* 3d ed. Boca Raton, Fla.: CRC Press, 2001. Overview of hazardous-waste management technologies with discussion concerning disposal of radioactive and biomedical wastes.

Cabaniss, Amy D., ed. *Handbook on Household Hazardous Waste.* Lanham, Md.: Government Institutes, 2008. Comprehensive discussion concerning household hazardous-waste disposal that encourages responsible disposal.

LaGrega, Michael D., Philip L. Buckingham, and Jeffrey C. Evans. 2d ed. *Hazardous Waste Management.* Long Grove, Ill.: Waveland Press, 2010. Text written for students interested in resolving hazardous-waste problems based on case studies and treatment solutions.

Miller, Jeffrey G., and Craig N. Johnston. *Law of Hazardous Waste Disposal and Remediation: Cases, Legislation, Regulations, Policies.* 2d ed. St. Paul, Minn.: Thomson/West, 2005. Introductory text discussing federal and state laws pertaining to hazardous-waste disposal and cleanup.

Spellman, Frank. *Transportation of Hazardous Materials Post-9/11.* Lanham, Md.: Rowman & Littlefield, 2007. Useful information concerning current hazardous materials, transportation regulations, and management of security threats.

Voyles, James K. *Managing Your Hazardous Wastes: A Step-by-Step RCRA Compliance Guide.* 2d ed. Rockville, Md.: Government Institutes, 2002. Guide for developing management plans that comply with laws regulating the generation and processing of hazardous wastes.

WEB SITES
Environmental Protection Agency
Wastes–Hazardous Wastes
http://www.epa.gov/epawaste/hazard/index.htm

Nuclear Regulatory Commission
Radioactive Waste
http://www.nrc.gov/waste.html

Occupational Safety and Health Administration
Safety and Health Topics: Hazardous Waste
http://www.osha.gov/SLTC/hazardouswaste/index.html

Status Clean
Hazardous Wastes
http://www.statusclean.com/waste-management/waste-disposal/hazardous-waste.aspx

See also: Hydrology and Hydrogeology; Sanitary Engineering; Sewage Engineering; Toxicology.

HEAT-EXCHANGER TECHNOLOGIES

FIELDS OF STUDY

Thermodynamics; fluid mechanics; heat transfer; mechanics of materials; calculus; differential equations.

SUMMARY

A heat exchanger transfers thermal energy from one flowing fluid to another. A car radiator transfers thermal energy from the engine-cooling water to the atmosphere. A nuclear power plant contains very large heat exchangers called steam generators, which transfer thermal energy out of the water that circulates through the reactor core and makes steam that drives the turbines. In some heat exchangers, the two fluids mix together, while in others, the fluids are separated by a solid surface such as a tube wall.

KEY TERMS AND CONCEPTS

- **Closed Heat Exchanger:** Exchanger in which the two fluids are separated by a solid surface.
- **Counterflow:** Fluids flowing in opposite directions.
- **Cross-Flow:** Fluids flowing perpendicular to each other.
- **Deaerator (Direct-Contact Heater):** Open heat exchanger in which steam is mixed with water to bring the water to its boiling point.
- **Open Heat Exchanger:** Exchanger in which the two fluids mix together and exit as a single stream.
- **Parallel Flow (Co-Current):** Both fluids flowing in the same direction.
- **Plate Heat Exchanger:** Exchanger in which corrugated flat metal sheets separate the two flowing fluids.
- **Shell:** Relatively large, usually cylindrical, enclosure with many small tubes running through it.
- **Shell And Tube Heat Exchanger:** Exchanger composed of many relatively small tubes running through a much larger enclosure called a shell.
- **Tube Sheet:** Flat plate with many holes drilled in it. Tubes are inserted through the holes.

DEFINITION AND BASIC PRINCIPLES

There are three modes of heat transfer. Conduction is the method of heat transfer within a solid. In closed heat exchangers, conduction is how thermal energy moves through the solid boundary that separates the two fluids. Convection is the method for transferring heat between a fluid and a solid surface. In a heat exchanger, heat moves out of the hotter fluid into the solid boundary by convection. That is also the way it moves from the solid boundary into the cooler fluid. The final mode of heat transfer is radiation. This is how the energy from the Sun is transmitted through space to Earth.

The simplest type of closed heat exchanger is composed of a small tube running inside a larger one. One fluid flows through the inner tube, while the other fluid flows in the annular space between the inner and outer tubes. In most applications, a double-pipe heat exchanger would be very long and narrow. It is usually more appropriate to make the outer tube large in diameter and have many small tubes inside it. Such a device is called a shell and tube heat exchanger. When one of the fluids is a gas, fins are often added to heat-exchanger tubes on the gas side, which is usually on the outside.

Plate heat exchangers consist of many thin sheets of metal. One fluid flows across one side of each sheet, and the other fluid flows across the other side.

BACKGROUND AND HISTORY

Boilers were probably the first important heat exchangers. One of the first documented boilers was invented by Hero of Alexandria in the first century. This device included a crude steam turbine, but Hero's engine was little more than a toy. The first truly useful boiler may have been invented by the Marquess of Worcester in about 1663. His boiler provided steam to drive a water pump. Further developments were made by British engineer Thomas Savery and British blacksmith Thomas Newcomen, though many people mistakenly believe that James Watt invented the steam engine. Watt invented the condenser, another kind of heat exchanger. Combining the condenser with existing engines made them much more efficient. Until the late nineteenth century boilers and condensers dominated the heat-exchanger scene.

The invention of the diesel engine in 1897 by Bavarian engineer Rudolf Diesel gave rise to the need for other heat exchangers: lubricating oil coolers, radiators, and fuel oil heaters.

During the twentieth century, heat exchangers grew rapidly in number, size, and variety. Plate heat exchangers were invented. The huge steam generators used in nuclear plants were produced. Highly specialized heat exchangers were developed for use in spacecraft.

How It Works

Thermal Calculations. There are two kinds of thermal calculations–design calculations and rating calculations. In design calculations, engineers know what rate of heat transfer is needed in a particular application. The dimensions of the heat exchanger that will satisfy the need must be determined. In rating calculations, an existing heat exchanger is to be used in a new situation. The rate of heat transfer that it will provide in this situation must be determined.

In both design and rating analyses, engineers must deal with the following resistances to heat transfer: the convection resistance between the hot fluid and the solid boundary, the conduction resistance of the solid boundary itself, and the convection resistance between the solid boundary and the cold fluid. The conduction resistance is easy to calculate. It depends only on the thickness of the boundary and the thermal conductivity of the boundary material. Calculation of the convection resistances is much more complicated. They depend on the velocities of the

fluid flows and on the properties of the fluids such as viscosity, thermal conductivity, heat capacity, and density. The geometries of the flow passages are also a factor.

Because the convection resistances are so complicated, they are usually determined by empirical methods. This means that research engineers have conducted many experiments, graphed the results, and found equations that represent the lines on their graphs. Other engineers use these graphs or equations to predict the convection resistances in their heat exchangers.

In liquids, the convection resistance is usually low, but in gases, it is usually high. In order to compensate for high convection resistance, fins are often installed on the gas side of heat-exchanger tubes. This can increase the amount of heat transfer area on the gas side by a factor of ten without significantly increasing the overall size of the heat exchanger.

Hydraulic Calculations. Fluid friction and turbulence within a heat exchanger cause the exit pressure of each fluid to be lower than its entrance pressure. It is desirable to minimize this pressure drop. If a fluid is made to flow by a pump, increased pressure drop will require more pumping power. As with convection resistance, pressure drop depends on many factors and is difficult to predict accurately. Empirical methods are again used. Generally, design changes that reduce the convection resistance will increase the pressure drop, so engineers must reach a compromise between these issues.

Strength Calculations. The pressure of the fluid flowing inside the tubes is often significantly different from the pressure of the fluid flowing around the outside. Engineers must ensure that the tubes are strong enough to withstand this pressure difference so the tubes do not burst. Similarly, the pressure of the fluid in the shell (outside the tubes) is often significantly different from the atmospheric pressure outside the shell. The shell must be strong enough to withstand this.

Fouling. In many applications, one or both fluids may cause corrosion of heat-exchanger tubes, and

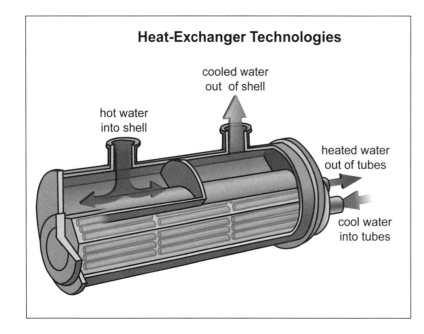

Heat-Exchanger Technologies

cooled water out of shell

hot water into shell

heated water out of tubes

cool water into tubes

they may deposit unwanted material on the tube surfaces. River water may deposit mud. Seawater may deposit barnacles and other biological contamination. The general term for all these things is fouling. The tubes may have a layer of fouling on the inside surface and another one on the outside. In addition to the two convection resistances and the conduction resistance, there may be two fouling resistances. When heat exchangers are designed, a reasonable allowance must be made for these fouling resistances.

APPLICATIONS AND PRODUCTS

Heat exchangers come in an amazing variety of shapes and sizes. They are used with fluids ranging from liquid metals to water and air. A home with hot-water heat has a heat exchanger in each room. They are called radiators, but they rely on convection, not radiation. A room air conditioner has two heat exchangers in it. One transfers heat from room air to the refrigerant, and the other transfers heat from the refrigerant to outside air. Cars with water-cooled engines have a heat exchanger to get rid of engine heat. It is called a radiator, but again it relies on convection.

Boilers. Boilers come in two basic types: fire tube and water tube. In both cases, the heat transfer is between the hot gases produced by combustion of fuel and water that is turning to steam. As the name suggests, a fire-tube boiler has very hot gas, not actually fire, inside the tubes. These tubes are submerged in water, which absorbs heat and turns into steam. In a water-tube boiler, water goes inside the tubes and hot gases pass around them. Water-tube boilers often include superheaters. These heat exchangers allow the steam to flow through tubes that are exposed to the hot combustion gases. As a result the final temperature of the steam may reach 1,000 degrees Fahrenheit or higher. An important and dangerous kind of fouling in boilers is called scale. Scale forms when minerals in the water come out of solution and form a layer of fouling on the hot tube surface. Scale is dangerous because it causes the tube metal behind it to get hotter. In high-performance boilers, this can cause a tube to overheat and burst.

Condensers. Many electric plants have generators driven by steam turbines. As steam leaves a turbine it is transformed back into liquid water in a condenser. This increases the efficiency of the system, and it recovers the mineral-free water for reuse. A typical condenser has thousands of tubes with cooling water

flowing inside them. Steam flows around the outsides, transfers heat, and turns back into liquid water. The cooling water may come from a river or ocean. When a source of a large amount of water is not available, cooling water may be recirculated through a cooling tower. Hot cooling water leaving the condenser is sprayed into a stream of atmospheric air. Some of it evaporates, which lowers the temperature of the remaining water. This remaining water can be reused as cooling water in the condenser. The water that evaporates must be replaced from some source, but the quantity of new water needed is much less than when cooling water is used only once. When river water or seawater is used for cooling, there may be significant fouling on the insides of the tubes. Because the steam leaving a turbine contains very small droplets of water moving at high speed, erosion on the outsides of the tubes is a problem. Eventually a hole may develop, and cooling water, which contains dissolved minerals, can leak into the condensing steam.

Steam Generators. In a pressurized-water nuclear power plant, there is a huge heat exchanger called a steam generator. The primary loop contains water under high pressure that circulates through the reactor core. This water, which does not boil, then moves to the steam generator, where it flows inside a large number of tubes. Secondary water at lower pressure surrounds these tubes. As the secondary water absorbs heat it turns into steam. This steam is used to drive the turbines. Steam generators are among the largest heat exchangers in existence.

Deaerators. Because the condensers in steam systems operate with internal pressures below one atmosphere, air may leak in. Some of this air dissolves in the water that forms as steam condenses. If this air remained in the water as it reached the boiler, rapid rusting of boiler surfaces would result. To prevent this, an open heat exchanger, called a deaerator, is installed. Water is sprayed into the deaerator as a fine mist, and steam is also admitted. As the steam and water mix, the water droplets are heated to their boiling point but not actually boiled. The solubility of air in water goes to zero as the water temperature approaches the boiling point, so nearly all air is forced out of solution in the deaerator. Once the air is in gaseous form, it is removed from the system.

Feedwater Heaters. Leaving the deaerator, the water in a steam plant is on its way to the boiler. The system can be made more efficient by preheating the water along the way. This is done in a feedwater heater. Steam is extracted from the steam turbines to serve as the heat source in feedwater heaters. Feedwater flows inside the tubes and steam flows around the tubes. Feedwater heaters are often multi-pass heat exchangers. This means that the feedwater passes back and forth through the heat exchanger several times. This makes the heat exchanger shorter and fatter, which is a more convenient shape.

Intercoolers. Many diesel engines have turbochargers that pressurize the air being fed to the cylinders. As air is compressed, its temperature rises. It is desirable to lower the temperature of the air before it enters the cylinders, because that means a greater mass of air can occupy the same space. More air in the cylinder means more fuel can be burned, and more power can be produced. An intercooler is a closed heat exchanger between the turbocharger and the engine cylinders. In this device, air passes around the tubes, and cooling water passes inside them. There are usually fins on the outsides of the tubes to provide increased heat transfer area.

Industrial air compressors also have intercoolers. These compressors are two-stage machines. That means air is compressed part way in one cylinder and the rest of the way in another. As with turbochargers, the first compression raises the air temperature. An intercooler is often installed between the cylinders. Compressed air flows through the intercooler tubes. Either atmospheric air or cooling water flows around the outside. Cooling the air before the second compression reduces the power required there.

IMPACT ON INDUSTRY

By 1880 the use of boilers was widespread, but there was little or no regulation of their design and construction. As a result, boiler explosions were commonplace, and many lives were lost in that way every year. The American Society of Mechanical Engineers (ASME) was founded that year to develop safety standards for boilers.

Early in the twenty-first century, an industry group called Tubular Exchanger Manufacturers Association (TEMA) published standards for shell-and-tube heat exchangers. This organization certifies heat exchangers that are built in accordance with its standards. Purchasers of heat exchangers with TEMA certification are assured that these devices are properly designed and safe to operate. More than twenty manufacturers are members of TEMA. Standards for shell-and-tube heat exchangers are also published by ASME, Heat Transfer Research, and the American Petroleum Institute.

Manufacturers of small heat exchangers include Exergy LLC and Springboard Manufacturing. These heat exchangers are small enough to hold in one's hand. Basco/Whitlock manufactures medium-size shell-and-tube heat exchangers from two inches to eight inches in diameter and eight inches to ninety-six inches in length. The shells are made of brass or stainless steel, and the tubes are made of copper, Admiralty metal (an alloy of copper, zinc, and tin), or stainless steel.

Westinghouse, Combustion Engineering, and Babcock & Wilcox have built many of the huge steam generators used in U.S. nuclear power plants. Heat Transfer Equipment Company manufactures large steel heat exchangers. It specializes in heat exchangers for the oil and gas industry. In Japan, Mitsubishi Heavy Industries has manufactured about one hundred steam generators for use in nuclear power plants there.

The Electric Power Research Institute, an organization composed of companies that generate electricity, funds and conducts research on all aspects of electric power generation, including the heat exchangers that are involved. Much of the basic research on convection resistance in heat exchangers was conducted in university laboratories during the middle of the twentieth century. Although this remains an active research area, the level of activity is lower in the early twenty-first century.

CAREERS AND COURSE WORK

Heat exchangers are usually designed by mechanical engineers who hold bachelor or master of science degrees in this field. Students of mechanical engineering take courses in advanced mathematics, mechanics of materials, thermodynamics, fluid mechanics, and heat transfer. An M.S. degree provides advanced understanding of the physical phenomena involved in heat exchangers. Research into the theory of heat transfer is normally carried out by mechanical engineers with doctoral degrees. They conduct research in laboratories, at universities, private research companies, or large corporations that build heat exchangers. As mentioned earlier, convection heat transfer calculations rely on equations that are

---

Fascinating Facts About Heat-Exchanger Technologies

- Heat exchangers that have seawater flowing through them usually have small zinc blocks in the flow. These blocks help protect the exchanger surfaces from corrosion.

- Steam condensers operate below atmospheric pressure. Any leakage that occurs is air going in rather than steam coming out. These heat exchangers have air-removal devices called air ejectors connected to them.

- Heat exchanger tubes are usually attached to tube sheets by "rolling," a process that expands the tube to make it fit tightly in its hole in the tube sheet.

- Shell and tube heat exchangers have baffle plates that direct the fluid flowing over the outsides of the tubes. The baffle plates also help support the tubes.

- Electronic sensing devices called eddy-current sensors are passed through the tubes of a shell heat exchanger to detect potential problems before tubes fail.

- Steam generators in nuclear power plants can be as tall as 70 feet and weigh as much as 800 tons when empty. These are among the largest heat exchangers in existence.

- The slang "crud" is said to be an acronym of Chalk River unidentified deposits, which refers to radioactive fouling found in the steam generators of the Chalk River Laboratories in Canada.

- Radiators in cars and home-heating systems make use of convection heat transfer, which refers to the transfer of heat into a fluid. In both cases, heat is transferred to air. They do not make significant use of radiation heat transfer.

- Heat exchangers in spacecraft make use of radiation heat transfer, because there is no air or water to absorb the heat.

---

derived from extensive experiments. Much research work continues to be devoted to improving the accuracy of these equations.

Construction of heat exchangers is executed by companies large and small. The work is carried out by skilled craftsmen using precise machine tools and other equipment. Machinists, welders, sheet-metal, and other highly trained workers are involved. Students who pursue such careers may begin with vocational-technical training at the high school level. They become apprentices to one of these trades. During apprenticeship, the workers receive formal training in classrooms and on-the-job training. As their skills develop they become journeymen and then master mechanics.

Workers who operate, maintain, and repair heat exchangers have a variety of backgrounds. Some have engineering or engineering technology degrees. Others have vocational-technical and on-the-job training. At nuclear power plants, the Nuclear Regulatory Commission requires a program of extensive testing of the vital heat exchangers. This is carried out by engineers with B.S. or M.S. degrees, assisted by skilled craftsmen.

SOCIAL CONTEXT AND FUTURE PROSPECTS

Although heat exchangers are not glamorous, they are an essential part of people's lives. Every home has several, as does every car and truck. Without heat exchangers, people would still be heating their homes with fireplaces, and engines of all types and sizes would not be possible. Heat exchangers are essential in all manner of industries. In particular, they play a key role in the generation of electricity.

The design of heat exchangers is based on empirical methods rather than basic principles. While empirical methods are reasonably effective, design from basic principles would be preferred. In the early twenty-first century, extensive research projects are under way with the goal of solving the very complicated equations that represent the basic principles of heat transfer. These projects make use of very powerful computers. As the cost of computers continues to drop and their power continues to increase, heat exchangers may come to be designed from basic principles.

Edwin G. Wiggins, B.S., M.S., Ph.D.

FURTHER READING

Babcock and Wilcox Company. *Steam: Its Generation and Use*. Reprint. Whitefish, Mont.: Kessinger Publishing, 2010. This easy-to-read book provides extensive information about all manner of fossil-fueled boilers as well as nuclear steam generators.

Blank, David A., Arthur E. Bock, and David J. Richardson. *Introduction to Naval Engineering*. 2d ed. Annapolis, Md.: Naval Institute Press, 2005. Intended for use by freshmen at the U.S. Naval Academy, this book provides a simple, nonmathematical discussion of heat exchangers used on naval ships.

McGeorge, H. D. *Marine Auxiliary Machinery*. 7th ed. Burlington, Mass.: Butterworth-Heinemann, 1995. Excellent descriptions of shell-and-tube heat exchangers, plate heat exchangers, and deaerators are provided in very readable form, including a simple presentation of heat exchanger theory.

Thurston, Robert Henry. *A History of the Growth of the Steam-Engine*. Ithaca, N.Y.: Cornell University Press, 1939. Although very old, this book is useful because it provides a comprehensive history of the development of steam engines, boilers, and condensers.

WEB SITES
American Petroleum Institute
http://www.api.org

American Society of Mechanical Engineers
http://www.asme.org

Electric Power Research Institute
http://my.epri.com/portal/server.pt?

Heat Transfer Research, Inc.
http://www.htri.net

Tubular Exchanger Manufacturers Association
http://tema.org

See also: Calculus; Fluid Dynamics; Mechanical Engineering.

HEAT PUMPS

FIELDS OF STUDY

Civil engineering; mechanical engineering; renewable-energy engineering; physics; geology; environmental science; heating, ventilation, and air-conditioning (HVAC); refrigeration; thermodynamics.

SUMMARY

A heat pump is a thermal energy exchanger that transfers energy from one medium to another in order to provide heating or cooling. This technology is rapidly developing as a renewable-energy technology to provide moderate space heating and cooling or water heating in residential, commercial, and institutional settings. The benefit of the heat-pump design over conventional energy systems is that it condenses and uses existing thermal energy, rather than fossil fuels, to generate new thermal energy. As a renewable-energy technology, heat pumps rely on the relatively constant temperature just below the Earth's surface to act as a source or sink for thermal energy.

KEY TERMS AND CONCEPTS

- **Closed-Loop Subsystem:** Earth connection subsystem in which a fluid, typically water or a refrigerant, is circulated continuously through a system and not released.
- **Coefficient Of Performance (COP):** Measure of energy performance applied to electric, vapor-compression heat pumps that describes the number of kilowatts of thermal energy output per kilowatt electricity input.
- **Earth Connection Subsystem:** Component of a ground source heat-pump system that collects thermal energy from or distributes thermal energy to the Earth below its surface; also known as a "loop."
- **Geoexchange:** Refers to ground-source heat-pump systems.
- **Heat Distribution Subsystem:** Component of a ground-source heat-pump system that collects thermal energy from or distributes thermal energy to the inside of a building.
- **Open-Loop Subsystem:** Earth-connection subsystem in which a fluid, typically water, is drawn in, circulated, and released.
- **Temperature Lift:** Increase in temperature achieved by a heat pump; measures the difference between the temperatures inside the evaporator and condenser.

DEFINITION AND BASIC PRINCIPLES

In the simplest sense, a heat pump takes in thermal energy at a low temperature (the source) and releases it at a higher temperature (the sink), working against the temperature gradient using a small amount of electricity or other energy. This thermal energy is carried between two spaces by a liquid refrigerant. The temperature at which a liquid will change into a gaseous state is related to the atmospheric pressure it is under. For example, water will normally change state from liquid to gas at 212 degrees Fahrenheit (F) at sea level, or under 1 atmosphere of pressure. However, under one-tenth of normal atmospheric pressure water will boil at 98 degrees F.

Thermal energy naturally moves down a temperature gradient, from a warmer medium to a cooler medium, in accordance with the second law of thermodynamics. This law characterizes the trend of the universe toward disorganization, or entropy, and explains how heat is transferred to or from the heat-pump system. While the refrigerant in a heat pump is in the sink, or evaporation stage, it absorbs ambient heat, which turns the low-pressure liquid into a higher-energy gas. This gas is then compressed to a higher pressure, and it releases that energy into the sink and condenses into a high-pressure liquid. These are the basic principles under which heat pumps, refrigerators, and air-conditioning units operate.

BACKGROUND AND HISTORY

The vapor-compression cycle was first used by French engineer Nicolas Léonard Sadi Carnot in 1824. Then in 1832, American inventor Jacob Perkins was the first to demonstrate a compression cooling technology that used ether as a refrigerant. But it was in 1852 that Scottish engineer William Thomson, also known as Lord Kelvin, conceptualized the first heat pump system, dubbed the "heat multiplier."

He recognized that this early predecessor of the air-source heat pump could provide both heating and cooling. Thomson estimated that his machine could generate an equal amount of heat as that produced through direct heating but using only 3 percent of the energy required.

In the early nineteenth century, skeptics maintained that electricity would never be a practical energy source for generating heat in large amounts. Just a handful of scientists looked at the possibility of converting electrical energy to mechanical energy as a means of generating heat, and then pumping that heat from a lower temperature to a higher temperature. But in 1927, British engineer T. G. N. Haldane demonstrated that the heat pump could operate both as a heating and a cooling apparatus using a vapor-compression cycle refrigerator. When it functioned normally, the refrigerator provided cooling, but when operated in reverse, it could provide heat.

HOW IT WORKS

There are two main types of heat pumps: vapor-compression pumps and absorption pumps. Vapor-compression heat pumps are the more common of the two.

Vapor-Compression Pumps. The vapor-compression refrigeration cycle removes heat from the air or water inside a building and enables this energy to be transferred to another medium. Vapor-compression pumps are filled with a heat-conducting refrigerant, such as Freon, and include four main components: a motor, a compressor, an evaporator, and a condenser. The motor uses electricity to draw the refrigerant in its vaporous form from the evaporator through the compressor, where it is compressed to increase pressure and temperature. The vapor then passes on to the condenser for cooling and condensing. At this point, the vapor surrenders much of the thermal energy it contains to the air or liquid it is designed to heat and returns to a liquid phase. The liquid refrigerant then continues through the expansion valve to the evaporator. The expansion valve allows only as much refrigerant to pass through as will vaporize. As this occurs, the pressure of the refrigerant falls rapidly and lowers its boiling point. The lower boiling point allows it to vaporize and accept thermal energy from the air or liquid. The refrigerant then moves on to the compressor to repeat the process.

Absorption Heat Pumps. Absorption heat pumps operate using thermal energy, rather than electrical energy, and function like an air-source heat pump. Just a small amount of electricity may be required by the solution pump. Fossil fuels, solar energy, or another energy source may be used to power the pump. Absorption heat-pump components consist of an absorber, a solution pump, a generator, an evaporator, a condenser, and an expansion valve. While vapor-compression heat pumps use only one fluid, the refrigerant, absorption heat pumps require a working fluid and an absorbent. The working fluid must have the ability to vaporize under operating conditions, and the absorbent must be able to absorb this vapor. Most systems employ either water as the working fluid and lithium bromide as the absorbent or ammonia as the working fluid and water as the absorbent. Inside the evaporator, low-pressure vapor from the working fluid is absorbed by the absorbent, resulting in the production of heat. The solution pump then increases the pressure of the resulting solution, and the heat added to the system by the external heat source causes the working fluid to boil out of the solution. The vaporous working fluid is cooled and condensed back into liquid form in the condenser, and the absorbent is diverted back to the absorber by the expansion valve. As in the vapor-compression heat pump, heat is accepted by the working fluid inside the evaporator and released inside the condenser; however, in an absorption heat pump, heat is also released inside the absorber.

APPLICATIONS AND PRODUCTS

Many people unknowingly use heat pumps every day. The refrigerator is probably the most common example of a heat pump. The refrigerant inside this type of heat pump is selected to have a vaporizing temperature that is below the desired temperature inside the refrigerator, so that heat from inside the refrigerator is captured by the refrigerant following the temperature gradient. The refrigerant flows to the evaporator, where is turns into a gas. But the refrigerant must also have a condensing temperature that is higher than room temperature, so that the refrigerant draws thermal energy away from the condenser and changes to a liquid, releasing the heat into the air surrounding the refrigerator. Water chillers, air conditioners, and dehumidifiers are other common forms of heat pumps that operate similarly. All of these types of heat pumps are designed to allow heat to move in only one direction.

Three kinds of heat-pump systems can be used for indoor space heating and cooling: air source, water source, and ground source. Since they must perform both heating and cooling, these systems are designed both to produce and receive thermal energy.

Air-Source Heat Pumps. Air-source systems may be used in place of central air-conditioning systems, as a means of cooling indoor air. These are typically installed on rooftops or other building areas where access to circulating air is available. In addition to the four main components of a heat pump, air-source heat pumps also have a four-way reversing valve that is capable of reversing the flow of refrigerant through the heat pump while maintaining the same direction of flow through the compressor. The expansion valve in an air-source heat pump is also modified to control bidirectional flow of refrigerant. When it is performing space cooling, an air-source heat pump operates exactly like an air conditioner. Refrigerant

---

Fascinating Facts About Heat Pumps

- When the temperature outside is 50 degrees Fahrenheit (F), the coefficient of performance (COP) of an air-source heat pump is approximately 3.3. When the temperature falls to 17 degrees F, the COP decreases to approximately 2.3, because the heat pump must perform more work to extract thermal energy from colder air.
- Because the outside air temperature in countries such as the United States, Canada, Japan, and Germany is more variable than the temperature underground or in lakes or ponds, ground-source heat pumps can operate in a temperature range closer to their optimal COP and yield the greatest energy savings.
- Electric vapor-compression heat pumps typically have a COP of 2.5 to 5.0, indicating that they produce up to five times as much energy as they require to operate.
- Integrating space heating and cooling and water-heating systems can reduce water-heating costs by up to 50 percent.
- The potential energy cost savings achieved by switching to a heat-pump system depends both on the energy performance of the new system and the fuel source used by the system currently in place.

---

under pressure leaves the compressor at a higher temperature than the outside air, causing heat to follow the temperature gradient from the coil to the outside air as the refrigerant condenses. When the refrigerant passes out of the condenser, it is in a pressurized liquid form, and the expansion valve releases only as much refrigerant as can be vaporized when it receives thermal energy from the indoor air. The pressure of the refrigerant falls as it passes through the expansion valve and its temperature decreases, causing it to receive more heat from the indoor air and vaporize the remaining refrigerant. Once the pressure of the refrigerant has completely fallen, the refrigerant passes through the reversing valve and is directed back to the compressor to repeat the process. If the direction of the reversing valve is switched to allow for heating, the process moves in the opposite direction. Air-source heat pump systems may be connected to water-heating systems to provide hot water during the cooling cycle.

One limitation of air-source heat pump systems is the need for relatively mild winter temperatures, between 25 and 35 degrees F. Otherwise, supplemental heating may be required. One solution may be to include an electric heating device on the air outflow inside the building.

Water-Source Heat Pumps. Water-source systems transfer thermal energy to and from water, rather than air. During the winter, a fossil-fuel-powered water heater produces heat that is condensed by the heat pump and distributed throughout the building by water-filled pipes. In warmer temperatures, the indoor air heats the water in the pipes and the warm water is pumped up to an evaporative cooling tower outside the building, typically located on the roof. The water inside the pipes remains at a steady temperature of between 60 and 90 degrees F.

Since water is more heat conductive than air because of its higher specific heat, these distribution systems require only 25 percent of the energy that would be required for an air-filled duct distribution system, and because water is denser than air, water-filled pipes take up less space inside a building than ducts. For these reasons, water-source systems are typically preferred in large, multistory buildings. Because many such buildings also have sprinkler systems, the distribution pipes may be integrated into the sprinkler system to reduce overall costs.

Ground-Source Heat Pumps. Ground-source heat pumps take advantage of the relatively consistent temperature underground or beneath the surface of bodies of water such as lakes and ponds as a source or sink for thermal energy. Ground-source systems include three major components: the Earth connection subsystem, the heat-distribution system, and the heat pump. The Earth connection subsystem, or "loops," consists of pipes buried approximately six to ten feet below the Earth's surface. Loops may be buried underground in the soil, placed in an underground aquifer, or located in a lake or pond. Loops may be oriented vertically or horizontally, based on the available space and the difficulty associated with vertical drilling or excavation. Earth connection subsystems may be open or closed to the surrounding soil, groundwater, or surface water. The heat-distribution system consists of air-filled ducts or fluid-filled pipes that transport thermal energy through a building, either to or from the heat pump at any given time.

Unlike air-source heat pumps, ground-source heat pumps are not exposed to fluctuating outdoor temperatures because the temperature below the Earth's surface remains relatively consistent year-round. Consequently, ground-source systems may be more appropriate where both heating and cooling is needed and seasonal temperatures fall below 25 degrees F to 35 degrees F. Supplemental heating systems may still be required to maintain desired indoor temperatures during extreme temperatures. Where loops are submerged in ponds or lakes, it is important to ensure that freezing will not occur.

Air cooling and water heating may be achieved simultaneously by connecting the heat-distribution system and heat pump to the hot-water system inside a building. Instead of transferring thermal energy from the heat-distribution system into the ground, this energy can be put to use for water heating. In this way, energy use is being reduced from both space cooling and water heating.

Rising energy prices and awareness about the environmental impacts of energy use have contributed to the popularity of heat-pump systems. Heat-pump systems offer a potentially lower-emission option for space heating and cooling and water heating because of their relative efficiency under suitable conditions and their reliance on electricity as an energy source. Where prevailing climatic conditions are moderate and electricity is a lower-emission energy source than

natural gas, heat-pump systems reduce greenhouse gas emissions from residential, commercial, or industrial buildings. Since the heat pump is using existing thermal energy rather than attempting to generate it from another source, such as fossil fuels, heat-pump systems require energy inputs only to operate the motor and compressor. For this reason, heat pumps require only 20 kilowatts (kW) to 40 kW of electricity to generate 100 kW of thermal energy; some highly efficient heat pumps use even less electricity for a similar heat output.

While it is often easier and more cost effective to incorporate a heat-pump system into a new building design, retrofits can also be tailored to existing site and building characteristics. Electric heat-pump systems tend to be most appropriate for single-family dwellings, while engine-driven systems can be developed for larger condominium-style, commercial, or institutional buildings.

Industrial Applications. Industrial applications of heat pumps are currently limited, but heat pumps may be used for space heating, process heating and cooling, water heating, steam generation, drying, evaporation, distillation, and concentration. Waste heat is typically captured for space heating, and recapture is minimal; however, heat reuse is common in drying, evaporation, and distillation.

Industrial heat pumps may use conventional heating from fossil fuels or waste-heat capture to provide space heating inside greenhouses or other facilities. These systems typically employ vapor-compression heat pumps.

Operations that require hot water for cleaning typically require water temperatures of between 100 degrees F and 190 degrees F. Heat-pump systems for these applications may simultaneously provide space cooling and water heating. Vapor-compression heat pumps are typically employed, but absorption heat pumps may be adapted to these uses. Vapor-compression heat-pump systems can also provide steam at various pressures for process water heating at temperatures between 210 degrees F and 390 degrees F.

Pulp and paper, lumber, and food processing industries may use heat pumps to provide drying and dehumidification at temperatures up to 210 degrees F. In these applications, heat pumps have demonstrated superior performance and product quality compared with conventional drying systems. Drying systems are typically closed and produce fewer odor emissions.

There are several types of heat pumps that are currently feasible for industrial applications, including mechanical vapor-compression systems, closed-cycle mechanical heat pumps, absorption heat pumps, heat transformers, and reverse Brayton-cycle heat pumps.

Heat pumps can provide valuable energy savings and lower costs associated with industrial energy use by capturing waste steam that is at a temperature too low to be useful for heating and increasing its temperature, allowing for reuse. The exact dollar value of energy savings can be calculated based on the temperature lift, the increase in temperature achieved by the heat pump. Achieving a higher lift requires a more powerful pump, which in turn requires a larger amount of energy.

IMPACT ON INDUSTRY

With the wide variety of heat-pump technology applications available around the world, there are countless businesses involved in engineering, manufacturing, and selling heat pumps; and an even greater number that design and install air-source, water-source, and ground-source heat-pump systems. Annual growth of the global market for ground-source heat pumps has grown by at least 10 percent since 2000. The majority of market expansion has occurred in North America and Europe. The United States boasts most of the world's installed ground-source heat-pump capacity, followed by Sweden, Germany, Switzerland, and Canada.

Government and University Research. The International Energy Association (IEA) has several annexes under its heat pump program dedicated specifically to studying and promoting applications of heat-pump systems. Annexes are country-specific activities that occur under the IEA banner. IEA activities focus primarily on ground-source heat pumps and include examination of barriers to heat-pump implementation.

The International Ground-Source Heat Pump Association (IGSHPA) is a nonprofit organization housed within Oklahoma State University in Stillwater. Through its affiliation with the IGSHPA, Oklahoma State University's Building and Environmental Thermal Systems Research Group is involved in considerable research focused on the development, improvement, and application of heat-pump technology.

A number of countries, including Austria, Canada, China, Denmark, Japan, the Netherlands, Norway, Sweden, Switzerland, and the United States, have also established national heat-pump research institutes or government programs supporting heat-pump research.

Natural Resources Canada provides online access to RETScreen International, a standardized renewable energy project analysis software program that features case studies and free ground-source heat-pump project modeling. The Canadian Office of Energy Efficiency provides information about heat-pump applications.

The Heat Pump and Thermal Storage Technology Center of Japan is a national research organization that is linked with the IEA through its research into heat-pump cooling. Its researchers explore thermal storage systems and technologies and actively promote adoption of heat-pump technologies. Japan's New Energy and Industrial Technology Development Organization has also been developed as a collaborative institution involving government, academia, and industry, with a mandate to advance environmental and renewable-energy issues. Offices are located in Japan as well as China, Europe, Thailand, India, and the United States.

In the United States, the Geo-Heat Center is operated out of the Oregon Institute of Technology and provides access to research, case studies, periodic newsletters, and a directory of heat-pump consultants and manufacturers. The institute also owns a database of known groundwater wells and springs with high potential for ground-source heat-pump applications, fluid chemistry characteristics of these sites, and known sites currently using ground-source heat pumps, which includes information for sixteen western states. The U.S. Department of Energy (DOE) has also conducted a significant amount of research focused on heat pumps and promotes residential, commercial, and institutional applications through its Geothermal Technologies Program.

Industry and Business. The Institute of Electrical and Electronics Engineers (IEEE) is an international, nonprofit, member-driven organization that works to advance electricity-related technology. The organization promotes innovation, develops standards, and provides members with newsletters and networking opportunities. Its membership currently includes more than 400,000 engineers, among whom are experts working in the field of heat pump technologies.

Industry associations for heat pump-related businesses have been established in Austria, Canada, China, Finland, France, Germany, Italy, Japan, the Netherlands, Romania, Sweden, Switzerland, the United Kingdom, and the United States

The Earth Energy Society of Canada is an organization that was established by a small group of industry members to promote ground-source heat-pump systems. The organization provides information about the technology, a list of members, and a directory of International Ground-Source Heat Pump Association-accredited system contractors located across the country.

The Canadian Geoexchange Coalition has both a national presence and provincial branches. The organization offers training and accreditation for system designers and installers, information about installation in new and existing buildings, links to government resources and grant programs, case studies, upcoming events, and a list of members.

The United Kingdom's Heat Pump Association is dedicated to providing information about domestic, commercial, and industrial heat-pump applications to the public, as well as case studies, and a list of members. The Ground-Source Heat Pump Association is a more specialized industry group also operating out of the United Kingdom that focuses on providing members with services such as news, events, and grant opportunities.

In the United States, the Geothermal Heat Pump Consortium encompasses all aspects of the ground-source heat-pump industry, including contractors, manufacturers, drillers, installers, engineers, and system designers. It maintains a list of professionals in Canada and the United States and provides information about case studies, current projects, incentive programs, and offers training and events.

CAREERS AND COURSE WORK

The design, manufacturing, and installation of heat-pump components and technologies often requires an academic background in mechanical engineering. Specific courses required may include thermodynamics, heat transfer, fluid dynamics, mechanical design, thermal systems design, control systems, and manufacturing processes. Professional licensure, achieved through a recognized engineering

association, such as the National Society of Professional Engineers, may be required for some positions. Regional, state, or federal requirements may dictate that an accredited professional engineer must approve the installation of ground-source heat-pump systems.

Many technical and trade schools offer programs geared toward installing renewable energy technology systems. Engineering systems and energy technology programs teach applied skills for ground-source heat-pump installation. Careers in heating, refrigeration, and air conditioning may require either a bachelor's degree in mechanical engineering or a technical diploma in a related field. Diploma programs in heating, refrigeration, and air conditioning enable individuals to install, service, maintain, and upgrade heat pumps and systems for these applications. Additional certificates may be required. A diploma in plumbing may be required for installing, servicing, or upgrading heat-distribution systems involving water-source or ground-source heat pumps. Where drilling is required, as in the case of an Earth connection subsystem, a diploma in drilling may be required for installing loops.

While many university or technical programs provide the foundations for understanding heat pumps and associated systems, industry associations offer additional, specialized training and accreditation. It is important to check regional, state, or federal requirements for professionals in the specific field of interest.

SOCIAL CONTEXT AND FUTURE PROSPECTS

Tighter environmental regulations for building construction and industrial processes are forecast to increase the use of heat-pump technologies to reduce energy use and greenhouse-gas emissions. However, it is important that the source of electricity used to power the heat pump does not generate more greenhouse-gas emissions on a life-cycle basis than the fossil fuels required to generate the same amount of heat in a conventional energy system. One advantage of heat-pump technologies is that they permit switching to electricity produced from renewable-energy sources, such as hydro or wind, as these energy sources become available, while a heating system that relies solely on fossil fuels, such as natural gas, does not. Therefore, as more renewable energy is supplied to the grid, greenhouse-gas emissions associated with heat-pump systems will continue to fall.

Residential, Commercial, and Institutional Buildings. Heat pumps can decrease building energy consumption by up to 70 percent in cold climates, such as those of Nordic countries. Leadership in Energy and Environmental Design (LEED) and net zero emissions facility design have led to the incorporation of heat pumps into green-building design. Retrofits to existing buildings and advanced heat-pump systems will present new challenges to heat-pump system designers and installers. For example, in member-countries of the international Organisation for Economic Co-operation and Development (OECD), new housing is not projected to grow rapidly, and the majority of opportunities for heat-pump systems are expected to be retrofits to existing buildings.

Historically, heat-pump applications have been limited by high initial cost, difficulty optimizing retrofitted heat-pump systems in existing buildings, and inadequate performance of air-source heat pumps in seasonally variable climates. Technological advancements in heat-pump engineering have made significant progress toward eliminating these issues. While the growing popularity of ground-source heat-pump systems has led to an increase in the number of installations and decreasing system and installation costs, research as of 2011 is focusing on further cost reduction and system optimization.

Ultrahigh efficiency heat pumps are being developed to achieve space cooling using a fraction of the energy of the heat pumps available in the second decade of the twenty-first century. As the standard of living continues to rise in the developing world, demand for space cooling is forecasted to increase; thus the market for ultrahigh efficiency heat pumps is projected to grow considerably as the technology matures.

Industry. The applicability of heat pumps in the industrial sector depends largely on the temperatures of the heat source and sink and the temperature lift required. Temperature lift is influenced by the design of the heat pump and the properties of the refrigerant. While previously developed refrigerants, such as Freon, performed well in technological applications, the longer-term environmental consequences of their use were unacceptable. Research has turned to developing improved refrigerants that are capable of producing greater temperature lift for industrial applications. Research from the International Energy Agency is focusing on developing refrigerants capable of producing

temperature lift in the range between 170 degrees and 300 degrees F.

Ngaio Hotte, B.Sc. M.F.R.E.

FURTHER READING

Egg, Jay, and Brian Howard. *Geothermal HVAC: Green Heating and Cooling.* New York: McGraw-Hill, 2011. Provides a guide for contractors and consumers about the benefits of commercial and residential heating, ventilation, and cooling using heat pumps. Includes practical information about the types of systems available, how they work, costs and efficiencies, and troubleshooting.

Herold, Keith E., Reinhard Radermacher, and Sanford A. Klein. *Absorption Chillers and Heat Pumps.* Boca Raton, Fla.: CRC Press, 1996. This text provides a detailed investigation of absorption chillers and heat pumps, with particular attention to thermodynamic properties and movement of working fluids.

Kavanaugh, Stephen P., and Kevin Rafferty. *Ground-Source Heat Pumps: Design of Geothermal Systems for Commercial and Institutional Buildings.* Atlanta: American Society of Heating, Refrigerating, and Air-Conditioning Engineers, 1997. Written and published by a respected professional organization, this guide is a resource for engineers working with ground-source and water-source heat pumps.

Langley, Billy C. *Heat Pump Technology.* 3d ed. Upper Saddle River, N.J.: Prentice Hall, 2002. This book serves as both a text and reference manual. Fundamental concepts of heat pump operation and service are written for a student audience.

Radermacher, Reinhard, and Yunho Hwang. *Vapor Compression Heat Pumps with Refrigerant Mixtures.* Boca Raton, Fla.: CRC Press, 2005. The author provides a more advanced examination of refrigerant mixtures and working fluids, including thermodynamic aspects, refrigerant cycles, and heat transfer.

Silberstein, Eugene. *Heat Pumps.* Clifton, N.Y.: Thomson/Delmar Learning, 2003. Designed as a reference text for technicians working with heat pumps, this text provides an introduction to vapor-compression heat-pump technology and applications. Includes several diagrams and uses straightforward explanations of components and processes.

U.S. Department of Energy, Energy Efficiency and Renewable Energy, Industrial Technologies Program. "Industrial Heat Pumps for Steam and Fuel Savings." June 2003. http://www1.eere.energy.

gov/industry/bestpractices/pdfs/heatpump.pdf . Explains heat-pump applications for industrial processes and includes diagrams, types of heat pumps, benefits, and typical applications by type.

WEB SITES
American Society of Heating, Refrigerating and Air-Conditioning Engineers
http://www.ashrae.org

Geothermal Technologies Program
http://www1.eere.energy.gov/geothermal/heat-pumps.html

Heat Pump Centre
http://www.heatpumpcentre.org/en

Institute of Electrical and Electronics Engineers (IEEE)
http://www.ieee.org

International Energy Agency
http://www.iea.org

National Society for Professional Engineers
http://www.nspe.org

U.S. Department of Energy
http://www.energy.gov

See also: Civil Engineering; Energy-Efficient Building; Engineering; Fluid Dynamics; Mechanical Engineering.

HEMATOLOGY

SUMMARY

Hematology is the study of blood and blood-related disorders. It involves the diagnosis and treatment of a wide variety of diseases including anemia, leukemia, cancer, and clotting and bleeding disorders. As knowledge of the human genome has grown, many genetic mutations have been linked to hematological disorders. This has resulted in new methods of detection and novel treatments for diseases once thought incurable.

KEY TERMS AND CONCEPTS

- **Anemia:** Decrease in hemoglobin or the volume of red blood cells, resulting in less oxygen being delivered to body tissues.
- **Automated Cell Counter:** Instrument that uses electric impedance or optical light scatter to rapidly count blood cells.
- **Flow Cytometer:** Instrument that can simultaneously measure multiple physical characteristics of a single cell as it flows in suspension through a measuring device.
- **Hematopoiesis:** Production, differentiation, and development of blood cells.
- **Hematopoietic Stem Cell:** Precursor cell that has the ability to replicate and proliferate into all the lymphoid and myeloid blood cell lines.
- **Hemoglobin:** Respiratory protein that is found in the cytoplasm of the red blood cells.
- **Hemoglobinopathy:** Genetic defect that results in either an amino acid substitution or diminished production of one of the protein chains of hemoglobin.
- **Hemostasis:** Process resulting in the balancing of bleeding and clotting within the circulatory system
- **Leukemia:** Disease characterized by the overproduction of immature or mature cells of various leukocyte types in the bone marrow or peripheral blood.
- **Lymphoma:** Disease characterized by malignant tumors of the lymph nodes and associated tissues or bone marrow.

DEFINITION AND BASIC PRINCIPLES

Hematology is the study of blood and its related disorders. The components of blood include the plasma (55 percent) and the formed elements, or cells (45 percent). Plasma is 91.5 percent water and 8.5 percent solutes, mainly the proteins, albumin, globulins, and fibrinogen. The cells can be subdivided into three types, erythrocytes (red blood cells), leukocytes (white blood cells), and platelets. The primary step in assessing hematologic function and disease processes is to examine the cellular elements, determining the percentage of each type of cell and its morphology.

Erythrocytes consist of a plasma membrane that surrounds a solution of proteins (mainly hemoglobin) and electrolytes. Mature red blood cells do not contain a nucleus. They are small, about 7-8 micrometers (μm) and shaped like a biconcave disk. Evaluation of red blood cells is important in the diagnosis and monitoring of anemia.

Platelets are small (1-4 μm) bits of cytoplasm that contain granules and no nucleus. Platelets are necessary for hemostasis. They have the ability to adhere, aggregate, and provide a surface for coagulation to occur. Platelets are important in bleeding and clotting disorders.

Leukocytes are white blood cells that are differentiated into five varieties, neutrophils, eosinophils, basophils, lymphocytes, and monocytes. Mature white blood cells contain a nucleus and often cytoplasmic granules. It is important to distinguish the different types of white blood cells and their relative amounts. Neutrophils have a segmented nucleus and are important in fighting infections. They also contain granules that secrete specific enzymes that aid in the killing of bacteria. Neutrophils play a key role in phagocytosis and inflammation. Eosinophils contain large granules that stain reddish orange. The number of eosinophils increases in cases of parasitic infections and some types of allergies. Basophils also

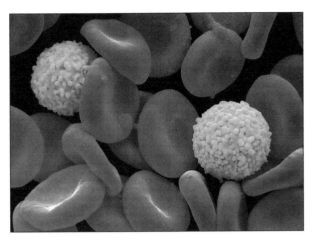

Human red and white blood cells. (Power and Syred/ Photo Researchers, Inc.)

contain granules; however, these stain a dark purple and contain the components heparin and histamine. Basophils play an important role in allergic reactions. Lymphocytes are the second most numerous white cells in the blood. They typically do not contain granules in the cytoplasm. Lymphocytes play an important role in immunity because of their ability to initiate an immune response and produce antibodies. Monocytes are the largest of the white blood cells. Often the monocyte's cytoplasm displays psuedopods. These cells are mobile and are important in fighting infections and removing foreign elements from the blood and tissues.

Hematopoiesis is the continual process of blood cell production and the development of the various cell lines. Blood cells are the progeny of a hematopoetic stem cell. Through the process of hematopoiesis, the body is supplied with ample blood cells of all lines. The hematopoetic stem cell can both replicate itself and differentiate into the mature cells found normally in the peripheral blood. The hematopoietic system includes the bone marrow, liver, spleen, lymph nodes, and thymus. Regulatory growth factors also play a role in hematopoiesis.

BACKGROUND AND HISTORY

The field of hematology came into existence in 1642 when Antoni van Leeuwenhoek visualized cells using a microscope that he invented. It was not until 1842 that French microbiologist Alexandre Donne discovered platelets. In 1845, German physician

Rudolf Virchow discovered an excess of white blood cells in a patient who had died from a condition he called leukemia (meaning white blood). A staining technique developed in 1877 by American scientist Paul Ehrlich allowed the visualization and differentiation of blood cells. French anatomist Louis-Charles Malassez invented the hemocytometer in 1874 and was able to quantitatively measure blood cells. The following year, French physician Georges Hayem developed a method for quantitatively measuring platelets using a hemocytometer.

Many diseases of the blood and their treatments, cures, and causes were discovered in the twentieth century. In 1925, American pediatrician Thomas Cooley described Cooley's anemia (thalassemia major). In the 1970's, American geneticist Janet Davison Rowley demonstrated the translocation of chromosomes 8 and 21 in acute myelogenous leukemia and of chromosomes 9 and 22 in chronic myelogenous leukemia. In 1972, American immunologist and geneticist Leonard Herzenberg invented the fluorescence-activated cell sorter (FACS), which aided the study of cancer. Argentine immunologist César Milstein and German biologist Georges Köhler used hybridization to develop monoclonal antibodies in 1975. These were followed by many other discoveries and treatments, some of which were the result of knowledge about the human genome.

HOW IT WORKS

Hematology diseases involve disorders of red blood cells, white blood cells, and platelets and hemostasis.

Disorders of Red Blood Cells. For the red blood cell to survive and function properly in the body, it must maintain a proper membrane, possess structurally correct and appropriately functioning hemoglobin, and have properly working metabolic pathways. Problems or defects in any of these areas will result in the red blood cell having a shortened life. Anemia results when the circulating red blood cells are unable to provide an adequate supply of oxygen to the tissues of the body. The many causes of anemia can be classified as nutritional deficiency (vitamin B12 or folic acid deficiency), blood loss, accelerated red cell destruction, hemoglobin defects (hereditary or acquired), and enzyme deficiencies.

Disorders of White Blood Cells. White blood cells perform a variety of functions in the body, including the destruction of bacteria, mediation of the inflammatory process, and production of antibodies or immunoglobulins. White blood cell disorders include diseases that affect the number of cells (quantitative defects) and those that affect the functioning of the cells (qualitative defects). White blood cell disorders range from slight inflammation to acute leukemia. Leukemia is a malignant disease that involves the hematopoietic tissue. Abnormal cells can be found in both the bone marrow and the peripheral blood. Leukemia is classified as either chronic or acute and can affect any of the white blood cell types, red blood cells, or platelets. It is important to distinguish between the different types of leukemias to provide the proper and most effective treatment.

Disorders of Platelets and Hemostasis. Hemostasis is the process by which the body stops bleeding and maintains the fluid state of the blood. In other words, it is a balance between bleeding and clotting. It involves the platelets, blood vessels, and specialized coagulation proteins. Disorders of this system include qualitative and quantitative platelet disorders, such as idiopathic thrombocytopenic purpura (ITP) and von Willebrand's disease. Some disorders that involve problems with coagulation proteins include factor V Leiden mutation and hemophilia A and B. The interaction between the platelets, clotting factors, and blood vessels plays an immensely important role in the functioning of the cardiovascular system. The field of hematology has been integral in developing modern cardiovascular therapies, including heart catheterization and the use of stents.

Disseminated intravascular coagulation (DIC) is another hemostasis disorder. It is a complex disorder that involves the development of small clots within the blood vessels and the dissolution of these clots. Platelets and the clotting factors often are consumed during this disease process, and intensive therapy is necessary to resolve its occurrence.

APPLICATIONS AND PRODUCTS

Automated Cell-Counting Instruments. Until the mid-1950's, cell counts were performed manually using a diluted fluid and a hemocytometer, and blood smears were viewed microscopically. Modern automated instruments can perform a complete blood count (CBC), which includes red blood cells, white blood cells, platelets, hemoglobin, hematocrit, and a five-part differential. This testing accounts for

the primary hematology testing performed for almost every disorder. The instruments use the principles of impedance and optical light scattering to enumerate and determine the characteristics of each cell. Using the impedance principle, the cells are passed through an electrically charged aperture. Because blood cells are poor conductors of electricity, they create a resistance that can be measured as a pulse. The number of pulses is equivalent to the number of blood cells. The height of each pulse is

---

Fascinating Facts About Hematology

- A mature red blood cell does not contain a nucleus and therefore possess no DNA. Red blood cells serve the vital function of supplying oxygen to the rest of the body.
- The shortest-lived cell in the blood, the polymorphoneutrophil, circulates for only seventy-two hours. Other cells in the blood, such as the monocyte/macrophages, can live for months or even years.
- Sickle cell anemia is a genetic defect that results in a single amino acid substitution in the protein chain of hemoglobin. Hemoglobin has four protein chains with more than 574 amino acids in a specific sequence.
- Iron-deficiency anemia is the most common anemia and affects an estimated 2 billion people worldwide, according to the World Heath Organization.
- Although leukemia can strike at any age, more than 50 percent of all cases of leukemia occur after the age of sixty-four.
- In 2010, an estimated 43,050 new cases of leukemia were diagnosed in the United States.
- Allogenic bone marrow transplants have been successful in curing chronic myelogenous leukemia. Long-term survival rates of as high as 78 percent have been reported. A majority of patients have successfully resumed their personal and professional lives.
- Hemophilia A is a bleeding disorder that affects men. This disorder was found in the Royal House of Stuart in Europe and Russia. Queen Victoria was later proved to be the carrier. The disorder, previously thought to be an absence of factor VIII, is actually a molecular defect in the factor that renders it unable to perform its clotting function.

---

equal to the cell's volume. Using the optical light scattering principle, each cell is passed through a beam of light (either optical or laser). Forward scatter and side scatter of the light is created by each cell. The forward scatter represents the size of the cell, and the side scatter represents the degree of complexity of the cells (cytoplasmic organelles can be assessed).

Flow Cytometry. Flow cytometry can detect molecules on the surface of a cell, making it possible to sort cells according to their surface composition. The molecule of interest is labeled using a fluorescent marker. Flow cytometry is used for immunophenotyping (identification of antigens on the cell surface), reticulocyte counting, and analysis of DNA. The information obtained about the subset of cells in flow cytometry is of critical use in the diagnosis, classification, and treatment of malignancies of mature lymphocytes, acute leukemia, and immunodeficiency disorders.

Cytogenic Analysis. In cytogenic analysis, cells are harvested and processed to visualize chromosomes in mitotically active cells. The cells are arrested in metaphase, and chromosome bands are visualized by various staining procedures. A similar technique, fluorescence in situ hybridization (FISH) may also be used. In this technique, fluorescent-labeled DNA probes are used to visualize specific chromosome centromeres (region on a chromosome joining two sister chromatids), whole arms, whole chromosomes, and individual genes. Chromosomes are visualized microscopically, and a karyotype can be constructed by using a video-computer-linked analysis system. Many chromosome aberrations are considered diagnostic or have significant prognostic implications for hematologic malignancies and solid tumors, such as chronic myelogenous leukemia, acute myelogenous leukemia, acute lymphoblastic leukemia, and lymphomas.

Chemotherapy. The goal of chemotherapy is to destroy all malignant cells within the bone marrow and allow the bone marrow to be repopulated with normal precursor cells. However, the drugs used for chemotherapy are not specific for leukemic cells, and many normal cells are also killed in the process. This results in the severe complications of bleeding, infections, and anemia.

Molecular-Targeted Therapy. Therapies that target specific genetic mutations can either silence the expression of a particular gene or reactivate a silenced gene. These types of therapies are better tolerated by patients than traditional chemotherapy and have been developed for chronic myelogenous leukemia and acute promyelocytic leukemia.

Bone Marrow Transplant. For a bone marrow transplant, drugs and radiation are used to first eradicate all leukemic cells that may be present. Then, bone marrow from a suitable closely matched donor is transplanted into the patient to provide a source of normal stem cells that can then repopulate the patient's bone marrow. Autologous bone marrow transplants are also an option. Some of the patient's bone marrow is removed while the patient is in remission. This specimen is treated to remove any residual leukemic cells and preserved. The patient is then treated to eradicate all leukemic cells and given back his or her own bone marrow. Many bone marrow transplants have been successful, and this procedure is being performed on an increasing basis.

Stem Cell Transplant. Stem cells can be found in either the bone marrow or the peripheral circulating blood. They also can be found in umbilical cord blood and fetal marrow and liver. Before stem cells can be harvested from a donor, mobilization of stem cells into the peripheral blood is stimulated by the use of cytokines. A process called apheresis is used to collect the stem cells from the donor's blood. These harvested stem cells are then injected into the patient. As with a bone marrow transplant, it is also possible to perform an autologous stem cell transplant. The transplanted patient will usually begin to produce new blood cells ten to twenty-one days after receiving the harvested stem cells.

IMPACT ON INDUSTRY

Industry that involves the field of hematology is mainly centered on its application, which is in the field of medicine and treating patients with hematological or cancerous diseases.

Medicine. Physicians may practice in hematology or oncology in government, nonprofit, or private institutions of medicine. Hematology is considered a specialty area of medicine. Usually patients are referred to a hematologist by their general physician. Hematologists also specialize in the area of oncology, since many cancers involve the blood and circulatory system.

Most large medical center will have a hematology/oncology department that is staffed with hematologists, oncology nurses, medical laboratory scientists, and radiology technologists. These professionals work together as a health care team to diagnose and provide treatment for hematological disorders.

Research. Hematology research is conducted in a variety of settings. Research is ongoing to understand basic cellular and molecular mechanisms, hematopoiesis, and the regulation of genes that control normal blood cell maturation and function. Other areas of research are stem cells, growth factors, synthetic hemoglobin, and drugs for the treatment of all types of cancers. For example, the U.S. Department of Health and Human Services funds a multifaceted research program through the National Institute of Diabetes and Digestive and Kidney Diseases.

Another area of medicine that relies heavily on hematology research is cardiology, particularly cardiovascular surgery and the treatment of cardiovascular diseases. Because the blood vessels are in contact with the blood at all times, these two areas are heavily reliant on each other.

Industry. The medical laboratory instrument, reagent, and pharmaceutical industries are all essential to the diagnosis and treatment of hematology disorders. Revenues from these industries are approaching $4.5 billion.

The United States leads the medical laboratory instrument industry, with an estimated 38 percent of the market. Asia, Europe, and Germany follow closely behind. A large area of development involves point-of-care instruments, which allow the patient and caregivers to monitor the patient's laboratory results at home or at the bedside.

The development of drugs for treatment is a market that continues to grow. Some examples of drugs developed for hematology treatments are recombinant erythropoietin (EPO), granulocyte-macrophage colony stimulating factor (G-CSF), and recombinant blood clotting factors. The market for pharmaceuticals for treatment of leukemia, anemia, and other blood disorders continues to grow as more diseases are diagnosed and as research provides information about the causes of these diseases.

CAREERS AND COURSE WORK

Courses in biology, genetics, molecular and cell biology, chemistry, immunology, hematology, and immunohematology are core courses for those interested in pursing a career in hematology. There is a need for professionals with hematology knowledge at all levels of careers.

Medical Laboratory Scientists. Medical laboratory scientists, previously known as medical technologists or clinical laboratory scientists, work in all areas of the clinical laboratory, including hematology. They perform the necessary tests to diagnose and treat diseases of the hematological systems. To become a medical laboratory scientist, one must obtain a bachelor's degree and complete an accredited medical laboratory science program. This can be done through a four-year course of study at most universities.

Cytotechnologists. Cytotechnologists are responsible for examining human cells under the microscope. They look for early signs of cancer and other diseases. A bachelor's degree and completion of an accredited cytotechnologist program are required.

Hematologists/Oncologists. Medical doctors specializing in hematology or oncology can diagnose and treat patients with diseases that affect the hematology system. They can work in private practice, for academic centers, or in governmental agencies. To become a hematologist requires a bachelor's degree and a medical degree. A residency and a fellowship in the area of hematology or oncology also are necessary.

Pathologists. Pathologists are physicians who examine tissues and are responsible for the accuracy of laboratory tests. The pathologist and the patient's other doctors consult on which tests to order, the interpretation of test results, and the appropriate treatments. A bachelor's degree, medical school, and a residency and a fellowship in pathology are required.

SOCIAL CONTEXT AND FUTURE PROSPECTS

Never in the history of medical science have the opportunities for advancement in the field of hematology been greater. Many genes have been characterized as disease specific or disease related. This creates the possibility of finding cures or treatments for many diseases that have plagued humankind for centuries.

With this newfound knowledge also comes the challenge of balancing the treatment of patients with new technologies and the ethical application of these treatments. Stem cell research has spurred some controversy as to how to ethically provide the stem cells needed for treatments. Another controversy arises when genetic testing for diseases or certain gene markers that are tested for as risk factors are used in the everyday practice of medicine.

Mary R. Muslow, B.S., M.H.S., M.T.(ASCP)S.C.

FURTHER READING

Carradice, Duncan, and Graham J. Lieschke. "Zebrafish in Hematology: Sushi or Science?" *Blood* 111, no. 7 (April, 2008): 3331-3342. Describes the potential of the use of zebrafish for the modeling of hematologic diseases.

Harmening, Denise M. *Clinical Hematology and Fundamentals of Hemostasis.* 5th ed. Philadelphia: F. A. Davis, 2009. Includes chapters on types of anemia, white blood cell disorders, hemostasis, and laboratory methods.

Herzenberg, Leonard A., and Leonore A. Herzenberg. "Genetics, FACS, Immunology, and Redox: A Tale of Two Lives Intertwined." *Annual Review of Immunology* 22 (2004) 1-31. The inventors of flow cytometry describe their lives and work.

McKenzie, Shirlyn B., and J. Lynne Williams. *Clinical Laboratory Hematology.* 2d ed. Upper Saddle River, N. J.: Pearson, 2010. Cover types of anemia, neoplastic hematologic disorders, hemostasis, and hematology procedures. Includes some excellent photomicrographs and illustrations.

Patlak, Margie. "Targeting Leukemia: From Bench to Bedside." *The Federation of American Societies for Experimental Biology* 16, no. 3 (March, 2002): 273. An interesting review of leukemia diagnosis and treatment.

WEB SITES

American Society of Hematology
http://www.hematology.org

American Society of Pediatric Hematology Oncology
http://www.aspho.org

Hematology/Oncology Pharmacy Association
http://www.hoparx.org

See also: Cardiology; Histology; Stem Cell Research and Technology.

HERBOLOGY

FIELDS OF STUDY

Medicine; nutrition; naturopathy; biochemistry; botany; pharmacy; physiology.

SUMMARY

Herbology is the study of plants for use in the prevention and treatment of health conditions and disease. Herbs have been used throughout history for medicinal purposes. Herbology is an essential part of traditional Chinese medicine and has been incorporated into Western naturopathy. Growing interest in natural remedies in North America has increased the popularity of training programs in herbology. Although some cultures have used herbal treatments for thousands of years, their efficacy and quality have been questioned, and some people have called for stricter regulation of herbal products.

KEY TERMS AND CONCEPTS

- **Adaptogen:** Herb that treats physical and emotional stress.
- **Alterative:** Herb that restores normal bodily function.
- **Astringent:** Tannin containing herb used for contraction of tissues.
- **Bitter:** Digestive herb.
- **Carminative:** Herb for relief of gas.
- **Cathartic:** Laxative herb.
- **Cholagogue/Choleretic:** Herb to increase bile production.
- **Decoction:** Tea made from the fibrous part of plants.
- **Demulcent:** Mucilaginous herb that treats mucous membranes.
- **Diaphoretci:** Herb to promote perspiration.
- **Emmenagogue:** Herb to stimulate menstruation.
- **Extract:** Concentrated liquid made from herbs.
- **Galactagogue:** Herb with sugar and nonsugar composition.
- **Hepatic:** Quality of supporting liver function.
- **Herbalist:** Practitioner trained in the medicinal use of plants.
- **Infusion:** Herbal tea made from the soft parts of plants such as flowers and leaves.
- **Tincture:** Concentrated liquid made with water, alcohol, and herbs.
- **Vunerary:** Herb to promote wound healing.

DEFINITION AND BASIC PRINCIPLES

Herbology, also known as phytotherapy, is the study of plants and plant extracts for medicinal use. Herbology is an integral part of Chinese traditional medicine, folk medicine, and naturopathic medicine. Herbology is often used by practitioners who also favor alternative medical treatments such as acupuncture. An herbalist is a person who specializes in the use of plants and plant extracts for the prevention and treatment of diseases and conditions. Some herbalists also use naturally occurring nonplant products such as minerals and animal products.

Herbology is based on the knowledge that phytochemicals (chemical substances within plants) have pharmaceutical properties. Many herbal remedies are based on traditional practices, and many scientific studies have been done to verify the properties of herbal products. In 1978, the German Commission E, a governmental regulatory agency, was formed to evaluate the safety and efficacy of herbs available for general use. The commission produced 380 monographs. Although the commission is no longer in existence, the monographs are available in *The Complete German Commission E Monographs* published by the American Botanical Council in 1998. These monographs are one of the most thorough collections of scientific data on the subject of herbology. Since the commission was disbanded, numerous scientific studies have been published by other groups regarding the effectiveness of various herbs in the treatment and prevention of disease.

As of 2010, herbalists were not required to be licensed in the United States and Canada. Training programs in herbology are available through a number of organizations. The American Herbalists Guild was founded in 1989 to serve as an educational organization for herbalists in the United States. The American Botanical Council is a nonprofit organization that was founded in 1988 to provide information to herbalists and consumers.

Herbal remedies are not regulated by the U.S. Food and Drug Administration (FDA) and are not investigated by the FDA unless there are reports of adverse reactions. A similar situation exists in Canada. As more and more herbal products have become available, reports of adverse reactions due to product misuse or product contaminants have been increasing. The National Center for Complementary and Alternative Medicine (NCCAM), a branch of the National Institutes of Health (NIH), takes a scientific approach. It provides research-based information on the safety and efficacy of herbs and offers training in herbology.

A wide variety of medical practitioners uses herbology. In addition to traditional herbalists, medical herbalists, natural healers, holistic medical doctors, naturopaths, and practitioners of alternative medicine (such as traditional Chinese medicine) use herbs. In a practice called complementary medicine, some conventional doctors and other medical professionals treat patients using herbology and other alternative therapies in addition to conventional Western medicine. According to the National Institutes of Health, in 2007, more than 38 percent of adults in the United States reported using complementary or alternative medicine, including herbs.

BACKGROUND AND HISTORY

Plants and plant extracts have been used in the prevention and treatment of medical conditions for many centuries. Ancient Egyptians used opium, garlic, and other plants, and the Old Testament includes references to the use of herbs. The use of herbology has been an integral part of traditional Chinese medicine for more than a thousand years.

Many pharmaceutical products are derived from plants. A well-known example is digitalis, which is derived from the foxglove plant and is used in the treatment of heart conditions. Another example is morphine, which was originally extracted from poppies. Aspirin is derived from salicin, which is present in the bark and leaves of the willow tree. As herbal remedies have become more popular, the manufacture and sale of herbal products including combinations of herbs in tablet or extract forms has increased. Many products are readily available at health food stores and pharmacies.

Although the practice of herbology relies on traditional knowledge, a number of efforts have been made to verify the efficacy and safety of herbal remedies. The most well-known source for scientific information about herbs is the monographs produced by the German Commission E. The National Center for Complementary and Alternative Medicine also provides information on and training in herbology.

HOW IT WORKS

Herbalists use herbal remedies made from plants. These remedies are based on the knowledge that phytochemicals have pharmacologic properties. These phytochemicals can be brought out by preparing the plant material in a variety of ways, depending on the remedy.

Herbal medicines are often prepared as a tea, which is properly called an infusion or decoction. A decoction is typically made by placing the stems, roots, or bark of a specific plant in water, and boiling the water until the volume is reduced. Infusions are made by steeping plant parts in hot or cold water. Infusions are usually made from the softer parts of plants such as the flowers or leaves. Decoctions and infusions extract water-soluble phytochemicals.

Herbal remedies may also be prepared as a tincture or extract. Tinctures are made by macerating the plant material in water and alcohol. The ratios of plant material, water, and alcohol will vary depending on the herbal remedy and dosages. Extracts will usually be more concentrated. For some extracts, glycerin is used as a solvent.

Tablets, lozenges, and dried herbs are also available. Some herbal products can also be found in creams or ointments.

Commercially available products are not regulated by the FDA, and the contents of some products have differed from the product label. Others have been contaminated with other herbs, metals, or pesticides. To ensure safety, NCCAM suggests that patients become informed about the herbal product, discuss it with their primary care physician, read the label carefully, and remember that natural does not always mean safe.

Consumers and practitioners should be aware that herbal products can cause adverse reactions or result in negative outcomes for a number of reasons. First, consumers or herbalists may not recognize the seriousness of a medical condition, which might require more aggressive medical treatment than herbs can provide. Second, problems can arise from allergic reactions, misidentification of herbal products, intentional or unintentional contamination of herbal products with heavy metals or other harmful substances, or mistakes in the dosage. Third, herbs and conventional medications may interact and exacerbate an existing condition. Finally, unsubstantiated herbal remedies may be totally ineffective and allow a condition to remain basically untreated.

Herbology is practiced in different ways according to the philosophy of treatment. Some dosage recommendations are derived from herbal traditions such as traditional Ayurvedic or Chinese medicine. Many books are available on the subject of herbal remedies, and some commercially available products will have recommended dosages on the label.

Although there are no standards for training in herbology, training programs are available through NCCAM and other organizations.

APPLICATIONS AND PRODUCTS

Traditional Chinese Medicine. Herbology involves the use of plants and plant extracts in the treatment and prevention of disease. The application of herbal remedies may depend on the medical tradition followed by the herbalist. For example, a herbalist following traditional Chinese medicine may recommend herbs to restore a person's body to a balanced state, based on an examination of aspects such as yin and yang, as well as qi (chi) and blood. Yin and yang are opposite aspects, defined as light and dark, hot and cold, and moist and dry, respectively. Qi is often described as energy, or a dynamic essence. Blood, transformed from the essence of food, is important as well. Chinese medicine also determines treatment according to the five elements–wood, fire, earth, metal, and water–which correspond to specific organs, senses, and bodily functions. Chinese herbal products range from specific foods such as wheat or grapes to combinations of herbs in a tablet or a tea.

Western Herbology. Some Western herbalists subdivide herbal remedies according to the organs or systems affected or the action of the herb. Herbal actions fall into several categories. Adaptogens are herbs used to treat emotional and physical stress. Tonics provide energy. Vulnerary herbs promote healing, and nervines work on the nervous system. Astringents contract tissues. Diaphoretics promote perspiration, and febrifuge is used to treat a fever. Alterative herbs restore normal function and may be further divided into hepatic, digestive, or antimicrobial. The gallbladder can be treated with a cholagogue or choleritic to promote secretion of bile. Bitters and carminative herbs are used to support digestion. Purgative and carthartic herbs are used for a laxative effect, and stomachic is

an agent to promote appetite. Demulcents are mucilaginous agents that sooth mucous membranes. Menstrual flow is enhanced by emmenagogues. Phytoestrogens enhance the estrogen system. Galactagogues promote lactation.

Applications. Herbology is used in a number of settings. Herbalists may be solo practitioners or may work in a health food store. Nutritionists or conventional physicians may incorporate herbology in their practices. Herbal remedies are readily available in pharmacies, and many pharmacists will have some knowledge about the benefits and dosing of herbal remedies. Herbal remedies are often used along with other forms of alternative medicine such as yoga, acupuncture, and meditation.

IMPACT ON INDUSTRY

Herbal products, used by more than 38 percent of Americans and millions of people worldwide, have become big business. Billions of dollars of herbal products are sold each year. According to some estimates, Americans spent more than $20 billion on vitamins, herbs, and other supplements in 2007. Health Canada reports that more than 71 percent of Canadians regularly take natural health products. Suppliers of herbal products range from small companies that specialize in a few products to major companies, including some major pharmaceutical companies. The growth in sales of herbs and supplements slowed after deaths were linked to the supplement ephedra, which was banned in 2004. However, herbal products continue to generate large revenues.

Organizations. Herbology organizations range from governmental agencies designed to educate consumers to associations made up of people in the industry. The National Center for Complementary and Alternative Medicine, part of the NIH, provides research-based information to consumers as well as training in herbology. The American Botanical Council provides information about herbal medicine to consumers and herbalists. The American Herbal Products Association represents growers, importers, manufacturers, marketers, and corporations in the herbal industry. The American Herbalists Guild represents herbalists specializing in the medicinal use of plants.

Governmental Regulation. According to the Dietary Supplement Health and Education Act of 1994, companies marketing supplements in the United States cannot make specific therapeutic claims about their products. Companies can make nutritional claims if they include a disclaimer that the claims have not been evaluated by the FDA. Reports of adverse reactions from herbal products are investigated by the FDA. In Canada, supplements can be treated as either foods or drugs. Supplements regulated as prescription drugs have a drug identification number (DIN) that appears on the product label. The National Health Products Regulations, which came into effect in January 1, 2004, defined natural health products as over-the-counter drugs. Those evaluated by Health Canada are given a natural product number (NPN).

Government Research. Research on herbal products has been conducted by governmental agencies such as the German Commission E and Canada's Natural Health Products Research Program (2003-2008). National Center for Complementary and Alternative Medicine funds a number of research programs and, in cooperation with the NIH's Office of Dietary Supplements, runs the Botanical Research Centers Program. Thousands of scientific articles have been published on herbal products and peer-reviewed articles are available through NIH's National Library of Medicine.

Fascinating Facts About Herbology

- Herbs have been used as medicine for thousands of years. Perhaps the most expensive herb is saffron, which costs between $500 and $1,000 per pound. Some herbalists recommend saffron for memory enhancement and antiaging.
- Garlic is both a common remedy and a common ingredient in foods. The ancient Greeks and Romans used garlic to ward off disease. Garlic, which may reduce cholesterol, has been used to prevent heart disease, cancer, colds, and the flu. It is considered both a vegetable and a herb and is a member of the onion family. Garlic has been used as a natural mosquito repellent and was once believed to ward off vampires.
- Red clover has been reported to have magical powers, including the ability to protect against witchcraft and evil spirits. Four-leaf variants have been used as lucky charms.
- A flavenoid called quercetin present in apples may protect the brain, lending support for the adage "An apple a day keeps the doctor away."

It can be challenging to study herbal remedies because of large variations in the practice of herbology, including variable dosing and the practice of using a mixture of herbs. Another problem is the placebo effect. In blind studies, up to 30 percent of those taking placebos (a substance that resembles the product but does not have the active ingredient) report benefits. In addition, some researchers feel that those who take herbs are more likely to have a healthier lifestyle than those who do not take herbal products. Researchers have produced conflicting reports about a number of products, including St. John's wort, which has been used to treat depression. Later data seem to indicate that St. John's wort may not be effective and can have serious drug interactions. The efficacy of ginkgo biloba, a common herb taken to combat memory loss, has also been called into question.

Industry and Business. Herbology involves a variety of sectors, both small and large. Herbology practices range from individuals working solo to those who are part of a larger alternative group practice. Herbal products are commonly sold in health food stores that specialize in the sales of herbs and supplements. These stores may be mom-and-pop establishments or branches of large corporate chains. Herbal products are also sold by several direct-sales companies.

The industry also contains companies that grow, import, manufacture, and distribute herbal products. This segment of the industry includes large and small companies dedicated to herbal and other supplements as well as major pharmaceutical companies with an herbal product line. The distribution and sales of these herbal products in small and large retail chain stores generate billions of dollars in revenue every year.

CAREERS AND COURSE WORK

For those wishing to become a practicing herbalist, programs are readily available in the United States and Canada. Herbology training usually involves studying botany, herbal medicine, and the history and philosophy of herbal medicine, and receiving specialized training in the use of herbs in pregnancy and childhood.

The Association of Accredited Naturopathic Medical Colleges consists of seven North American schools that provide naturopathic training. These schools offer a variety of programs leading to a degree or certificate, including a bachelor of science in herbal sciences, a certificate in Chinese herbal medicine, and a doctor of naturopathic medicine degree. The requirements for entry and completion of these programs vary depending on the chosen degree. These naturopathic programs are built around the six fundamental principles of healing: the body has an inherent ability to create, maintain, and restore health (the healing power of nature), the physician should not interfere with this healing power (first do no harm), the cause of an illness must be identified for its treatment, the physician should heal the whole person, the physician should educate and encourage the patient, and the physician should help the patient achieve health through preventative medicine. There are also a variety of online programs available for training in herbology.

For careers in the business side of herbology, business training or industry experience may be helpful. Researchers in herbology often have advanced degrees in related fields such as biochemistry, botany, chemistry, or physiology. Grants are available through the National Center for Complementary and Alternative Medicine for additional training in herbology.

SOCIAL CONTEXT AND FUTURE PROSPECTS

The use of plants and plant extracts for therapeutic treatment and prevention of disease will continue to be practiced worldwide. With the development of government organizations, degree programs, and medical school courses in alternative medicine, it is likely that someday mainstream medicine will incorporate the study of herbology.

Consumer interest in herbal products continues to increase and to spur growth in the herbal industry. As the number of herbal products expands and more and more people use them, governments are likely to increase oversight of these products to ensure consumer safety and prevent fraudulent practices. The FDA may introduce licensing or other measures to control the production and use of herbal products.

Continued research into herbal remedies is likely to lead to improved understanding of the benefits of herbs as well as possible side effects and interactions. The phytochemicals responsible for the herbal effects will probably lead to the development of conventional pharmaceuticals as well.

Ellen E. Anderson Penno, M.D., M.S., F.R.C.S.C., A.B.O.

FURTHER READING

Blumenthol, Mark, et al., eds. *The Complete German Commission E Monographs: Therapeutic Guide to Herbal Medicines*. Austin, Tex.: American Botanical Council, 1998. Still considered a valuable resource for science-based information about herbs for therapeutic use.

Boon, Heather, and Michael Smith. *Fifty-five Most Common Medicinal Herbs*. 2d ed. Toronto: Robert Rose, 2009. A detailed description of the fifty most commonly used herbs. Includes an introduction with information about herbal preparations and legislation of herbal products.

Hurley, Dan. *Natural Causes: Death, Lies, and Politics in America's Vitamin and Herbal Supplement Industry*. New York: Broadway Books, 2006. Hurley criticizes the supplements industry for its excesses and discusses the politics behind regulatory decisions.

PDR for Nonprescription Drugs, Dietary Supplements, and Herbs. 30th ed. Montvale, N.J.: Thomson Healthcare, 2008. The physician's desk reference provides information on dietary supplements and herbs.

Tierra, Michael. *The Way of Chinese Herbs*. New York: Pocket Books, 1998. A very thorough compendium of Chinese herbal remedies with descriptions, including the philosophy of treatment underlying the Chinese herbal tradition.

WEB SITES

American Botanical Council
http://abc.herbalgram.org

American Herbalists Guild
http://www.americanherbalistsguild.com

American Herbal Products Association
http://www.ahpa.org

Health Canada
Natural Health Products
http://www.hc-sc.gc.ca

Herb Research Foundation
http://www.herbs.org

National Center for Complementary and Alternative Medicine, NIH
http://www.nccam.nih.gov

Office of Dietary Supplements, NIH
http://ods.od.nih.gov

See also: Horticulture; Pharmacology.

HISTOLOGY

FIELDS OF STUDY

Microscopy; cell biology; embryology; anatomy and physiology; histopathology; histochemistry, immunohistochemistry; hematology, oncology, tissue transplantation.

SUMMARY

Histology is an interdisciplinary branch of science that focuses on the structure and function of normal and diseased tissues of the human body using microscopy and staining techniques. It studies comparative morphology of tissues, changes in tissues and organs during embryonic development, evolutionary changes of structure, and function of tissues in different species. In clinical medicine, histology is used as a diagnostic tool to understand and treat pathological developments in the body tissues. In forensics, histology is employed to understand the degenerative events in injured and dead tissues (autopsies) and to determine the cause of death in criminal investigations.

KEY TERMS AND CONCEPTS

- **Autopsy:** Medical test that involves the extraction of cells or tissues from a corpse to evaluate any disease or injury and determine a cause of death. It is usually performed by a specialized medical doctor called a pathologist.
- **Biopsy:** Medical test that involves the extraction of cells or living tissues for observation and analysis of structural and functional abnormalities by methods such as microscopy or histochemistry.
- **Cytology:** Science that focuses on the fine structure and function of cells.
- **Embryology:** Study of the development of an embryo from fertilization to the fetus stage.
- **Hematology:** Study of blood cell formation (hematopoiesis), the organs that form blood cells, and diseases of the blood.
- **Histochemistry:** Use of chemical reactions between specific laboratory reagents and components within tissue biopsy material for diagnostic purposes.

- **Histology Staining:** Use of particular staining reagents to stain tissue sections for structural characterization of cells and tissues.
- **Histopathology:** Subdiscipline of histology that focuses on development of disease at a tissue level. It uses biopsies to diagnose and evaluate disease progression.
- **Immunohistochemistry:** Use of fluorescent-labeled antibodies on tissue sections to identify the location of particular structures (antigens) to which such antibodies would specifically bind.
- **Microscopy:** Study of objects too small to be seen with the naked eye, using a magnifying instrument called a microscope.
- **Microtome:** Device that slices ultra-thin sections of tissues that can be used for microscopic observation and analysis.
- **Oncology:** Branch of medicine that studies tumors (cancers).

DEFINITION AND BASIC PRINCIPLES

Histology studies the morphology of cells and tissues of the human body. It is sometimes called microscopic anatomy because it looks at the structure and function of the human body at the microscopic level. In the body, individual cells are organized in tissues, which form organs and organ systems to perform complex functions to maintain homeostasis. Although histology as a discipline is descriptive in many ways, it also involves a great deal of analysis because it focuses on structure and function relationships under normal and pathological conditions. The human body is made up of epithelial, connective, muscle, and nervous tissue types, but there are structural variations within each of these four groups. Changes in the body homeostasis are directly reflected in the changes–from temporary to irreversible–of the structure and function of tissues. Signs of infection, autoimmunity, aging, and malignant growths can be observed in various types of tissue by histological analysis of tissue probes. A trained histologist is capable of drawing a diagnostic conclusion based on such assessment alone or in combination with other methods.

Histology is based on the preparation of stained samples of tissue (from autopsies, biopsies, and cell

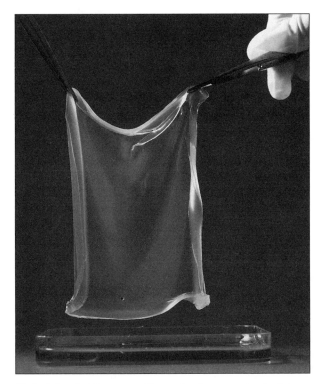

Technician holding an epidermal strip grown in the laboratory. Epidermal cells make up the epidermis, the tough outer layer of the skin. When cultured under the correct conditions, these human cells form epidermal strips that can be used to treat skin diseases such as vitiligo, a pigmentation disorder. (Mauro Fermariello/Photo Researchers, Inc.)

cultures) and their microscopic analysis. The development of various staining techniques and tools for specimen preparation and microscopy have strengthened histology as a diagnostic approach in biomedical research and clinical medicine. The microscope evolved from a simple set of magnifying lenses to highly sophisticated optical (light), scanning, and electron microscopes that can use computer control to create high-resolution digital images for researchers, doctors, clinicians, and educators.

BACKGROUND AND HISTORY

The word "histology" comes from the Greek *histos* (mast, or tissue) and *logie* (study) and means the study of tissues. Descriptions of tissues can be found in the works of the ancient Greek philosopher Aristotle, the eleventh-century Persian physician Avicenna, and the sixteenth-century German physician Andreas Vesalius, long before the microscope was invented. The English natural philosopher Robert Hooke was the first to introduce the term "cell" in 1665, while studying structures of a cork tissue using a magnifying device, a simple microscope. In 1674, Dutch scientist Antoni van Leeuwenhoek, using a microscope of his own invention, observed living organisms. He discovered that an animal cell is structurally organized and has a nucleus. The term "histology" was introduced by August Mayer in 1819, and in the nineteenth century, histology was established as an academic discipline. Studies of animal tissues were conducted by such prominent scientists as Marcello Malpighi, Camillo Golgi, Caspar Friedrich Wolff, and many others.

Histology developed much faster after around 1838, when microscopes with a greater magnification became available. Theodor Schwann developed the cell theory, which states that living organisms are made of similarly structured units, or cells, and that each individual cell exhibits characteristics of life. Rudolf Virchow, regarded as the father of modern pathology, contributed to the cell theory by stating that cells divide to produce daughter cells. In 1906, the Nobel Prize in Physiology or Medicine was awarded to Golgi for his outstanding work on the histology of the nervous system. Modern histology is a multidisciplinary branch of biomedical science. The results of histological observations contribute to the understanding of mechanisms of diseases and ways to treat diseases.

HOW IT WORKS

Preparation of Tissue Samples. Histology and histopathology examine tissue samples using microscopy; therefore, preparation of high-quality samples is an important step toward obtaining reliable results. For light microscopy, a sample of tissue is acquired, processed so that it can be thinly sliced or sectioned using a microtome, and stained with a specific dye. The tissue section is then analyzed using a microscope, and the results are interpreted, usually by comparison to normal tissue samples. A research scientist would examine how his experimental protocol affected the tissue, while a physician would analyze biopsied tissue to observe disease progression and determine the correct treatment. All steps of this process must be performed according to certain standards to avoid misinterpretation.

There are many methods of studying extracted tissues, and the choice of methodology depends on the type of tissue and the final objective. Because tissues degenerate quickly, it is important to preserve them as soon as possible, using a chemical process called fixation. The chemicals preserve the cells in the tissue without changing their structure. Individual protocols for fixation of different tissues vary and are the product of experimentation by many histologists. After fixation, tissue samples are treated to remove water so that they can be saturated with a waxy substance such as paraffin, which solidifies at room temperature. The paraffin-saturated tissue specimen is then sliced into sections using a microtome. The sections are mounted on a glass slide, treated to dissolve the paraffin, and stained with dye. The thickness of the tissue sections can vary from 0.002 to 0.02 millimeters (mm). Certain modifications to this process are made to prepare bone tissue, which is very hard because of its high mineral content, and adipose tissue, which has a high fat content. Staining is an important step, because stains have different chemical affinities for various cell and tissue components. For instance, the same kind of dye might stain the nucleus, cytoplasm, and membrane structures differently. These differing absorption rates make it possible to highlight particular structures of interest. The specimen must be thoroughly rinsed to remove any unbound stain and avoid nonspecific staining.

Electron microscopy employs completely different techniques for preparation of specimens and allows observation of fine cell and tissue structures under very high magnification. A glass or diamond knife is used with an ultramicrotome to produce ultra-thin sections less than 0.001 mm thick. In some situations, extracted tissues can be frozen and sliced immediately, then stained and analyzed without lengthy tissue processing. Such tissue sections can deteriorate rapidly but are used to quickly analyze biopsies during surgery or in applications that do not require a detailed structural analysis.

Microscopy. The most common tool in examining cells and cell structures in tissue samples is the microscope. Optical microscopy uses visible light, which passes thought the object (a tissue section mounted on a glass slide) and the lenses to create a magnified image. The maximum resolution that can be achieved with light microscopes is 0.2 micrometers (μm). Transmission electron microscopes send a beam of electrons through the sample and allow much higher resolution, around 0.05 namometers (nm), but they require complex tissue processing and good technical skills to produce usable specimens on an ultramicrotome. Scanning electron microscopes produce a three-dimensional image of the cells and tissues. Electron microscopes are not suitable for routine or rapid observations of tissue samples.

Tissue Culture. Cell and tissue cultures can be grown on an artificial growth medium in a laboratory by providing adequate growing conditions in an incubator. Cultures are used to study living cells and tissues under experimental conditions. Tissue cultures are used to study tissue physiology, mechanisms of cell differentiation and development, cell-cell interaction, and gene regulation. Cell cultures are also used as a testing system for drug development. For example, fibroblasts, a cell type of connective tissue, grows well in the laboratory, providing a homogenous, live tissue material for drug testing.

Fluorescent Staining. Certain methods of tissue staining differentially detect the presence of particular chemical components in cells. One approach is to use a fluorescent label to trace components inside the cell. Fluorescent staining not only attests to the presence or absence of certain components in cells but also provides their precise localization in a particular cell or tissue type. For example, immunochemistry uses fluorescent antibodies that specifically bind with the target structures in the tissue so that the labeled antibodies will highlight the presence of the structures of interest in the analyzed specimen. Some applications of these techniques are staining of virus-infected cells in tissues and detection of immune-complex deposition in the kidneys and skin of patients with lupus.

APPLICATIONS AND PRODUCTS

Progress in histology has been dependent on the development of microscopy. Many companies are developing microscopes with various technical specifications because microscopes are being used not only in the biomedical field but also in fields such as engineering, optical physics, and biotechnology. Industry leaders such as Olympus and Carl Zeiss offer high-quality optics, superior construction, and high-image contrast to satisfy the growing demands of researchers. The companies develop and produce instruments, software, and accessories for microscope systems for use in industrial settings and clinical and research laboratories. Advances in technology have resulted in the creation of stereo microscopy, virtual microscopy, and total internal reflection fluorescence (TIRF) microscopy.

Fluorescence Microscopy. Improvements in fluorescence observation have resulted in images that are twice as bright as conventional fluorescence images. Fluorescence microscopy is used for staining and

observation of tuberculosis bacteria and of tissues from pulmonary adenocarcinoma and breast cancer. The Carl Zeiss company, a member of the Stop Tuberculosis Initiative, provided the Primo Star iLED fluorescence microscope at a lower price to seventy-four countries with a high incidence of the disease. The microscope delivers up to four times faster detection of tuberculosis than can be achieved with traditional techniques.

NanoZoomer Digital Pathology is a system for scanning glass sides of tissue and rapidly converting them to high-resolution digital slides, known as virtual slides. By magnifying or shrinking a chosen area while changing the focus on a monitor screen, researchers can view intracellular structures in fine detail. This system of virtual microscopy can be employed for toxicology tests, gene-expression analysis, and protein localization. The great advantage of digital slides is that they eliminate the worry of the fluorescence fading and degrading the image quality. Digital slide technology enables researchers to archive entire slides at high resolution, copy and edit them, and easily share them with others, allowing them to be used as the focal point for discussions.

Stereo microscopy is used by cutting-edge biological and medical laboratories that require the most effective imaging and observation of a large quantity of live specimens. It achieves the world's highest zoom ratio of 1:16.4, enabling remarkably sharp three-dimensional imaging and considerably enhanced specimen manipulation.

Live microscopes, such as Olympus's VivaView fluorescence incubator microscope, achieve the dual objectives of culturing cells under monitored conditions and simultaneously observing the cultured cells under a microscope. These microscopes allows cell analysis in the most optimized environment.

Total internal reflection fluorescence (TIRF) microscopes are employed to study a diverse phenomena, including cell transport, signaling, replication, motility, adhesion, and migration; cell membranes and transport; the structure of ribonucleic acid (RNA); neurotransmitters; and virology.

Industrial microscopy offers a wide range of both industrial microscopes and microimaging systems for metrology, semiconductor wafer manufacturing, quality assurance, advanced materials analysis, metallography, and other precision applications in various industrial sectors.

Fascinating Facts About Histology

- Antoni van Leeuwenhoek was the first person to observe human red blood cells. The preserved tissue samples that he used are part of the collections of the Royal Society of London. Leeuwenhoek was not a trained scientist but a merchant who made microscopes for a hobby.

- Olympus's VivaView fluorescence incubator microscope allows researchers to observe cultured cells growing in real time. This combination of a microscope and an incubator results in high-quality live-cell imaging with the freedom to control growth conditions of cells and use the cells for a significantly longer time than was previously possible.

- A transmission electron microscope allows researchers to see parts of individual human chromosomes, single proteins, and DNA.

- The life span of red blood cells is between 100 and 120 days, after which the cells undergo apoptosis (programmed cell death).

- There are about 100 trillion cells in the human body.

- A Pap smear (also known as a Papanicolaou test) involves microscopic examination of cells scraped from stratified squamous epithelium of the vagina and cervix of the uterus. It is performed for early detection of a precancerous condition or cancer. Collected cells are smeared on a microscope slide and sent to a laboratory for analysis.

- Tissue engineering is a technology in which living tissues are combined with synthetic materials to grow new tissues in the laboratory. Laboratory-grown versions of skin, cartilage, and bone are developed by using biodegradable synthetic materials or natural collagen fibers as a scaffolding system to immobilize and grow cells. The new tissue is then implanted into the patient.

Sample Preparation. Isolation of single cells and cell groups, as well as biomolecules, from a heterogeneous tissue is crucial for preservation of the material. Therefore, the proper preparation and staining of high-purity samples is critical for production of reliable results. Laser microdissection with optical tweezers, a technology developed by Carl Zeiss, delivers absolute purity in sample isolation with maximum preservation of the material and without affecting the viability of live cells. The technology combines microdissection and advanced imaging and incorporates digital camera technology. Major achievements of this technology include instant viewing, editing, and sharing of images over a network from any remote location. It is employed for specimen capture and for isolation of DNA, RNA, and proteins. Leica Microsystems was the first company to automate microtomy (microtome sectioning) and the first to introduce an integrated workstation for staining and cover slipping. The system does not require constant supervision, which provides an enormous time-saving advantage.

Clinical and Educational Applications. Histology is used in clinical, research, and educational settings. Biomedical research and clinical testing centers use fluorescent tags in staining to identify components of interest. For instance, flow cytometry analysis uses fluorescent-labeled antibodies as markers to perform rapid counts of blood cell types for clinical analysis. Hospitals, universities, and research centers provide training, education, and workshops on various protocols for specimen preparation as well as digital slide scanning for researchers and medical professionals. Histology is used in the manufacture of prepared tissue slides for biology, anatomy, and physiology courses at high schools and colleges.

Diagnosis of Diseases. Microscopic analysis of tissue samples is used to diagnose and classify many types of cancer, including cervical, breast, prostate, pancreatic, skin, and blood cancers, as well as noncancerous diseases. Histology tests can reveal if a tumor is cancerous, and if it is, they can determine the stage of progression, which leads to appropriate clinical treatment and an overall prognosis. Hematology analysis (of the blood cells) is usually the first step in the diagnostic process for cancer and noncancerous diseases.

IMPACT ON INDUSTRY

Histology, a valuable tool in biomedical research and clinical diagnostics, is funded through federal and private sources. The governmental agencies that provide most of the funding for biomedical research in universities and research institutes are the National Institutes of Health (NIH) and the National Science Foundation (NSF). Biomedical research funding increased from $37.1 billion in 1994 to $94.3 billion in 2003 and doubled when adjusted for inflation. By 2007, biomedical research funding had increased to $101.1 billion. Funding for this area of research continues to grow steadily.

Manufacturers that develop innovative microscopes, microtomes for tissue sectioning, staining reagents, and the glassware used by biomedical researchers and clinical laboratories are part of a multibillion-dollar industry. These manufacturers provide histology-related equipment to hospital and academic research centers, diagnostic laboratories, biotechnology companies, and other biomedical companies. Every major hospital has histology facilities that test and interpret samples.

CAREERS AND COURSE WORK

Specialists in histology can choose from numerous training programs. Histologists are in demand, and job prospects are excellent. To become a histologist, an individual must study microscopy, anatomy, physiology, microbiology, and chemistry, and master laboratory and diagnostic techniques. Some histology students obtain special training in molecular biology so that they can isolate DNA and RNA from human tissue.

Careers in histology can take different paths. Histology specialists can work in hospitals, medical laboratories, and universities as technicians, assistants, and researchers, or they can become instructors or professors at universities. As of 2010, the average yearly salary for a histology technician in the United States was about $46,000. University professors receive between $40,000 and $120,000 per year. Histology technicians and assistants work in hospitals and laboratories helping with clinical and laboratory procedures. A bachelor's degree program in histology is required for histology technicians. For histology assistants, a two-year associate's degree in medical assisting is sufficient. Advanced degrees such as a M.D. or Ph.D. are necessary for becoming a university professor. At universities, histology specialists may divide their time between research and teaching.

SOCIAL CONTEXT AND FUTURE PROSPECTS

Histology is a field of continuous growth. Many areas of medicine–such as oncology and regenerative medicine–depend on and require knowledge of histology. Histological parameters such as tumor size and morphologic characteristics of tissues are the most important diagnostic factors for cancer. Tissue engineering, a subspecialty of regenerative medicine, deals with repairing and replacing tissues (such as skin, bone, cartilage, or blood vessels) in part or in their entirety. Examples of tissue engineering include artificial bladders and edible artificial animal muscle tissue (artificial meat). Many materials, both natural and synthetic (including carbon nanotubes), are being actively investigated for use in tissue engineering. Scientists are also pursuing stem cell development and manipulation for tissue regeneration. Interest in this area has expanded, leading to the establishment of the California Institute for Regenerative Medicine in 2004.

Elvira R. Eivazova, Ph.D.

FURTHER READING

Croft, William J. *Under the Microscope: A Brief History of Microscopy.* Hackensack, N.J.: World Scientific, 2006. Traces the microscope from early beginnings to modern instruments, discussing how each works.

Hewitson, Tim D., and Ian A. Darby, eds. *Histology Protocols.* New York: Humana Press, 2010. This laboratory manual looks at tissue preparation and staining, with explanations of complex procedures.

Ovalle, William K., and Patrick C. Nahirney. *Netter's Essential Histology.* Philadelphia: Saunders/Elsevier, 2008. This atlas covers cells and tissues and the major bodily systems. Features a great collection of images by Frank H. Netter.

Ross, Michael H., and Pawlina Wojciech. *Histology: A Text and Atlas–With Correlated Cell and Molecular Biology.* 6th ed. Philadelphia: Wolters Kluwer/Lippincott Williams & Wilkins Health, 2011. Contains great illustrations, easy to follow diagrams. Recommended for medical, health professions, and undergraduate biology students

Tortora, Gerard J., and Bryan Derrickson. *Principles of Anatomy and Physiology.* 12th ed. Hoboken, N.J.: John Wiley & Sons, 2009. Excellent anatomy and physiology textbook with great illustrations and clinical references.

WEB SITES

American Society for Clinical Pathology
http://www.ascp.org

California Institute for Regenerative Medicine
http://www.cirm.ca.gov

Loyola University Medical Education Network
Histology
http://www.lumen.luc.edu/lumen/MedEd/Histo/frames/histo_frames.html

National Society for Histotechnology
http://www.nsh.org

Olympus America
Microscopy Resource Center
http://www.olympusmicro.com

See also: Artificial Organs; Bionics and Biomedical Engineering; Biosynthetics; Cell and Tissue Engineering; Electron Microscopy; Hematology; Microscopy; Pathology; Stem Cell Research and Technology.

HOLOGRAPHIC TECHNOLOGY

FIELDS OF STUDY

Optics; photography; computer science; laser technology; chemistry; electrical engineering; electronics; mathematics; and physics.

SUMMARY

Holographic technology employs beams of light to record information and then rebuilds that information so the reconstruction appears three-dimensional. Unlike photography, which traditionally produces fixed two-dimensional images, holography re-creates the lighting from the original scene and results in a hologram that can be viewed from different angles and perspectives as if the observer were seeing the original scene. The technology, which was greatly improved with the invention of the laser, is used in various fields such as product packaging, consumer electronics, medical imaging, security, architecture, geology, and cosmology.

KEY TERMS AND CONCEPTS

- **Diffracted Light:** Light that is modified by bending when it passes through narrow openings or around objects.
- **Emulsion:** Light-sensitive mixture that forms a coating on photographic film, plates, or paper.
- **Incandescent Light:** Light that is white, glowing, or luminous as a result of being heated.
- **Monochromatic Light:** 1. Having or appearing to have one color or hue. 2. Consisting of or composed of radiation of a single wavelength.
- **Object Beam:** Beam of light that first reaches the object and is then reflected off the object.
- **Reference Beam:** Beam of light that does not reflect off an object but interferes with the object beam to produce an interference pattern on the emulsion.

DEFINITION AND BASIC PRINCIPLES

Holography is a technique that uses interference and diffraction of light to record a likeness and then rebuild and illuminate the likeness. Holograms use coherent light, which contains waves that are aligned with one another. Beams of coherent light interfere with one another as the image is recorded and stored, thus producing interference patterns. When the image is re-illuminated, diffracted light allows the resulting hologram to appear three dimensional. Unlike photography, which produces a fixed image, holography re-creates the light of the original scene and yields a hologram, which can be viewed from different angles and different perspectives just as if the original subject were still present.

Several basic types of holograms can be produced. A transmission hologram requires an observer to see the image through light as it passes through the hologram. A rainbow hologram is a special kind of transmission hologram: Colors change as the observer moves his or her head. This type of transmission hologram can also be viewed in white light, such as that produced by an incandescent lightbulb. A reflection hologram can also be viewed in white light. This type allows the observer to see the image with light reflected off the surface of the hologram. The holographic stereogram uses attributes of both holography and photography. Industry and art utilize the basic types of holograms as well as create new and advanced technologies and applications.

BACKGROUND AND HISTORY

Around 1947, Hungarian-born physicist Dennis Gabor, while attempting to improve the electron microscope, discovered the basics of holography. Early efforts by scientists to develop the technique were restricted by the use of the mercury arc lamp as a monochromatic light source. This inferior light source contributed to the poor quality of holograms, and the field advanced little throughout the next decade. Laser light was introduced in the 1960's and was considered stable and coherent. Coherent light contains waves that are aligned with one another and is well suited for high-quality holograms. Subsequently, discoveries and innovations in the field began to increase and accelerate.

In 1960, American physicist Theodore Maiman of Hughes Research Laboratories developed the pulsed ruby laser. This laser used rubies to operate and generated powerful bursts of light lasting only nanoseconds. This laser, which acted much like a camera's flashbulb, became ideal for capturing images of moving objects or people.

In 1962, deciding to improve upon Gabor's technique, scientists Emmett Leith and Juris Upatnieks, working at the University of Michigan, produced images of three-dimensional (3-D) objects–a toy bird and train. These were the first transmission holograms and required an observer to see the image through light as it passed through the holograms.

Additionally in 1962, Russian scientist Yuri N. Denisyuk combined his own work with the color photography work of French physicist Gabriel Lippmann. This resulted in a reflection hologram that could be viewed with white light reflecting off the surface of a hologram. Reflection holograms did not need laser light to be viewed.

In 1968 electrical engineer Stephen Benton developed the rainbow hologram–as the observer moves his head, he sees the spectrum of color as in a rainbow. This type of hologram can also be viewed in white light.

Holographic art appeared in exhibits beginning in the late 1960's and early 1970's, and holographic portraits, made with the pulsed ruby laser, found some favor beginning in the 1980's. Advances in the field have continued, and many varied types of holograms are utilized in many different areas of science and technology, while artistic applications have lagged in comparison.

Fascinating Facts About Holographic Technology

- Larry Siebert of the Conductron Corporation produced the first pulsed hologram of a person when he created a self-portrait on October 31, 1967.
- The Cranbrook Academy of Art in Michigan held the first holographic art exhibition in 1968.
- Dennis Gabor received the 1971 Nobel Prize in Physics for his discovery of and work in holography.
- Kiss II, a well-known multiplex hologram, was produced in 1974 by Lloyd Cross. A 3-D image of Pam Brazier floats in the air–she blows a kiss and winks to passersby.
- *National Geographic* crafted an entire magazine cover made of an embossed hologram in 1988.
- The first holoprinter was created by Dutch Holographic Laboratory in 1992.

HOW IT WORKS

A 3-D subject captured by conventional photography becomes stored on a medium, such as photographic film, as a two-dimensional (2-D) scene. Information about the intensity of the light from a static scene is acquired, but information about the path of the light is not recorded. Holographic creation captures information about the light, including the path, and the whole field of light is recorded.

A beam of light first reaches the object from a light source. Wavelengths of coherent light, such as laser light, leave the light source "in phase" (in sync) and are known collectively as an object beam. These waves reach the object, are scattered, and then are interfered with when a reference beam from the same light source is introduced. A pattern occurs from the reference beam interfering with the object waves. This interference pattern is recorded on the emulsion. Re-illumination of the hologram with the reference beam results in the reconstruction of the object light wave, and a 3-D image appears.

Light Sources. An incandescent bulb generates light in a host of different wavelengths, whereas a laser produces monochromatic wavelengths of the same frequency. Laser light, also referred to as coherent light, is used most often to create holograms. The helium-neon laser is the most commonly recognized type.

Types of lasers include all gas-phase iodine, argon, carbon dioxide, carbon monoxide, chemical oxygen-iodine, helium-neon, and many others.

To produce wavelengths in color, the most frequently used lasers are the helium-neon (for red) and the argon-ion (for blue and green). Lasers at one time were expensive and sometimes difficult to obtain, but modern-day lasers can be relatively inexpensive and easier to use for recording holograms.

Recording Materials. Light-sensitive materials such as photographic films and plates, the first resources used for recording holograms, still prove useful. Since the color of light is determined by its wavelength, varying emulsions on the film that are sensitive to different wavelengths can be utilized to record information about scene colors. However, many different types of materials have proven valuable in various applications.

Other recording materials include dichromated gelatin, elastomers, photoreactive polymers, photochromics, photorefractive crystals, photoresists, photothermoplastics, and silver-halide sensitized gelatin.

APPLICATIONS AND PRODUCTS

Art. Holographic art, prevalent in the 1960's through the 1980's, still exists, although fewer artists practice holography solely. Many artistic creations contain holographic components. A modicum of schools and universities teach holographic art.

Digital Holography. Digital holography is one of the fastest-growing realms and has applications in the artistic, scientific, and technological communities. Computer processing of digital holograms lends an advantage, as a separate light source is not needed for re-illumination.

Digital holography first began to appear in the late 1970's. The process initially involved two steps: the writing of a string of digital images onto film and then converting the images into a hologram. Around 1988, holographer Ken Haines invented a process of creating digital holograms in one step.

Digital holographic microscopy (DHM) can be utilized noninvasively to study changes in the cells of living tissue subjected to simulated microgravity. Information is captured by a digital camera and processed by software.

Display. Different types of holograms can be displayed in store windows, as visual aids to accompany lectures or presentations, in museums, at art, science, or technology exhibits, in schools, libraries, or at home as simple decorations hung on a wall and lit by spotlights.

Embossed Holograms. Embossed holograms, which are special kinds of rainbow holograms, can be duplicated and mass produced. These holograms can be used as means of authentication on credit cards and driver's licenses as well as for decorative use on wrapping paper, book covers, magazine covers, bumper stickers, greeting cards, stickers, and product packaging.

Holograms in Medicine and Biology. The field of dentistry provided a setting for an early application in medical holography. Creating holograms of dental casts markedly reduced the space needed to store dental records for Britain's National Health Service. Holograms have also proved useful in regular dental practice and dentistry training.

The use of various types of holograms prove beneficial for viewing sections of living and nonliving tissue, preparing joint replacement devices, noninvasive scrutiny of tumors or suspected tumors, and viewing the human eye. A volume-multiplexed hologram can be used in medical-scanning applications.

Moving Holograms. Holographic movies created for entertaining audiences in a cinema are in development. While moving holograms can be made, limitations exist for the production of motion pictures. Somewhat more promising is the field of holographic video and possibly television.

Security. A recurring issue in world trade is that of counterfeit goods. Vendors increasingly rely on special holograms embedded in product packaging to combat the problem. The creation of complex brand images using holographic technology can offer a degree of brand protection for almost any product, including pharmaceuticals.

Security holography garners a large segment of the market. However, makers of security holograms, whose designs are utilized for authentication of bank notes, credit cards, and driver's licenses, face the perplexing challenge of counterfeit security images. As time progresses, these images become increasingly easier to fake; therefore, this area of industry must continually create newer and more complex holographic techniques to stay ahead of deceptive practices.

Stereograms. Holographic stereograms, unique and divergent, use attributes of both holography and photography. Makers of stereograms have the potential of creating both very large and moving images. Stereograms can be produced in color and also processed by a computer.

Non-Optical Holography. Types of holography exist that use waves other than light. Some examples include acoustical holography, which operates with sound waves; atomic holography, which is used in applications with atomic beams; and electron holography, which utilizes electron waves.

IMPACT ON INDUSTRY

As time progresses, holography increasingly appears in more and varied industries, even branching into non-optical fields. In addition to those mentioned above, applications can also be found in industries such as aeronautics, architecture, automotive, the environment and weather, cosmology, geology, law enforcement, solar energy, and video-game design and programming. As of 2011, commercial, technological, and scientific applications outnumber applications in fine art.

Government Research. United States Army and Air Force research offices have funded holographic work leading to the development of technologies including 3-D mapping, risk reduction for helicopter flight and landing, and improving visualization and interface technology during military operations.

Acoustic holography and sonar technology have been used in research by the United States Navy, which has also worked with holographic cameras capable of operating underwater.

Schools and Universities. A few schools are breaking ground with new developments in optics, holographic television, holographic sensors, and recording materials. Some companies have collaborated with a few progressive schools and universities, providing art students with innovative holographic systems for the purpose of experimental and creative work. Such work evolves the field of holography and opens doors to future applications, which can ultimately influence industry. Among the schools that have made valuable contributions to the holographic frontier as well as to the many related disciplines are Adelphi University; Cambridge University and De Montfort University in the United Kingdom; Korean National University of Art and Holocenter Korea; Academy of Media Arts (KHM) Cologne in Germany; Kun Shan University in Taiwan; University of Arizona; and University of New South Wales College of Art in Australia.

Major Corporations. Holographic applications in industry have become, over time, increasingly diverse and valuable. Among the global corporations that have made significant contributions to the industry are Colour Holographic and De La Rue in the United Kingdom, Dutch Holographic Laboratory in the Netherlands, Geola in Italy, Holographics North in Vermont, Illumina in California, Rabbitholes Media in Canada, SAIC in Virginia, and Zebra Imaging in Texas.

CAREERS AND COURSE WORK

Outside of fine arts, careers that include holographic applications are varied. While studies should be concentrated in a particular industry or area of interest, students should have a basic foundation in chemistry, electronics, mathematics, and physics. Advanced courses in these subjects should be taken according to the requirements of the overall industry of interest. Course work in computer science, software engineering, electrical engineering, laser technology, optics, and photography should also be given consideration. Students should expect to earn at least a bachelor of science in their chosen field. For advanced career positions, generally a master's or doctorate degree will be required.

SOCIAL CONTEXT AND FUTURE PROSPECTS

Holography in one form or another is prevalent in modern society, whether as a security feature on a credit card, a component of a medical technique, or a colorful wrapping paper. Holograms have been interwoven into daily life and will likely continue to increase their impact in the future.

Next-generation holographic storage devices have been developed, setting the stage for companies to compete for future markets. Data is stored on the surface of DVDs; however, devices have been invented to store holographic data within a disk. The significantly enlarged storage capacity is appealing for customers with large storage needs who can afford the expensive disks and drives, but some companies are also interested in targeting an even larger market by revising existing technology. Possible modification of current technology, such as DVD players, could potentially result in less expensive methods of playing 3-D data.

Upcoming technology for recording and displaying 3-D images in another format is that of 3-D television. This idea has been around for awhile; however, advances have been slow but potentially promising.

Glenda Griffin, B.A., M.L.S.

FURTHER READING

Ackermann, Gerhard K., and Jürgen Eichler. *Holography: A Practical Approach.* Weinheim, Germany: Wiley-VCH, 2007. Based on university laboratory courses, and contains more than 100 problems with solutions. Also discusses new developments in holography.

Hariharan, P. *Basics of Holography.* Cambridge, England: Cambridge University Press, 2002. This resource introduces the basics of holography for students of science and engineering as well as for people with an interest in holography and a basic understanding of physics.

Harper, Gavin D. J. *Holography Projects for the Evil Genius.* New York: McGraw Hill, 2010. Explains the basics of holography and provides do-it-yourself projects with easily accessible materials. Glossary, index, and supplier's index included.

Johnston, Sean F. "Absorbing New Subjects: Holography as an Analog of Photography." *Physics in Perspective* vol. 8, issue 2 (2006): 164-188. Provides the history and background in the development of holography and examines the cultural influences on the field.

Saxby, Graham. *Practical Holography*. 3d ed. Bristol, England: Institute of Physics Publishing, 2004. Explains the basics of holography and provides easy steps to making holograms. Offers ideas on creating a holographic studio.

Yaroslavsky, Leonid. *Digital Holography and Digital Image Processing: Principles, Methods, Algorithms*. Norwell, Mass.: Kluwer Academic, 2010. A basic introduction to digital holography including valuable information on signal processing.

WEB SITES
American Institute of Physics
http://www.aip.org

International Hologram Manufacturer's Association
http://www.ihma.org

National Science Foundation
http://www.nsf.gov

See also: Computer Science; Electronics and Electronic Engineering; Laser Technologies; Optics; Photography; Software Engineering.

HORTICULTURE

FIELDS OF STUDY

Plant science; crop science; botany; agronomy; agriculture; biochemistry; ecology; entomology; genetics; landscape architecture; soils engineering.

SUMMARY

Horticulture is both a science and an art and includes the propagation and cultivation of plants often associated with gardening. Trees, bushes, grasses, and fungi are typical naturally occurring plant materials. Cultivated plants include not only trees, shrubs, grasses, and fungi but also vegetables, fruits, nuts, herbs, spices, ornamentals, and flowers. Although horticulture has existed since ancient times, modern scientific advances in the field have resulted in many practical uses of plant materials, including the development of pharmaceuticals. Genetics plays a significant role in modern-day horticultural practices, especially in areas such as food production.

KEY TERMS AND CONCEPTS

- **Botany:** Branch of biology that involves the scientific study of plants.
- **Genetic Engineering:** Technique used to modify the genes of a plant so that it has a desired trait.
- **Hydroponics:** Method of growing plants in water, without soil, by adding nutrients to the water.
- **Micropropagation:** Cloning of multiple plants from the tissues of another plant.
- **Plant Biotechnology:** Using biological systems to genetically engineer hardier plants and enhance their productivity.
- **Plant Structures:** Parts of plants used most often in propagation, including runners, tubers, corms, bulbs, and rhizomes.
- **Propagation:** Sexual production of plants from seeds or asexual production through cuttings, grafts, and tissue cultures.
- **Xeriscaping:** Method of growing gardens in arid conditions by using drought-resistant plants and water conservation techniques.

DEFINITION AND BASIC PRINCIPLES

Horticulture is the art of aesthetically arranging plant materials and the science of breeding, propagating, and growing plants. Horticulture requires a knowledge of landscape design, plant science, and individual plants. The ability to identify plants and their growing needs is necessary for determining how to use plant materials in landscape designs and also in plant breeding and propagation.

Plants are classified as herbaceous or woody. Woody plants include deciduous and evergreen trees, shrubs, and smaller plant materials. Deciduous plants become dormant at the end of the growing season, and plants classified as evergreens will retain their foliage–needles and broad leaves–throughout the year.

Horticulture is not the same as agriculture, which involves growing crops on a much larger scale than the typical horticultural venture and may also include animal production. Horticulture can be pursued either as a hobby or as a commercial pursuit. Those involved in horticulture need to be versed in many disciplines. Design, planning, graphics, and construction engineering are important for the art of horticulture, and a background in math and biochemistry are needed for horticultural science.

BACKGROUND AND HISTORY

Horticulture is associated with ancient botany, which began as early as the fourth century B.C.E. Plants were selected and domesticated, and early propagation practices such as grafting can be seen in artwork found in ancient Egyptian tombs. Classical philosophers, such as Aristotle, defined plant types and developed some of their early uses for food, drugs, and fuel. Although an interest in botanical sciences waned during the Middle Ages, monasteries were known for their gardening prowess. During the eighteenth and nineteenth centuries, gardening became important as an art form, especially in Europe and Asia, and horticulture as a science began to expand.

One of the first modern horticulturalists in the United States was Liberty Hyde Bailey, a graduate of Michigan Agricultural College. As is true of many horticulturalists, Bailey wrote many texts on botany and horticulture. Bailey also worked with President Theodore Roosevelt's administration to improve rural life through horticulture. Early landscape architects, such as Frederick Law Olmstead, were also instrumental in advancing the field of horticulture.

How It Works

A horticulturalist must be cognizant of soil, light, temperature, and water conditions to propagate and cultivate plants. Horticulture begins with the preparation of soil through tilling and fertilization, unless the plant is grown without soil as in hydroponics. Light and temperature can be controlled by beginning plant cultivation inside greenhouses and transplanting plants to the outdoors when light and temperature are optimal for growth. Installation of water collectors and irrigation systems, which have been part of horticulture since ancient times, is also necessary. Xeriscaping with modified drought-resistant plant materials, however, has reduced the reliance on water in some locations. Control of pests and diseases is also important, although many pest- and disease-resistant plant materials exist because of modern technology. Some horticulturalists no longer rely on chemical fertilizers and pesticides, and organic gardening is becoming profitable because of the public demand for naturally grown foods. Horticulture may also involve modern harvesting and handling practices that use more mechanical methods as opposed to manual labor.

Propagation. Propagation techniques vary from microscopic techniques such as gene splicing and

Separating plant tissue from culture for micropropagation. (Nigel Cattlin/Holt Studios/Photo Researchers, Inc.)

producing tissue cultures to plant breeding and employing natural and cultivated plant structures and cuttings to generate new plants. The most common form of plant propagation is with seeds. Vegetables and flowers are typically propagated in this manner.

Clonal propagation can be done naturally or achieved through scientific methods involving plant structures rather than seeds. Some examples of natural clonal propagation through plant structures include strawberry runners, potato tubers, crocus corms, daffodils bulbs, and Calla lily rhizomes. Scientific propagation can be as simple as placing cuttings of shoots, roots, and leaves into water or moist soil. Shoots, such as willows, will regenerate roots. Roots, such as sweet potatoes, will grow new shoots, and leaves, such as African violets, will grow both new shoots and roots.

More complex clonal propagation methods include grafting techniques that physically join the parts from two plants together (often used in the propagation of fruit trees) and micropropagation, which requires the horticulturalist to grow plant cells and tissues artificially. Micropropagation is used in the production of modified plants that may have desirable qualities such as pest and disease resistance. It enables the production of a greater number of hardier plants in a shorter amount of time than could be produced from seeds or other propagation methods in the same amount of time.

Cultivation. Cultivation involves the direct planting of seeds and plant structures or the transplanting of plants that were started elsewhere. Cultivation of plant materials requires pruning and weeding to enhance future growth and improve on a plant's appearance. Ornamental cultivation techniques might include training plants so that they are reoriented to better conditions, such as light. Dwarfing is an ornamental cultivation practice that is often used in Asian cultures. Ornamental horticulture is important for those involved in floriculture and landscape architecture.

Applications and Products

Plants are everywhere and are necessary to sustain life. The practice of horticulture, therefore, has multiple applications that are important to society. In addition, modern technology has resulted in a greater diversity in plant-based products.

Food Production. Plant biotechnology includes genetic engineering, which allows the exchange of genes between plant species, usually ones that are closely related. More than 70 percent of processed foods contain some ingredients that have been genetically engineered, and biotech crops, such as corn and soybeans, are on the rise. New horticultural technologies and research are especially important in global regions, where horticultural production is restricted because of marginal climate conditions, poor soils, lack of water for irrigation, and reduced land resources for cultivation. DNA plays a major role in plant breeding of both quantitative and qualitative traits. Developing nations must especially depend on horticultural research to breed plants with pest- and drought-resistant characteristics to expand productivity on less acreage.

The global food market is changing. The ability to transport food products longer distances and the improved shelf life of such food are the result of horticultural advances that have opened up new markets for foods that may be exported by small producers in developing nations. Moreover, the methods of producing food are changing, and horticulturalists are finding ways to meet society's desire for more natural and organic food products while sustaining and protecting the environment.

Pharmaceuticals. Herbal remedies and medicinal plants have been used by healing practitioners since ancient times. Although scientists have questioned the curative powers of some herbs, horticulture continues to play an important role in the breeding of plants used as alternative medicines and in the development of plant-based mainstream drugs. Horticulturalists are also involved in synthesizing plant compounds into new medicines. Rain-forest plants have become especially important in this research because of the many naturally occurring chemical defenses against pests and diseases that these plants exhibit. Horticulturalists are on the forefront in conducting research concerning the curative properties of naturally occurring chemical compounds found in plants. These chemicals, known as pytochemicals, may provide new drugs for fighting cancer.

Flavors and Fragrances. The cultivation of plants such as flowers and fruits is necessary to produce many diverse products. Among these are cosmetics, perfumes, and food flavorings. Plant breeding and genetic engineering have enhanced the quality of plant-based flavorings, thus making some foods more palatable.

Farms and Nurseries. Horticultural practices play an important role in any enterprise that involves the growing of plants for commercial, public, or private uses. Tree farms range from farms specializing in the production of Christmas trees to those providing trees for forest conservation, which protects the environment by providing flood control and carbon sequestration. Other farms may produce vegetables, flowers, or turf, and golf course, park, and garden designers depend on farms and nurseries to produce plant materials to implement their plans.

Energy Production. Horticulture plays a role in the production and breeding of plants that can be converted into biofuels. Although the biofuel industry is in its infancy, substantial global research is taking place. For example, Clemson University is involved in research to turn discarded peaches into hydrogen gas, and nations such as India are producing biodiesel from nonedible plant oil seeds. Sugarcane and maize have also been used to produce ethanol. Because diverse plants can be used to produce biofuels, developing nations may be able to create biofuels cheaply by using naturally growing plant materials and thus create new industries to sustain population growth. However, horticulturalists must ensure that sufficient plants are available for both food production and these emerging biofuel industries.

Education. Development of healthy eating habits has become an important goal in many developed nations, such as the United States, where obesity is on the increase. Educators are employing gardening and horticulture in their lesson plans to teach children about nutrition. In less-developed countries, where poor nutrition is caused by a lack of horticultural crops, international organizations such as the United Nations provide education not only on advanced production and cultivation practices including irrigation and pest control but also on food safety, conservation, and marketing.

Therapy. Horticulture and gardening are therapeutic for those with learning, mental health, and physical disabilities. The American Horticultural Therapy Association is one of several organizations involved in this growing practice. Horticultural therapists incorporate gardens into wellness programs, and this type of therapy has become popular in programs to treat alcohol and substance abusers.

IMPACT ON INDUSTRY

The Global Horticulture Initiative, known as GlobalHort, is an international consortium of institutions, governmental agencies, and nongovernmental organizations involved in horticultural research, training, and information sharing. Located in both Belgium and Tanzania, GlobalHort focuses on improving international health and nutrition through the production and promotion of various horticultural crops. Partners include the Food and Agriculture Organization of the United Nations, the World Health Organization, and the World Bank.

Government Agencies. In the United States, one of the primary agencies involved in horticultural research and education is the U.S. Department of Agriculture (USDA). State and local governments also have active horticultural research and public education programs, including those associated with county and university agricultural extension services. The USDA is in partnership with many researchers at land-grant universities and sponsors the National Agricultural Library, which collaborates with public, private, and nonprofit agencies to provide education, often concerning horticultural aspects of food production.

University Research. Much of horticultural research is conducted at universities. In the United States, many of these institutions are state universities that started as land-grant colleges. University research is concentrated on new plant science, especially related to propagation and cultivation. Many land-grant universities have botanical gardens that provide ample opportunities for field research with new plant materials and research trials to enhance plant characteristics. Often university research is conducted at experimental stations; research ranges from agroforestry experimentation at the University of Missouri to urban horticulture research at the University of California Cooperative Extension.

Institutions and Organizations. Great Britain's Royal Horticultural Society, which has focused on gardening for about 150 years, continues to provide horticultural advice and education, especially through its published journal. Horticultural research and education has been provided internationally since 1864 by the International Society for Horticultural Science, made up of scientists representing about 150 nations. The American Society for Horticultural Science provides a similar service in the United States.

Within the Society of Chemical Industry, an international private organization with offices in the United States, United Kingdom, and Australia, is a horticultural group concerned with transferring horticultural science into business practices, especially those associated with agriculture and food. In the United States, the American Horticultural Society is an important organization for providing private gardeners with horticultural information. Online networking is possible through the nonprofit organization, Society for Advancement of Horticulture, which connects researchers, growers, and commercial enterprises associated with horticulture.

Growers and Producers. Commercial and private growers and producers associate with one another to learn the latest technologies in horticulture and to share information. Growers and producers cultivate a variety of plant materials not limited to commercial produce, such as vegetables, fruits, nuts, and decorative trees, shrubs, grasses, and flowers.

Fascinating Facts About Horticulture

- Approximately 25 percent of Western medicines and drugs are derived from plants. The U.S. National Cancer Institute believes that more than 70 percent of tropical plants have medicinal properties that could help fight cancers.

- The oldest known living tree is a Norway spruce found in Sweden. Although the exposed part of the tree is only 13 feet tall, scientists believe its root system is about 9,550 years old.

- Most plants contain both male and female reproductive parts, thus simplifying pollination. Some plants, however, such as hollies, have separate male and female plants, and the female plant cannot produce red berries unless a male plant, which produces the pollen, is nearby so that pollination can take place.

- Plants such as ferns do not flower and must reproduce through spores, which are located on the underside of the plant's fronds.

- Bamboo not only is the world's tallest grass but also grows quickly. The plant can reach heights of 130 feet and grow 4 feet in a day.

- There are more than one hundred species of carnivorous plants that capture insects and digest them for their nutrients.

Some horticultural specialties include growing grapes and berries for wine, aquatic gardening, and producing unusual species such as carnivorous plants, cacti and succulents, and bonsai and dwarf varieties. Flower growers often specialize in one species such as roses, orchids, or lilies.

Related Industry and Business Sectors. Growers and producers must depend on industry to supply them with new horticultural products and the latest scientific advances, including seeds and plant structures, machinery, greenhouse and irrigation systems, and fertilizers and pesticides. Many of the newer companies are associated with the growth in organic and natural plant production. Once commercial products–everything from vegetables to flowers–are produced, the growers and producers must depend on the business sector to process plant products safely and to market them.

CAREERS AND COURSE WORK

Horticultural careers include gardening and landscaping, which may require little education, and scientific careers, such as plant breeding and genetic engineering, for which advanced degrees are required. A bachelor of science degree in horticulture or a related field prepares a student for careers including plant and crop cultivator and plant disease or pesticide specialist. Governmental careers such as extension work are also popular. A dual bachelor's degree in business administration and horticulture would support work in a greenhouse, nursery, garden center, or golf course. Those interested in garden and landscape design might seek a bachelor's degree in landscape architecture.

Students interested in plant biotechnology, one of the fastest growing fields, can work in research laboratories as technicians or research scientists. Many who have gained a doctorate degree in biotechnology work in educational settings and continue to conduct horticultural research. Biotechnologists are sought by governmental agencies and private industry, especially companies associated with food production and pharmaceuticals.

Basic courses for students seeking a career in horticulture include plant biology, botany, and plant materials. More advanced course work includes entomology, plant physiology, breeding and propagation, plant nutrition, and plant pathology. Specialized educational courses might include floriculture for those

who want to work with flowers or arboriculture for those interested in caring for trees. Students interested in an advanced degree, such as biotechnology, must expand their course work to include biochemistry, genetics, and molecular science.

SOCIAL CONTEXT AND FUTURE PROSPECTS

As the worldwide population continues to increase, horticultural advances through biotechnology can benefit society by ensuring an adequate food supply. Genetically engineered plants are more pest and disease resistant, and modified plants that withstand cold temperatures and are more tolerant of drought and salt allow for longer growing seasons and more numerous habitats. Plants are being engineered to have superior nutrient properties that could reduce the quantity of food necessary to nourish the world's growing population. Furthermore, many of the plants created through biotechnology and genetic engineering are playing a greater role in the production of beneficial medicines and drugs; some experts predict that society may one day have access to edible vaccines.

However, these scientific advances are raising concerns. Scientists are debating whether genetically engineered foods could produce gene mutations in organisms that consume them, create new food allergens, or cause environmental damage because of the antibiotic resistance of genetically engineered foods. Cross-pollination has also become an issue, as it cannot be prevented or stopped. Some fear the creation of super weeds that cannot be controlled after naturally occurring weeds cross-pollinate with genetically modified pesticide-resistant plants.

Carol A. Rolf, B.S.L.A., M.Ed., M.B.A., J.D.

FURTHER READING

Adams, Charles R., Katherine M. Bamford, and Michael P. Early. *Principles of Horticulture.* 5th ed. Burlington, Mass.: Butterworth-Heinemann, 2008. Introductory text dealing with all aspects of horticulture care and propagation, pest control, and soils, water, and nutrition.

DiSabato-Aust, Tracy. *The Well-Tended Perennial Garden: Planting and Pruning Techniques.* 2d ed. Portland, Oreg.: Timber Press, 2006. Useful guide for planning a perennial garden with a section for notes and personal documentation.

Garner, Jerry. *Careers in Horticulture and Botany.* 2d ed. New York: McGraw-Hill, 2006. Lists many job opportunities in the field of horticulture and describes the necessary education and training.

Pauly, Philip J. *Fruits and Plains: The Horticultural Transformation of America.* Cambridge, Mass.: Harvard University Press, 2008. Discusses how horticultural practice in the United States helped shape the nation.

Reiley, Edward H., and Carroll L. Shry, Jr. *Introductory Horticulture.* 7th ed. Florence, Ky.: Delmar Cengage Learning, 2006. Easy-to-read textbook that includes information on the latest technologies and on finding a job in horticulture.

Toogood, Alan. *American Horticultural Society Plant Propagation: The Fully Illustrated Plant-by-Plant Manual of Practical Techniques.* New York: DK Adult, 1999. Good reference for propagating more than 1,500 plants from existing specimens with illustrations and explanations of plant biology.

WEB SITES
American Horticultural Society
http://www.ahs.org

American Horticultural Therapy Association
http://www.ahta.org

American Society for Horticultural Science
http://www.ashs.org

Global Horticulture Initiative
http://www.globalhort.org

International Society for Horticultural Science
http://www.ishs.org

U.S. Department of Agriculture
Horticultural Crops Research Unit
http://www.ars.usda.gov/main/site_main.htm?modecode=53-58-10-00

See also: Agricultural Science; Agroforestry; Herbology; Hydroponics; Landscape Architecture and Engineering; Landscape Ecology; Pharmacology; Silviculture; Soil Science.

HUMAN-COMPUTER INTERACTION

FIELDS OF STUDY

Computer science; computer graphics; software engineering; systems analysis; graphic design; industrial design; ergonomics; mechanical engineering; information science; information architecture; robotics; artificial intelligence; cognitive science; psychology; social psychology; linguistics; neurobiology; psychophysics; social neuroscience; anthropology; scientific computing; data visualization; typography; anthropometrics.

SUMMARY

Human-computer interaction (HCI) is a field concerned with the study, design, implementation, evaluation, and improvement of the ways in which human beings use or interact with computer systems. The importance of human-computer interaction within the field of computer science has grown in tandem with technology's potential to help people accomplish an increasing number and variety of personal, professional, and social goals. For example, the development of user-friendly interactive computer interfaces, Web sites, games, home appliances, office equipment, art installations, and information distribution systems such as advertising and public awareness campaigns are all applications that fall within the realm of HCI.

KEY TERMS AND CONCEPTS

- **Accessibility:** Extent to which an interface can be used by people with visual, auditory, cognitive, or physical impairments.
- **Direct Manipulation:** Interacting with a graphic representation of an object to accomplish a task.
- **Ethnography:** Process of observing, interviewing, and analyzing the activities of people in their everyday environments in order to gain the perspective of a user.
- **Graphical User Interface (GUI):** Means of interacting with an electronic device based on images rather than text.
- **Heuristic:** Guideline or rule of thumb used to quickly evaluate an interface or product.

- **Information Architecture:** Study and practice of organizing data so that they can be found and used efficiently.
- **Likert Scale:** Method of ascribing quantitative value to qualitative data, used in questionnaires; typically asks respondents to rate how much they agree or disagree with a statement.
- **Ubiquitous Computing:** Model of computing that sees technology as being fully integrated into every aspect of daily life rather than limited to the functionality of a machine on a desktop.
- **Usability:** Extent to which a product can easily, efficiently, and effectively be used to achieve a certain goal.
- **Wayfinding:** Ways in which users orient themselves and navigate from place to place within an interface.
- **Widget:** Interactive component on a Web site, such as one that allows the user to click through a list of options and choose one.
- **Wire Frame:** Skeleton version of a Web site that leaves out visual design elements and focuses on how pages will be linked.

DEFINITION AND BASIC PRINCIPLES

Human-computer interaction is an interdisciplinary science with the primary goal of harnessing the full potential of computer and communication systems for the benefit of individuals and groups. HCI researchers design and implement innovative interactive technologies that are not only useful but also easy and pleasurable to use and anticipate and satisfy the specific needs of the user. The study of HCI has applications throughout every realm of modern life, including work, education, communications, health care, and recreation.

The fundamental philosophy that guides HCI is the principle of user-centered design. This philosophy proposes that the development of any product or interface should be driven by the needs of the person or people who will ultimately use it, rather than by any design considerations that center around the object itself. A key element of usability is affordance, the notion that the appearance of any interactive element should suggest the ways in which it can be manipulated. For example, the use of shadowing around a button on a Web site might help make it look three-dimensional, thus suggesting that it can be pushed or clicked. Visibility is closely related to affordance; it is the notion that the function of all the controls with which a user interacts should be clearly mapped to their effects. For example, a label such as "Volume Up" beneath a button might indicate exactly what it does. Various protocols facilitate the creation of highly usable applications. A cornerstone of HCI is iterative design, a method of development that uses repeated cycles of feedback and analysis to improve each prototype version of a product, instead of simply creating a single design and launching it immediately. To learn more about the people who will eventually use a product and how they will use it, designers also make use of ethnographic field studies and usability tests.

BACKGROUND AND HISTORY

Before the advent of the personal computer, those who interacted with computers were largely technology specialists. In the 1980's, however, more and more individual users began making use of software such as word-processing programs, computer games, and spreadsheets. HCI as a field emerged from the growing need to redesign such tools to make them practical and useful to ordinary people with no technical training. The first HCI researchers came from a variety of related fields: cognitive science, psychology, computer graphics, human factors (the study of how human capabilities affect the design of mechanical systems), and technology. Among the thinkers and researchers whose ideas have shaped the formation of HCI as a science are John M. Carroll, best known for his theory of minimalism (an approach to instruction that emphasizes real-life applications and the chunking of new material into logical parts), and Adele Goldberg, whose work on early software interfaces at the Palo Alto Research Center (PARC) was instrumental in the development of the modern graphical user interface.

In the early days of HCI, the notion of usability was simply defined as the degree to which a computer system was easy and effective to use. However, usability has come to encompass a number of other qualities, including whether an interface is enjoyable, encourages creativity, relieves tension, anticipates points of confusion, and facilitates the combined efforts of multiple users. In addition, there has been a shift in HCI away from a reliance on theoretical findings from cognitive science and toward a more hands-on approach that prioritizes field studies and usability testing by real participants.

HOW IT WORKS

Input and Output Devices. The essential goal of HCI is to improve the ways in which information is transferred between a user and the machine he or she is using. Input and output devices are the basic tools HCI researchers and professionals use for this purpose. The more sophisticated the interaction between input and output devices–the more complex the feedback loop between the two directions of information flow–the more the human user will be able to accomplish with the machine.

An input device is any tool that delivers data of some kind from a human to a machine. The most familiar input devices are the ones associated with personal computers: keyboards and mice. Other commonly used devices include joysticks, trackballs, pen styluses, and tablets. Still more unconventional or elaborate input devices might take the shape of head gear designed to track the movements of a user's head and neck, video cameras that track the movements of a user's eyes, skin sensors that detect changes in body temperature or heart rate, wearable gloves that precisely track hand gestures, or automatic speech recognition devices that translate spoken commands into instructions that a machine can understand. Some input devices, such as the sensors that open automatic doors at the fronts of banks or supermarkets, are designed to record information passively, without the user having to take any action.

An output device is any tool that delivers information from a machine to a human. Again, the most familiar output devices are those associated with personal computers: monitors, flat-panel displays, and audio speakers. Other output devices include wearable head-mounted displays or goggles that provide visual feedback directly in front of the user's field of vision and full-body suits that provide tactile feedback to the user in the form of pressure.

Perceptual-Motor Interaction. When HCI theorists speak about perceptual-motor interaction, what they are referring to is the notion that users' perceptions—the information they gather from the machine—are inextricably linked to their physical actions, or how they relate to the machine. Computer systems can take advantage of this by using both input and output devices to provide feedback about the user's actions that will help him or her make the next move. For example, a word on a Web site may change in color when a user hovers the mouse over it, indicating that it is a functional link. A joystick being used in a racing game may exert what feels like muscular tension or pressure against the user's hand in response to the device being steered to the left or right. Ideally, any feedback a system gives a user should be aligned to the physical direction in which he or she is moving an input device. For example, the direction in which a cursor moves on screen should be the same as the direction in which the user is moving the mouse. This is known as kinesthetic correspondence.

Another technique HCI researchers have devised to facilitate the feedback loop between a user's perceptions and actions is known as augmented reality. With this approach, rather than providing the user with data from a single source, the output device projects digital information, such as labels, descriptions, charts, and outlines, on the physical world. When an engineer is looking at a complex mechanical system, for example, the display might show what each part in the system is called and enable him or her to call up additional troubleshooting or repair information.

APPLICATIONS AND PRODUCTS

Computers. At one time, interacting with a personal computer required knowing how to use a command-line interface in which the user typed in instructions–often worded in abstract technical language–for a computer to execute. A graphical user interface, based on HCI principles, supplements or replaces text-based commands with visual elements such as icons, labels, windows, widgets, menus, and control buttons. These elements are controlled using a physical pointing device such as a mouse. For instance, a user may use a mouse to open, close, or resize a window or to pull down a list of options in a menu in order to select one. The major advantage graphical user interfaces have over text-based interfaces is that they make completing tasks far simpler and more intuitive. Using graphic images rather than text reduces the amount of time it takes to interpret and use a control, even for a novice user. This enables users to focus on the task at hand rather than to spend time figuring out how to manipulate the technology itself. For instance, rather than having to recall and then correctly type in a complicated command, a user can print a particular file by selecting its name in a window, opening it, and clicking on an icon designed to look like a printer. Similarly, rather than choosing options from a menu in order to open a certain file within an application, a user might drag and drop the icon for the file onto the icon for the application.

Besides helping individuals navigate through and execute commands in operating systems, software engineers also use HCI principles to increase the usability of specific computer programs. One example is the way pop-up windows appear in the word-processing program Microsoft Word when a user types in the salutation in a letter or the beginning item in a list. The program is designed to recognize the user's task, anticipate the needs of that task, and offer assistance with formatting customized to that particular kind of writing.

Consumer Appliances. Besides computers, a host of consumer appliances use aspects of HCI design to improve usability. Graphic icons are ubiquitous parts of the interfaces commonly found on cameras, stereos, microwave ovens, refrigerators, and televisions. Smartphones such as Apple's iPhone rely on the same graphic displays and direct manipulation techniques as used in full-sized computers. Many also add extra tactile, or haptic, dimensions of usability such as touchscreen keyboards and the ability to rotate windows on the device by physically rotating the device itself in space. Entertainment products such as video game consoles have moved away from keyboard and joystick interfaces, which may not have kinesthetic correspondence, toward far more sophisticated controls. The hand-held device that accompanies the Nintendo Wii, for instance, allows players to control the motions of avatars within a game through the natural movements of their own bodies. Finally, HCI research influences the physical design of many household devices. For example, a plug for an appliance designed with the user in mind might be deliberately shaped so that it can be inserted into an outlet in any orientation, based on the understanding that a user may have to fit several plugs into a limited amount of space, and many appliances have bulky plugs that take up a lot of room.

Increasingly, HCI research is helping appliance designers move toward multimodal user interfaces. These are systems that engage the whole array of human senses and physical capabilities, match particular tasks to the modalities that are the easiest and most effective for people to use, and respond in tangible ways to the actions and behaviors of users. Multimodal interfaces combine input devices for collecting data from the human user (such as video cameras, sound recording devices, and pressure sensors) with software tools that use statistical analysis or artificial intelligence to interpret these data (such as natural language processing programs and computer vision applications). For example, a multimodal interface for a GPS system installed in an automobile might allow the user to simply speak the name of a destination aloud rather than having to type it in while driving. The system might use auditory processing of the user's voice as well as visual processing of his or her lip movements to more accurately interpret speech. It might also use a camera to closely follow the movements of the user's eyes, tracking his or her gaze from one part of the screen to another and using this information to helpfully zoom in on particular parts of the map or automatically select a particular item in a menu.

Workplace Information Systems. HCI research plays an important role in many products that enable people to perform workplace tasks more effectively. For example, experimental computer systems are being designed for air traffic control that will increase safety and efficiency. Such systems work by collecting data about the operator's pupil size, facial expression, heart rate, and the forward momentum and intensity of his or her mouse movements and clicks. This information helps the computer interpret the operator's behavior and state of mind and respond accordingly. When an airplane drifts slightly off its course, the system analyzes the operator's physical modalities. If his or her gaze travels quickly over the relevant area of the screen, with no change in pupil size or mouse click intensity, the computer might conclude that the operator has missed the anomaly and attempt to draw attention to it by using a flashing light or an alarm.

Other common workplace applications of HCI include products that are designed to facilitate communication and collaboration between team members, such as instant messaging programs, wikis (collaboratively edited Web sites), and videoconferencing tools. In addition, HCI principles have contributed to many project management tools that enable groups to

---

Fascinating Facts About Human-Computer Interaction

- Children working with scientists at a University of Maryland HCI laboratory created their own toy: a set of animal blocks that each plays a recorded factoid about an animal and reveal its name. When the blocks are jumbled up, they create a nonsensical animal using syllables from each animal's name.

- Roomba, a robotic vacuum cleaner, appeals so well to users' emotions (through features such as "cute" chirping alert noises when the robot finds itself stuck in a corner) that many people find themselves naming and talking to their robots.

- To print a file using a command-line interface rather than a graphical user interface, a user would have to type in the instruction "print [/d:Printer] [Drive:] [Path] FileName [. . .]."

- HCI researchers have coined the term critical incident to describe times when a piece of technology makes a person feel frustrated, angry, or confused. Some 80 percent of computer users say frustrating experiences with their machines have led them to shout expletives out loud.

- One common application of HCI research in Japan is smart toilets with automatic seat warmers and speakers that emit flushing sounds to mask bathroom noises.

- The mouse was invented in 1964 and began its life as a simple wooden box with rolling wheels.

---

schedule and track the progress they are making on a shared task or to make changes to common documents without overriding someone else's work.

Education and Training. Schools, museums, and businesses all make use of HCI principles when designing educational and training curricula for students, visitors, and staff. For example, many school districts are moving away from printed textbooks and toward interactive electronic programs that target a variety of information-processing modalities through multimedia. Unlike paper and pencil worksheets, such programs also provide instant feedback, making it easier for students to learn and understand new concepts. Businesses use similar programs to train employees in such areas as the use of new software and the company's policies on issues of workplace ethics. Many art and science museums have installed electronic kiosks with touchscreens that visitors can use to learn more about a particular exhibit. HCI principles underlie the design of such kiosks. For example, rather than using a text-heavy interface, the screen on an interactive kiosk at a science museum might display video of a museum staff member talking to the visitor about each available option.

IMPACT ON INDUSTRY

Cutting-edge HCI research is taking place at technology centers all over the globe, but Japan is perhaps the country in which usability has become most established as a cornerstone of product design. With robotic receptionists so realistically human and efficient that they are almost capable of fooling visitors, Japan is a world leader in the goal of making technology ever more able to anticipate the needs of the user.

Government and University Research. Although many HCI technologies have emerged from privately funded corporations, both government institutes and university research laboratories are equally important contributors to the field. For example, the United States' Defense Advanced Research Projects Agency (DARPA) invests heavily in military-related HCI applications such as simulation equipment used to train soldiers for the battlefield and airplane pilots for flying. The number of universities that have set up academic research institutes focusing on HCI has been growing rapidly. Among them are Carnegie Mellon University, Northeastern University, Stanford University, and the University of Maryland.

Industry and Business. The major industry sectors related to HCI are those that directly involve the development of technological devices for personal and business use, including computer and software manufacturers and electronic appliance manufacturers. Companies such as Apple, Microsoft, IBM, Hewlett-Packard, and Xerox are among the major corporations whose research and development departments have been at the forefront of new HCI applications for many years. However, because the use of technology of all forms is so deeply embedded in contemporary life, HCI design principles have come to influence virtually every other commercial sector.

Marketers of all stripes, for instance, use HCI theories to create Web sites that make browsing for, choosing, and purchasing products as easy as possible. Amazon.com's one-click ordering system is a good example of an HCI-influenced application. It shortens the distance between the decision to buy something and the point of purchase to a single click, making it less likely that a customer will be turned off by a lengthy ordering process or change his or her mind halfway through the process. Similarly, several travel Web sites have begun using a search interface that allows users to type in their desired parameters (departure and destination airports, dates of travel) just once, then simply click on a button to use the same search terms with any of a dozen different airline Web sites–a far less laborious task than visiting each of those sites individually and using their internal search functionalities. Increasingly, it has become common for businesses to create site elements that are not immediately related to the purchase of a product but that add value to customers' experiences on the Web site and develop brand loyalty. For instance, a company that sells cooking and baking equipment for home cooks might add a forum to their site in which users can share recipes and discuss the foods they love. By including on their site a piece of technology that allows potential customers to build a virtual community around a shared interest, the company can encourage repeat visits and future sales.

CAREERS AND COURSE WORK

The paths toward becoming a HCI professional are extraordinarily varied. Bachelor's degrees in cognitive science, neuroscience, computer science, graphic design, psychology, engineering, art, and many other fields could serve as appropriate preparation for a career as someone who uses HCI principles. No matter which concentration an aspiring HCI researcher or student chooses, it is important to acquire basic programming skills, a broad understanding of human psychophysiology, and some practical experience or training with either graphic design or product design. Common areas of work include developing Web sites; computer operating systems; interfaces for consumer appliances such as cell phones, printers, or cameras; and educational materials such as interactive employee training courses, advertising campaigns, or any other applications that demand accessible, learnable, and usable computer systems. Although a graduate degree is not required for entering the field, many universities offer specialized master's programs in HCI.

SOCIAL CONTEXT AND FUTURE PROSPECTS

As HCI moves forward with research into multimodal interfaces and ubiquitous computing, notion of the computer as an object separate from the user may eventually be relegated to the archives of technological history, to be replaced by wearable machine interfaces that can be worn like clothing on the user's head, arm, or torso. Virtual reality interfaces have been developed that are capable of immersing the user in a 360-degree space that looks, sounds, feels, and perhaps even smells like a real environment–and with which they can interact naturally and intuitively, using their whole bodies. As the capacity to measure the physical properties of human beings becomes ever more sophisticated, input devices may grow more and more sensitive; it is possible to envision a future, for instance, in which a machine might "listen in" to the synaptic firings of the neurons in a user's brain and respond accordingly. Indeed, it is not beyond the realm of possibility that a means could be found of stimulating a user's neurons to produce direct visual or auditory sensations. The future of HCI research may be wide open, but its essential place in the workplace, home, recreational spaces, and the broader human culture is assured.

M. Lee, B.A., M.A.

FURTHER READING

Bainbridge, William Sims, ed. *Berkshire Encyclopedia of Human-Computer Interaction: When Science Fiction Becomes Fact.* 2 vols. Great Barrington, Mass.: Berkshire, 2004. Contains more than a hundred HCI topics, including gesture recognition, natural-language processing, and education. Each article includes further reading listings and may contain sidebars, figures, tables, or photographs.

Helander, Martin. *A Guide to Human Factors and Ergonomics.* 2d ed. Boca Raton, Fla.: CRC Press/ Taylor & Francis, 2006. Discusses the cognitive and physical aspects of human capabilities that inform product development, and includes an appendix examining the use of checklists to improve safety and effectiveness in human factors design.

Sears, Andrew, and Julie A. Jacko, eds. *The Human-Computer Interaction Handbook: Fundamentals, Evolving Technologies, and Emerging Applications.* 2d ed. New York: Lawrence Erlbaum Associates, 2008. Prominent researchers in the field address issues in design, development, testing, and evaluation. Contains hundreds of explanatory figures and tables.

Sharp, Heken, Yvonne Rogers, and Jenny Preece. *Interaction Design: Beyond Human-Computer Interaction.* 2d ed. Hoboken, N.J.: John Wiley & Sons, 2007. Each chapter on interaction design contains an outline, summary, subsections, and text boxes highlighting case studies and other points of focus. Includes a comprehensive bibliography.

Thatcher, Jim, et al. *Web Accessibility: Web Standards and Regulatory Compliance.* New York: Springer, 2006. Presents an overview of laws, policies, and technical standards for creating accessible Web sites.

Tufte, Edward R. *The Visual Display of Quantitative Information.* 2d ed. Cheshire, Conn.: Graphics Press, 2007. A seminal work on ways in which complex information can be clearly and simply displayed in visual forms such as graphics, maps, charts, and tables.

WEB SITES

Human Factors and Ergonomics Society
Internet Technical Group
http://www.internettg.org

Special Interest Group on Human Computer Interaction
http://www.sigchi.org

See also: Computer Engineering; Computer Graphics; Computer Science; Electronics and Electronic Engineering; Graphics Technologies; Information Technology; Internet and Web Engineering; Wireless Technologies and Communication.

HUMAN GENETIC ENGINEERING

FIELDS OF STUDY

Biotechnology; genetics; cytogenetics; molecular genetics; biochemical genetics; genomics; population genetics; developmental genetics; clinical genetics; genetic counseling.

SUMMARY

Human genetic engineering is a branch of genetic engineering focusing on the understanding of human genes to produce applications that can improve human life. Genes, formulated by DNA (deoxyribonucleic acid), determine genotype–the complete genetic information carried by an individual, even if not expressed. Visible human characteristics, by contrast, are formed as the result of human genes interacting with the environment and are called the phenotype. Human genetic engineering aims to alter genotypes to cause changes in phenotypes; also, and more often, the knowledge of human genetics is used to engineer products, such as medications, that can cure or improve the quality of human life by addressing genetic disorders. Many of the applications of what is now known as human genetic engineering arose out of the mapping of the human genome during the Human Genome Project, completed in 2003.

KEY TERMS AND CONCEPTS

- **Bioinformatics:** Science of compiling and managing genetic and other biology data using computers, requisite in human genome research.
- **Biologics:** Medicines produced using genes and genetic manipulations.
- **Clones:** Genetically identical living organisms produced via genetic engineering.
- **DNA (Deoxyribonucleic Acid):** Molecule, found in all living organisms, that by reproducing itself allows for the inheritance of characteristics from one generation to the next.
- **Dysmorphology:** Abnormal physical development resulting from a genetic disorder.
- **Forensic Genetics:** Application of genetics, particularly DNA technology, to the analysis of evidence used in criminal cases and paternity testing.

- **Gene:** Specific DNA sequence that codes for a specific protein.
- **Gene Therapy:** Use of a viral or other vector to incorporate new DNA into a person's cells with the objective of alleviating or treating the symptoms of a disease or condition.
- **Genetic Screening:** Use of the techniques of genetics research to determine a person's risk of developing, or his or her status as a carrier of, a disease or other disorder.
- **Gene Transfer:** Using a viral or other vector to incorporate new DNA into a person's cells; used in gene therapy.
- **Genetic Testing:** Process of investigating a specific individual or population of people to detect the presence of genetic defects.
- **Genomics:** Branch of genetics dealing with the study of the genetic sequences of organisms, including the human being.
- **Pharmacogenomics:** Branch of human medical genetics that evaluates how an individual's genetic makeup influences his or her response to drugs.
- **Proteomics:** Study of how proteins are expressed in different types of cells, tissues, and organs.
- **Recombinant DNA:** DNA that has been transferred from one cell to another. Genes are recombined from a human chromosome to another cell, usually from bacteria; if the transferred human genes code for insulin, the bacteria accepting the transferred genes will now produce human insulin.
- **Stem Cell:** Progenitor cell that has the capability to become a more specialized cell, such as a kidney, liver, or heart cell. Once a cell becomes a specific kind of cell, it cannot change to another type of cell.
- **Toxicogenomics:** Science of evaluating ways in which genomes respond to chemical and other pollutants in the environment.

DEFINITION AND BASIC PRINCIPLES

Human genetic engineering is the science and technology of manipulating or changing human genes to alter or control visible characteristics of a human newborn or adult. Genes, formulated by DNA (deoxyribonucleic acid), are called the genotype. Visible human traits or characteristics, formed from the interaction of genes and the environment, are called the phenotype. Human genetic engineering aims to change genotypes to cause change in the phenotype. To understand human genetic engineering capabilities, it is important to understand basic genetic principles.

BACKGROUND AND HISTORY

Human genetic engineering is a scientific endeavor, and as such this field builds on the information and knowledge gained from the decades of experimentation accomplished in years past. Without this foundation, human genetic engineering would not exist. This foundation brings the prospect of human cloning and the use of human genetics for therapeutic purposes.

A keystone event in modern genetics occurred in 1953, when American biologist James D. Watson and English physicist Francis Crick deduced the double-helical structure of DNA. This structural information enabled effective study of how genetic material codes for life. In 1968, the DNA genetic code was deciphered. Armed with this important genetic information, geneticists undertook the first recombinant DNA experiments on bacteria in 1973. The ambitious Human Genome Project started in 1990, with the goal of mapping out the entire human genetic sequence. In June, 2000, the first working draft of the human genetic sequence was produced from the efforts of this project. April, 2003, saw the announcement of the first complete human genetic sequence–breakthrough information for human genetic engineering.

Cloning, a subdiscipline of bioengineering, is the reproduction of genetically identical living organisms. In 1996, the first mammal was cloned, a sheep named Dolly. Other animals have been cloned since this pioneering event, including a bull in 1999 and a pig in 2000. The year 2003 saw the cloning of a mule, a horse, and a rat, followed by the cloning of a dog in 2005. Attempts at pet cloning have occurred: John Sperling, a wealthy and influential American educator, has funded pet-cloning projects, and researchers at Texas A&M University successfully cloned a cat in 2002. Commercial attempts at pet cloning started in April, 2004, with a company called Genetics Savings and Clone (now defunct) offering pet gene banking and cloning. Korean researchers published claims of successful human embryonic cloning in 2004, but these claims were later retracted because of fabricated data and other problems with the research. In May, 2010, the journal *Science* reported that scientists J. Craig Venter, Clyde Hutchison III, and Hamilton Smith had created a living creature in the laboratory. This new life-form, a bacterium, was artificially produced using genetic-engineering techniques. It had no ancestor and it reproduced, a key ability of living organisms.

Milestones in other areas of human genetic engineering—with more practicable and practical results—occurred in the areas of medicine, pharmaceuticals, forensics, and even psychology, as identified below in Applications and Products. Many, if not most, of these blossomed shortly after the mapping of the human genome was completed in 2003. Many more will be developed as scientists and researchers continue to investigate the data that were gathered through that monumental accomplishment.

HOW IT WORKS

Until the middle of the eighteenth century, most biologists believed in spontaneous generation—that life arises from combinations of decaying matter, as if flies arose from garbage. It is now known that DNA genetically codes for many physical characteristics.

The story of genetics starts with DNA and ends with protein. DNA, the genetic material found in the nucleus of every cell, codes for (that is, creates instructions for the building of) various proteins by means of components of the DNA molecule called nucleotides, which form the building blocks of DNA. The nucleotides establish the code. Proteins make up the structural elements of the body, including collagen, ligaments, tendons, and muscles; some hormones, such as insulin, are made of protein as well. Perhaps most important, however, are the protein enzymes.

All the enzymes in the body are made up of protein and protein alone. Enzymes are key because they accelerate chemical reactions. Thousands of chemical reactions occur in human bodies all the time. Protein enzymes catalyze all these reactions. DNA, by dictating the production of enzymes, controls these chemical reactions.

DNA dictates which proteins are produced by living and staying in the cell nucleus during the entire protein-making process. Much like a general in a command center, the DNA sends out orders but does not leave the nucleus. DNA is made up of nucleotide

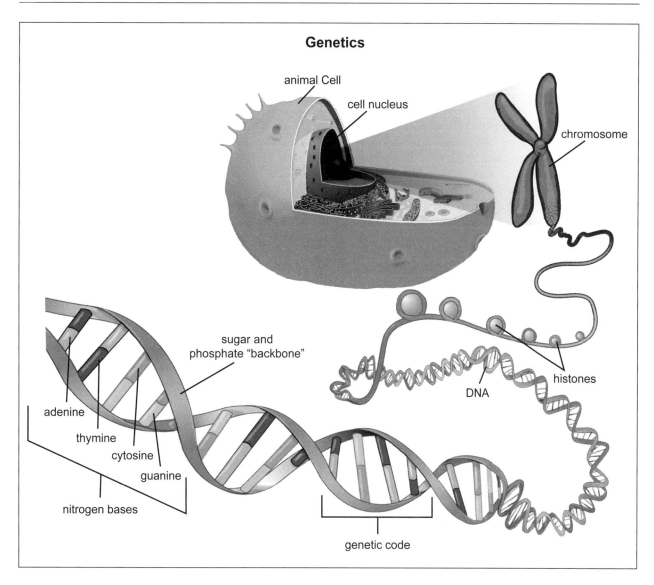

Genetics

animal Cell

cell nucleus

chromosome

sugar and phosphate "backbone"

adenine

thymine

cytosine

guanine

nitrogen bases

histones

DNA

genetic code

bases, and the first step in protein production involves the reading of the nucleotide code in the DNA. When the nucleotides are read, a strand of messenger ribonucleic acid (mRNA) is produced and sent from the nucleus into the cytoplasm of the cell. The DNA stays in the nucleus, while the mRNA leaves the nucleus. This process of reading the DNA nucleotide code and producing mRNA is called transcription.

The next step involves reading the nucleotide code on the mRNA in the cytoplasm of the cell. The cytoplasm is the liquid environment inside the cell where all the cellular organelles float. Transfer RNA (tRNA) reads the code on the mRNA, and this process is called translation. tRNA is called transfer because it transfers a specific amino acid when it reads the appropriate code on the mRNA. The basic building blocks of protein are called amino acids. About twenty-two different amino acids build all the various proteins the body uses. It is like the English alphabet: Twenty-six letters and vowels comprise the English alphabet, and combining these various letters and vowels results in tens of thousands of different combinations and all the various words in the English language. Likewise, the body uses the different amino acids to form the tens of thousands of different proteins in the body.

During translation, a specific amino acid is coded for and carried by the transfer RNA to a ribosome. Ribosomes (along with the rough endoplasmic reticulum) are where the cell's proteins are produced. Along ribosomes, different amino acids are transported by tRNA and linked, forming a protein molecule. Some proteins may be only eighty or ninety amino acids long, whereas others, such as hemoglobin, may have more than 300 amino acids as their amino acid backbone.

The way DNA codes for all this involves the nucleotide bases that make up DNA. Four nucleotide bases make up DNA: adenine, cytosine, guanine, and thymine. Adenine will chemically bind with thymine, and cytosine always chemically binds with guanine. When DNA is transcribed to form mRNA, if the nucleotide sequence in the DNA reads cytosine-cytosine-guanine, these nucleotide bases will code for guanine-guanine-cytosine in the mRNA. Then when mRNA is translated by tRNA, the code goes back to the original DNA code, cytosine-cytosine-guanine. Cytosine-cytosine-guanine can code for a specific amino acid, and in that fashion DNA codes for the amino acid sequence of all protein molecules. The nucleotide base sequence in the mRNA is called the codon and the complimentary base sequence found in tRNA is called the anticodon.

The example of how Dolly the sheep was cloned demonstrates how genetic engineering in mammals works, and, hence, how human cloning could work. Cloning is an ultimate example of genetic engineering because cloning produces an entire living organism via genetic engineering. Dolly was a Finn Dorset sheep, which is all white. A Blackface ewe, named because of the distinctive black face these sheep have, was used as an egg donor and as a surrogate mother.

Cells taken from a Finn Dorset ewe were grown in a tissue culture. An egg cell, from the Blackface ewe, had the nucleus removed. The nucleus contained the genes and DNA. The nucleus and genetic information from the Finn Dorset ewe were placed in the enucleated Blackface ewe egg cell. The Blackface ewe egg cell, now containing genetic information from the Finn Dorset ewe, was placed in the uterus of the Blackface ewe after an electric pulse is applied to stimulate growth and duplication of the cells. The Blackface ewe gave birth to Dolly, the all-white Finn Dorset ewe. The newborn Finn Dorset ewe was an identical genetic copy of the Finn Dorset ewe originally used to harvest the genetic information found in the nucleus.

Recombinant DNA refers to DNA transfer from one cell to another. In human genetic engineering, genes transfer from a human chromosome to another cell, usually bacteria. If the transferred human genes code for insulin, the bacteria accepting the transferred genes will now produce human insulin.

In this process, the desired genes are isolated and removed from the human cell. Bacterial cells have small, circular strips of DNA called plasmids. These circular plasmids are removed from the bacterial cells and opened up. Various enzymes are used to cut the human DNA and bacterial DNA sequences at specific points. Restriction enzymes cut the original DNA in specific locations. DNA ligase pastes strips of DNA together. Scientists mix isolated human genes with the opened bacterial plasmids, along with DNA ligase. The human genes are spliced into the bacterial plasmid and the circle of genetic information in the bacterial plasmid closes.

The bacterial plasmid with the spliced human genes is now called a vector. The plasmid vectors are taken up by the bacterial cells. Once inside the bacterial cells, the bacteria multiply and reproduce the spliced human genes. Whatever specific human genes were selected for splicing, for example, human insulin genes, are now functioning in the reproduced bacteria, and human insulin is harvested from the bacterial clones.

APPLICATIONS AND PRODUCTS

Early Medical Applications. Recombinant DNA techniques are remarkable biological life adaptations, and many medicines based on this technology are used. Medications generated through human genetic engineering techniques have been in use since 1982, with the production of human insulin using recombinant DNA techniques. Human genes are inserted into a bacterial host that then makes human insulin. Prior to this type of genetic manipulation, diabetics needing insulin had to rely on insulin harvested from pigs or cows. Genetic techniques produce human growth hormone, previously only available from human cadavers. A genetically engineered hepatitis B vaccine has been in use since 1987.

Since these first human genetic medicines and vaccines, many types of biological products have been introduced or are under current investigation and development. These new medicinal products are called biologics to distinguish them from chemically synthesized medicines. Genes and genetic manipulations produce biologics. Major types of biologics include hormones, antibodies, and cell-receptor proteins.

Insulin and human growth hormone, discussed above, are classic protein hormones produced with recombinant DNA technology. The immune system produces protein antibodies that attack disease causing-agents such bacteria and viruses. Genetic antibody production interferes with or attacks entities associated with diseases such as psoriatic arthritis and Crohn's disease. Recombinant DNA technology produces proteins binding with specialized white blood cells to reduce inflammation associated with rheumatoid arthritis.

Bioinformatics. The purpose of bioinformatics is to help organize, store, and analyze genetic biological information in a rapid and precise manner, dictated by the need to be able to access genetic information quickly. In the United States the online database that provides access to these gene sequences is called GenBank, which is under the purview of the National Center for Biotechnology Information (NCBI) and has been made available on the Internet. In addition to human genome sequence records, GenBank provides genome information about plants, bacteria, and animals other than humans.

Proteomics. Bioinformatics provides the basis for all modern studies of human genetics, including analyzing genes and gene sequences, determining gene functions, and detecting faulty genes. The study of genes and their functions is called proteomics, which involves the comparative study of protein expression. That is, it studies the metabolic and morphological relationship between the protein encoded within the genome and how that protein works. Geneticists are now classifying proteins into families, superfamilies, and folds according to their configuration, enzymatic activity, and sequence. Ultimately proteomics will complete the picture of the genetic structure and functioning of all human genes.

Toxicogenomics. Another newly developing field that relies on bioinformatics is the study of toxicogenomics, which is concerned with how human genes respond to toxins. As of 2011, this field is specifically concerned with evaluating how environmental factors

---

Fascinating Facts About Human Genetic Engineering

- In 2011, researchers at Brigham and Women's Hospital in Boston identified a self-renewing human lung stem cell. This particular stem cell is also able to form and integrate a variety of biological lung structures, including bronchi, alveoli, and pulmonary vessels.

- DNA fingerprints, used in forensic genetics, are made by using enzymes splitting the genetic sequence up into patterns unique to an individual. DNA fingerprinting used to require test tubes of blood for analysis. Polymerase chain reaction technology now allows the reproduction of DNA (and subsequent analysis) from a sample as small as dried saliva from the back of a stamp.

- Synthetic biology aims to produce life-forms. Life-forms reproduce, by definition. An invention such the atomic bomb cannot reproduce itself, but biologics can.

- Dolly the sheep, the first cloned mammal, lived six years, half the normal life expectancy. The sheep originating the cloned DNA was six years old at the time of donation, raising speculation about the genetic age of the donor DNA.

- Genetic engineering techniques could resurrect life for extinct animals, such as mammoths, or extinct human species, such as Neanderthals. Although "farming" DNA that is viable for such use is difficult, theoretically such feats could be accomplished. A film that was based on this possibility is *Jurassic Park* (1993), based on a novel of the same name by Michael Crichton.

---

negatively interact with mRNA translation, resulting in disease or dysfunction.

Gene Testing. In a gene-testing protocol, a sample of blood or body fluids is examined to detect a genetic anomaly such as the transposition of part of a chromosome or an altered sequence of the bases that comprise a specific gene, either of which can lead to a genetically based disorder or disease. As of 2011, more than 600 tests are available to detect malfunctioning or nonfunctioning genes. Most gene tests have focused on various types of human cancers, but other tests are being developed to detect genetic deficiencies that cause or exacerbate infectious and vascular diseases.

The emphasis on the relationship between genetics and cancer lies in the fact that all human cancers are genetically triggered or have a genetic basis. Some cancers are inherited as mutations, but most result from random genetic mutations that occur in specific cells, often precipitated by viral infections or environmental factors not yet well understood.

At least four types of genetic problems have been identified in human cancers. The normal function of oncogenes, for example, is to signal the start of cell division. However, when mutations occur or oncogenes are overexpressed, the cells keep on dividing, leading to rapid growth of cell masses. The genetic inheritance of certain kinds of breast and ovarian cancers results from the nonfunctioning tumor-suppressor genes that normally stop cell division. When genetically altered tumor-suppressor genes are unable to stop cell division, cancer results. Conversely, the genes that cause inheritance of colon cancer result from the failure of DNA repair genes to correct mutations properly. The accumulation of mutations in these "proofreading" genes makes them inefficient or less efficient, and cells continue to replicate, producing a tumor mass.

If a gene screening reveals a genetic problem several options may be available, including gene therapy and genetic counseling. If the detected genetic anomaly results in disease, then pharmacogenomics holds promise of patient-specific drug treatment.

Gene Therapy. The science of gene therapy uses recombinant DNA technology to cure diseases or disorders that have a genetic basis. Still in its experimental stages, gene therapy may include procedures to replace a defective gene, repair a defective gene, or introduce healthy genes to supplement, complement, or augment the function of nonfunctional or malfunctioning genes. Several hundred protocols are being used in gene-therapy trials, and many more are under development. As of 2011, trials are focusing on two major types of gene therapy, somatic cell gene therapy and germ-line gene therapy.

Somatic cell gene therapy concentrates on altering a defective gene or genes in human body cells in an attempt to prevent or lessen the debilitating impact of a disease or other genetic disorder. Some examples of somatic cell gene therapy protocols now being tested include ones for adenosine deaminase (ADA) deficiency, cystic fibrosis, lung cancer, brain tumors, ovarian cancer, and AIDS.

In somatic cell gene therapy a sample of the patient's cells may be removed and treated and then reintegrated into body tissue carrying the corrected gene. An alternative somatic cell therapy is called gene replacement, which typically involves insertion of a normally functioning gene. Some experimental delivery methods for gene insertion include use of retroviral vectors and adenovirus vectors. These viral vectors are used because they are readily able to insert their genomes into host cells. Hence, adding the needed (or corrective) gene segment to the viral genome guarantees delivery into the cell's nuclear interior. Nonviral delivery vectors that are being investigated for gene replacement include liposome fat bodies, human artificial chromosomes, and naked DNA (free DNA, or DNA that is not enclosed in a viral particle or any other "package").

Another type of somatic gene therapy involves blocking gene activity, whereby potentially harmful genes such as those that cause Marfan syndrome and Huntington's disease are disabled or destroyed. Two types of gene-blocking therapies being investigated include the use of antisense molecules that target and bind to the mRNA produced by the gene, thereby preventing its translation, and the use of specially developed ribozymes that can target and cleave gene sequences that contain the unwanted mutation.

Germ-line therapy is concerned with altering the genetics of male and female reproductive cells (gametes) as well as other body cells. Because germ-line therapy will alter the individual's genes as well as those of his or her offspring, both concepts and protocols are still very controversial. Some aspects of germ-line therapy now being explored include human cloning and genetic enhancement.

Clinical Genetics. Clinical genetics is that branch of medical genetics involved in the direct clinical care of people afflicted with diseases caused by genetic disorders. Clinical genetics involves diagnosis, counseling, management, and support. Genetic counseling is a part of clinical genetics directly concerned with medical management, risk determination and options, and decisions regarding reproduction of afflicted individuals. Support services are an integral feature of all genetic counseling themes.

Clinical genetics begins with an accurate diagnosis that recognizes a specific, underlying genetic cause of a physical or biochemical defect following guidelines outlined by the National Institutes of

Health (NIH) Counseling Development Conference. Clinical practice includes several hundred genetic tests that are able to detect mutations such as those associated with breast and colon cancers, muscular dystrophy, cystic fibrosis, sickle-cell disease, and Huntington's disease.

Genetic counseling follows clinical diagnosis and focuses initially on explaining the risk factors and human problems associated with the genetic disorder. Both the afflicted individual and family members are involved in all counseling procedures. Important components include a frank discussion of risks, of options such as preventive operations, and of options involved with regard to reproduction. All reproductive options are described along with their potential consequences, but genetic counseling is a support service rather than a directive mode. That is, it does not include recommendations. Instead, its ultimate mission is to help both the afflicted individuals and their families recognize and cope with the immediate and future implications of the genetic disorder.

Pharmacogenomics. That branch of human medical genetics dealing with the correlation of specific drugs to fit specific diseases in individuals is called pharmacogenomics. This field recognizes that individuals may metabolically respond differentially to therapeutic medicines based on their genetic makeup. It is anticipated that testing human genome data will greatly speed the development of new drugs that not only target specific diseases but also will be tailored to the specific genetics of patients.

Forensic Genetics. Forensic genetics is the use of human genetics in criminal or paternity cases. For example, DNA testing on blood, saliva, or other tissue can be used to determine the source of evidence, such as blood stains or semen, left at a crime scene. Forensic DNA analysis is also used to determine paternity and other kinship. Finally, with the increasing use of forensic genetics since the 1990's, some incarcerated prisoners have been released after it was clearly determined that they could not possibly have been guilty of crimes they were convicted of, as DNA evidence eliminated them from suspicion.

Potential for Human Cloning. Human therapeutic cloning involves the production of cloned human embryos, with the idea of harvesting embryonic stem cells. The hope is that the stem cells can be grown into a wide variety of cells to replace or repair organs, such as liver, kidney, or heart cells. Although human cloning has not yet reached this potential, future applications could offer identically matched kidneys for people with failing kidneys or even a genetically duplicate heart for someone in severe heart failure.

IMPACT ON INDUSTRY

The many applications of our knowledge of human genetics have had a revolutionary impact on industry, spawning entirely new industries as well as new product lines. Three main areas on which these industries focus is research, or the analysis of the human genome to identify new and potentially profitable areas of investigation; pharmaceutical applications–many of the large pharmaceutical companies produce medications based on human genetic research, such as the aforementioned synthetic insulin hormones; and finally, genetic counseling, which the National Society of Genetic Counselors defines as the practice of "health care professionals with specialized graduate training in the areas of medical genetics and counseling" who "work as members of a health care team, providing information and support to families who have members with birth defects or genetic disorders and to families who may be at risk for a variety of inherited conditions."

Genentech, established in 1976, is considered the founder of the biotechnology industry, with special focus on human genetics. Its stated mission is "using human genetic information to discover, develop, manufacture, and commercialize medicines to treat patients with serious or life-threatening medical conditions . . . to create, produce, and market innovative solutions of high quality for unmet medical needs." Other leading companies in the field include Amgen, Genzyme, Gilead Sciences, Biogen Idec, Cephalon, and MedImmune in the United States; Merck Serono in Switzerland; CSL in Australia; and UCB in Belgium.

CAREERS AND COURSE WORK

A variety of career paths exist for people interested in genetics. Educational requirements vary from bachelor's to doctorate degrees. Research scientists, working at major universities, companies, and government agencies such as the NIH usually need a Ph.D. or M.D. A clinical geneticist who treats patients typically completes medical school and specialty training. Geneticists or research scientists usually have Ph.D. training in fields such as molecular genetics, cytogenetics, or the burgeoning field of synthetic biology. Genetic counselors complete a master's degree program in genetic counseling. Research assistants in genetics laboratories usually have a master's degrees in genetics or biological sciences. Genetic laboratory technicians work in forensic or research labs with a bachelor's degree in science-related fields.

Select engineering schools, such as the University of Michigan, have biomedical engineering programs. Biomedical engineers interface with medical device and drug industries. Biomedical engineers will play an important role bringing genetic therapies into clinical practice. Tissue engineering involves the manipulation of genes to affect changes in phenotypes, the essence of genetic engineering.

SOCIAL CONTEXT AND FUTURE PROSPECTS

The Human Genome Project painstakingly mapped out the human DNA sequence in 2003, after a decade and a half of meticulous multicenter collaboration. Genetic databases are now rapidly filling with genetic detail because of technological advances in the speed of analyzing DNA sequences. While the speed of this analysis has increased considerably, the price of such investigations has dropped significantly. Genetic databases currently hold information on a wide variety of life-forms, and significant amounts of new generic information is added frequently.

The DNA sequencing found in genetic databases provides the burgeoning field of synthetic biology with important basic information needed for human genetic engineering. This information can be used for modeling and as supply depots for the mixing and matching of genes. As the speed of genetic analysis has increased significantly and the price of genetic investigations has dropped considerably, the process of DNA synthesis is much less expensive and faster than it was in the beginning of the twenty-first century.

More genetic information, faster artificial DNA synthesis, and significant technological cost savings result in more feasible human genetic engineering projects. Genes are the stuff of life, and the field is on the verge of changing life and even making new life-forms, via genetic engineering. How and what changes are made will present significant bioethical and societal challenges, along with potentially fantastic and beneficial results.

Richard P. Capriccioso, M.D.

FURTHER READING

Andrews, Lori B. *The Clone Age: Adventures in the New World of Reproductive Technology.* New York: Henry Holt, 1999. A lawyer specializing in reproductive technology, Andrews examines the legal ramifications of human cloning, from privacy to property rights.

Baudrillard, Jean. *The Vital Illusion.* Edited by Julia Witwer. New York: Columbia University Press, 2000. A sociological perspective on what human cloning means to the idea of what it means to be human.

Capriccioso, Richard P. "Genetic Testing." In *Salem Health: Cancer,* edited by Jeffrey A. Knight. Pasadena, Calif.: Salem Press, 2009. A comprehensive overview of genetic testing covering different types of genetic tests with a review of the science behind the testing.

The Economist. "Artificial Lifeforms: Genesis Redux." 395, no. 8683 (May 20, 2010): 81-83. Informative article on synthetic biology and the creation of a new form of life in the laboratory.

Hartwell, Leland, et al. *Genetics: From Genes to Genomes.* 4th ed. New York: McGraw-Hill, 2011. A comprehensive textbook on genetics, including human genetics discussed in a comparative context.

Hekimi, Siegfried, ed. *The Molecular Genetics of Aging: Results and Problems in Cell Differentiation.* Berlin: Springer, 2000. Examines various genetic aspects of the aging process. Illustrated.

Jorde, Lynn B., John C. Carey, and Michael J. Bamshad. *Medical Genetics.* 4th ed. Philadelphia: Mosby, 2010. Provides both an introduction to the field of human genetics with chapters on clinical aspects of human genetics, such as gene therapy, genetic screening, and genetic counseling.

Lewis, Ricki. *Human Genetics: Concepts and Applications.* 9th ed. New York: McGraw-Hill, 2010. This textbook provides a broad overview of human genetics and genomics.

Pasternak, Jack J. *An Introduction to Human Molecular Genetics: Mechanisms of Inherited Diseases.* 2d ed. Hoboken, N.J.: Wiley-Liss, 2005. Discusses treatment advances, fundamental molecular mechanisms that govern human inherited diseases, the interactions of genes and their products, and the consequences of these mechanisms on disease states in major organ systems such as muscles, the nervous system, and the eyes. Also addresses cancer and mitochondrial disorders.

Rudin, Norah, and Keith Inman. *An Introduction to Forensic DNA Analysis.* 2d ed. Boca Raton, Fla.: CRC Press, 2002. An overview of many DNA typing techniques, along with numerous examples and a discussion of legal implications.

Shostak, Stanley. *Becoming Immortal: Combining Cloning and Stem-Cell Therapy.* Albany: State University of New York Press, 2002. Examines the question of whether human beings are equipped for potential immortality.

Wilson, Edward O. *On Human Nature.* 1978. Cambridge, Mass.: Harvard University Press, 2004. A look at the significance of biology and genetics for the way people understand human behaviors, including aggression, sex, and altruism, and the institution of religion.

WEB SITES

American Society of Human Genetics (ASHG)
http://www.ashg.org

Center for Genetics and Society
http://www.geneticsandsociety.org

Genetics Education Center
University of Kansas Medical Center
http://www.kumc.edu/gec

Human Genome Project
http://www.ornl.gov/sci/techresources/Human_Genome/home.shtml

National Institutes of Health
Stem Cell Information
http://stemcells.nih.gov/info/basics

National Society of Genetic Counselors
http://www.nsgc.org

See also: Bioengineering; Cell and Tissue Engineering; Cloning; DNA Analysis; DNA Sequencing; Genetically Modified Food Production; Genetic Engineering; Proteomics and Protein Engineering.

HYBRID VEHICLE TECHNOLOGIES

FIELDS OF STUDY

Electrical engineering; mechanical engineering; chemistry; ecology.

SUMMARY

Hybrid vehicle technologies use shared systems of electrical and gas power to create ecologically sustainable industrial and passenger vehicles. With both types of vehicles, the main goals are to reduce hazardous emissions and conserve fuel consumption.

KEY TERMS AND CONCEPTS

- **Driving Range:** Distances, both in city driving and open-road driving, that can be covered by HEVs before refueling or recharging is necessary.
- **Fuel Cells:** Very advanced form of electrochemical cell (instead of a conventional battery) that produces electricity following a reaction between an externally supplied fuel, frequently hydrogen, which becomes positively charged ions following contact with an oxidant, frequently oxygen or chlorine, which "strips" its electrons.
- **Internal Combustion Engine (ICE):** Produces power by the pressure of expanding fossil fuel gases after they are ignited in cylinder chambers.
- **Lithium-Ion Battery:** Lightweight rechargeable battery with energy-efficient qualities tied to the fact that lithium ions travel from the negative to positive electrode when the battery discharges, returning to the negative during charging.
- **Parallel System Hybrid:** Main power comes from the gas engine; electric motor steps in only for power to accelerate.
- **Series System Hybrid:** Power is delivered to the wheels from the electric motor; the gas engine's role is to generate electricity.
- **Third-Generation Hybrid:** Improved vehicle models using newer technology after the previous (first and second) generations.

DEFINITION AND BASIC PRINCIPLES

As the word "hybrid" suggests, hybrid vehicle technology seeks to develop an automobile (or, more broadly defined, any power-driven mechanical system) using power from at least two different sources. Before and during the first decade of the twenty-first century, hybrid technology emphasized the combination of an internal combustion engine working with an electric motor component.

BACKGROUND AND HISTORY

Before technological development of what is now called a hybrid vehicle, the automobile industry, by necessity, had to have two existing forms of motor energy to hybridize–namely internal combustion in combination with some form of electric power. Early versions of cars driven with electric motors emerged in the 1890's and seemed destined to compete very seriously with both gasoline (internal combustion engines) and steam engines at the turn of the twentieth century.

Although development of commercially attractive hybrid vehicles would not occur until the middle of the twentieth century, the Austrian engineer Ferdinand Porsche made a first-series hybrid automobile in 1900. Within a short time, however, the commercial attractiveness of mass-produced internal combustion engines became the force that dominated the automobile industry for more than a half century. Experimentation with hybrid technology as it could be applied to other forms of transport, especially motorcycles, however, continued throughout this early period.

By the 1970's the main emerging goal of hybrid car engineering was to reduce exhaust emissions; conservation of fuel was a secondary consideration. This situation changed when, in the wake of the 1973 Arab-Israeli War, many petroleum-producing countries supporting the Arab cause cut exports drastically, causing a nationwide gas shortage and worldwide fears that oil would be used as a political weapon.

Until 1975 government support for research and development of hybrid cars was tied to the Environmental Protection Agency (EPA). In that year (and after at least two unsatisfactory results of EPA-supported hybrid car projects), this role was shifted to the Energy Research and Development Administration, which later became the Department of Energy (DOE).

During the decade that followed the introduction of Honda's Insight hybrid car in 1999, the most widely recognized commercially marketed hybrid automobile was Toyota's Prius. Despite some setbacks in sales in 2010 following major recalls connected with (among other less dangerous problems) the malfunctioning anti-lock braking system and accelerator devices, the third-generation Prius still held a strong position in total hybrid car sales globally going into 2011.

HOW IT WORKS

"Integrated motor assist," a common layperson's engineering phrase borrowed from Honda's late-1990's technology, suggests a simple explanation of how a hybrid vehicle works. The well-known relationship between the electrical starter motor and the gas-driven engine in an internal combustion engine (ICE) car provides a (technically incomplete) analogy: The electric starter motor takes the load needed to turn the crankcase (and the wheels if gears are engaged) until the ICE itself kicks in. This overly general analogy could be carried further by including the alternator in the system, since it relieves the battery of the job of supplying constant electricity to the running engine (recharging the battery at the same time).

In a hybrid system, however, power from the electric motor (or the gas engine) enters and leaves the drivetrain as the demand for power to move the vehicle increases or decreases. To obtain optimum results in terms of carbon dioxide emissions and overall

The engine compartment of a Chevrolet Volt hybrid gas/electric car. (Spencer Grant/Photo Researchers, Inc.)

fuel efficiency, the power train of most hybrid vehicles is designed to depend on a relatively small internal combustion engine with various forms of rechargeable electrical energy. Although petroleum-driven ICE's are commonly used, hybrid car engineering is not limited to petroleum. Ethanol, biodiesel, and natural gas have also been used.

In a parallel hybrid, the electric motor and ICE are installed so that they can power the vehicle either individually or together. These power sources are integrated by automatically controlled clutches. For electric driving, the clutch between the ICE and the gearbox is disengaged, while the clutch connecting the electric motor to the gearbox is engaged. A typical situation requiring simultaneous operation of the ICE and the electric motor would be for rapid acceleration (as in passing) or in climbing hills. Reliance on the electric motor would happen only when the car is braking, coasting, or advancing on level surfaces.

It is extremely important to note that one of the most vital challenges for researchers involved in hybrid-vehicle technology has to do with variable options for supplying electricity to the system. It is far too simple to say that the electrical motor is run by a rechargeable battery, since a wide range of batteries (and alternatives to batteries) exists. A primary and obvious concern, of course, will always be reducing battery weight. To this aim, several carmakers, including Ford, have developed highly effective first-, second-, and even third-generation lithium-ion batteries. Many engineers predict that, in the future, hydrogen-driven fuel cells will play a bigger role in the electrical components of hybrids.

Selection of the basic source of electrical power ties in with corollary issues such as calculation of the driving range (time elapsed and distances covered before the electrical system must be recharged) and optimal technologies for recharging. The simplest scenario for recharging, which is an early direct borrowing from pure-electric car technology, involves plugging into a household outlet (either 110 volt or 220 volt) overnight. But, hybrid-car engineers have developed several more sophisticated methods. One is a "sub-hybrid" procedure, which uses very lightweight fuel cells, mentioned above, in combination with conventional batteries (the latter being recharged by the fuel cells while the vehicle is underway). Research engineers continue to look at any number of ways to tweak energy and power sources from different phases of hybrid vehicle operation. One example, which has been used in Honda's Insight, is a process that temporarily converts the motor into a generator when the car does not require application of the accelerator. Other channels are being investigated for tapping kinetic-energy recovery during different phases of simple mechanical operation of hybrid vehicles.

APPLICATIONS AND PRODUCTS

Some countries, especially Japan, have begun to use the principle of the hybrid engine for heavy-duty transport or construction-equipment needs, as well as hybrid systems for diesel road graders and new forms of diesel-powered industrial cranes. Hybrid medium-power commercial vehicles, especially urban trolleys and buses, have been manufactured, mainly in Europe and Japan. Important for broad ecological planning, several countries, including China and Japan, have incorporated hybrid (diesel combined with electric) technology into their programs for rail transport. The biggest potential consumer market for hybrid technology, however, is probably in the private automobile sector.

By the second decade of the twenty-first century, a wide variety of commercially produced hybrid automobiles were on European, Asian, and American markets. Among U.S. manufacturers, Ford has developed the increasingly popular Escape, and General Motors produces about five models ranging from Chevrolet's economical Volt to Cadillac's more expensive Escalade. Japanese manufacturers Nissan, Honda, and Toyota have introduced at least one, three, and two standard hybrid models, respectively, to which one should add Lexus's RX semi-luxury and technologically more advanced series of cars. Korea's Hyundai Elantra and Germany's Volkswagen Golf also competed for some share of the market.

One of the chief attractions to Toyota's hybrid technology has usually been its primary goal of using electric motors in as many operational phases as possible. The closest sales competitor in the United States to the Prius (mainly for mileage efficiency) was Chevrolet's Volt.

At the outset of 2011, Lexus launched an ambitious campaign to attract attention to what it called its full hybrid technology (as compared with mild hybrid) in its high-end RX models. A main feature of the full hybrid system, according to Lexus, is a combination of both series and parallel hybrid power in

one vehicle. Such a combination aims at transferring a variable but continuously optimum ratio of gas-engine and electric-motor power to the car. Another advance claimed by Lexus's full hybrid over parallel hybrids is its reliance on the electric motor only at lower speeds.

Early in 2011, Mercedes-Benz also announced its intention to capture more sales of high-end hybrids by dedicating, over a three-year period, more research to improve the technology used in its S400 model. Audi, a somewhat latecomer, unveiled plans for its first hybrid, the Q5, to appear in European markets late in 2011.

As fuel alternatives continue to be added to the ICE components of HEVs, advanced fuel-cell technology could transform the technological field that supplies electrical energy to the combined system.

IMPACT ON INDUSTRY

Given the factors of added cost associated with designing and producing hybrid vehicles, private companies, both manufacturers and research institutions, are more likely to enter the field if they can receive some form of governmental financial assistance. This can be in the form of direct subsidies, tax reductions, or grants. Similarly, institutions of higher learning, both public and private, frequently seek outside sources of funding for specially targeted research activities. Major national laboratories that are not tied to academic institutions, such as the Argonne National Laboratory near Chicago, also submit research proposals for grants. Private foundations that favor ecological research, particularly reduction of carbon dioxide emissions and alternative methods for producing electrical energy, can also be approached for seed money.

In the United States in 2010, hybrid vehicles made up only 3 percent of the total car sales. If this is a valid indicator, only a much bigger sales potential is likely to induce manufacturers to fund research that could bring to the market more fuel-efficient hybrid-technology cars at increasingly attractive prices. Major vested interests on the potentially negative side are: the gigantic ICE automobile industry itself, which is resistant to changes that require major new investment costs, and the fossil-fuel (petroleum) industry, which holds a near monopoly on fuels supplying power to automobiles, including diesel, and, in some cases, ethanol, around the world.

Future expansion in the number of hybrid cars might also cause important changes in the nature of equipment needed for various aspects of hybrid refueling and recharging–equipment that will eventually have to be made available for commercial distribution. At an even more local level, hiring specialized and, perhaps, higher-paid mechanics capable of dealing with the more advanced technical components of hybrid vehicles may also become part of automotive shops' planning for as-yet unpredictable new directions in their business.

CAREERS AND COURSE WORK

Academic preparation for careers tied to HEV technology is, of course, closely tied to the fields of electrical and mechanical engineering and, perhaps to a lesser degree, chemistry. All of these fields demand course work at the undergraduate level to develop familiarity not only with engineering principles but with basic sciences and mathematics used, especially those used by physicists. Beyond a bachelor's degree, graduate-level preparation would include continuation of all of the above subjects at more advanced levels, plus an eventual choice for specialization, based on the assumption that some subfields of engineering are more relevant to HEV technology than others.

The most obvious employment possibilities for engineers interested in HEV technology is with actual manufacturers of automobiles or heavy equipment. Depending on the applicant's academic background, employment with manufacturing firms can range from hands-on engineering applications to more conceptually based research and design functions.

Employment openings in research may be found with a wide variety of private- sector firms, some involving studies of environmental impact, others embedded in actual hybrid-engineering technology. These are too numerous to list here, but one outstanding example of a major private firm that is engaged on an international level in environmentally sustainable technology linked to hybrid vehicle research is ABB. ABB grew from late-nineteenth-century origins in electrical lighting and generator manufacturing in Sweden (ASEA), merging in 1987 with the Swiss firm Brown Boveri. ABB carries on operations in many locations throughout the world.

Internationally known U.S. firm Argonne National Laboratory not only produces research data but also serves as a training ground for engineers who either move on to work with smaller ecology-sensitive engineering enterprises or enter government agencies and university research programs.

Finally, employment with government agencies, especially the EPA, the DOE, and Department of Transportation, represents a viable alternative for applicants with requisite advanced engineering and managerial training.

Fascinating Facts About Hybrid Vehicle Technologies

- The basic technology that is being used, albeit in perfected form, in post-2000 hybrid vehicles was first used to manufacture a working hybrid car more than one hundred years ago.
- Consideration of total weight of a hybrid vehicle is so important that engineers devote major attention to possible innovations for any and all hybrid electric vehicle (HEV) components. The most obvious component that undergoes changes from one generation of HEV to the next involves ever-more-efficient modes of supplying electrical energy.
- Technological use of hybrid power systems need not, probably will not, be limited to land transportation. It is possible that aviation–a transport sector that went from conventional to jet engines in the middle of the 20th century–could become considerably more economical and ecological by a combination of power sources.
- Many major cities, especially in Europe (most notably Paris) have fleets of municipal bicycles that can be checked in and out for inner-city use by individuals physically capable and desirous of peddling. It is to be hoped that a next stage–hourly rentals of small HEVs, especially those with major electrical power sources–will follow when production of "basic" (markedly less expensive) hybrid vehicles becomes feasible.
- As more and more sophisticated hybrid-power procedures are developed, the possibility of using different forms of fuel, and eventually bypassing dependence on petroleum, and even biofuels, is a long-range goal of hybrid technology.
- Using regenerative braking, engineers are able to recover electric energy from the magnetic field created when braking results and store it in the HEV's battery for future use.

SOCIAL CONTEXT AND FUTURE PROSPECTS

Although obvious ecological advantages can result as more and more buyers of new vehicles opt for hybrid cars, a variety of potentially negative socioeconomic factors could come into play, certainly over the short to medium term. The higher sales price of hybrids that were available toward the end of 2010 already raised the question of consumer ability (or willingness) to pay more at the outset for fuel-economy savings that would have to be spread out over a fairly long time frame–possibly even longer than the owner kept the vehicle. It is nearly impossible to predict the number of potential buyers whose statistically lower purchasing ability prevents them from paying higher prices for hybrids. Continued unwillingness or inability to purchase hybrids would mean that a proportionally large number of used older-model ICE's (or brand-new models of older-technology vehicles) would remain on the roads. This socioeconomic potentiality remains linked, of course, to any investment strategies under consideration by industrial producers of cars.

How is one to know which companies worldwide are developing new, economically attractive applications for forthcoming hybrid cars?

The European digital news service EIN News, established in the mid-1990's, provides (among dozens of other categories of information) a specific subsection on hybrid vehicle technology and marketing events, including exhibitions, to its subscribers.

Subscribers from all over the world can obtain up-to-date information on hybrid technology from the Detroit publication *Automotive News* and *Automotive News Europe* and *Automotive News China*. There are, of course, many different marketing congresses (popularly labeled automotive shows) all over the globe, where the latest hybrid technology is introduced and different manufacturers' models can be compared.

In the United States, the Society of Automotive Engineers (SAE) is an important source of up-to-date information for ongoing hybrid vehicle research for both engineering specialists and well-informed general readers.

Byron D. Cannon, M.A., Ph.D.

FURTHER READING

Bethscheider-Kieser, Ulrich. *Green Designed: Future Cars.* Ludwigsburg, Germany: Avedition, 2008. Presents European estimates of technologies that should be compared with the hybrid gas-electric approach to fuel economy.

Clemens, Kevin. *The Crooked Mile: Through Peak Oil, Hybrid Cars and Global Climate Change to Reach a Brighter Future.* Lake Elmo, Minn.: Demontreville Press, 2009. As the title suggests, issues of hybrid car technology need to be placed in a very broad ecological context, where even bigger issues (downward decline in world oil reserves, climate change) may necessitate emphasis on new possible technological solutions.

Lim, Kieran. *Hybrid Cars, Fuel-cell Buses and Sustainable Energy Use.* North Melbourne, Australia: Royal Australian Chemical Institute, 2004. Provides an idea of technologies and programs imagined in other countries.

Society of Automotive Engineers. *1994 Hybrid Electric Vehicle Challenge.* Warrendale, Penn.: Society of Automotive Engineers, 1995. Reports published by thirty American and Canadian college and university engineering laboratories on their respective HEV research programs.

WEB SITES

Electric Auto Association
http://www.electricauto.org

Electric Drive Transportation Association
http://www.electricdrive.org

Society of Automotive Engineers
http://www.sae.org

U.S. Department of Energy
Clean Cities
http://www1.eere.energy.gov/cleancities

See also: Biofuels and Synthetic Fuels; Electrical Engineering; Electric Automobile Technology; Mechanical Engineering.

HYDRAULIC ENGINEERING

FIELDS OF STUDY

Physics; mathematics; computer programming; numerical analysis and modeling; material science; civil engineering; water resources engineering; fluid mechanics; hydrostatics; fluid kinematics; hydrodynamics; hydraulic structures; reservoir operations; dam design; open-channel flow; channel design; bridge design; river navigation; coastal engineering; water supply; hydraulic transients; pipeline design; storm drainage; irrigation; water reclamation and recycling; sanitary engineering; environmental engineering; hydraulic machinery; hydroelectric power; pump design.

SUMMARY

Hydraulic engineering is a branch of civil engineering concerned with the properties, flow, control, and uses of water. Its applications are in the fields of water supply, sewerage evacuation, water recycling, flood management, irrigation, and the generation of electricity. Hydraulic engineering is an essential element in the design of many civil and environmental engineering projects and structures, such as water distribution systems, wastewater management systems, drainage systems, dams, hydraulic turbines, channels, canals, bridges, dikes, levees, weirs, tanks, pumps, and valves.

KEY TERMS AND CONCEPTS

- **Froude Number:** Dimensionless number for open-channel flow, defined as the ratio of the velocity to the square root of the product of the hydraulic depth and the gravitational acceleration. The Froude number equals 1 for critical flow.
- **Gravity Flow:** Flow of water with a free surface, as in open channels or partially full pipes.
- **Ideal Fluid Flow:** Hypothetical flow that assumes a frictionless flow and no fluid viscosity; also known as inviscid flow.
- **Incompressible Fluid:** Fluid that assumes constant fluid density; applies to fluids such as water and oil, but not gases, which are compressible.
- **Laminar Flow:** Streamlined flow in a pipe in which a fluid particle follows an observable path. It occurs at low velocities, in high-viscosity fluids, and at low Reynolds numbers, typically less than 2,000.
- **Pressure Flow:** Flow under pressure, such as water in a pipe flowing at capacity.
- **Real Fluid Flow:** Flow that assumes frictional (viscous) effects; also known as viscid flow.
- **Reynolds Number:** Dimensionless number defined by the product of the pipe diameter and the flow velocity, divided by the fluid's kinematic viscosity.
- **Steady Flow:** Flow that remains constant over time at every point.
- **Subcritical Flow:** Relatively low-velocity and high-depth flow in an open channel. Its Froude number is smaller than 1.
- **Supercritical Flow:** Relatively high-velocity and small-depth flow in an open channel. Its Froude number is larger than unity.
- **Turbulent Flow:** Flow characterized by the irregular motion of particles following erratic paths. It may be found in streams that appear to be flowing very smoothly and in swirls and large eddies caused by a disturbance source.
- **Uniform Flow:** Flow that occurs when the cross section and the velocity remain constant along the channel or pipe.
- **Unsteady Flow:** Flow that changes with time; also known as hydraulic transient flow.
- **Varied Flow:** Flow that occurs when the free water surface and the velocity vary along the flow path; also known as nonuniform flow.

DEFINITION AND BASIC PRINCIPLES

Hydraulic engineering is a branch of civil engineering that focuses on the flow of water and its role in civil engineering projects. The principles of hydraulic engineering are rooted in fluid mechanics. The conservation of mass principle (or the continuity principle) is the cornerstone of hydraulic analysis and design. It states that the mass going into a control volume within fixed boundaries is equal to the rate of increase of mass within the same control volume. For an incompressible fluid with fixed boundaries, such as water flowing through a pipe, the continuity equation is simplified to state that the inflow rate is equal to the outflow rate. For unsteady flow in a channel or a reservoir, the continuity principle states that the flow rate into a control volume minus the outflow rate is equal to the time rate of change of storage within the control volume.

Energy is always conserved, according to the first law of thermodynamics, which states that energy can neither be created nor be destroyed. Also, all forms of energy are equivalent. In fluid mechanics, there are mainly three forms of head (energy expressed in unit of length). First, the potential head is equal to the elevation of the water particle above an arbitrary datum. Second, the pressure head is proportional to the water pressure. Third, the kinetic head is proportional to the square of the velocity. Therefore, the conservation of energy principle states that the potential, pressure, and kinetic heads of water entering a control volume, plus the head gained from any pumps in the control volume, are equal to the potential, pressure, and kinetic heads of water exiting the control volume, plus the friction loss head and any head lost in the system, such as the head lost in a turbine to generate electricity.

Hydraulic engineering deals with water quantity (flow, velocity, and volume) and not water quality, which falls under sanitary and environmental engineering. However, hydraulic engineering is an essential element in designing sanitary engineering facilities such as wastewater-treatment plants.

Hydraulic engineering is often mistakenly thought to be petroleum engineering, which deals with the flow of natural gas and oil in pipelines, or the branch of mechanical engineering that deals with a vehicle's engine, gas pump, and hydraulic breaking system. The only machines that are of concern to hydraulic engineers are hydraulic turbines and water pumps.

BACKGROUND AND HISTORY

Irrigation and water supply projects were built by ancient civilizations long before mathematicians defined the governing principles of fluid mechanics. In the Andes Mountains in Peru, remains of irrigation canals were found, radiocarbon dating from the fourth millennium B.C.E. The first dam for which there are reliable records was built before 4000 B.C.E. on the Nile River in Memphis in ancient Egypt. Egyptians built dams and dikes to divert the Nile's floodwaters into irrigation canals. Mesopotamia (now Iraq and western Iran) has low rainfall and is supplied with surface water by two major rivers, the Tigris and

the Euphrates, which are much smaller than the Nile but have more dramatic floods in the spring. Mesopotamian engineers, concerned about water storage and flood control as well as irrigation, built diversion dams and large weirs to create reservoirs and to supply canals that carried water for long distances. In the Indus Valley civilization (now Pakistan and northwestern India), sophisticated irrigation and storage systems were developed.

One of the most impressive dams of ancient times is near Marib, an ancient city in Yemen. The 1,600-foot-long dam was built of masonry strengthened by copper around 600 B.C.E. It holds back some of the annual floodwaters coming down the valley and diverts the rest of that water out of sluice gates and into a canal system. The same sort of diversion dam system was independently built in Arizona by the Hohokam civilization around the second or third century C.E.

In the Szechwan region of ancient China, the Dujiangyan irrigation system was built around 250 B.C.E. and still supplies water in modern times. By the second century C.E., the Chinese used chain pumps, which lifted water from lower to higher elevations, powered by hydraulic waterwheels, manual foot pedals, or rotating mechanical wheels pulled by oxen.

The Minoan civilization developed an aqueduct system in 1500 B.C.E. to convey water in tubular conduits in the city of Knossos in Crete. Roman aqueducts were built to carry water from large distances to Rome and other cities in the empire. Of the 800 miles of aqueducts in Rome, only 29 miles were above ground. The Romans kept most of their aqueducts underground to protect their water from enemies and diseases spread by animals.

The Muslim agricultural revolution flourished during the Islamic golden age in various parts of Asia and Africa, as well as in Europe. Islamic hydraulic engineers built water management technological complexes, consisting of dams, canals, screw pumps, and *norias*, which are wheels that lift water from a river into an aqueduct.

The Swiss mathematician Daniel Bernoulli published *Hydrodynamica* (1738; *Hydrodynamics by Daniel Bernoulli*, 1968), applying the discoveries of Sir Isaac Newton and Gottfried Wilhelm Leibniz in mathematics and physics to fluid systems. In 1752, Leonhard Euler, Bernoulli's colleague, developed the more generalized form of the energy equation.

In 1843, Adhémar-Jean-Claude Barré de Saint Venant developed the most general form of the differential equations describing the motion of fluids, known as the Saint Venant equations. They are sometimes called Navier-Stokes equations after Claude-Louis Navier and Sir George Gabriel Stokes, who were working on them around the same time.

The German scientist Ludwig Prandtl and his students studied the interactions between fluids and solids between 1900 and 1930, thus developing the boundary layer theory, which theoretically explains the drag or friction between pipe walls and a fluid.

How It Works

Properties of Water. Density and viscosity are important properties in fluid mechanics. The density of a fluid is its mass per unit volume. When the temperature or pressure of water changes significantly, its density variation remains negligible. Therefore, water is assumed to be incompressible. Viscosity, on the other hand, is the measure of a fluid's resistance to shear or deformation. Heavy oil is more viscous than water, whereas air is less viscous than water. The viscosity of water increases with reduced temperatures. For instance, the viscosity of water at its freezing point is six times its viscosity at its boiling temperature. Therefore, a flow of colder water assumes higher friction.

Hydrostatics. Hydrostatics is a subdiscipline of fluid mechanics that examines the pressures in water at rest and the forces on floating bodies or bodies submerged in water. When water is at rest, as in a tank or a large reservoir, it does not experience shear stresses; therefore, only normal pressure is present. When the pressure is uniform over the surface of a body in water, the total force applied on the body is a product of its surface area times the pressure. The direction of the force is perpendicular (normal) to the surface. Hydrostatic pressure forces can be mathematically determined on any shape. Buoyancy, for instance, is the upward vertical force applied on floating bodies (such as boats) or submerged ones (such as submarines). Hydraulic engineers use

---

Fascinating Facts About Hydraulic Engineering

- An interesting application of Bernoulli's theorem outside the field of hydraulic engineering is the lift force on an airplane wing. The air velocity over the longer, top side of the wing is faster than the velocity along the shorter underside. Bernoulli's theorem proves that the pressure on the top of the wing is lower than the pressure on the bottom, which results in a net upward force that lifts the plane in the air.

- Hoover Dam, built between 1931 and 1936, is a 726-foot-tall concrete arch-gravity dam on the border between Nevada and Arizona. It is the second tallest dam in the United States after the 770-foot-tall Oroville Dam in California, built between 1961 and 1968. The world's tallest dam is the 984-foot-tall Nurek Dam in Tajikistan, an earth-fill embankment dam constructed between 1961 and 1980.

- One of the first recorded dam failures was in Grenoble, France, in 1219. Since 1865, twenty-nine dam failures have been recorded worldwide, ten of which occurred in the United States.

- The 305-foot-tall Teton Dam in Idaho became the highest dam to fail in June, 1976, a few months after construction was complete. Spring runoff had filled the reservoir with 80 million gallons of water, which gushed through the mostly evacuated downstream towns, killing eleven people and causing $1 billion in damages.

- Hurricane Katrina hit New Orleans in August, 2005, causing the failure of levees and flood walls at about fifty locations. Millions of gallons of water flooded 85 percent of the coastal city, and more than 1,800 people died.

- A tidal bore is a moving hydraulic jump that occurs when the incoming high tide forms a wave that travels up a river against the direction of the flow. Depending on the water level in the river, the tidal bore can vary from an undular wave front to a shock wave that resembles a wall of water.

- A water hammer is hydraulic transient in a pipe, characterized by dangerously large pressure fluctuations caused by the sudden shutdown of a pump or the rapid opening or closing of a valve. The high-velocity wave travels back and forth in the pipe, causing the pressure to fluctuate from large positive values that could burst the pipe to very low negative values that could cause the walls of the pipe to collapse.

hydrostatics to compute the forces on submerged gates in reservoirs and detention basins.

Fluid Kinematics. Water flowing at a steady rate in a constant-diameter pipe has a constant average velocity. The viscosity of water introduces shear stresses between particles that move at different velocities. The velocity of the particle adjacent to the wall of the pipe is zero. The velocity increases for particles away from the wall, and it reaches its maximum at the center of the pipe for a particular flow rate or pipe discharge. The velocity profile in a pipe has a parabolic shape. Hydraulic engineers use the average velocity of the velocity profile distribution, which is the flow rate over the cross-sectional area of the pipe.

Bernoulli's Theorem. When friction is negligible and there are no hydraulic machines, the conservation of energy principle is reduced to Bernoulli's equation, which has many applications in pressurized flow and open-channel flow when it is safe to neglect the losses.

APPLICATIONS AND PRODUCTS

Water Distribution Systems. A water distribution network consists of pipes and several of the following components: reservoirs, pumps, elevated storage tanks, valves, and other appurtenances such as surge tanks or standpipes. Regardless of its size and complexity, a water distribution system serves the purpose of transferring water from one or more sources to customers. There are raw and treated water systems. A raw water network transmits water from a storage reservoir to treatment plants via large pipes, also called transmission mains. The purpose of a treated water network is to move water from a water-treatment plant and distribute it to water retailers through transmission mains or directly to municipal and industrial customers through smaller distribution mains.

Some water distribution systems are branched, whereas others are looped. The latter type offers more reliability in case of a pipe failure. The hydraulic engineering problem is to compute the steady velocity or flow rate in each pipe and the pressure at each junction node by solving a large set of continuity equations and nonlinear energy equations that characterize the network. The steady solution of a branched network is easily obtained mathematically; however, the looped network initially

offered challenges to engineers. In 1936, American structural engineer Hardy Cross developed a simplified method that tackled networks formed of only pipes. In the 1970's and 1980's, three other categories of numerical methods were developed to provide solutions for complex networks with pumps and valves. In 1996, engineer Habib A. Basha and his colleagues offered a perturbation solution to the nonlinear set of equations in a direct, mathematical fashion, thus eliminating the risk of divergent numerical solutions.

Hydraulic Transients in Pipes. Unsteady flow in pipe networks can be gradual; therefore, it can be modeled as a series of steady solutions in an extended period simulation, mostly useful for water-quality analysis. However, abrupt changes in a valve position, a sudden shutoff of a pump because of power failure, or a rapid change in demand could cause a hydraulic transient or a water hammer that travels back and forth in the system at high speed, causing large pressure fluctuations that could cause pipe rupture or collapse.

The solution of the quasi-linear partial differential equations that govern the hydraulic transient problem is more challenging than the steady network solution. The Russian scientist Nikolai Zhukovsky offered a simplified arithmetic solution in 1904. Many other methods–graphical, algebraic, wave-plane analysis, implicit, and linear methods, as well as the method of characteristics–were introduced between the 1950's and 1990's. In 1996, Basha and his colleagues published another paper solving the hydraulic transient problem in a direct, noniterative fashion, using the mathematical concept of perturbation.

Open-Channel Flow. Unlike pressure flow in full pipes, which is typical for water distribution systems, flow in channels, rivers, and partially full pipes is called gravity flow. Pipes in wastewater evacuation and drainage systems usually flow partially full with a free water surface that is subject to atmospheric pressure. This is the case for human-built canals and channels (earth or concrete lined) and natural creeks and rivers.

The velocity in an open channel depends on the area of the cross section, the length of the wetted perimeter, the bed slope, and the roughness of the channel bed and sides. A roughness factor is estimated empirically and usually accounts for the material, the vegetation, and the meandering in the channel.

Open-channel flow can be characterized as steady or unsteady. It also can be uniform or varied flow, which could be gradually or rapidly varied flow. A famous example of rapidly varied flow is the hydraulic jump.

When high-energy water, gushing at a high velocity and a shallow depth, encounters a hump, an obstruction, or a channel with a milder slope, it cannot sustain its supercritical flow (characterized by a Froude number larger than 1). It dissipates most of its energy through a hydraulic jump, which is a highly turbulent transition to a calmer flow (subcritical flow with a Froude number less than 1) at a higher depth and a much lower velocity. One way to solve for the depths and velocities upstream and downstream of the hydraulic jump is by applying the conservation of momentum principle, the third principle of fluid mechanics and hydraulic engineering. The hydraulic jump is a very effective energy dissipater that is used in the designs of spillways.

Hydraulic Structures. Many types of hydraulic structures are built in small or large civil engineering projects. The most notable by its size and cost is the dam. A dam is built over a creek or a river, forming a reservoir in a canyon. Water is released through an outlet structure into a pipeline for water supply or into the river or creek for groundwater recharge and environmental reasons (sustainability of the biological life in the river downstream). During a large flood, the reservoir fills up and water can flow into a side overflow spillway–which protects the integrity of the face of the dam from overtopping–and into the river.

The four major types of dams are gravity, arch, buttress, and earth. Dams are designed to hold the immense water pressure applied on their upstream face. The pressure increases as the water elevation in the reservoir rises.

Hydraulic Machinery. Hydraulic turbines transform the drop in pressure (head) into electric power. Also, pumps take electric power and transform it into water head, thereby moving the flow in a pipe to a higher elevation.

There are two types of turbines, impulse and reaction. The reaction turbine is based on the steam-powered device that was developed in Egypt in the first century C.E. by Hero of Alexandria. A simple example of a reaction turbine is the rotating lawn sprinkler.

Pumps are classified into two main categories, centrifugal and axial flow. Pumps have many industrial, municipal, and household uses, such as boosting the flow in a water distribution system or pumping water from a groundwater well.

IMPACT ON INDUSTRY

The vast field of hydraulics has applications ranging from household plumbing to the largest civil engineering projects. The field has been integral to the history of humankind. The development of irrigation, water supply systems, and flood protection shaped the evolution of societies.

Hydraulic engineering is not a new field. Its governing principles were established starting in the eighteenth century and were refined through the twentieth century. Modern advances in the industry have been mainly the development of commercial software that supports designers in their modeling.

Since the 1980's, water distribution systems software has been evolving. Software can handle steady flow, extended period simulations, and hydraulic transients. Wastewater system and storm drainage software are also being developed.

One-dimensional open-channel software can be used for modeling flow in channels and rivers and even to simulate flooding of the banks. However, two-dimensional software is better for modeling flow in floodplains and estuaries, although most software still has convergence problems. Three-dimensional software is used for modeling the flow in bays and lakes.

CAREERS AND COURSE WORK

Undergraduate students majoring in civil or environmental engineering usually take several core courses in hydraulic engineering, including fluid mechanics, water resources, and fluid mechanics laboratory. Advanced studies in hydraulic engineering lead to a master of science or a doctoral degree. Students with a bachelor's degree in a science or another engineering specialty could pursue an advanced degree in hydraulic engineering, but they may need to take several undergraduate level courses before starting the graduate program.

Graduates with a bachelor's degree in civil engineering or advanced degrees in hydraulics can work for private design firms that compete to be chosen to work on the planning and design phases of large governmental hydraulic engineering projects. They can work for construction companies that bid on governmental projects to build structures and facilities that include hydraulic elements, or for water utility companies, whether private or public.

To teach or conduct research at a university or a research laboratory requires a doctoral degree in one of the branches of hydraulic engineering.

SOCIAL CONTEXT AND FUTURE PROSPECTS

In the twenty-first century, hydraulic engineering has become closely tied to environmental engineering. Reservoir operators plan and vary water releases to keep downstream creeks wet, thus protecting the biological life in the ecosystem.

Clean energy is the way to ensure sustainability of the planet's resources. Hydroelectric power generation is a form of clean energy. Energy generated by ocean waves is a developing and promising field, although wave power technologies still face technical challenges.

Bassam Kassab, B.Eng., M.Eng., M.Sc.

FURTHER READING

Basha, Habib A. "Nonlinear Reservoir Routing: Particular Analytical Solution." *Journal of Hydraulic Engineering* 120, no. 5 (May, 1994): 624-632. Presents a mathematical solution for the flood routing equations in reservoirs.

Basha, Habib A., and W. El-Asmar. "The Fracture Flow Equation and Its Perturbation Solution." *Water Resources Research* 39, no. 12 (December, 2003): 1365. Shows that the perturbation method could be used not only on steady and transient water distribution problems but also on any nonlinear problem.

Boulos, Paul F. "H2ONET Hydraulic Modeling." *Journal of Water Supply Quarterly, Water Works Association of the Republic of China (Taiwan)* 16, no. 1 (February, 1997): 17-29. Introduces the use of modeling software (the H2ONET) in water distribution networks.

Boulos, Paul F., Kevin E. Lansey, and Bryan W. Karney. *Comprehensive Water Distribution Systems Analysis Handbook for Engineers and Planners.* 2d ed. Pasadena, Calif.: MWH Soft Press, 2006. Includes chapters on master planning and water-quality simulation.

Chow, Ven Te. *Open-Channel Hydraulics.* 1959. Reprint. Caldwell, N.J.: Blackburn, 2008. Contains chapters on uniform flow, varied flow, and unsteady flow.

Finnemore, E. John, and Joseph B. Franzini. *Fluid Mechanics With Engineering Applications.* 10th international ed. Boston: McGraw-Hill, 2009. Features chapters on kinematics, energy principles, hydrodynamics, and forces on immersed bodies.

Walski, Thomas M. *Advanced Water Distribution Modeling and Management.* Waterbury, Conn.: Haestad Methods, 2003. Examines water distribution modeling, has a chapter on hydraulic transients.

WEB SITES
American Society of Civil Engineers
http://www.asce.org

International Association of Hydro-Environment Engineering and Research
http://www.iahr.net

International Association of Hydrological Sciences
http://iahs.info

United States Society on Dams
http://www.ussdams.org

U.S. Bureau of Reclamation
Waterways and Concrete Dams Group
http://www.usbr.gov/pmts/waterways

See also: Civil Engineering; Flood-Control Technology; Fluid Dynamics; Hydrology and Hydrogeology; Water Supply Systems.

HYDROELECTRIC POWER PLANTS

FIELDS OF STUDY

Biology; civil engineering; earth science; electrical engineering; electronics; environmental science; geology; hydrogeology; mechanical engineering; physics; natural resource planning; seismology; urban planning

SUMMARY

Hydroelectric power plants produce electricity using water, a renewable resource. The power plants convert the energy in flowing water into electricity that can supply the needs of an entire city or supplement the power available for a region or other area on the power grid. To produce the electricity, water collects behind a dam before flowing through a turbine. As the water flows through the turbine, a generator uses magnets to create an electromagnetic field and then electricity. There is little pollution created in the process, but there is an impact on the ecosystem at the site of the plant.

KEY TERMS AND CONCEPTS

- **Dam:** Any barrier that curtails the flow of water.
- **Excitor:** Sends an electrical current to the rotor as the turbine turns.
- **Generator:** The part of the turbine that uses magnets to produce electricity.
- **Intake:** The penstock that brings water into the powerhouse.
- **Outflow:** The used water moved through pipes to reenter the river downstream from the dam.
- **Penstock:** Closed conduit or pipe that uses gravity to bring water to the powerhouse and into the turbine.
- **Powerhouse:** A facility for the generation of electricity.
- **Power Lines:** The wires that carry electricity from the powerhouse.
- **Reservoir:** A large natural or human-made lake used to contain water for drinking or power.
- **Rotor:** The series of electromagnets that spin inside the stator.
- **Stator:** The tightly wound coil of copper wire inside the generator.
- **Tailraces:** The pipelines that carry the outflow.
- **Transformer:** Located inside the powerhouse, it takes the alternating current and converts it to higher voltage currency.
- **Turbine:** A rotary engine that extracts energy from a fluid flow and converts it into usable energy.
- **Turbine Blades:** The parts of the turbine that spin when water passes over them as it falls over the dam.
- **Wicket Gate:** Controls the amount of water flowing from the penstock through the turbine.

DEFINITION AND BASIC PRINCIPLES

Hydroelectric power plants take the stored energy of water in a reservoir and convert it into electricity. Swiftly moving water is brought into the powerhouse and then moved through the turbine engine. As the water passes by the turbine, it causes the blades of the turbine to spin. These blades in turn cause a series of rotors with magnets mounted on rotors inside the generator to rotate past copper coils. The magnets and coiled copper wire act as giant electromagnets and produce an alternating current. The alternating current, or AC, is then converted into a higher voltage of current before exiting the powerhouse by way of the power lines that carry the power to the electric grid.

The amount of electricity generated depends upon the volume of water flow from the reservoir. The larger the reservoir, the greater the volume flow and the greater the amount of electricity produced, as the greater the volume flow, the more quickly the magnets will spin within the coil.

For a hydroelectric power plant to produce a steady flow of electricity, a steady source of water must flow into the reservoir. To ensure this, some hydroelectric power plants have an upper reservoir and a lower reservoir. Water from the upper reservoir is used to produce electricity and then channeled into a lower reservoir, rather than back into the river downstream of the dam. During off-peak hours, water from the lower reservoir is pumped back up to the upper reservoir to be used again as a source of flowing fluid for the turbines. In the absence of a lower reservoir, water is directed into the river downstream of the dam.

Current technology calls for large quantities of water moving at a high rate of speed. Newer technologies that use the kinetic properties of water without building dams, like the waterwheels of years ago, are also being explored.

BACKGROUND AND HISTORY

Civilizations have been using water as a power source for thousands of years. From the ancient Greeks, who used water wheels to replace manual labor in the grinding of wheat into flower, to the Romans, who created floating mills when under siege by the Goths in 537 C.E., water has been used to get work done. Because they had fewer slave workers than the Romans, the ancient Chinese used waterpower to their advantage throughout the empire. Water, for example, was used to power the bellows used in iron casting.

During the medieval period in Europe, waterpower grew in prominence. Because of the Black Death (the plague) and the shortage of labor that resulted, it was critical to find an inexpensive way to grind wheat to sustain the surviving population. The number of mills increased dramatically, with some of the mills built to take advantage of tidal changes.

Another important development was the use of waterpower by a religious order. The Cistercian monastic order lived in rural areas, where they perfected the use of hydropower for milling, woodcutting, and olive crushing. They did not use dams, but rather placed their waterwheels in swiftly moving water without impeding the flow of that water. They also used the water for washing and for sewage disposal. As the order moved through Europe, the technology traveled with them, until waterpower next came to prominence during the Industrial Revolution. At that time, swiftly running streams were used to provide power for a variety of manufacturing processes before the widespread use of fossil fuel.

Hydroelectric power plants in the United States have often been constructed to take advantage of the force of moving water while also solving flooding or other water-related problems. Typically the dams have been among the largest concrete structures built in the area. They also have altered the ecosystem by controlling the flow of water downstream while flooding the area upstream of the dam. Because of this, a great deal of opposition to new construction accompanies any dam proposal; people debate issues

of water rights and the environmental impact of dams in general.

HOW IT WORKS

The process of converting the energy from a flowing liquid, such as water, into electricity that can be used to supply the power needs of a city or to increase the amount of electricity available on a local or regional grid is actually simple, as long as the conditions are right.

Swift Water Flow. These conditions include the existence of a steady supply of swiftly moving water. The water can flow swiftly as a result of the pressure on the water from the reservoir as the water enters the intake, or penstock, area of the powerhouse located near the base of the dam. Alternately, it can flow swiftly as a result of being released over the top of the dam to spill into the penstock at a high speed. Either way, it is essential the water moves rapidly when it flows through the turbine, although it is not necessary that the water first be held in a reservoir created by a dam.

The turbine has blades that are turned by the flowing water. The more rapidly the water flows, the more rapidly the blades spin. The spinning motion of the blades in turn causes the magnets in the generator, mounted on rotors, to spin inside coils of copper wire. This spinning creates an electromagnetic field that produces alternating current that is then converted to a higher voltage current in the transformer. The higher voltage current can be stored but is typically transmitted over power lines to become part of the power grid serving a city or region.

A typical powerhouse will have water entering through multiple penstocks to flow through one of several turbines mounted under generators. Once the water has flowed past the turbine blades, it will reenter the flow of the river downstream of the dam. The water can also be channeled into a lower reservoir to be pumped back to the upper reservoir for reuse in generating additional electricity. Whether or not this will be done depends on the amount of water available in the reservoir and the expediency of pumping the water back to the higher reservoir.

The placement of the hydroelectric power plant is vitally important. Because a steady flow of water must be moving at a high speed, the dam needs to be located on a river with a reliable supply of water. Gravity also plays an important part in the generation

The Dalles, a hydroelectric dam. Columbia River Gorge, Oregon. (Peter Skinner/Photo Researchers, Inc.)

of hydroelectric power because water picks up speed as it moves from a higher level to a lower level. Thus, dams are often built in areas where there is a natural downward flow.

Reservoirs. The construction of a dam often also includes building a reservoir that covers hundreds of acres of land. When that land is flooded, wildlife will lose their habitat and plant life will be ruined. Reservoir construction may also require the relocation of a significant number of people. Because of this, a study of the impact of the dam is often an important part of the planning process, and not all dams are built at the optimal site.

Land Integrity. Finally, the dam site must be one that can bear the weight of the dam and the water that will accumulate. It must also be a site without significant seismic activity or the likelihood of such activity. To ensure this, a thorough geological study of the site is necessary before construction begins.

APPLICATIONS AND PRODUCTS

Hydroelectric power plants supply power that is used for many purposes. Most plants supply power to an existing power grid. Once part of the grid, the power is allocated to the area of greatest need (along with power from other sources). It is possible, though not common, to have a dedicated hydroelectric power plant, one that is created specifically to meet the needs of an individual or factory.

The main application of hydroelectric power plants is to ensure a steady supply of electricity through a process that uses a renewable resource

with little pollution. Hydroelectric power plants do have an environmental impact, however, and that needs to be taken into consideration.

Hydroelectric power plants have had a significant impact on industry through the ages. Originally a simple replacement for slave labor, the use of hydroelectric power peaked before the widespread use of fossil fuels as power sources. It is possible for a dedicated waterwheel to power a simple manufacturing process such as milling, on a small scale.

The largest hydroelectric power plant in the world is the Three Gorges Dam in the People's Republic of China. It has a capacity of 22,500 MW (megawatts) and is visible from space. The largest hydroelectric power installation in the United States, and the fifth largest in the world, is the Grand Coulee. Located in Washington State, the power plant produces electricity and is also used to irrigate the land around it. One of the largest concrete structures in the world, the Grand Coulee has a capacity of 6,809 MW. Brazil, Venezuela, Russia, and Canada are also home to hydroelectric power plants with significant capacity.

IMPACT ON INDUSTRY

Hydroelectric power is considered a green power source. Water is a renewable resource that is plentiful on the North American continent. As a result, several government agencies are actively involved in managing and promoting hydroelectric power plants and projects.

The U.S. Department of the Interior's Bureau of Reclamation is responsible for water operations in the western United States. The region includes the Pacific Northwest, upper and lower Colorado River, and the mid-Pacific. Washington, Oregon, Idaho, California, Nevada, Utah, Arkansas, New Mexico, and portions of Wyoming, Colorado, and Texas are under their jurisdiction. The Hydropower Technical Services Group provides specialized technical knowledge to hydroelectric power engineers as they work to increase energy production at their own sites.

The U.S. Department of Energy's Water Power Program is designed to optimize existing hydroelectric power plant production while promoting construction of new plants and hydropower technologies. The agency maintains an online database with information and data about hydroelectric power installations, plans, and new technologies. The database also includes information about technologies in use or under development in the United States and abroad. It can be used by anyone and allows anyone to submit a technology or project idea for consideration.

The U.S. Geological Survey (USGS) also collects information about water resources throughout the United States. The USGA's Web site highlights water science by state and includes links to a full range of publications and reports on water resources, quality, and use. As governments worldwide move toward renewable sources of energy, hydroelectric power and associated technologies will remain at the forefront.

CAREERS AND COURSEWORK

Hydroelectric power and the technology behind existing and new forms of hydroelectric power are growing fields. Study in the fields of engineering, hydrogeology, civil engineering, environmental science, and seismology are just a few areas of study that will lead to careers in this industry.

Impact and feasibility studies must be performed before construction can begin on a new facility. These studies include the expert opinions of geologists, seismologists, civil engineers, and environmental scientists about the quality of the sites under consideration and the type of facility that can be built and maintained there. The studies also detail the environmental impact of the facility under consideration. Decisions about the use of the water before and after the generation of power will also be made. The size of the facility will also be determined once the experts figure the desired output and uses of the water before and after the electricity is produced.

After approval, civil engineers oversee the construction of the dam and penstock while electrical engineers work to install the turbines, generators, and transformers. Specialists in the installation and operation of the turbines, generators, and transformers will be on hand throughout the construction phase. When construction is completed, professionals ensure that the hydroelectric power plant runs at optimal efficiency.

Engineers inspect the integrity of the dam at frequent intervals. Naturalists ensure that any fish ladders or other equipment function appropriately and are in operation at the times that are essential to the upriver journey of the fish. Naturalists also oversee the health of the fish population and the reservoir.

SOCIAL CONTEXT AND FUTURE PROSPECTS

Hydroelectric power plants are important sources of renewable energy. Producing the electricity generated by these plants does not result in significant levels of pollution. It also does not consume resources that take centuries to replenish, does not require labor-intensive or costly processes, and does not damage the environment.

The dam site and its reservoir, however, do affect the local ecosystem. The land lost to the reservoir through flooding is likely already home to many species of animals and plants and may include towns or villages, all of which will be displaced or destroyed by the dam project. It is possible the reservoir will form a wetlands area at the shoreline. It also is possible that this will not occur, resulting instead in areas of stagnant water that are not hospitable to wildlife.

Fascinating Facts About Hydroelectric Power Plants

- Hydroelectric power plants provide power from a renewable resource at low cost while generating little pollution.
- Construction of a dam for a hydroelectric power plant often alters, both positively and negatively, the surrounding ecosystem.
- The first hydroelectric power plant in the United States began operation in Appleton, Wisconsin, in 1882.
- Hydroelectric power is used around the world, providing almost 25 percent of the world's electricity.
- China generates more hydroelectricity than any other country.
- The largest hydroelectric power plant in the United States is located on the Columbia River in northern Washington State at the Grand Coulee Dam.
- The Three Gorges Dam project in China displaced 1.4 million people.
- The Three Gorges Dam is so large that it creates its own seismic activity.
- The Three Gorges Dam is one of only a few human-made structures visible from space.
- Hydroelectric power can be used anywhere there is falling water.
- Many dams incorporate fish lifts to help fish such as the American shad or the salmon return upstream to spawn.

Furthermore, the dams built for hydroelectric power plants, for example, cut off access to the spawning grounds of anadromous, migratory salmon or the American shad. These fish must return upriver to lay their eggs. To facilitate the return journey, fish lifts, ladders, or elevators are in place at many dams. These structures, which help fish move upstream, also help to avoid the disruption of the local habitat.

As the importance of renewable energy sources becomes incontrovertible, greater demand for hydroelectric power can be anticipated. As the cost in terms of loss of habitat gains greater appreciation, the call to protect existing wildlife and vegetation can also be expected. With existing technology, these goals are not easily met simultaneously.

The challenge for the next generation of hydroelectric power professionals will be to modify this existing technology or explore the use of alternatives that do not require such a large footprint. Using water as it moves, without impeding its flow, is one possible way that hydroelectric power plants can better coexist with the populations they serve.

Gina Hagler, M.B.A.

FURTHER READING

Gevorkian, Peter. *Sustainable Energy System Engineering.* New York: McGraw-Hill, 2007. Presents a variety of green energy solutions that includes hydroelectric power.

Hicks, Tyler. *Handbook of Energy Engineering Calculations.* New York: McGraw-Hill, 2011. Covers such topics as combustion of fossil fuels, alternative power sources, and hydroelectric power.

Monroe, James S., and Reed Wicander. *The Changing Earth: Exploring Geology and Evolution.* Belmont, Calif.: Thomson Higher Education, 2006. An excellent overview of the role of geology in general, with several chapters on energy sources such as hydroelectric power.

Nag, P. K. *Power Plant Engineering.* New Delhi, India: Tata McGraw-Hill, 2008. A thorough discussion of power generation from a variety of sources, including hydroelectric power. Also presents considerations for appropriate site selection.

WEB SITES

Tennessee Valley Authority "Hydroelectric Power."
http://www.tva.gov/power/hydro.htm.

U.S. Department of Energy, Water Power Program
http://www1.eere.energy.gov/water.

U.S. Department of the Interior "Reclamation: Managing Water in the West—Hydroelectric Power."
http://www.usbr.gov/power/edu/pamphlet.pdf.

U.S. Geological Survey "Water Resources of the United States."
http://water.usgs.gov.

See also: Architecture and Architectural Engineering; Civil Engineering; Earthquake Engineering; Ecological Engineering; Electrical Engineering; Energy Storage Technologies; Engineering Seismology; Environmental Engineering; Fossil Fuel Power Plants; Landscape Ecology; Land-Use Management; Water Supply Systems; Wildlife Conservation; Wind Power Technologies.

HYDROLOGY AND HYDROGEOLOGY

FIELDS OF STUDY

Hydrology; hydrogeology; physics; physical geography; geology; chemistry; biology; fluid mechanics; statistics; water resources; groundwater; water supply; precipitation; fluvial processes; earth sciences; aquifers; water quality; climatology; meteorology; evapotranspiration.

SUMMARY

Hydrology is a broad interdisciplinary science that includes the hydrologic cycle and global distribution of water in both solid and liquid form in the atmosphere, oceans, lakes, streams, and subsurface formations. Hydrogeology is a subset of hydrology that focuses on groundwater and related geologic factors governing its distribution and magnitude. The shortages of water and the increasing pollution in many countries have heightened concern about availability and quality of water for many people in the world.

KEY TERMS AND CONCEPTS

- **Aquifer:** Water-bearing geologic formation of saturated permeable materials.
- **Base Flow:** Groundwater discharge moving into receiving streams.
- **Discharge:** Amount of water flowing through a stream cross section, measured in cubic feet per second or cubic meters per second.
- **Evapotranspiration:** Combined loss of water from evaporation from soils, streams, and lakes, plus plant transpiration.
- **Exotic River:** Stream that flows in an arid region from water coming from distant well-watered uplands.
- **Groundwater:** Subsurface water in the saturated zone that varies with the type of geological formation.
- **Hydrologic Cycle:** Total transfer and storage of all of the readily available water on the Earth as it moves through the gaseous, liquid, and solid states.
- **Watershed:** Area of land that has a drainage network and is separated from other watersheds by a drainage divide; also known as a drainage basin.

DEFINITION AND BASIC PRINCIPLES

Hydrology is the science of water, a unique substance that affects all life on Earth. Although water could exist on Earth without life, life could not exist without water. Water is the most abundant liquid on Earth, covering 71 percent of the Earth's surface in its liquid and solid forms. In humans, water constitutes about 92 percent of blood plasma, 80 percent of muscle tissue, 60 percent of red blood cells, and more than half of most other tissues.

Hydrology is a part of many scientific disciplines. The study of water in the atmosphere involves the fields of climatology and meteorology. The study of the hydrosphere includes the fields of physical geography, potamology (rivers), glaciology, cryology (snow, ice, and frozen ground), and limnology (lakes). The study of the lithosphere (the topmost rock layer of the Earth) includes the fields of hydrogeology (groundwater location, movement, and magnitude), geomorphology (the science of surface processes and landforms on the Earth), and limnology (the science of lakes). Given the importance of water to plants and animals, hydrology includes the fields of silviculture (forestry), plant ecology, and hydrobiology (the biology of bodies of freshwater such as lakes).

Other fields related to hydrology by virtue of their strong connection to water resources include watershed management, potable water supply, wastewater treatment, irrigation, water law, political science (water policy), economics (costs of water projects), drainage, flood control, hydropower, salinity control and treatment, erosion and aspects of sediment control, navigation, lake and inland fisheries, and recreational uses of water.

BACKGROUND AND HISTORY

Credit for developing part of the hydrologic cycle, a fundamental component of hydrology, goes to Marcus Vitruvius Pollio, a Roman engineer and writer during the reign of the emperor Augustus, who developed a theory that groundwater is mostly recharged by precipitation infiltrating the ground. This early theory was buttressed by Leonardo da Vinci and Bernard Palissy during the sixteenth century. The seventeenth century was a period of measurement, when scientists studied precipitation, evaporation, and stream discharge in the Seine River in France. Hydraulic studies developed during the eighteenth century. In the nineteenth century, the active area of investigation was experimental hydrology, particularly in stream-flow measurement and in groundwater. The U.S. government created several important agencies, including the U.S. Army Corps of Engineers in 1802, the U.S. Geological Survey in 1879, and the U.S. Weather Bureau (now the National Weather Service) in 1891.

The nineteenth and twentieth centuries witnessed the increasing use of statistical and theoretical analysis in hydrologic studies. One example of this was the development of the bed-load function in sedimentation research in 1950 by Hans Albert Einstein, the son of Albert Einstein. Research has benefited from the increasing use of computers that can handle larger amounts of data in shorter periods of time. Several types of sophisticated statistical packages can assist in the analysis of increasingly complicated studies in hydrology and hydrogeology.

How It Works

The Hydrologic Cycle. The never-ending circulation of water and water vapor over the Earth is called the hydrologic cycle. This continuous circulation affects all three parts of the global system: the water spreading over the Earth's surface (the hydrosphere), the gaseous envelope above the hydrosphere (the atmosphere), and the rock layer beneath the hydrosphere (the lithosphere). The Sun's energy and gravity power this circulation that has no beginning or end.

The oceans cover 71 percent of the Earth's surface and account for 86 percent of the moisture in the atmosphere. The evaporated water that is transported into the atmosphere can travel tens to hundreds of miles before it is returned to the Earth in the form of rain, hail, sleet, snow, or ice. When precipitation gets closer to the surface of the Earth, the water may be intercepted and transpired by vegetation, or it can reach the ground surface and eventually flow into streams or simply infiltrate into the ground. A large portion of the water that reaches plants and the runoff flowing in streams is evaporated back into the atmosphere. A portion of the water that infiltrates into the ground may penetrate to deeper layers in the Earth to become groundwater. In turn, this groundwater may return to the streams as the base-flow component of runoff, which will eventually flow into the oceans and evaporate back into the atmosphere to complete the hydrologic cycle.

Urbanization and Stream Flow. The expansion of cities into open spaces outside the metropolitan area has strongly affected local streams. As impervious cover—in the form of houses, roads, driveways, and large parking areas for shopping malls and office buildings—increases, larger and larger areas of previously water-penetrable surfaces become impervious. Depending on local land-use regulations, impervious cover can easily approach or exceed 80 percent of the total area. The immediate effect of this high impervious percentage is to reduce the amount of water that can infiltrate into the ground and result in increased overland flow.

Storm drainage systems can also affect stream flow, as runoff is deliberately directed into nearby streams. This rapid exiting of water from the increased impervious area can quickly reduce the lag time between precipitation input and flood runoff. The resultant increase in the stream hydrograph invariably gives rise to peak discharge flows that result in local and regional flooding.

In the light of flooding problems associated with an increase in impervious cover, some counties and states have required new developments, particularly in suburban areas, to give up part of their site for detention basins. These structures are designed to detain, for varying amounts of time, the excess runoff that the new buildings and roads on the site will generate. The basins can reduce flooding and also provide an opportunity for sediment to settle out and thereby improve the quality of the water that moves downstream.

Applications and Products

Flow-Duration Curves. One example of the type of analysis that is commonly employed in hydrogeologic research is to study the variability of stream flow in watersheds that have lithologic heterogeneity (differences in their rock formations). The physical attributes of watersheds affect stream-flow variability. Some formations found on coastal plains have large amounts of sand with high infiltration rates and thereby have high groundwater yields. Other geologic formations, such as basalt, diabase, and granite, have low infiltration rates and consequently have low yields of groundwater. Indeed, the differences in

Fascinating Facts About Hydrology and Hydrogeology

- Unlike most substances, which expand and decrease in density as they are heated, water reaches its highest density at 39.2 degrees Fahrenheit. Thus, ice, which is less dense than water, floats.
- Water in lakes and oceans freezes from the surface downward; this permits water circulation to continue under the ice so that fish can survive.
- Water has the highest specific heat of all of the common substances, and its huge heat capacity has an equalizing effect on the Earth's climate. Therefore, at comparable latitudes, coastal areas will have milder climates than interior areas.
- Water boils at 212 degrees Fahrenheit at sea-level pressure. It has one of the highest boiling points of any fluid on Earth.
- Almost all fluids experience an increase in viscosity as the pressure increases, but the viscosity of water decreases as the pressure increases. This explains why water under high pressure in water distribution systems is able to flow rather than dribble out of a kitchen tap.
- The surface tension of water is double or triple that of most common liquids because of hydrogen bonding. As a result, some insects can walk on water, and steel needles will float.
- Surface tension (called cohesion) and the natural tendency of water to wet solid surfaces (called adhesion) result in capillarity, which allows water to climb a tube or wall. If water had a weaker surface tension (and consequently weaker capillary forces), water in the soil would not be able to overcome gravity, and plant life would die.
- The oceans hold the vast majority of the world's water (97 percent). About 2 percent is frozen in icecaps and glaciers. This means that almost all the water in the world is either salty or frozen.
- The average volume of all of the rivers on Earth amounts to only 0.0001 percent of the total water on the planet.

water yield between formations can approach an order of magnitude.

One useful technique that developed in the twentieth century was to use the low-flow and high-flow ends of flow-duration curves to provide useful information about the hydrogeologic characteristics of

any watershed. The flow-duration curve is a cumulative frequency curve that shows the percentage of time that specified stream discharges were equaled or exceeded. The values are plotted on logarithmic-probability graph paper with discharge in cubic feet per second on the y-axis (ordinate) and the frequency that specified discharges were exceeded in percent on the x-axis (abscissa). The slope of the curve provides a measure of temporal variability; the steeper the curve, the more variable the value plotted. Steeply sloping curves indicate a flashy stream, where the flow is mostly derived from direct runoff, indicating minimal groundwater storage. As a result, that watershed has limited potential for ample groundwater supplies.

Water-Quality Issues. Concern is growing about the release of pharmaceuticals from manufacturing facilities into surface waters. In a 2004-2009 study, the U.S. Geological Survey (USGS) found that the water released into surface waters by two wastewater-treatment plants in New York that received 20 percent of their wastewater from nearby pharmaceutical plants had concentrations of drugs that were ten to one thousand times higher than the water released from twenty-three other plants in the United States that were not treating any waste from pharmaceutical plants. A sampling of the maximum concentrations in the outflows from the two New York plants were 3,800 parts per billion (ppb) for the muscle relaxer metaxalone and 1,700 ppb and 400 ppb, respectively, for oxycodone and methadone, both opioid pain relievers. In stark contrast, the twenty-three plants that were not receiving any wastewater effluent from pharmaceutical facilities reported drug concentrations of less than 1 ppb for these drugs.

Treated effluent from wastewater-treatment plants is routinely discharged into streams that flow downstream to one or more water-treatment plants that distribute potable water to their service areas. A prime example is New Orleans, located close to the mouth of the Mississippi River, the largest drainage basin in North America, and downstream from many wastewater-treatment facilities.

The problem is that water containing a growing number of pharmaceuticals is entering wastewater-treatment plants that are not equipped to remove them. The issue is compounded by the comparable lack of techniques available to water supply treatment facilities. This water-quality issue will most probably increase in importance.

Specific Capacity. The determination of specific capacity is a useful procedure to evaluate the magnitude of the expected yield of a well. It is obtained by simply dividing the tested well yield in gallons per minute by the drawdown of the well in feet (gpm/ft). Drawdown is a measure of how much the water table is lowered as a well is pumped. The purpose of this test is to ensure that the well can sustain a required minimum yield over the long term.

Pump tests for residential wells should take at least four hours, although some communities require six hours or more. Most states require a minimum pumping rate of 0.5 gpm/ft for residential use. In addition, the original static level of the water table should recover twenty-four hours after end of the test. For large-scale commercial and industrial users, many states have more stringent standards for pumping, such as a minimum testing period of at least forty-eight hours.

Specific capacity values can vary over several orders of magnitude, from less than 1 gpm/ft to more than 100 gpm/ft, depending on the type of geologic formations present. High permeability and porosity usually result in high specific capacity values, and the converse is expected if the formations are either poorly fractured or have low permeability and porosity.

Water Use. The U.S. Geological Survey (USGS) began estimating water use every five years beginning in 1950. Its reports present a large amount of data on a wide selection of water-use categories. The USGS collects information from all fifty states in the United States, as well as the District of Columbia, Puerto Rico, and the U.S. Virgin Islands.

Some water-use categories have changed over the years, but the overall data collected are very useful. For example, the public supply category pertains to water that is furnished to at least 25 people or that serves a minimum of fifteen connections. The distributed water category includes domestic, commercial, and industrial users and contains estimates about system losses, such as leaks and the flushing of pipes. About 258 million people (86 percent of the total population) depend on public water for household needs. Surface water (streams and lakes) accounts for about two-thirds of the public water supply, and the remaining one-third comes from groundwater sources. In New Jersey, the most densely populated state in the nation, 11 percent of the population uses their own wells for water. This relatively large percentage has remained about the same since about the 1980's. This situation presumably reflects large-lot zoning and the high cost of bringing in water from distant suppliers to isolated clusters of a few homes.

Water Disputes. Conflicts over the world's water resources have been numerous and lengthy. The list of water conflicts goes back several thousand years, and the growing disparity between well-watered and poorly watered countries means future conflicts are likely.

The hydrologic imbalance in water supplies is physically based, but social, economic, and political factors play an important role. Countries that have an abundance of headwater streams can build large dams and use the water for irrigation, resulting in less flow to downstream countries. For example, the headwaters of the Tigris and Euphrates rivers start in Turkey, and any diminishment in flow through Syria and Iraq en route to the Persian Gulf would obviously affect downstream agriculture. Another well-known example is the almost total dependence of Egypt on the Nile River, which begins in Ethiopia (Blue Nile), and the lakes Albert and Victoria (White Nile) in east-central Africa. A large diversion of the waters of the Nile by the upstream states would have a substantial impact on Egypt.

Continued growth in irrigation and population in the semi-arid portion of the southwestern United States has led to serious problems on the Colorado River. The drainage area of the Colorado is 246,000 square miles and flows for 1,450 miles from its headwaters in the Rocky Mountains in Colorado, Wyoming, and New Mexico through Utah, Arizona, Nevada, and California before emptying into the Gulf of California in Mexico. Although the seven states and Mexico have made various agreements pertaining to water allocation, numerous problems have developed and are likely to worsen in future years, as the initial allocation in the early twentieth century used an average flow that was based on precipitation values that were above normal. The allocations would have to be reduced when drier or even more normal precipitation cycles return, resulting in manifold problems to large users in the basin.

IMPACT ON INDUSTRY

The private and governmental organizations that deal with hydrology range from large-scale international and national government agencies and associations to small groups focused on local issues. The increase in both the size and the number of these organizations reflects the attention that must be given to the study of problems and issues related to water and water use.

Government and University Research. The International Association of Hydrological Sciences issues a series of publications that convey the results of hydrological research and practice to all interested parties. Research covers all water-related aspects from as many countries as possible.

The U.S. government has twelve agencies that deal with water issues. The oldest agency is the U.S. Army Corps of Engineers, established in 1802. Its major activities are flood control and the improvement of navigable waters by dredging and related means. Other activities include wetlands protection. The U.S. Geological Survey (established 1879) is charged with monitoring and examining the geological and mineral resources of the nation. In 1889, the agency established the first gauging station for measuring stream flow in New Mexico; it has come to manage, with some state participation, more than 7,600 sites in the United States. Additional responsibilities include hydrogeologic investigations, water-quality analyses, and water-use monitoring. The U.S. Bureau of Reclamation in the Department of the Interior was established in 1902 with a mission to build irrigation projects, such as the Hoover and Grand Coulee dams, in the drier lands west of the 100th meridian that include seventeen Western states. The Environmental Protection Agency was established in 1970 with the responsibility for protecting and improving water quality in the United States. It has the power to issue fines and set standards for drinking water, wastewater discharge, and levels of pollution in rivers.

Professional and Educational Organizations. The American Institute of Hydrology is a nonprofit scientific and educational organization that offers certification to professionals in all fields of hydrology, including groundwater, surface water, and water quality. The certification process includes examinations and information regarding the professional and academic experience of the applicant. The American Water Resources Association provides a forum for education, professional development, and information covering many related disciplines. It also sponsors meetings in different parts of the country on various hydrologic

topics. The National Groundwater Association, which is focused on groundwater-related topics, holds technical meetings in different locations. The Association of American Geographers hosts professional meetings at varied locations that include specialty sessions in many water-related fields such as groundwater, water supply, and wastewater issues, and the application of new techniques in geographic information systems that pertain to water.

Trade Associations. The Water Quality Association is an international trade association for household and commercial water-quality improvement issues. The American Water Works Association, started in 1881, has become one of the largest organizations of water professionals, including scientists, manufacturers, and water-treatment plant personnel and managers. Its membership includes more than 4,600 treatment plants that are responsible for the delivery of potable water to 180 million people in North America. The Water Environment Federation, formed in 1928, is a nonprofit organization that provides educational meetings and technical material for wastewater-treatment plant operators and managers.

CAREERS AND COURSE WORK

Given the interaction that water has with a wide array of disciplines and subdisciplines, anyone entering the field of hydrology can pursue a variety of paths to become proficient in the subject. Although few academic institutions have a hydrology department, many offer courses in hydrology within academic departments such as civil engineering, geology, physical geography, and environmental science.

The course path for students is determined by their interests. For example, students interested in hydrogeology would most likely major in geology and possibly minor in geography, so that they could take courses in physical geography, cartography, and geographic information systems. Students interested in water-quality issues would be drawn to a major in environmental science, with chemistry and biology as suitable minors. Other useful courses include statistics, economics, civil engineering, mathematics, meteorology, and climatology.

Employment opportunities in hydrology include positions in federal, state, and local government, state water project associations (such as the Central Arizona Project), bi-state river basin commissions (such as the Delaware River Basin Commission), industry, professional associations, and nonprofit watershed associations that act as guardians for their local drainage area. Also, teaching and research positions are available at colleges and universities.

SOCIAL CONTEXT AND FUTURE PROSPECTS

Providing sufficient clean water for the growing population of the world is vital to human survival. The growing importance of an adequate water supply is demonstrated by the increasing number of books, articles, meetings, and commentaries on the topic. The Pacific Institute, established in 1987 in Oakland, California, focuses on water issues, including water and human health, controversies over large dams, freshwater conflicts, climate change and water resources, new water laws and institutions, and efficient urban water use. It conducts research; publishes reports, including the biennial series *The World's Water*; and works with stakeholders to develop solutions.

The future prospects for hydrologic research and technological change are good. For example, desalination plants are being considered for a variety of coastal locations in areas such as the Persian Gulf region, where alternative sources of water are very limited, and in the Los Angeles-San Diego area, where additional imports of water from the Colorado River would encounter opposition from other users. Another technologic advance is the invention in Israel of a type of drip irrigation designed to deal with water scarcity and salinity problems in water-short areas. This irrigation method is used in more than half of the irrigated land in Israel and its use is spreading in California, particularly for high-value crops such as orchards and vineyards.

Unlike most other resources on Earth, water is renewable. Technology and innovative thinking can play important roles in ensuring adequate supplies of clean water for a growing population. For example, the recognition that storm-water runoff, gray water (from dishwashers and washing machines), and reclaimed wastewater could be used for landscape irrigation and some industrial processes resulted in significant water conservation.

Robert M. Hordon, B.A., M.A., Ph.D.

FURTHER READING

Cech, Thomas V. *Principles of Water Resources: History, Development, Management, and Policy*. 3d ed. Hoboken, N.J.: John Wiley & Sons, 2010. A highly readable and very informative text containing a wide variety of useful and well-written chapters on water resources.

Fetter, Charles W. *Applied Hydrogeology*. 4th ed. Englewood Cliffs, N.J.: Prentice Hall, 2001. One of the standard texts in the field, it has good coverage of groundwater topics along with pertinent case studies.

Gleick, Peter H., et al. *The World's Water, 2008-2009: The Biennial Report on Freshwater Resources*. Washington, D.C.: Island Press, 2009. Contains a series of short informative chapters followed by numerous and detailed tables covering a wide range of useful data.

Glennon, Robert. *Water Follies: Groundwater Pumping and the Fate of America's Fresh Waters*. Washington, D.C.: Island Press, 2002. An interesting nontechnical and readable discussion of a variety of groundwater problems in different areas of the United States resulting from misguided direction.

Powell, James L. *Dead Pool: Lake Powell, Global Warming, and the Future of Water in the West*. 2008. Reprint. Berkeley: University of California Press, 2010. A detailed but readable book on the water problems in the southwestern United States.

Spellman, Frank R. *The Science of Water: Concepts and Applications*. 2d ed. Boca Raton, Fla.: CRC Press, 2008. Contains many useful chapters on water biology, ecology, chemistry, and water pollution and treatment.

Strahler, Alan. *Introducing Physical Geography*. 5th ed. Hoboken, N.J.: John Wiley & Sons, 2011. An excellent textbook with diagrams, maps, and photographs that cover the fields of weather and climate, fresh water of the continents, and landforms that are made by running water.

WEB SITES

American Water Works Association
http://www.awra.org

International Association of Hydrological Sciences
http://iahs.info

National Groundwater Association
http://www.ngwa.org

U.S. Army Corps of Engineers
http://www.usace.army.mil

U.S. Bureau of Reclamation
http://www.usbr.gov

U.S. Geological Survey
Water Resources
http://www.usgs.gov/water

See also: Climatology; Coastal Engineering; Fisheries Science; Flood-Control Technology; Hydroelectric Power Plants; Meteorology; Ocean and Tidal Energy Technologies; Oceanography; Water-Pollution Control; Water Supply Systems.

HYDROPONICS

FIELDS OF STUDY

Horticulture; agriculture; agricultural engineering; agronomy; botany; biology; biochemistry; biological engineering; chemistry; plant physiology; microbiology; fisheries and allied aquacultures; food technology; food science; nutrition; floriculture; hydrology; genetics; agricultural economics; rural sociology; industrial engineering; electrical engineering; mechanical engineering; computer science; physics; robotics.

SUMMARY

Hydroponics uses varied scientific and technological processes to cultivate plants without soils usually associated with agriculture. Throughout history, humans have delivered nutrients directly to plant roots with water. Based on that principle, modern hydroponics has diverse applications for commercial and utilitarian agriculture. Hydroponics supplies food to military personnel in places where agricultural resources are limited because of climate and terrain. Astronauts eat fresh vegetables grown with hydroponics in space. Food security is bolstered by the availability of substantial yields year-round assured by hydroponics, providing people access to nutrients and relief from hunger. Agribusinesses sell hydroponic crops and equipment to consumers. Many scientific educational curricula incorporate hydroponic lessons.

KEY TERMS AND CONCEPTS

- **Aeroponics:** Nutritional material sprayed in fine droplets onto roots.
- **Aquaponics:** Using hydroponic techniques for plant cultivation simultaneously with fisheries production.
- **Controlled Environment Agriculture:** Cultivation providing plants optimum growing temperatures, moisture, and aeration.
- **Macronutrients:** Elements that plants require significant quantities of to thrive, including nitrogen, phosphorus, sulfur, potassium, calcium, and magnesium.

- **Micronutrients:** Elements that plants need in lesser amounts such as iron, chlorine, manganese, boron, zinc, copper, and molybdenum.
- **Nutrient Solution:** Combination of water, minerals, and elements mixed in specific proportions for each plant type.
- **Rock Wool:** Inorganic fibers created by melting limestone, coke, and volcanic rock at extremely high temperatures to enhance water retention.
- **Substrate:** Material used to support roots.

DEFINITION AND BASIC PRINCIPLES

Hydroponics is the scientific use of chemicals, organic and inorganic materials, and technology to grow plants independently of soil. Solutions composed of water and dissolved minerals and elements provide essential macronutrients and micronutrients and supplement oxygen and light necessary for plant growth. Plant roots absorb nutrients that are supplied through various methods. Some hydroponic systems involve suspending roots into liquid solutions. Other hydroponic techniques periodically wash or spray roots with solutions. Methods also utilize containers filled with substrates, such as gravel, where roots are flooded with nutrient solutions. Hydroponics is practiced both in greenhouses, where temperatures and lighting can be regulated, and outdoors, where milder climates pose few natural detriments to plants.

Hydroponics enables agriculturists to grow crops continually without relying on weather, precipitation, and other factors associated with natural growing seasons. These systems permit agricultural production in otherwise unsuitable settings for crop cultivation such as congested cities, deserts, and mountains. Hydroponics is convenient, producing foods in all seasons. Plants can be grown closely together because root growth does not spread like soil-based plant roots extending to seek nutrients and water. Agriculturists can grow crops that are not indigenous to areas, such as tropical fruits. Growth typically occurs more quickly with hydroponics than in soil, because plants invest energy in maturing rather than competing for resources, resulting in large yields. Many hydroponic systems recycle water not absorbed by roots to use for other purposes. Crops cultivated with hydroponic systems are usually safer for consumers than field-grown crops because their exposure to soil-transmitted diseases has been minimized.

Negative aspects of hydroponics include costs associated with acquiring equipment and supplies. Automation and computerized systems require substantial investments in machinery, software, and training personnel to operate them.

BACKGROUND AND HISTORY

Records indicate people cultivated plants using water instead of soil in ancient Mesopotamia, Egypt, and Rome. Aztecs in Mexico innovated floating barges, *chinampas*, for growing food because they lacked suitable agricultural land. In the mid-nineteenth century, German botanist Julius von Sachs and German agrochemist Wilhelm Knop experimented with combining minerals with water to nourish plants.

During the 1920's, University of California, Berkeley, plant nutrition professor Dennis R. Hoagland studied how roots absorb nutrients. William F. Gericke, also a professor at the University of California, Berkeley, cultivated tomatoes in tanks of mineral-rich solution and discussed his research in a February, 1937, *Science* article, noting colleague William A. Setchell referred to that process as hydroponics, representing Greek vocabulary: *hydro* (water) and *ponos* (work).

In the 1940's, the United States military utilized hydroponics to provide sustenance to World War II soldiers in the Pacific. Oil companies built hydroponic gardens on Caribbean islands to feed employees extracting natural resources in that region. The United States Army established a hydroponics branch to supply troops serving in the Korean War in the early 1950's, growing eight million pounds of food.

By mid-century, researchers began incorporating plastic equipment in hydroponic systems. Engineers innovated better pumps and devices, automating some hydroponic processes with computers. By the 1970's new methods included drip-irrigation systems and the nutrient film technique (NFT) created by English scientist Allen Cooper, which helped commercial hydroponics expand globally. The United Nations' Food and Agriculture Organization (FAO) funded hydroponic programs in areas experiencing food crises. Scientists continued devising new techniques, such as aerohydroponics, in the late twentieth century.

HOW IT WORKS

Hydroponics. Hydroponic processes represent examples of controlled environment agriculture in which plant cultivation involves technology such as greenhouses enabling growers to stabilize conditions. Most hydroponic systems function with basic components that supply oxygen and nutrients necessary to sustain plants until they have matured for harvest. Electrical or solar-powered lights, fans, heaters, and pumps regulate temperatures, ventilation for plant respiration, water flow, and photosynthesis impacting plant growth. Each hydroponic system incorporates variations of equipment and methods according to growers' resources and goals. Styrofoam, wood, glass, and plastic are materials used to construct hydroponic systems.

Basic hydroponic procedures involve placing seeds in substrates that consist of organic materials such as coconut fibers, rice hulls, sawdust, peat moss or inorganic mediums including gravel, pumice, perlite, rock wool, or vermiculite. After roots emerge during germination, growers keep seedlings in substrates or remove them depending on which hydroponic method is selected for cultivation. Roots undergo varying durations of exposure to nutrient solutions to absorb macronutrients and micronutrients. Most hydroponic processes utilize either an open, or non-recirculating, system, or a closed system, referred to as recirculating, depending on whether nutrient solution contacts roots once and is discarded or is kept for consistent or repeated use.

Hydroponic greenhouse. Plants grown using mineral nutrient solutions, in water, without soil. (Tom Myers/ Photo Researchers, Inc.)

Water-Culture Techniques. These hydroponic methods, which are frequently used to cultivate plants that quickly attain maturity, involve roots constantly being suspended in a nutrient solution. Water-culture hydroponic techniques are often utilized to grow lettuce crops. For the raft culture technique, growers place plants on platforms drilled with holes to pull roots through so roots can be submerged in pools of nutrient solution on which the platforms float. In the dynamic root floating technique, roots closest to the plant are kept dry so they can supply oxygen to the plant. The lower roots are constantly exposed to nutrient solutions and absorb those minerals and elements to nourish the plant.

Pumps and air stones oxygenate nutrient solutions so that roots are aerated. Lighting is essential for plants to undergo photosynthesis above the solution surface. Growers monitor nutrient solutions' pH levels and the presence of any algae, which might harm roots, interfere with their adsorption of nutrients, and impede plant growth. Growers also replenish fluids lost to evaporation.

Nutrient Solution Culture (NSC) Methods. Several forms of NSC are utilized to feed plants. Continuous-flow NSC involves nutrient solutions being poured into a trough and constantly moving through roots. Nutrient solutions contact roots less frequently in intermittent-flow NSC. The drip NSC technique delivers nutrient solutions through tubing and emitters that dispense water on the substrate near roots. Some drip systems recycle nutrient solution. The wick system utilizes strings that extend from substrates to a reservoir filled with nutrient solution.

In the ebb-and-flow method, nutrient solution contacts roots in cycles after flooding trays containing roots and substrates then draining and returning to a tank to store for additional delivery. Timers control pump mechanisms, which move nutrient solutions. Aquaponics systems transport water from ponds or greenhouses where fish tanks are kept to greenhouses where plants are grown so that wastes from the fish can provide nutrients for plants.

Nutrient Film Technique (NFT). This closed system continually pumps nutrient solutions into a channel placed at an angle in which roots hang under plants supported from above by platforms or other equipment. No substrates are used. The solution contacting roots is delivered as a watery film to assure roots will receive sufficient oxygen. The

---

Fascinating Facts About Hydroponics

- Diners eating at restaurants in Disney World can purchase vegetables, including squash and tomatoes, resembling Mickey Mouse, which are specially cultivated with hydroponics using plastic molds in Epcot Center's Land Pavilion.
- Mukesh Ambani, the world's fifth wealthiest person in 2010 according to *Forbes*, cultivates food with hydroponic systems placed throughout his twenty-seven-story tall, $2 billion Mumbai, India, residence. He values hydroponic plants' role in conserving energy by cooling that structure through adsorption of solar heat.
- Japanese agriculturists grow Koshihikari rice using hydroponic techniques in a renovated, subterranean bank vault equipped with fans to ventilate the underground paddy.
- OrganiTech in Israel utilizes automation technology to cultivate vegetables unable to thrive naturally in the country's arid conditions. The company's Grow-tech 2000, a hydroponic machine, uses sensors that alert robots when to plant and harvest crops.
- Hydroponic technology at McMurdo Station, located on Ross Island in Antarctica, produces about 250 pounds of vegetables and herbs monthly.
- In 1992, teacher Vonneke Miller and her students at Peterson Middle School in Sunnyvale, California, built a hydroponics laboratory they called ASTRO 1 because it resembled a spacecraft capsule. The vegetables they cultivated, including tomatoes and lettuce, were incorporated into meals prepared in the cafeteria.

---

hydroponic trough system uses a reservoir, which includes a filtering device to strain contaminants from nutrient solutions. Resembling NFT methods, aeroponics does not rely on substrates. Sprayers attached to timers continually dispense nutrient solutions on roots suspended in air below plants.

APPLICATIONS AND PRODUCTS

Nutrition. Hydroponics enables ample production of food supplies that meet nutritional needs for vitamins, antioxidants, and amino acids crucial for maintaining people's bodies. These techniques aid hunger relief in arid regions where climate change is associated with expanding desertification and loss of arable land, threatening food security. Hydroponic cultivation provides both rural and urban populations access to affordable, fresh, healthy food despite loss of access to traditional agricultural supplies due to political, economic, or military crises; natural disasters; or famines. Various hydroponic techniques can be applied to produce crops with increased levels of calcium, potassium, and other elements essential to sustain health. Hydroponic processes can be designed to grow food with appealing tastes, textures, and appearances.

Agribusiness. Hydroponics generates profits for commercial sellers of crops, manufacturers of hydroponic equipment, nutrient solutions, and supplies and wholesalers and retailers that distribute hydroponic merchandise to consumers. Agribusinesses create and market hydroponic greenhouses of varying sizes, including small growing containers such as AeroGarden for use inside homes, to consumers. Many florists grow stock cultivated with hydroponics at their stores. Internationally, the number of hydroponic businesses has expanded on all continents except Antarctica, contributing to countries' economies. By 2008, Advanced Nutrients, one of the most successful hydroponic businesses internationally, sold its merchandise to customers from forty-one countries. Some hydroponic companies develop and sell smart phone applications to perform hydroponic functions, such as General Hydroponics' calculator for preparing nutrient solutions.

Education. Students at various levels, from elementary through graduate school, often study hydroponics in science classes. Some courses may discuss hydroponics to explain basic scientific principles such as how roots absorb nutrients, while others may focus on special topics such as genetics. Students frequently investigate aspects of hydroponics for science-fair competitions or projects for the Future Farmers of America. Teachers instructing Advanced Placement biology courses often encourage students to develop hydroponic systems to comprehend concepts associated with plant growth and nutrition. Some school cafeterias use foods grown on their campuses or students sell products cultivated with hydroponic techniques for fund-raisers. Universities sometimes award funds to students' innovative hydroponic projects, especially those with humanitarian applications. The

Denver Botanic Gardens and other botanical centers offer hydroponic classes.

Military and Exploration. Military troops benefit from the establishment of hydroponic systems near bases and battlefields to produce fresh vegetables for rations regardless of soil and climate conditions in those areas. Hydroponic applications for military usage enable crews on vessels undergoing lengthy sea voyages to grow foods when they are between ports. Veterans with hydroponic experience or who complete Veterans Sustainable Agriculture Training or similar programs are often sought out for employment in that field. The ability to grow foods without soil nourishes people traveling by submarine, whether for military or scientific reasons. Workers in remote locations, such as off-shore oil and natural-gas rigs, eat meals incorporating hydroponic produce grown at those sites.

Scientists conducting research at Antarctic stations rely on hydroponics for sustenance and as a method to recycle, purify, and store water. The South Pole Food Growth Chamber, designed by the Controlled Environment Agriculture Center at the University of Arizona, uses NFT methods and is automated with an Argus climate-control system. The National Aeronautics and Space Administration (NASA) funds projects such as Controlled Ecological Life Support Systems (CELSS), in which hydroponic plants remove carbon dioxide and pollutants while producing food on spacecraft. Researchers are investigating using hydroponics for future missions of long duration.

Urban Planning. Some twenty-first-century architecture incorporates hydroponics as a strategy to feed increasing populations, particularly in urban areas. Rooftops are popular sites for hydroponic systems in places where land is unavailable for gardens. These urban farms grow large yields of basic vegetable crops and supply fresh produce to residents who might otherwise not have access to those foods. Vertical farming techniques inspired proposals applying hydroponics. In New York City, Dickson Despommier, a Columbia University microbiologist, introduced his idea to renovate almost two thousand empty structures with hydroponic equipment. The Seoul Commune 2026 in South Korea presented another vertical farming proposal. This project will involve covering skyscrapers, some fifty floors high, with supports for plants. Nutrients delivered by fog machines and irrigation technology to roots will nourish plants growing on those garden buildings.

Tourism. Some hotels, especially in exotic locales, apply hydroponics for agricultural and aesthetic uses. On Anguilla in the Caribbean, the CuisinArt Resort and Spa grows hydroponic herbs, vegetables, and flowers. Guests can tour areas with hydroponic equipment to see where food served in the hotel's restaurant is grown. These businesses sometimes sell hydroponic products, often identified by resorts' brands, to cruise ships docking nearby or to markets. Visitors can tour hydroponic displays at the Hampshire Hydroponicum in England and Epcot's Land Pavilion in Orlando, Florida. Zoos occasionally utilize hydroponic processes to cultivate grain and grass to feed animals.

IMPACT ON INDUSTRY

Organic Monitor declared the organic agriculture industry, including hydroponics, generated $50.9 billion globally in 2008. Hydroponic systems used for production of legal crops represented part of the $26.6 billion of organic products sold during 2009 in North America. In 2006, hydroponic retail sales totaled $55 million, a $40-million increase since 2002. Internationally, scientists and engineers, working for universities, industries, and governments, conduct research to improve the hydroponic industry for humanitarian, utilitarian, and commercial applications by manufacturers, businesses, and agriculturists.

Government and University Research. Most hydroponic research is supported with funds from academic institutions, governmental agencies, or professional organizations. Hydroponic projects conducted at land-grant universities typically occur at experiment stations where scientists and engineers contact experts in other fields so they can incorporate interdisciplinary approaches to resolving problems and innovating new technologies. Academic and research laboratories host studies investigating procedures to advance hydroponics by designing plants and equipment compatible with hydroponic processes. Some hydroponics researchers utilize computer models to test hypotheses. Scientists can observe distant hydroponic facilities such as those on Antarctica with Internet and satellite technology to gather data.

The Ohio Agricultural and Research Development Center initiated notable research in the early twenty-first century regarding hydroponics used to grow lettuce. Scientists perform experiments relevant to hydroponic systems designs, substrates, and rate of nutrient flow to determine how those factors impact yields. Horticulturists study which plant species are best suited for cultivation using hydroponic methods. Some genetics researchers at the University of Wisconsin focused on developing smaller versions of plants, such as dwarf wheat, to grow in limited spaces. Scientists also investigate increasing maturation of plants so they can be harvested more quickly. Starting in 2008, Pakistan's government funded BioBlitz, a hydroponic program that grows 50,000 kilograms of tomatoes weekly on five acres. Officials projected hydroponics could enable Pakistan to earn $500 million annually by cultivating 1,000 acres.

Environmental impact of hydroponics is a universal concern among researchers. The United States Environmental Protection Agency (EPA) promoted hydroponic research to increase use of inorganic substrates to minimize agriculturists' reliance on chemicals in fields. New York's State Energy Office encouraged hydroponic cultivation to decrease the amount of energy needed to grow and transport food.

Industry and Business. Agribusinesses invest in hydroponic research to enhance the profitability of their companies by offering improved technology and products that appeal to customers, whether individuals, governments, or corporations.

Industry experts project that the hydroponic industry will continue growing in the twenty-first century. This is concurrent with the market for organic foods, which increased eleven percent from 2008 to 2009. The hydroponic industry will expand to produce equipment for new applications as they are envisioned and introduced. Businesses will recognize investment opportunities hydroponics systems offer because that technology can expand food production beyond the limited quantities grown from natural resources.

Partnerships link hydroponic agriculture with existing food manufacturing methods. For example, in India, Himalya International, which manufactures frozen foods, contracts with the British company New World Paradigm to produce crops using hydroponic systems for Himalya International to process for distribution to markets. New World Paradigm also supplies hydroponic produce to food manufacturers in Europe and North America. Alliances between companies, such as American Hydroponics and Rimol

Greenhouse Systems, are financially benefited by centralizing hydroponics resources for consumers to purchase.

Successful hydroponic technology manufacturers include Orbitec, which created the Biomass Production System that can be used to grow dwarf plants developed by researchers. The Orbitec hydroponics technology regulates nutrients, moisture, lighting, and temperatures. DuPont researchers innovated materials that are biodegradable for growing plants hydroponically. Companies design and produce greenhouses of various sizes and shapes, including geodesic domes. Manufacturers build machinery incorporating computerized functions to perform tasks. OrganiTech developed the automated hydroponic Grow-tech 2000 equipment, securing revenues of $3.5 million in 2005, including a $2.73 million contract to construct a 100,000-square-foot greenhouse in Russia.

CAREERS AND COURSE WORK

Students interested in professions associated with hydroponics can complete diverse educational programs to pursue their career goals. Many entry-level hydroponic positions are available to people with high school educations or associate's degrees. Some employers seeking qualified workers to build and maintain hydroponic systems expect candidates to have completed basic horticultural courses at technical schools, community colleges, or universities, preparing them to cultivate plants and assemble equipment. Experience working for landscaping businesses, farms, or other positions that involve tending plants enhances one's employability. One can sometimes find available positions at gardening businesses that use hydroponics to grow crops and ornamental plants to sell to consumers and markets. Resorts, botanical gardens, and theme parks hire people with educational and work experience to establish and maintain hydroponic gardens.

Government, academic, and industrial employers that staff scientific and technological positions focusing on hydroponics usually require the minimum of a bachelor of science degree in a related field. Candidates can acquire basic knowledge for plant cultivation by studying horticulture, botany, agriculture, or subjects applicable to hydroponics. Those seeking research positions typically need to earn advanced degrees—a master's or doctorate—in relevant

subjects, acquiring expertise that will benefit the quality of their employers' services and hydroponic products. Agricultural engineering, computer science, or robotics courses prepare employees for positions designing hydroponic structures, machinery, and automation software. Candidates with advanced education or hydroponic experience have credentials for many positions as administrators or educators in schools, experiment stations, extension services, or government agencies.

SOCIAL CONTEXT AND FUTURE PROSPECTS

Throughout the twenty-first century, hydroponics will continue to provide humanitarian and commercial benefits. The Hydroponic Merchants Association stated in 2004 that hydroponic greenhouses grew 55,000 acres of vegetables internationally, of which 5,800 acres in North America produced tomatoes, peppers, and cucumbers valued at $2.4 billion. That organization estimated the hydroponic industry will continue growing ten percent yearly because of increasing demand and advances in hydroponic technology. Industry experts suggest that hydroponics, universal to diverse cultures, will continue to expand for several reasons, including depletion of arable lands caused by natural disasters and global warming, expenses associated with machinery and operation of conventional agriculture; and public disapproval of bioengineering associated with field crops.

The early twenty-first-century economic recession motivated consumers to use hydroponic equipment because many cannot afford produce sold in stores. Some domestic hydroponic growers sell their products to earn money while they are unemployed or to supplement incomes. Many people practice hydroponics when grocery stores in their communities close because of financial problems, resulting in those consumers lacking access to fresh food. Hydroponics presents food-security solutions to the increasing population, which is estimated to reach nine billion people by 2050. Some experts speculate hydroponics will eventually surpass mainstream agriculture to produce the most food worldwide.

Elizabeth D. Schafer, Ph.D.

FURTHER READING

Despommier, Dickson. *The Vertical Farm: Feeding the World in the Twenty-First Century*. New York: St. Martin's Press, 2010. Emphasizing humanitarian and environmental benefits, narrative promotes development of urban hydroponic farms to increase food supplies.

Giacomelli, Gene A., et al. "CEA in Antarctica: Growing Vegetables on 'the Ice.'" *Resource: Engineering & Technology for a Sustainable World* 13, no. 1 (January/February, 2006): 3-5. Discusses how hydroponic systems serve both scientists in polar regions and researchers monitoring those processes remotely with communications technology.

Hansen, Robert, Jeff Balduff, and Harold Keener. "Development and Operation of a Hydroponic Lettuce Research Laboratory." *Resource: Engineering & Technology for a Sustainable World* 17, no. 4 (July/August, 2010): 4-7. Text and photographs document Ohio Agricultural and Research Development Center cultivation experiments' growth sequence, results, and impact on growers.

Jones, J. Benton, Jr. *Hydroponics: A Practical Guide for the Soilless Grower.* 2d ed. Boca Raton, Fla.: CRC Press, 2005. Agronomist addresses diverse aspects of cultivating crops with nutrient solutions. Chapter concentrating on hydroponics in

education outlines ideas and resources for projects; includes CD-ROM.

Lloyd, Marion. "Gardens of Hope on the Rooftops of Rio." *The Chronicle of Higher Education* 52, no. 8 (October 14, 2005): A56. Describes hydroponic work conducted by students in a Brazilian shantytown to teach residents how to grow fresh, nutritious foods.

WEB SITES

Food and Agriculture Organization of the United Nations
http://www.fao.org

National Aeronautics and Space Administration
Farming for the Future
http://www.nasa.gov/missions/science/biofarming.html

National Plant Data Center
http://npdc.usda.gov

See also: Agricultural Science; Agronomy; Food Science; Horticulture.

HYPNOSIS

FIELDS OF STUDY

Psychology; humanistic psychology; psychoanalysis; neuroscience; philosophy of mind; placebo research; altered states of consciousness; complementary medicine; alternative healing; ethno-psychotherapy; shamanism.

SUMMARY

Hypnosis, a subfield of psychology, studies the influence of hypnotherapy (therapy undertaken in hypnosis) and hypnosis (as an altered state, social role, or response expectance) on people, especially its impact on psychological and physical health. Hypnosis is believed to give access to the unconscious mind, and there is empirical evidence that in highly susceptible subjects, hypnosis is successful as a therapeutic treatment (or as adjunct treatment) against a variety of psychological and psychosomatic disorders, such as post-traumatic stress disorder and depression. Hypnotherapy has also proven to be helpful as a substitute for anesthetics, for pain reduction in general, and for relaxation. Finally, hypnosis is used in the context of sports, education, advertising, and marketing.

KEY TERMS AND CONCEPTS

- **Age Regression and Progression:** Attempts to focus the mental attention of a subject on an earlier or later age, which can be therapeutically constructive in the context of experiencing roles, elucidating expectations, and testing consequences but does not necessarily produce real-life memories.
- **Altered State of Consciousness (ASC):** Mental state different from wakefulness and sleeping, includes drug-induced and trance states, as well as those brought about by mental, physical, and holistic techniques; also known as altered state of awareness or altered state of mind.
- **Hypnotherapy:** Form of psychotherapy or counseling that employs hypnosis.
- **Induction:** First set of suggestions in the hypnotic process, inducing relaxation, concentration, and the establishment of hypnotic rapport, or the definition of the roles of the hypnotist and subject; also known as hypnotic induction.
- **Nocebo Effect:** Unpleasant or harmful effect brought about by a placebo or a medical intervention that simulates the actual procedures.
- **Placebo Effect:** Positive therapeutic effect caused by a simulated medical intervention.
- **Rapport:** Subjective experience of consciously or unconsciously feeling trust or an emotional affinity with another person; also known as hypnotic rapport.
- **Suggestion:** Guidance of thoughts, imaginations, feelings, and behaviors of one person by a different person (hetero-suggestion) or by the same person (auto-suggestion, self-suggestion); also known as hypnotic suggestion.
- **Susceptibility:** Ability to be hypnotized or to respond to hypnotic suggestion, as measured by such scales as the Stanford Hypnotic Susceptibility Scale and the Harvard Group Scale of Hypnotic Suggestibility.
- **Trance:** Altered state of consciousness achievable through hypnosis, meditation, prayer, shamanistic rituals, or drug use; also known as hypnotic trance.
- **Unconscious:** Domain of the mind that is not conscious, of which people are unaware during the normal waking state.

DEFINITION AND BASIC PRINCIPLES

Hypnosis is a wakeful state in which the attention is focused on one or several issues by diminished peripheral awareness and heightened suggestibility, usually induced by suggestions (hypnotic induction). Although the word "hypnosis" is derived from the Greek word for sleep (*hypnos*) and a hypnotized person might at times appear to be asleep, neurological research has revealed that brain waves during hypnosis do not resemble those of sleep. Hypnosis is an altered state of consciousness and a specific interactive situation with voluntarily assumed, defined roles in which a subject follows the suggestions of a hypnotist (in hetero-hypnosis) or the subject's own suggestions (in self-hypnosis). Some researchers think that a hypnotic state can occur without suggestion, as part of everyday life when people become extremely focused on a particular issue, for example, during concentrated learning or the creative process. The depth and the success of hypnosis are determined by psychological factors such as positive motivation, an appropriate attitude, expectations, susceptibility, and an active imagination.

Hypnotherapy is hypnosis in a psychotherapeutic or counseling setting, with goals such as stress management, pain reduction, and the modification of attitudes, habits, or behavior.

BACKGROUND AND HISTORY

Hypnosis as psychosocial phenomenon is as old as human culture; images from several cultures show trances in the context of what are probably religious rituals, in some cases, possibly induced for medical reasons. Hypnosis is similar to some forms of trance brought about by eastern meditative techniques and religious or shamanistic rituals in diverse traditional cultures.

In the eighteenth century, the German physician Franz Anton Mesmer (from whose name comes the word "mesmerize") invented a treatment dubbed "animal magnetism." The Scottish physician and surgeon James Braid developed a treatment known as neuro-hypnotism, or hypnotism, which shared some features with Mesmer's technique. Braid is the first advocate of hypnosis and hypnotherapy to gain scientific acceptance. In France, neurologists Jean-Martin Charcot and Hippolyte Bernheim conducted research and developed clinical forms of hypnosis. Austrian neurologist and psychiatrist Sigmund Freud was trained by Charcot and initially was enthusiastic about hypnotherapy, but he later abandoned it in favor of his own psychoanalytic approaches.

In the twentieth century, psychiatrist Milton H. Erickson, who founded the American Society of Clinical Hypnosis, was an advocate of hypnotherapy. Erickson's approaches were both innovative and controversial, according to his collaborator, André Muller Weitzenhoffer, one of the most prolific hypnosis researchers of the second half of the twentieth century.

HOW IT WORKS

In the first hypnosis or hypnotherapy session, the subject is informed about the basics of hypnosis. Each session usually begins with an introduction in which the subject, in most cases, will be asked to recline and to relax. This is followed by the first inductive suggestions (impressions of gravity, feelings of heaviness, and the like), followed by further suggestions that guide the subject toward becoming more relaxed but also more alert and focused toward his or her inner impressions, images, and imagination. The methods vary considerably depending on the therapist's training and philosophical worldview, the subject's aims or problems, and the respective circumstances. For example, induction can use the eye-fixation method, whereby the subject is told to keep his or her eyes fixed on a certain object such as the hypnotist's finger. Various other induction methods employ one or more senses and the imagination. According to the altered state theory, induction helps the subject transfer into an altered state of awareness or consciousness, and social role and response expectancy theories view the induction as a means of defining the roles of the client and hypnotist, increasing expectations, focusing attention, and increasing concentration. Posthypnotic suggestions given during the session are intended to trigger or support the subject's therapy goals, usually to change a behavior or alter an attitude in daily life. Hypnosis can be conducted with individuals and groups. Before the end of the session, the hypnotist conducts an exit procedure, in which any suggestions that are not posthypnotic are taken back and the subject is gradually brought back to a normal condition. The session can end with a review.

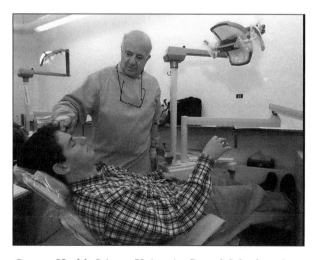

Oregon Health Science University Dental School professor Dr. J. Henry Clarke, right, hypnotizes a student. Dr. Clarke has used hypnosis instead of Novocaine on patients to remove teeth, and to perform root canals and other procedures. (AP Photo)

Altered State and Dissociation Theories. Braid, Erickson, and Weitzenhoffer, among many others, believed that hypnosis is an altered state of consciousness or an altered state of awareness. American psychologist Ernest Hilgard was of the opinion that consciousness is dissociated and that parallel streams of consciousness coexist and have certain degrees of autonomy (for example, one feeling pain and the other not). Therefore, the coordination and the emphasis of such streams of consciousness can be altered by suggestions. For example, a feeling of pain can be suggested to have less gravity, while a pleasant feeling can be emphasized. Additionally, many experts believe that hypnosis can give access to more remote, subordinated, or covered streams of the consciousness.

Social Role and Response Expectancy Theories. Social role theories (like those of American psychologist Theodore R. Sarbin), sociocognitive theories, and response expectancy theories emphasize the similarity of hypnosis and a placebo. Empirical evidence and meta-analyses suggest that the two main parameters that contribute to the effect of hypnosis in a significant way are the willingness to act socially compliant and an imaginative suggestibility (about one-third of subjects respond to imaginative suggestions and social pressure). Hypnotic suggestibility is not correlated with intelligence, social position, willpower, motivation, gender, introversion, extroversion, or credulity. Researchers such as psychology professor Irving Kirsch think that the effect of hypnosis, as well as that of placebos, is grounded in a kind of self-fulfilling prophecy, namely that largely subjects experience what they expect to experience. In patients with depression, the difference between a placebo and an antidepressant drug is not clinically significant.

Research in this area attempts to prove that the altered state theories are wrong, but that the effects of hypnosis are nevertheless real and that subjects do not fake the effects of hypnosis, as social role and response expectancy theorists believe. A number of researchers hold that both altered state and social role/response expectancy theories are correct and responsible for the effect of hypnosis. Additionally, it can be argued that a hypnotic state is a deeper form of hypnosis, following a nonaltered or less altered state in which suggestions are given and taken according to the social role/response expectancy theory.

APPLICATIONS AND PRODUCTS

Hypnotherapy, Psychotherapy, and Counseling. In clinical psychology, hypnosis, as an adjunct method, and hypnotherapy, as a stand-alone treatment, are successfully used to deal with pain reduction, psychosomatic symptoms, obsessive-compulsive disorder, post-traumatic stress disorder, anxiety disorders, and depression. Less successful is the use of hypnosis or hypnotherapy to treat problem habits such as excessive drinking, eating, and smoking, which are not manageable by self-control. Hypnosis is also used for pain and stress management, for self-improvement, and to change behavior and attitudes.

Hypnotic Analgesia. Hypnosis and shamanistic trance rituals have been used as analgesia. Hypnosis has been employed successfully to achieve relaxation and reduce anxiety, fear, discomfort, and pain before and during childbirth, in dental settings, and also during minor surgery. Hypnosis does not reduce the physical reception of pain, but its perception can be manipulated by hypnotic suggestions, whether administered by a hypnotist or the self.

Nonclinical Applications. Hypnosis and hypnotic suggestions (self- and hetero-) have been successfully used to cope with stage fright, to reduce stress levels, and to increase the degree of concentration and focus. They can also intensify relaxation and concentration in the context of creative arts, education, and sports. Research has been conducted on applications for military intelligence, investigations, and forensics, but there is no scientific evidence that such applications are of value. A number of business applications, for advertising, marketing, and improving sales, have been created; however, such applications are ethically questionable. Interest in raising athletic performance levels, losing weight, or quitting smoking has resulted in the proliferation of self-hypnosis products, usually in the form of CDs, DVDs, or books. However, self-hypnosis is best learned from a qualified practitioner. Stage hypnosis is usually considered to be neither of therapeutic interest nor an overly important issue in academic research. Leaving aside stage hypnosis, most of the applications still take place in the medical or clinical field.

IMPACT ON INDUSTRY

Hypnosis Research and Clinical Organizations. Hypnotherapy is practiced and researched all over the world. Hypnosis and hypnotherapy organizations exist at the international, national, and regional levels. They include the International Society of Hypnosis, the American Psychological Association's psychological hypnosis division, and the European Society of Hypnosis. Established scholarly journals include the Society for Clinical and Experimental Hypnosis's *International Journal of Clinical and Experimental Hypnosis* and the American Society of Clinical Hypnosis's *American Journal of Clinical Hypnosis*. A number of medical schools and psychology departments at established universities such as Stanford University, Harvard University, and the University of California, Berkeley, deal with hypnosis in research and education.

Dubious Fields. Although the science-based hypnosis has advanced, an increasing number of institutes for hypnotherapy-training exist in more or less grey areas. These programs might promise anything from turning a person into a hypnotherapist after a weekend of training to turning an individual into a happy person in one evening. Therefore, anyone seeking training in hypnosis should seek out reputable organizations or institutes affiliated with them. For example, the American Society of Clinical Hypnosis offers professionals training courses that are approved by the American Psychological Association. Also, stage or show hypnosis generally is the work of actors with no actual hypnosis taking place, but if actual hypnosis were performed, it could endanger the health of participants, as it can pose, for example, cardiovascular or psychological risks.

CAREERS AND COURSE WORK

Students pursuing careers as hypnotherapists should study counseling, psychology, or medicine to become counselors, psychologists, physicians, or psychiatrists and specialize or subspecialize in hypnotherapy. Physicians, dentists, educators, social workers, nurses, counselors, psychologists, and psychiatrists can also complete diverse courses and programs to acquire the practical qualifications for using hypnosis as stand-alone or adjunct treatment.

Courses or programs in hypnosis should be part of or affiliated with accredited universities or reputable hypnosis societies. For higher positions in education, research, and the clinical field, a master's or even a doctoral degree is recommended. If such a career path is intended, the thesis or dissertation should focus on a particular issue very closely related to hypnosis, undertaken in a traditional field such as

---

Fascinating Facts About Hypnosis

- In Greek mythology Hypnos, the god of sleep, is the brother of Thanatos, the god of death. Morpheus, the god of dreams, is Hypnos's son.
- The famous German writer Thomas Mann, in his 1930 novel *Mario und der Zauberer* (*Mario and the Magician*, 1930), explores how hypnosis relates to the mesmerizing power of Fascist political leaders.
- In a 1956 article, professor Frank A. Pattie claimed that Franz Anton Mesmer, one of the fathers of hypnosis, plagiarized his dissertation from a work by the English physician Richard Mead, an acquaintance of Sir Isaac Newton.
- American psychiatrist Milton H. Erickson (1901-1980), sometimes referred to as Mr. Hypnosis, was born tone-deaf and color-blind, and he attributed much of his heightened sensitivity to altered modes of sensory-perceptual functioning, body dynamics, and kinesthetic cues to his innate infirmities.
- Despite a lack of scientific evidence, many subjects claim to have past-life experiences under hypnosis; experiments undertaken by psychology professor Nicholas Spanos in the 1980's suggest that such past-life memories reflect social constructions.
- According to Guinness World Records, in 1987, the German stage hypnotist Manfred Knoke was able to hypnotize 1,811 people in the city of Bochum in a six-day period.
- In 1995, the German magazine *Der Spiegel* reported that International Society of Hypnosis president Walter Bongartz, appearing on a television show featuring Manfred Knoke and his feats, pleaded for a law to prohibit such stage-hypnosis spectacles, given the medical risks that they could pose.
- Dream Theater, a progressive metal band, released a concept album, *Metropolis Pt. 2: Scenes from a Memory* (1999), which tells the story of a man who undergoes hypnosis to experience the mystery of his past lives.

---

medicine, nursing science, psychology, human biology, anthropology, ethnology, philosophy, or any other field in which a scientific approach to hypnosis is possible and credible.

SOCIAL CONTEXT AND FUTURE PROSPECTS

Hypnosis still receives a lot of attention in the discourse concerning memory recovery. Debate exists as to in which circumstances, under which conditions, and how reliably forgotten memories of past events, especially traumatic experiences, can be "recovered" through hypnosis. The more research conducted on hypnosis and more empirical data collected, the more effectively and appropriately hypnotherapy can be used and the more it will gain acceptance in mainstream medicine.

The question of whether hypnosis is a social role/ expected response or a true altered state will not be conclusively answered until neuroscience advances and anthropologists use neurological tools to study shamanist-cultic and religious trance rituals. Anthropology and sociology have made it evident that the use of hypnosis as therapy is not a European invention but rather a phenomenon that can be traced back to therapeutic shamanistic rituals and religious trances in various parts of the world.

Another issue concerns the role of the subject's imagination in the curative powers of hypnosis. A subject in a hypnosis show is similar to the subject of a traditional cultic healing ritual in that the involvement and participation of the public is taken for granted. If both subjects experience a curative effect, then hypnosis is acting like a placebo, and its actual therapeutic benefits are questionable. The subject's own power of imagination and its neurological, biochemical, physiological, psychological, and holistic effects may be what is producing the curative effect.

Certain scholars hold that quite a number of everyday settings–such as intensive educational settings, artistic performances, and mass political events–have a hypnotic character. Research in this field will bring to light how business applications of hypnosis are possible. The ethical problem in such contexts is that subjects should not be hypnotized against their own will. Critics hold that this has already been done in the sphere of marketing.

From a feminist perspective, it can be argued that hypnosis cements and even perpetuates patriarchal structures, since most of the well-known hypnotherapists were men and the hypnotherapist exerts a kind of dominance over the patient or client, unlike as in a guided imagery setting in which the relationship is less hierarchical and less suggestive. Therefore, a unique feminist approach to hypnosis is also the subject of research.

Roman Meinhold, M.A., Ph.D.

FURTHER READING

Erickson, Milton H. *The Wisdom of Milton H. Erickson.* Compiled by Ronald A. Havens. 1985. Reprint. Williston, Vt.: Crown House, 2003. Erickson's thoughts on hypnosis and psychotherapy.

Kirsch, Irving, Steven J. Lynn, and Judith W. Rhue, eds. *The Handbook of Clinical Hypnosis.* 2d ed. Washington, D.C.: American Psychological Association, 2010. Broad introduction to hypnosis covering how it is used clinically to treat conditions and disorders.

Nash, Michael R., and Amanda J. Barnier, eds. *The Oxford Handbook of Hypnosis: Theory, Research and Practice.* New York: Oxford University Press, 2008. Presents both academic theory and practical approaches. Contains name and subject indexes.

Pattie, Frank A. *Mesmer and Animal Magnetism: A Chapter in the History of Medicine.* Hamilton, N.Y.: Edmonston, 1994. Mesmer specialist Pattie provides a thoughtful biography of Mesmer that includes the probable source of his thought.

Pintar, Judith, and Steven Jay Lynn. *Hypnosis: A Brief History.* Malden, Mass.: Wiley-Blackwell, 2008. Offers a compact but critical overview of the origins and the history of hypnosis, including accounts on debates within psychology, references and index.

Temes, Roberta. *The Complete Idiot's Guide to Hypnosis.* 2d ed. Indianapolis, Ind.: Alpha Books, 2004. Provides the basics of hypnosis. Contains a glossary, an index, and suggestions for further reading.

WEB SITES

American Psychological Association
Division 30: Society of Psychological Hypnosis
http://www.apa.org/about/division/div30.aspx

American Society of Clinical Hypnosis
http://www.asch.net

European Society of Hypnosis
http://www.esh-hypnosis.eu

International Society of Hypnosis
http://www.ish-hypnosis.org

Society for Clinical and Experimental Hypnosis
https://netforum.avectra.com/eWeb/StartPage.aspx?Site=SCEH

See also: Psychiatry; Somnology.

I

IMMUNOLOGY AND VACCINATION

FIELDS OF STUDY

Microbiology; general biology; cell biology; chemistry; molecular biology; biochemistry; biophysics; microbial genetics; immunology; genetic engineering.

SUMMARY

The function of vaccination is to induce immunity in humans and other animals for the purpose of providing protection against disease-causing organisms. Vaccination is generally carried out through the injection of attenuated or inactivated microorganisms such as bacteria or viruses, or the inactivated toxins produced by bacteria. The first vaccinations were directed against smallpox during the eighteenth century and involved the use of cowpox virus, similar but not identical to the virus that caused smallpox. Modern vaccinations often use purified components of the organism rather than the entire bacterium or virus, producing similar immunity without the danger of side effects or illness.

KEY TERMS AND CONCEPTS

- **Acellular Vaccine:** Vaccine prepared from subunits or genes from the agent rather than the whole bacterium or virus.
- **Antibody:** Protein produced by B lymphocytes in response to foreign molecules.
- **Antigen:** Commonly a protein, but any molecule perceived as being foreign to the body.
- **B Lymphocyte:** Type of white blood cell that produces and secretes antibodies.
- **Cellular Immunity:** Immunity based on activation of macrophages, natural killer cells, and antigen-specific cytotoxic T-lymphocytes.
- **Complement:** Pathway triggered by antigen/antibody complexes that consists of proteins that augment the immune response.
- **Humoral Immunity:** Immunity based on soluble proteins such as antibodies found in blood and body fluids.
- **Monoclonal Antibodies (mAB):** Antibodies produced by a clone of B lymphocytes, all of which are identical.
- **Phagocytosis:** Ingestion and digestion of material, including bacteria, by white blood cells.
- **T Lymphocyte:** White cell that matures in the thymus gland (indicated by the T) and that regulates the immune response.

DEFINITION AND BASIC PRINCIPLES

Immunology is the science that studies the reactions of immune cells within the body to foreign molecules referred to as antigens. The majority of immune cells are represented by populations of white blood cells, or leukocytes, found circulating within the bloodstream and lymphatic system. Although all immune cells originate and mature largely within the bone marrow, they undergo differentiation into highly specialized categories. Monocytes and neutrophiles are phagocytic cells circulating in the bloodstream and lymphatic system that function to ingest and digest both foreign antigens from outside the body and old or dying cells within the body. Monocytes mature within tissues and organs such as the spleen and lymph nodes into a class of cells called macrophage, transport the ingested antigens to the cell surface, and present the digested molecules to a second class of white cells called B and T lymphocytes. Only those lymphocytes that express a receptor specific for the presented antigen will respond, with B cells differentiating into plasma cells that secrete antibodies directed against the original antigen.

The principle underlying vaccination is that administration of a killed or attenuated form of bacterium, bacterial toxin, or virus will result in production of antibodies against the target, producing immunity in the individual against the bacterial or viral antigens. If the person is later exposed to the same microorganism or toxin, the presence of pre-existing antibodies will protect against infection or poisoning.

BACKGROUND AND HISTORY

Immunitas originally referred to the freedom from taxes among ancient Romans. The medical concept of freedom from disease—immunity—was described by the Greek historian Thucydides, who in his description of the plague (actually probably a typhoid fever epidemic) in Athens in 430 B.C.E. noted that individuals who survived "were never attacked twice."

An understanding of the cellular basis for immunity was not reached until late in the nineteenth century. However, the principle of prior exposure to a disease resulting in immunity had been known since about 1000 C.E., when the Chinese carried out a practice called variolation. In this practice, a person inhaled dried crust from the pocks that developed on smallpox victims. If he or she developed only a mild form of the disease, the most common outcome, the person was immune for life. The practice traveled to Eastern Europe and then England by the early eighteenth century. Although variolation was generally successful, sometimes it resulted in the person contracting smallpox.

The first successful active immunization is ascribed to Edward Jenner, an English country physician who in the 1790's tested a belief common among local dairymaids—that prior exposure to a mild cowpox infection of the udder on a cow provided immunization against smallpox. Beginning in 1796, Jenner carried out tests in which he intentionally infected people by applying cowpox "lymph" obtained from a lesion to small slits cut in the arms of volunteers. During a subsequent epidemic, none of the inoculated individuals developed smallpox. Jenner called the practice "vaccination," from *vacca*, Latin for cow.

HOW IT WORKS

An understanding of the cellular mechanism underlying successful vaccinations did not begin until the late nineteenth century and was the outcome of both a scientific and a nationalistic rivalry between French and German scientists. The major proponent of a cellular theory of immunity was the Russian scientist Élie Metchnikoff. While studying the differentiation of cells in animals such as starfish larvae, Metchnikoff observed that insertion of a wooden splinter into the larvae resulted in the infiltration of both large and small white blood cells. He called these macrophage, "large eaters," and microphage, "small eaters." Microphage later became known as neutrophils. Metchnikoff subsequently joined the laboratory of French scientist Louis Pasteur, where he became a proponent of the cellular theory of immunity.

The competing theory was defined by the German school, and became known as humoral immunity. Robert Koch, Emil von Behring, and their associates noted that blood plasma obtained from animals previously exposed to etiological agents of disease or to bacterial toxins could directly kill bacteria or neutralize these toxins. Behring and Paul Ehrlich applied their discovery in the development of the first vaccines against diphtheria. Soluble proteins, including antibodies, became the basis for humoral immunity.

It was not until the mid-twentieth century that the basis for immunization was established as a combination of both cellular and humoral immunity. Phagocytosis is indeed carried out by several classes of white blood cells, while antibodies are produced by a class of white cells called B lymphocytes. The actual immune mechanisms involve a complex interaction between these classes of cells and their soluble products in which the phagocytic cell presents the digested

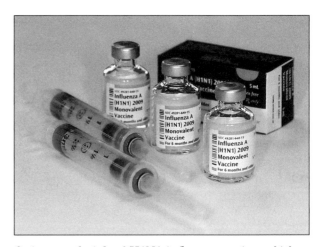

Syringes and vials of H1N1 influenza vaccine, which recently became available in the fall of 2009 to immunize people against swine flu. (Scott Camazine/Photo Researchers, Inc.)

antigen on its surface to the appropriate lympho-cyte. The end result is that B lymphocytes mature and differentiate into an end-stage antibody-pro-ducing factory called a plasma cell. Each plasma cell produces a single type of antibody, selected on the basis of possessing a receptor specific for the antigen presented by the phagocyte.

Vaccine Production. Vaccine production is based on activation of lymphocytes through exposure to bacterial or viral antigens (proteins), the result of which is the production of antibodies or the stimu-lation of phagocytic cells. Vaccines have historically been produced by three major mechanisms: use of inactivated or killed microorganisms, use of attenu-ated or cross-reacting organisms, or use of purified portions of microorganisms in recombinant vaccines. The smallpox vaccine is an example of a cross-re-acting organism; the cowpox virus is similar enough to smallpox that the immune response is protective against both.

Most viral vaccines have used attenuated strains of the original virus, selected either by passage through nonhuman animals or cells or by artificial selection on the basis of avirulence, an inability to cause dis-ease. The strains of poliovirus vaccines developed by Albert Bruce Sabin, as well as vaccines against rabies, chickenpox, measles, and mumps all consist of atten-uated viruses. The polio vaccine developed by Jonas Salk is a formalin-killed virus. A later generation of viral vaccines, those directed against viruses such as hepatitis B and human papillomavirus (which causes warts and cervical cancer), are subunit types con-sisting of surface proteins obtained from the virus, which through DNA recombination are linked to harmless carrier proteins. Vaccines directed against tetanus toxin are similar to those originally devel-oped by Behring and Ehrlich against diphtheria toxin. The toxin is chemically modified and injected.

The principle behind all vaccinations is the same. Exposure to the agent results in an immune response within the individual. Antibodies are produced, and cellular immunity is activated. The response is already in place in the event of future exposure to the same organism or toxin. In most cases, immunity is long-lasting, though periodic boosters are recom-mended to ensure a proper level of immunity.

Autoimmune Disease. In principle the immune re-sponse is directed only against foreign agents that could potentially cause disease. However, in certain circumstances, alterations in immune regulation take place, and antibodies are produced against the per-son's own tissues. The precise molecular mechanism that triggers autoimmune function is unclear. Some diseases run in families or are gender specific (women are more likely to contract certain autoimmune dis-eases), and other illnesses may be triggered by cross-reaction with viral or bacterial antigens. The tissue involved depends on the type of autoantibody pro-duced, but the mechanisms for damage are similar.

Autoimmune diseases are placed into two major categories, organ specific or systemic, reflecting the sites or systems involved. Examples of organ-specific diseases include type 1 diabetes, in which the B cells of the pancreatic islets of Langerhans are targeted; Crohn's disease, a form of inflammatory bowel dis-ease; and multiple sclerosis, characterized by inflam-mation of tissue in the central nervous system. Sys-temic autoimmune diseases include systemic lupus erythematosus, in which antigen/antibody com-plexes lodge in different organs, and rheumatoid arthritis, characterized by immune complexes that lodge in joints or bone. Although the type of anti-body may differ in autoimmune diseases, pathologies are similar in that each activates the complement pathway, components of which include degradative enzymes that contribute to inflammation and tissue destruction.

APPLICATIONS AND PRODUCTS

Vaccine Production. Historically, vaccines fell into two categories: live vaccines in which the agent was altered so as to be unable to cause disease but still able to replicate in the human host, triggering the immune response, and killed vaccines in which the organism was identical to the parent strain but un-able to replicate. Each had advantages. Live vaccines produced a greater response and often a lifelong im-munity, and killed vaccines would not result in rever-sion to the wild strain, causing disease. For example, before they were discontinued in 1990, the Sabin strains of attenuated poliovirus had a reversion rate of about one in one million persons inoculated in the United States, resulting in about ten vaccine-associ-ated cases of polio per year.

Live, or attenuated, vaccines were originally cre-ated by passage in nonhuman animals or in cell cul-tures in a laboratory. This was particularly true for vaccines for viruses, including those against rabies, polio, measles, mumps, rubella, and chickenpox. Because viruses develop random mutations, variant strains were selected on the basis of sensitivity to pH (acidity-alkalinity) or to elevated temperatures (fever), or for their inability to infect certain tissues. The Sabin poliovirus strains represent a prototype of attenuated viruses, being both temperature sensitive and incapable of infecting tissue in the central nervous system. Later methods of developing attenuated strains have involved active modification of viral genetic material or creation of recombinant viruses in which those genes necessary for replication have been deleted.

Most viruses can be grown in cell culture for vaccine production. Animal cells are easy to maintain in the laboratory, and viruses for vaccines can be grown to necessary concentrations. Influenza viruses are exceptions, which is one reason quick production of yearly influenza vaccines has been difficult. The influenza virus genome consists of eight individual segments; coinfection of cells with two different strains, often involving viruses from two different species such as humans and birds, routinely creates a new recombinant strain that is not recognized by the human immune system. Influenza viruses do not grow well in cell culture, so vaccines must be produced using viruses grown in eggs. The lead time necessary to produce sufficient quantities of vaccine for the influenza season, which begins in the fall, is about six months. Therefore, health agencies such as the World Health Organization and the Centers for Disease Control and Prevention must decide in late winter which strains are most likely to produce an outbreak later that year.

Monoclonal Antibodies (mAB). Exposure to antigens such as those found on bacteria or viruses triggers the production of a large number of different antibodies, each of which is specific for a particular molecular determinant on an organism. In the 1960's, it was discovered that persons with multiple myeloma, a cancer of plasma cells, produced large quantities of homogeneous antibodies.

British scientists Georges J. F. Köhler and César Milstein found that because myeloma cells, like those of most cancers, are immortalized, they could artificially fuse myeloma cells with immune cells of known specificity to produce a clone of "immortal" cells producing identical antibodies; because these cells represented a clone, the product became known as monoclonal antibodies.

Because antibodies, in theory, can be generated in the laboratory against any target, monoclonal antibodies can be used as a probe for detection of any cellular molecule. Initial applications used monoclonal antibodies for detection of cell surface proteins for identification of cell types or the maturation stage of cells during differentiation. Because these surface proteins exhibited clustering, they became known as cluster of differentiation (CD) proteins. Nearly two hundred cluster of differentiation proteins are now known.

The ability of monoclonal antibodies to bind surfaces on specific cells has led to their use in the diagnosis or treatment of certain types of cancers. Immuno-conjugates are prepared by chemically attaching a toxin or radioisotope to a monoclonal antibody and injecting the molecule into a patient. Binding of the conjugated monoclonal antibody to the tumor cell results in killing of the target. Although in theory immunotherapy could be applied to many forms of cancers, most tumors do not express proteins unique to that type of cancer.

IMPACT ON INDUSTRY

Most basic research in immunology has been carried out in university and medical school laboratories, but only industry has the capability to carry out large-scale production and testing of vaccines. Several factors have had a significant impact on the ability of industry to maintain this work. Vaccine production represents a significant expense, ultimately reflected in the cost to the consumer. New or untested vaccines are particularly expensive, with no guarantee of a return on investment for the company that produces them.

The second factor that has had a negative impact on industry is the cost of litigation in response to real or imagined injury to the vaccine recipient. In court cases through the 1960's, the benefit associated with vaccines was generally considered to override the known risk. However, beginning in the 1970's, courts increasingly found vaccine manufacturers liable for damages that may or may not have been associated with their products. The vaccines administered in response to an outbreak of swine influenza in 1976 resulted in more than 4,000 claims and awards of damages exceeding $72 million. In this case, the government assumed liability, but in most cases, the manufacturer had to assume the loss, and payouts exceeded the profits associated with the sale of vaccines. For example, Lederle (later part of Pfizer) was forced to pay claims against its Sabin poliovirus vaccine and diphtheria-tetanus-pertussis (DTP) vaccine that were two hundred times greater than their sales. The cost of the DPT vaccine rose 10,000 percent by 1986.

The problem of cost has been directly reflected in the loss of American companies involved in vaccine production. In 1967, thirty-seven companies in the United States manufactured vaccines, but by 1984, less than half of those companies continued to manufacture vaccines, and by 2005, only three companies remained in the business. During the same period, the actual number of licensed vaccines fell from 380 to fewer than 40. The reluctance of companies to engage in vaccine production and the long production

---

Fascinating Facts About Immunology and Vaccination

- Edward Jenner, the English physician who developed the first vaccine against smallpox, refused an offer by Captain James Cook of the HMS *Endeavour* to accompany the captain on his next trip to Pacific islands during the 1770's.

- Bishop Cotton Mather of Boston is notorious for his condemnation of witchcraft, but in the 1720's, he was the primary proponent of variolation.

- During the 1918 influenza epidemic, the bacterium *Haemophilus influenzae* was mistakenly thought to be the etiological agent of the disease. A vaccine was produced against it, which had no effect on the epidemic.

- Louis Pasteur's rabies vaccine was first tested on nine-year-old Joseph Meister in 1885. In 1940, Meister, then a caretaker at the Pasteur Institute, committed suicide rather than open Pasteur's tomb for the Nazis.

- The attenuated Edmonston strain of measles virus developed by physician John Enders was isolated in 1954 from an eleven-year-old boy, David Edmonston, son of a Bethesda mathematician. As an adult, Edmonston participated in the 1963 civil rights march on Washington and later became a schoolteacher in Mississippi.

- The JL strains of the mumps virus vaccine were developed by physician Maurice Hilleman from an isolate obtained in 1963 from his six-year-old daughter, Jeryl Lynn.

---

time involved has resulted in several shortages of influenza vaccines for the flu season in the twenty-first century. During the H1N1 flu (swine flu) pandemic of 2009, the vaccine was in short supply and was given only to those judged most in need of it. As of 2010, only ten companies worldwide were licensed to produce influenza vaccines, and only two, Medimmune-Avirion and Sanofi Pasteur, a French-owned company, had plants in the United States.

CAREERS AND COURSE WORK

As is true for most careers in medical science, students generally begin by earning a bachelor of science degree in a field such as biology or biochemistry. The undergraduate program should include courses in general biology, chemistry (particularly organic chemistry), microbiology, genetics, and biochemistry. An understanding of the human immune system is vital, so courses in immunology, virology, and pathogenic microbiology should be included.

The bachelor of science degree is sufficient for an entry-level position, but advanced training is necessary for someone wishing to be more than a technician. Historically, most research in the field of immunology has taken place in universities, often in association with medical schools or research institutes. A student wishing to pursue such research most commonly enters a graduate program in which faculty members are carrying out studies in a relevant area. The laboratory director may focus his or her research in development of a recombinant vaccine. The work often involves initial testing of efficacy and safety in nonhuman animals. A doctorate is necessary for working at the level of university faculty or laboratory director.

Development and marketing of any prospective vaccine requires a significant source of funding, which may be available through government grants but more likely involves funding from pharmaceutical companies. A master's degree in an area of science such as chemistry or biochemistry is the minimal requirement in industry for vaccine research and development, while a doctorate is preferred.

SOCIAL CONTEXT AND FUTURE PROSPECTS

The effective control of most childhood infectious diseases by the end of the twentieth century has caused the fear of such diseases to all but disappear among most modern populations. As some segments of European, British, and American populations have grown up in a time in which childhood infectious diseases appeared to be a thing of the past, many of these people do not fully understand the devastating nature of these diseases and question the value of vaccines. Also, the sheer number of recommended vaccinations has created concern among parents, some of whom are afraid their children's immune systems could be overwhelmed, perhaps resulting in autism. Although no evidence for a link between autism and vaccination has been found, some parents still believe that such a link exists.

Future immunizations are likely to rely less on whole virus or bacterial vaccines and more on acellular or subunit vaccines. Side effects resulting from the pertussis vaccine, generally mild fever or inflammation but occasionally a more serious problem, led to the development of an acellular pertussis vaccine using only bacterial proteins. Similar vaccines, some containing only genetic information for production of viral proteins, are likely to be used against other diseases in the future.

Immunization against some agents such as influenza viruses, which undergo yearly changes, will probably involve some form of combination vaccines, incorporating proteins that are common to most major strains of the virus. The simplicity of world travel in the twenty-first century means scientists must take a worldview of new strains, as an outbreak in a few countries can rapidly develop into a worldwide pandemic.

The ability of the human immunodeficiency virus (HIV) to undergo rapid mutations, even within the same individual, means a vaccine against the acquired immunodeficiency syndrome (AIDS) remains unlikely in the foreseeable future.

Richard Adler, Ph.D.

FURTHER READING

Allen, Arthur. *Vaccine.* New York: W. W. Norton, 2007. Discusses the history of vaccine development, from eighteenth century variolation to modern times, and the controversies that have surrounded vaccination.

Heller, Jacob. *The Vaccine Narrative.* Nashville, Tenn.: Vanderbilt University Press, 2008. Tells how four of the major vaccines of the twentieth century were developed.

Link, Kurt. *The Vaccine Controversy: The History, Use, and Safety of Vaccinations.* Westport, Conn.: Praeger, 2005. Short synopses of the history behind most major vaccines, from smallpox to acellular vaccines.

Plotkin, Stanley A., Walter A. Orenstein, and Paul A. Offit, eds. *Vaccines.* 5th ed. Philadelphia: Saunders/Elsevier, 2008. An excellent source for understanding the history of vaccine development against most major agents.

Tauber, Alfred. "Metchnikoff and the Phagocytosis Theory." *Molecular Cell Biology* 4 (November, 2003): 897-901. Discussion of Metchnikoff's discovery of phagocytosis. Includes original illustrations.

Williams, Tony. *The Pox and the Covenant.* Naperville, Ill.: Sourcebooks, 2010. Story of Cotton Mather and variolation in Boston during the 1720's.

WEB SITES

American Association of Immunologists
http://www.aai.org

Centers for Disease Control and Prevention
Vaccines and Immunization
http://www.cdc.gov/vaccines

World Health Organization
Immunization Surveillance, Assessment, and Monitoring
http://www.who.int/immunization_monitoring/en

See also: Epidemiology; Genetic Engineering; Human Genetic Engineering; Pathology; Pharmacology; Virology.

INDUSTRIAL FERMENTATION

FIELDS OF STUDY

Biology; microbiology; biochemistry; organic chemistry; biotechnology; bioprocess engineering; chemical engineering.

SUMMARY

Industrial fermentation is an interdisciplinary science that applies principles associated with biology and engineering. The biological aspect focuses on microbiology and biochemistry. The engineering aspect applies fluid dynamics and materials engineering. Industrial fermentation is associated primarily with the commercial exploitation of microorganisms on a large scale. The microbes used may be natural species, mutants, or microorganisms that have been genetically engineered. Many products of considerable economic value are derived from industrial fermentation processes. Common products such as antibiotics, cheese, pickles, wine, beer, biofuels, vitamins, amino acids, solvents, and biological insecticides and pesticides are produced via industrial fermentation.

KEY TERMS AND CONCEPTS

- **Antioxidant:** Chemical that prevents the oxidation of other chemicals.
- **Biomass:** Mass of organisms; traditionally, this term refers to the biomass of plants and microorganisms.
- **Bioreactor:** Apparatus for cell growth with practical purposes under controlled conditions. Bioreactors are closed systems and vary in size from the small laboratory scale (five to ten milliliters) to the large industrial scale (more than 500,000 liters).
- **Enzymes:** Biological catalysts made of proteins.
- **Fermentation:** In biology, the metabolic reactions necessary to generate energy in living (mainly microbial) cells; in industry, any large industrial process based on living things is called fermentation.
- **Fermenter:** Type of traditional bioreactor (stirred or nonstirred tanks) where cell fermentation takes place. Fermenters can be operated as continuous or batch-culture systems. In a continuous culture, nutrients are continuously fed into the fermentation vessel. This allows the cells to ferment indefinitely.
- **Probiotics:** Microorganisms and substances that promote the development of healthy intestinal microbial communities.
- **Substrate:** Molecule that is broken down by fermentation.

DEFINITION AND BASIC PRINCIPLES

Industrial fermentation is the use of living organisms (mainly microorganisms), typically on a large scale, to produce commercial products or to carry out important chemical transformations. The goal of industrial fermentation is to improve biochemical or physiological processes that microbes are capable of performing while yielding the highest quality and quantity of a particular product. The development of fermentation processes requires knowledge from disciplines such as microbiology, biochemistry, genetics, chemistry, chemical and bioprocess engineering, mathematics, and computer science. The major microorganisms used in industrial fermentation are fungi (such as yeast) and bacteria. Fermentation is performed in large fermenters or other bioreactors often of several thousand liters in volume. Industrial fermentation is a part of many industries, including microbiology, food, pharmaceutical, biotechnology, and chemical.

BACKGROUND AND HISTORY

Traditional fermentations such as those for making bread, cheese, yogurt, vinegar, beer, and wine had been used by people for thousands of years before its microbial nature was understood. Brewing beer was one of the first applications of fermentation in ancient Egypt as long as 10,000 years ago. The exact origins of dairy products are unknown—it may have been as early as 8000 B.C.E. It was probably nomadic Turkish tribes in Central Asia who invented cheese and yogurt making. Traditionally, dairy fermentation was a means of milk preservation. The scientific understanding of fermentation began only in the nineteenth century after French scientist Louis Pasteur published the results of his studies on the microbial nature of wine making.

The first industrial fermentation bioprocesses based on knowledge of microbes appeared in the

Pharmaceutical technician working at one of a row of fermentation units, or bioreactors. (Maximilian Stock Ltd./ Photo Researchers, Inc.)

early twentieth century. Russian biochemist Chaim Weizmann is considered to be the father of industrial fermentation. Weizmann used the bacterium *Clostridium acetobutylicum* for the production of acetone from starch in 1916. Acetone was used to make explosives during World War I.

Significant growth of this field began in the middle of the twentieth century, when the fermentation process for the large-scale production of antibiotic penicillin was developed. The goal of industrial-scale production of penicillin during World War II led to development of fermenters by engineers that were working together with biologists from the pharmaceutical company Pfizer. The fungus *Penicillium* grows and produces an antibiotic much more effectively under controlled conditions inside the fermenter. Continuous progress in industrial fermentation technology in the twentieth century has followed the development of genetic engineering. Genetic engineering allows

gene transfer between species and creates possibilities to generate new products from genetically modified microorganisms that are grown in fermenters.

The twenty-first century has been characterized by the introduction of biofuels, which are made by industrial fermentation processes. Once again, past and future developments in fermentation technology require contributions from a wide range of disciplines, including microbiology, genetics, biochemistry, chemistry, engineering, mathematics, and computer science.

How It Works

Industrial fermentation is based on microbial metabolism. Microbes produce different kinds of substances that they used for growth and maintenance of their cells. These substances can be useful for humans. The goal of industrial fermentation technology is to enhance the microbial production of useful substances.

Process of Fermentation. In biology, fermentation is a process of harvesting energy of organic molecules in oxygen-free conditions. Sugars are a prime example of what can be fermented, although, there are many other organic molecules that can be used. Different fermentations are known and are categorized by the substrate metabolized or the type of the product.

In industry, any large microbiological process is called fermentation. Thus, the term fermentation has a different meaning than in biology. Most industrial fermentations require oxygen.

Industrial Fermentation Organisms. Different organisms, such as bacteria, fungi, and plant and animal cells, are used in industrial fermentation processes. An industrial fermentation organism must produce the product of interest in high yield, grow rapidly on inexpensive culture media available in bulk quantities, be open to genetic manipulation, and be nonpathogenic (does not cause any diseases).

Fermentation Media. To make a desired product by fermentation, microorganisms need nutrients (substrates). Nutrients for microbial growth are known as media. Most fermentation requires liquid media or broth. General media components include carbon, nitrogen, oxygen, and hydrogen in the form of organic or inorganic compounds. Other minor or trace elements must also be supplied, for example, iron, phosphorus, or sulfur.

Fermentation Systems. Industrial fermentation takes place in fermenters, which are also called bioreactors.

Fermenters are closed vessels (to avoid microbial contamination) that reach vast volumes, as many as several hundred thousand liters. Designed by engineers, the main purpose of a fermenter is to provide controllable conditions for growth of microbial cells or other cells. Parameters such as pH, temperature, nutrients, fluid flow, and other variables are controlled. There are two kinds of fermenters, those for anaerobic processes (oxygen-free) and aerobic processes. Aerobic fermentation is the most common in industry. Anaerobic fermenters can be as simple as stainless-steel tanks or barrels. Aerobic fermenters are more complicated. The most critical part in these systems is aeration. In a large-scale fermenter transfer of oxygen is very important. Oxygen transfer and dispersion are provided by stirring with impellers or oxygen (air) sparging.

Fermentation Control and Monitoring. Industrial fermentation control is very important to ensure that organisms behave properly. In most cases computers are used for controlling and monitoring the fermentation process. Computers control temperature, pH, cell density, oxygen concentration, level of nutrients, and product concentration.

APPLICATIONS AND PRODUCTS

There are a wide range of industrial fermentation products and applications.

Food, Beverages, Food Additives, and Supplements. Industrial fermentation plays a major role in the production of food. Food products traditionally made by fermentation include dairy products (cheeses, sour cream, yogurt, and kefir); food additives and supplements (flavors, proteins, vitamins, and carotenoids); alcoholic beverages (beer, wines, and distilled spirits); plant products (bread, coffee, soy sauce, tofu, sauerkraut); and fermented meat and fish (pepperoni and salami).

Industrial Fermentation. The primary and largest industry revolves around food products. Milk from cows, sheep, goats, and horses have traditionally been used for the production of fermented dairy products. These products include cheese, sour cream, kefir, and yogurt. More recently so-called probiotics appeared and have been marketed as health-food drinks. Dairy products are produced via fermentation using lactic bacteria such as *Lactobacillus acidophilus* and *Bifidobacterium*. Fungi are also involved in making some cheeses. Fermentation produces lactic acid and other flavors and aroma compounds that make dairy products taste good.

Many products of industrial fermentation are added into food as flavors, vitamins, colors, preservatives, and antioxidants. These products are more desirable than food additives produced chemically. Many of the vitamins are made by microbial fermentations including thiamine (vitamin B_1), riboflavin (vitamin B_2), cobalamin (vitamin B_{12}), and vitamin C (ascorbic acid). Vitamin C is not only a vitamin but is also an important antioxidant that helps to prevent heart diseases. Carotenoids are another effective antioxidant. They are also used as a natural food color for butter and ice cream. Carotenoids are red, orange, and yellow pigments produced by bacteria, algae, and plants.

Food preservatives are yet another product of industrial fermentation. Organic acids, particularly lactic and citric acids, are extensively used as food preservatives. Some of these preservatives (such as citric acid) are used as flavoring agents. A mixture of two bacterial species (*Lactobacillus* and *Streptococcus*) is usually used for industrial production of lactic acid. The mold *Aspergillus niger* is used for citric acid manufacturing. Another common preservative is the protein nisin. Nisin is produced via fermentation by the bacterium *Lactococcus lactis*. It is employed in the dairy industry especially for production of processed cheese.

Antibiotics and Other Health Care Products. Antibiotics are chemicals that are produced by fungi and bacteria that kill or inhibit the growth of other microbes. They are the second most significant product of industrial fermentation. Most antibiotics are generated by molds or bacteria called actinomycetes. More than 4,000 antibiotics have been isolated from microorganisms, but only about 50 are produced regularly. Among them, beta-lactams, such as penicillins and tetracyclines, are most common. Penicillin is produced by the mold *Penicillium chrysogenum* via corn fermentation in bioreactors of up to 200,000 liters.

The other major health care products produced with the help of industrial fermentation are bacterial vaccines, therapeutic proteins, steroids, and gene therapy vectors. There are two categories of bacterial vaccines: living and inactivated vaccines. Living vaccines consist of weakened, also known as attenuated, bacteria. Examples of living vaccines include those for diseases such as anthrax, which is caused by *Bacillus anthracis*, and typhoid fever, which is caused by *Salmonella typhi*. Inactivated vaccines are composed of bacterial cells or their parts that have been inactivated by heat or formaldehyde. Examples of these vaccines are those for meningitis, whooping cough, and cholera. Vaccine production takes place in fermenters no bigger than 1,000 liters in volume. It requires highly controlled operations to avoid the release of bacteria into the environment. All exhaust gases pass through sterilization processes.

Therapeutic proteins include growth hormone, insulin, wound-healing factors, and interferon. Previously, such compounds were made from animal tissues and were very expensive to manufacture. Genetic engineering now allows their production by fermentation from bacteria. Human growth hormone is synthesized in the human brain and controls growth. Too little growth hormone can cause some cases of dwarfism. The American company Genentech started production of human growth hormone from genetically modified *Escherichia coli* by fermentation in 1985. Insulin is an animal and human hormone that is involved in the regulation of blood sugar. The body's inability to make sufficient insulin causes diabetes. Insulin extracted from pigs had been used to treat diabetes, but it has been replaced by insulin produced by industrial fermentation from genetically modified bacteria.

Chemicals. Numerous chemicals, such as amino acids, polymers, organic acids (citric, acetic, and lactic), and bioinsecticides are produced by industrial fermentation. Amino acids are used as a food and animal feed, as well as in the pharmaceutical, cosmetic, and chemical industries. Bacteria such as *Micrococcus luteus* and *Corynebacterium glutamicum* are used for industrial fermentation to produce chemicals. Bacterial toxins are effective against different insects. Since the 1960's, preparations of the bacteria *Bacillus thuringiensis* have been produced by fermentation as a biological insecticide.

Enzymes. Enzymes are used in many industries as catalysts. Microorganisms are the favored source for industrial enzymes. Seventy percent of these enzymes are made from *Bacillus* bacteria via fermentation. Most commercial microbial enzymes are hydrolases, which break down different organic molecules such as proteins and lipids. The enzyme glucose isomerase is important in the production of fructose syrups from corn and is widely used in the food industry.

Biomass Production. During biomass production by fermentation, the cells produced are the products.

Fascinating Facts About Industrial Fermentation

- Yeast (*Saccharomyces cerevisiae*) was one of the earliest domesticated organisms. It was used by ancient Egyptians for making bread and beer. People used yeast for thousands of years without knowing about it until the nineteenth century.
- The holes in some cheeses, such as Swiss cheese, are a result of gas production during bacterial fermentation. Bacteria *Propionibacterium* are used during ripening (aging) of cheese to add flavor.
- The most common mushroom in grocery stores is *Agaricus bisporus*. Since 1810, it has been cultivated using wheat straw and horse manure as food sources. *Agaricus* cultivation represents an example of the most efficient conversion of plant and animal wastes into edible food.
- Vinegar is produced via industrial fermentation of alcohol (ethanol) by acetic acid bacteria. These bacteria are not welcomed by winemakers since bacteria tend to spoil wine by converting it into vinegar. The word vinegar is derived from the French words *vin* (wine) and *egre* (sour).
- San Francisco has extended its recycling program to include fermentation of dog waste into methane. Methane is used to fuel electrical generators to produce electricity and to heat homes. San Francisco's dog population is 120,000, and this initiative promises to generate a significant amount of fuel while reducing a huge amount of waste.
- Chaim Weizmann, who is considered to be the father of industrial fermentation, became the first president of Israel. Weizmann's life path from the industrial fermentation to becoming a head of state illustrates the importance of industrial fermentation.

Biomass is used for four purposes: as a source of protein for human food or animal feed, in industry as fermentation starter cultures, in agriculture as a pesticide or fertilizer, and as a fuel source.

One major product of this application of industrial fermentation is baker's yeast biomass. Baker's yeast is required for making bread, bakery products, beer, wine, ethanol, microbial media, vitamins, animal feed, and biochemicals for research. Yeast is produced in large aerated fermenters of up to 200,000 liters. Molasses is used as a nutrient source for the cells. Yeast is recovered from fermentation liquid by centrifugation and then dried. It can then be sold as compressed yeast cakes or dry yeast.

Many bacteria have been considered as potential sources of protein to fulfill the food needs in some countries of the world. As of 2011, only a few species are cultivated around the world as a source of food and feed. Among them, cyanobacteria are the most popular. The protein level of *Cyanobacterium Spirulina* can be as high as those found in meat, nuts, and soybeans, from 50 to 70 percent. This cyanobacterium has been used as a human food for millennia in Asia, parts of Africa, and in Mexico.

Apart from yeast and bacteria, people are also using the biomass of algae. Algae are a source of animal feed, plant fertilizer, chemicals, and biodiesel. Because light is necessary to grow algae, the biomass is produced in open ponds or in transparent tubular glass or plastic bioreactors, called photobioreactors.

Biofuels. Industrial fermentation is used in the production of biofuels, mainly ethanol and biogas. These two biofuels are produced by the action of microorganisms in bioreactors. Fermentation can also be used for generation of biodiesel, butanol, and biohydrogen. Biofuels are considered, by many, as a future substitute for fossil fuels. Pollution from fossil fuels affects public health and causes global climate change due to the release of carbon dioxide (CO_2). Using biofuels as an energy source generates less pollutants and little or no CO_2.

Production of ethanol is a process based on fungal or bacterial fermentation of a variety of materials. In the United States, most of the ethanol is produced by yeast (fungal) fermentation of sugar from cornstarch. Sugar is extracted using enzymes, and then yeast cells convert the sugar into ethanol and CO_2. Ethanol is separated from the fermentation broth by distillation. Brazil, the second largest ethanol producer after the United States, uses sugarcane fermentation to generate ethanol. The Brazilian production of ethanol from sugarcane is more efficient than the American corn-based ethanol.

Biogas is produced during the anaerobic (non-oxygen) fermentation of organic matter by communities of microorganisms (bacteria and *Archaea*). There are different types of biogas. One type contains a mixture of methane (50 to 75 percent) and CO_2. Another type is composed primarily of nitrogen, hydrogen, and carbon monoxide (CO) with trace amounts of methane. Methane is generated by microorganisms called *Archaea* and is an integral part of their metabolism. For practical use, biogas is generated from wastewater, animal waste, and "gas wells" in landfills.

IMPACT ON INDUSTRY

Industrial fermentation plays a major role in a number of multibillion-dollar industries, including food, pharmaceutical, microbiological, biotechnological, chemical, and biofuel. The United States maintains a dominant position in the world in industrial fermentation. In fact, the first really large-scale industrial fermentation process to produce the antibiotic penicillin was developed in the United States. Many other developed and developing countries use industrial fermentation to produce varieties of products. Some countries have made industrial fermentation a prime national interest. Brazil, for example, is using the industrial fermentation for ethanol production. This country has the largest and most successful ethanol for fuel program in the world. As the result of this successful program, Brazil reached complete self-sufficiency in energy supply in 2006.

Government and University Research. Governmental agencies such as the National Science Foundation (NSF), the United States Department of Energy (DOE), and the United States Department of Agriculture provide funding for research in the industrial fermentation area. Currently, a vast majority of research is concentrated on biofuel generation by microorganisms and environmental applications.

Industry and Business. Scientists in the industry traditionally carry out a significant load of research in fermentation area. Companies such as Pfizer and Merck were pioneers in industrial production of the first antibiotic penicillin produced by fungal fermentation. A significant proportion of industrial fermentation research in industry has been directed to health care products (such as antibiotics).

Major Corporations. Examples of major corporations in food and beverage industries are Kraft Foods, Dannon, Coors, Guinness, and Anheuser-Busch. In the biofuel industry, major companies are Archer Daniels Midland, Poet Energy, Abengoa Bioenergy, and VeraSun Energy.

CAREERS AND COURSE WORK

There are several career options for people who

are interested in being trained in industrial fermentation. Food, biotechnology, microbiology, pharmaceutical, chemical, and biofuel companies are the biggest employers in the area. Students who are interested in conducting research in industrial fermentation can find jobs in university, government, and industry laboratories.

When choosing a career in industrial fermentation, one should be prepared for an interdisciplinary science. Students should obtain skills in microbiology, molecular biology, bioengineering, plant biology, organic chemistry, biochemistry, agriculture, bioprocess engineering, and chemical engineering.

Most professionals in industrial fermentation have a bachelor's degree in biology, microbiology, or biotechnology. Individuals who have managerial responsibilities often have a master's or doctorate in biology, microbiology, fermentation, molecular biology, biochemistry, biotechnology, bioprocess or chemical engineering, or genetics. Some universities, including the University of California, Davis, and Oregon State University, offer degrees in fermentation.

A career in industrial fermentation presents a variety of work options such as research, process development, production, technical services, or quality control. Some industrial fermentation specialists may be considered as genetic engineers (using DNA techniques to modify living organisms), while others are classified as bioprocess or chemical engineers (optimizing bioreactors and biochemical pathways for a desired product).

SOCIAL CONTEXT AND FUTURE PROSPECTS

Industrial fermentation plays a major role in providing food, chemicals, and fuels. End users are consumers, farmers, medical doctors, and industrialists. Industrial fermentation is changing the course of history. People have made food by fermentation for centuries. In the twentieth century, the development of antibiotics and their production by industrial fermentation had the most significant impact on the practice of medicine than any other development. The growth of the industrial fermentation field is continuing rapidly. Since the beginning of the twenty-first century, industrial fermentation underwent an unprecedented growth and expansion due to biofuel introduction. This record growth is particularly visible in the U.S. ethanol industry. In 1980, the U.S. ethanol industry produced 175 million gallons of ethanol by fermentation, and in 2009, 10.6 billion gallons.

The role of industrial fermentation in human society is likely to expand in the future because of increasing requirements for resources.

Sergei A. Markov, Ph.D.

FURTHER READING

Bailey, James E., and David F. Ollis. *Biochemical Engineering Fundamentals.* 2d ed. New York: McGraw-Hill, 1986. Classic textbook on biochemical engineering.

Bourgaize, David, Thomas R. Jewell, and Rodolfo G. Buiser. *Biotechnology: Demystifying the Concepts.* San Francisco: Benjamin/Cummings, 2000. Classic text on biotechnology.

Doran, Pauline M. *Bioprocess Engineering Principles.* San Diego: Academic Press, 1995. Describes various bioreactors and fermenters used for industrial fermentation.

Glazer, Alexander N., and Hiroshi Nikaido. *Microbial Biotechnology: Fundamentals of Applied Microbiology.* 2d ed. New York: Cambridge University Press, 2007. In-depth analysis of application of microorganisms in industrial fermentation.

Lydersen, Bjorn K., Nancy A. D'Elia, and Kim L. Nelson, eds. *Bioprocess Engineering: Systems, Equipment and Facilities.* New York: John Wiley & Sons, 1994. Describes equipment and facilities for industrial fermentation.

Wright, Richard T., and Dorothy F. Boorse. *Environmental Science: Toward a Sustainable Future.* 11th ed. Boston: Benjamin/Cummings, 2011. This textbook describes several bioprocesses used in waste treatment and pollution control.

WEB SITES

Biotechnology Industry Organization
http://www.bio.org

Society of Bioscience and Technology
http://www.socbioscience.org

United States Department of Agriculture Biotechnology
http://www.usda.gov/wps/portal/usda/
usdahome?navid=BIOTECH

See also: Chemical Engineering; Computer Science; Engineering; Food Preservation; Food Science.

INDUSTRIAL POLLUTION CONTROL

FIELDS OF STUDY

Chemistry; meteorology; engineering; environmental studies; marine biology.

SUMMARY

Industries that have contributed historically to problems of air, water, and soil pollution are too many to name. Some obvious examples are mining and metallurgy, pharmaceuticals, and all industries processing petroleum. Because different industries use different methods to produce their products, different approaches must be adopted and adapted for controlling pollutants that result.

KEY TERMS AND CONCEPTS

- **Acid Rain:** Occurs when air pollutants rise and collect in clouds, falling later to Earth in the form of chemically charged rain.
- **Adsorption:** Chemical process in which gases, liquids, or any dissolved substances adhere to a solid body next to it or in which it is contained.
- **Digestion:** Biochemical decomposition of organic materials to obtain different, less polluting, organic compounds.
- **Effluent:** Any liquid waste product leaving a source of pollution.
- **Entrainment:** Droplets of mist entering the air when bubbles form in boiling liquids; such droplets can carry polluting particles contained in the liquid source.
- **Sludge Seeding:** Injection of chemically active additives to sewage (or other biodegradable waste) to speed up the process of decomposition.

DEFINITION AND BASIC PRINCIPLES

Because various industries use chemicals or fossil fuels, by-products of such processes may occur as waste materials that are either emitted into the air or evacuated by means of polluted water drainage, sometimes into streams, lakes, or groundwater. Control of such pollutants requires both general legal guidelines, provided by the Environmental Protection Agency (EPA), and control or recycling procedures responding to the specific nature of the environmental pollution problem identified.

BACKGROUND AND HISTORY

Ultimately, industrial pollution goes back to the origins of the Industrial Revolution in Europe and its migration to the United States in the nineteenth century. In the early Industrial Revolution, methods of production were considerably closer to nature than they would become when heavy industries, most driven by fossil fuel energy, became predominant. Although water-driven industrial machines, particularly in the developing textile industry, might have wasted or diverted water, truly harmful effects (mainly from increases in chemical additives) were slow in coming.

Legislative actions in the United States aiming at control of industrial pollution— starting with concerns about water quality—have been numerous and diverse. Early measures, the Rivers and Harbors Appropriation Act of 1899 and organization of the Smoke Prevention Association in 1907, were followed by a very broad but important precedent: the Public Health Service Act of 1912. Over ensuing decades (except during the heavy emergency production period of World War II), particular industrial sectors became targets of pollution control laws, beginning with the Oil Pollution Act of 1924. Following World War II a major turning point came with the Water Pollution Control Act (WPCA) of 1948. In 1972, after multiple revisions beginning in 1961, when responsibility for the WPCA shifted to the new Department of Health, Education and Welfare, it became known as the Clean Water Act (CWA). This legislation established federally supervised procedures (originally to be overseen by the U.S. Surgeon General) for safeguarding the quality of water in public reservoirs, but also "interstate waters and [their] tributaries" as well as the "sanitary condition of surface and underground waters."

The CWA came in the wake of President Kennedy's "Special Message to the Congress on Natural Resources" on February 23, 1961. After this date it was clear that questions of water pollution needed to be considered along with industrial pollution in the air and in surface and subsurface soils. In 1970, Congress created the EPA. At the same time, President Nixon signed an executive order establishing the National Industrial Pollution Control Council (NIPCC). This special council, which included several representatives of private industry, was to help coordinate responsibilities of the new EPA, with specific emphasis on matters of industrial pollution.

The Pollution Prevention Act of 1990 represented a landmark in industrial pollution control. It shifted control emphasis from efforts to reduce the quantity of pollutants released by manufacturers to efforts to reduce pollution in earlier stages of industrial processing. This goal appeared in Section 6604, which created a new EPA Office of Pollution Prevention. Proponents of the 1990 laws argued that—in the long run—manufacturers stood to save money through "source reduction," both by reducing quantities of expensive (and polluting) materials and replacing them by less costly nonpolluting substitute materials. Manufacturers could also avoid looming liability costs (in the form of fines). The 1990 law also encouraged finding ways to treat industrial waste products so that they might be recycled, either in internal production operations or for use by other manufacturers.

How It Works

Methods of pollution control differ from industry to industry. The selected industries below demonstrate several approaches to pollution problems in different key sectors.

Mining and Metallurgy. Issues of pollution occur in all mining and metallurgical industries. Iron mining and steel production operations face problems widely shared with many other industrial subsectors. Pollution in the steel industry begins at mining sites. Water that has been used to treat extracted iron ore— containing iron sulfide and iron pyrite—has to be disposed of somehow. If oxidation of such materials occurs, harmful secondary chemicals, including sulfuric acid and carbon dioxide, result. Acid water requires chemical neutralization, usually with limestone additives, followed by careful disposal of the resulting sludge.

The later stage leading to steel—when iron ore is reduced in blast furnaces—presents a different set of pollution control problems. Three pollutants are produced at this point. Air and water is polluted when a mixture of coke, iron ore, and calcium carbonate is blasted in extremely hot furnaces. Once the process yields a percentage of molten metal, masses of solid waste need to be disposed of.

It may be possible to extract some useful substances, mainly hydrocarbons, through condensation, and some coke oven gases can be recycled for fuel needs elsewhere. The net effect of blast furnace operations, however, remains a menace to the environment. Efforts have been made to treat iron ore without coke-burning blast furnaces. These include direct reduction of crushed iron ore using a hydrogen additive, which produces pig iron that is transformable into steel in less polluting electric furnaces. Another method—called the HIsmelt process— uses force injected coal and oxygen to pre-reduce iron ore at an earlier stage.

Other problems arise when open hearth (or electric) furnaces and rolling mills are used to produce steel itself. Dangerous substances come from chemical by-products (called mill scale—a layer of metal oxides on new steel) in the rolling process. The steel industry has increasingly converted mill scale sludge into a concentrate that is about 70 percent iron, at the same time removing oil contaminants that—in earlier mill scale disposal processes—escaped into the air.

Control of pollution in the giant copper industry involves similar chemical by-products. Methods of control, however, particularly avoiding the need for high- temperature smelting, are different. Copper can be obtained by solvent-extraction means: Weak acid is percolated, both through ore and copper-bearing waste materials that would otherwise enter slag heaps. Copper ores containing sulfides, however, require new forms of treatment that use environmentally friendly bacterial additions.

Cement Industry. The cement industry ranks ninth in the list of major polluters in the United States; it is also a major (sixth-ranking) consumer of energy. After crushing limestone, shale, and sand components, mixed materials pass through high-temperature kilns. Here, a major pollutant, kiln dust, composed of at least eight polluting chemicals, enters the air.

Developing higher-efficiency machinery to filter kiln dust can help reduce its harmful chemical content. The cement industry does, however, have other commercially attractive options for disposal of this potential pollutant. Certain (not all) agricultural soils, for example, can be fertilized with kiln dust and it can also be used as a cattle grain supplement.

Pharmaceutical Industry. As a subbranch of the total chemical industry, the pharmaceutical industry must be as closely supervised as a potential polluter as it is for potentially dangerous contributions to commercial medicine. Some key processes, particularly fermentation of organic substances that yield antibiotics, produce harmful waste products. Some of these can be neutralized by careful calculation of biological additives to what are essentially pharmaceutical septic tanks.

In cases where wastes contain highly concentrated chemicals, neutralization (or adjustment of base-acid pH ratios) can be obtained via inorganic chemical additives. It should be noted, however, that such procedures—unless they are carefully managed by chemical experts—can produce other forms of risk, in some cases leading to fires or explosions.

APPLICATIONS AND PRODUCTS

Because polluting chemicals are frequently combatted by using other chemicals that neutralize their noxious effects, a number of potentially beneficial products may eventually come out of pollution control research. Any ongoing results of such specialized research are known to specialized chemists and as such remain beyond the ken of the general public.

By contrast, the public is becoming increasingly aware of much less high-tech applications that can make recycling not only ecologically beneficial but a significant contribution to middle levels of the economy. Most obvious are efforts to recycle paper products. Even though waste paper is biodegradable and therefore not highly polluting itself, industrial processes for reusing wastepaper not only save trees, they produce less pollution.

A similar observation applies to the recycling of plastic materials, all of which are in one way or another by-products of petroleum. Even the simplest methods of recycling plastics, therefore, promise reduced levels of industrial pollution linked to extraction and processing of crude oil.

In terms of higher levels of technology, each time a major pollution problem occurs, scientists and engineers find themselves recruited to develop effective ways to deal with the specifics of each occurrence. The variety of submarine applications that were tried (some for the first time) during the months-long British Petroleum (BP) *Deepwater Horizon* oil spill in the Gulf of Mexico in 2010 is a prime example of this challenge. The oil spill, which began after a violent explosion and fire destabilized the drilling platform

Technicians operating a vacuum recovery system during a training exercise. This equipment is used in the event of an oil spill and other water pollution events. The engine (left) creates a vacuum that enables oily water and debris to be sucked into the silver hopper (upper right). (Paul Rapson/ Photo Researchers, Inc.)

itself, illustrated another characteristic of pollution control demands: Highly specialized technical applications are used to prevent fire-producing disasters, but, if such fires occur, they must be controlled and extinguished with a minimum of pollution.

Finally, the BP oil spill, and even more dramatically, the 2011 nuclear reactor catastrophe in Fukushima, Japan, that followed the earthquake provided the public with unprecedented exposure to diverse scientific applications, some of which admittedly fell short of their stated goals.

IMPACT ON INDUSTRY

Pollution control obviously comes with a price tag. At each stage of new legislation the question of how costs can be reduced is raised. In recent years, major efforts have been made to recycle a number of industrially produced metals, particularly scrap copper. In the United States, more than half the copper used each year comes from recycling. Although reuse means less pollution, any eventual reduction in demand for products made from new raw materials can have economic repercussions for key segments of the industrial sector.

If one adopts a longer-term perspective, there is a real possibility that a number of key U.S. industries—faced with the necessity of revamping their mode of production to meet stricter pollution standards—may either cease operations altogether, or restructure their operations by resorting to a form of outsourcing to industrial sites in other countries where antipollution legislation is less stringent.

One of the biggest questions that loomed as the twenty-first century began was how to balance the risk of widespread ecological damage with what appears to be a binding necessity on a global level: expanding activities of the oil drilling and oil refining industry. The impact of the 2010 BP oil spill in the Gulf of Mexico is a major example of this quandary.

On the one hand, the enormous economic costs incurred to control a spill of this magnitude is in most cases billed to the private industrial concerns responsible for the pollution. On the other hand, if and when such procedures achieve the longer-term goal of taking measures to prevent similar troubles on the same or similar sites, the petroleum industry in particular seems to hold a position of bargaining with various agencies over terms governing their future operations. Such terms might include resumption of operations in the same location with similar equipment, even if public opinion seems skeptical of pampering the petroleum industry.

Needless to say, it is unlikely that large-scale petroleum companies would willingly accept government-imposed cutbacks in the scale of their operations, especially in areas where known oil reserves promise continuing profits.

Increasingly stringent demands by governments for costly preventive measures may, however, have the effect of raising production costs. In such cases, the petroleum industry's decision to pass such costs on to consumers (not only private consumers, but other fossil-fuel-dependent industries) could produce a spiraling impact, with consequences that would have to be measured carefully— if only to maintain necessary levels of stability in all sectors of the economy.

CAREERS AND COURSE WORK

Anyone seeking a career in industrial pollution control should plan his or her course work so that essential scientific fields, especially chemistry and biology, are covered at the undergraduate level and narrowed to specific subareas in graduate course work. Students nearing the end of their university training may qualify for a number of federally funded short- or medium-term or summer internships, especially with the EPA. Although many openings are

for clerical work, these temporary posts can provide valuable practical training in areas such as pollution policy analysis. Where applicants possess fairly advanced technical skills, work on EPA in-field projects is a possibility.

The EPA is the most important governmental agency involved in industrial pollution matters. It is headquartered in Washington, D.C., and has ten regional offices, all of which are equipped with laboratories staffed with trained experts, many of whom specialize in industrial pollution treatment.

One branch of the EPA, the Office of Chemical Safety and Pollution Prevention, offers an example of dovetailing between different specialized fields of training and employment. Its functions involve toxic

---

Fascinating Facts About Industrial Pollution Control

- So-called carbon taxes require industries that release CO_2 into the air to pay a supplemental tax to help fund vital environmental-preservation measures. A carbon tax was introduced in 2010 on the coal industry in India. The first European country to impose a carbon tax, in 1990, was Finland. At least ten countries in Europe had some form of carbon tax by 2010.

- In unusual cases, such as in a court case involving Iowa farmers, local private interests may prefer continuation of what amounts to industrial pollution. In the case noted above, local farmers successfully challenged EPA sanctions against a cement factory the dust of which was polluting the air—but depositing free fertilizing chemicals on surrounding farmlands.

- Despite continuing concerns about carbon dioxide emissions from automobile exhaust, much higher totals of carbon dioxide gas enter the atmosphere from industrial pollution.

- Industrial pollution affects not only the local regions in which industries are found. Pollutants spread rapidly in very wide patterns, either in the atmosphere or in underground water systems.

- Increasingly, methods are being found to recycle industrial waste products that can be useful and even sold commercially. This is sometimes done by means of additional chemical treatment. Plastic and paper recycling have become part of daily consumer habits all over the globe.

substances (including pesticide manufacture), food and drug safety, and pollution prevention in general.

Needless to say, almost every industry, heavy and light, hires personnel trained to trace sources of pollution in their production process and to devise methods to control it. In the automobile industry, such control concerns are present at two levels: factory hazardous waste and lowering exhaust emissions from new models.

Most local governments staff offices responsible for pollution control. Among many possible examples, one can cite local air pollution control districts, such as the one created as early as 1971 in California's Glenn County (near Sacramento), which maintains close association with the state's department of agriculture. One of its key concerns is to maintain informational networks with local industries that may need technical consultation based on specific local resources of mixed private and governmental agencies.

SOCIAL CONTEXT AND FUTURE PROSPECTS

In many formerly dominant industrial countries, especially the United States and European countries such as England, France, and Germany, economic and technological changes have made certain industrial subsectors much less important than they were up to the second half of the twentieth century. Metallurgy in general is an example of such changing conditions.

Even though some formerly very polluting heavy industries have reduced operations (or closed down entirely), lingering residue from earlier years of maximum production poses major health and safety problems for locally surrounding communities. Public reactions to such ecological dangers have in some cases led to citizen-based movements to oblige responsible powers, whether owners of industries themselves or governmental agencies, to clean up residual pollution.

A number of such movements could be cited, some in distant countries such as India (the Vasundhara research and policy advocacy group, for example, which supports initiatives to protect local communities from pollution and natural resource depletion) and Vietnam (see "Further Reading"). More and more cases are surfacing in the United States. Some more or less spontaneous local organizations depend on the media to gain support. An example appeared in *The New York Times* of April 4,

2011: An interfaith local movement pit itself against a defunct chromium- producing plant in Jersey City, New Jersey. Other more structured bodies, such Nexleaf Analytics (a private group working with the Center for Embedded Networked Sensing at University of California, Los Angeles), seek to use advanced technology to calculate levels of danger from various pollutants.

In most developed and developing societies, emphasis on recycling programs has begun to have an effect, not only on basic consumer patterns (either by reducing or eliminating entirely the use of certain industrial products that may end up as harmful additions to waste-disposal programs), but also on industries themselves. Examples of the latter may involve increased attention to ways in which polluting industrial by-products can be used in followup procedures that actually convert them to useful (and, in some cases, commercially valuable) by-products.

As technology advances, it is to be hoped that increasingly effective methods will be found to control (or even better, prevent) industrial pollution. For example, research should yield wider use of good chemicals to reduce the polluting effects of bad chemicals that must be used by some industrial manufacturers. Although most contributors to a special 1994-1995 issue of the *Georgia Law Review* dedicated to technical aspects of prevention (see "Further Reading"), a main suggestion by all was that future cooperation between scientists, lawmakers, and industrial manufacturers is becoming increasingly critical.

Such cooperation depends, however, on legislators' willingness to remain firm in the face of pressures to water down control regulations. An example of this problem appeared in *The New York Times* of April 16, 2011, citing Republican Party efforts to cut back both the EPA's budget and governmental regulations controlling industrial wastewater disposal. Opponents of EPA controls argued that the extra costs involved represented an additional burden to industries and might force them to cut back economically critical operations.

Byron D. Cannon, M.A., Ph.D.

FURTHER READING

Bishop, Paul L. *Pollution Prevention: Fundamentals and Practice.* Boston: McGraw-Hill, 2000. A study emphasizing prevention before pollution becomes a major problem involving more costly controls.

Eckenfelder, W. Wesley. *Industrial Water Pollution Control.* 3d ed. Boston: McGraw-Hill, 2000. Covers diverse aspects of the most common and, along with air pollution, potentially disastrous aspects of industrial waste control, including not only pollution of waterways but also groundwater reserves.

Hirschorn, Joel S. "Pollution Prevention Comes of Age." *Georgia Law Review* 29 (1994-1995): 325-347. This is a general review of what was, after the Pollution Prevention Act of 1990, a relatively new emphasis on methods to stop pollution at early to midpoint stages of industrial production, as well as finding ways to produce reusable products from industrial waste.

Lund, Herbert F., ed. *Industrial Pollution Control Handbook.* New York: McGraw-Hill, 1971. Although somewhat dated, this volume contains very useful technical information on different control processes in different industrial sectors.

Rodgers, William H., Jr. "The National Industrial Pollution Control Council: Advise or Collude?" *Boston College Law Review* vol. 13, issue 4 (1972): 719-747. A highly qualified professional familiar with the workings of government and industry wrote this criticism of the administrative steps to coordinate pollution control agencies.

Sell, Nancy J. *Industrial Pollution Control: Issues and Techniques.* 2d ed. New York: Van Nostrand Reinhold, 1992. This general study of different industrial pollution sectors continues to be widely cited in more recent literature for its comprehensive and accurate analysis.

WEB SITES

Air and Waste Management Association
http://www.awma.org

National Pollution Prevention Roundtable
http://www.p2.org

Worldwide Pollution Control Association
http://wpca.info

See also: Engineering; Meteorology; Thermal Pollution Control; Water-Pollution Control.

INFORMATION TECHNOLOGY

FIELDS OF STUDY

Information systems; computer science; software engineering; systems engineering; networking; computer security; Internet and Web engineering; computer engineering; computer programming; usability engineering; business; mobile computing; project management; information assurance; computer-aided design and manufacturing; artificial intelligence; knowledge engineering; mathematics; robotics.

SUMMARY

Information technology is a new discipline with ties to computer science, information systems, and software engineering. In general, information technology includes any expertise that helps create, modify, store, manage, or communicate information. It encompasses networking, systems management, program development, computer hardware, interface design, information assurance, systems integration, database management, and Web technologies. Information technology places a special emphasis on solving user issues, including helping everyone learn to use computers in a secure and socially responsible way.

KEY TERMS AND CONCEPTS

- **Application Software:** Software, such as word processors, used by individuals at home and at work.
- **Cloud Computing:** Software developed to execute on multiple servers and delivered by a Web service.
- **Computer Hardware:** Computing devices, networking equipment, and peripherals, including smartphones, digital cameras, and laser printers.
- **Human-Computer Interaction:** Ways in which humans use or interact with computer systems.
- **Information Assurance:** Discipline that not only secures information but also guarantees its safety and accuracy.
- **Information Management:** Storage, processing, and display of information that supports its use by individuals and businesses.
- **Networking:** Sharing of computer data and processing through hardware and software.

- **Programming:** Development of customized applications, using a special language such as Java.
- **Software Integration:** Process of building complex application programs from components developed by others.
- **Systems Integration:** Technique used to manage the computing systems of companies that use multiple types of software, such as an Oracle database and Microsoft Office.
- **Web Systems:** Database-driven Web applications, such as an online catalog, as well as simple Web sites.

DEFINITION AND BASIC PRINCIPLES

Information technology is a discipline that stresses systems management, use of computer applications, and end-user services. Although information technology professionals need a good understanding of networking, program development, computer hardware, systems development, software engineering, Web site design, and database management, they do not need as complete an understanding of the theory of computer science as computer scientists do. Information technology professionals need a solid understanding of systems development and project management but do not need as extensive a background in this as information systems professionals. In contrast, information technology professionals need better interpersonal skills than computer science and information systems workers, as they often do extensive work with end-users.

During the second half of the twentieth century, the world moved from the industrial age to the computer age, culminating in the development of the World Wide Web in 1990. The huge success of the Web as a means of communications marked a transition from the computer age, with an emphasis on technology, to the information age, with an emphasis on how technology enhances the use of information. The Web is used in many ways to enhance the use and transfer of information, including telephone service, social networking, e-mail, teleconferencing, and even radio and television programs. Information technology contains a set of tools that make it easier to create, organize, manage, exchange, and use information.

BACKGROUND AND HISTORY

The first computers were developed during World War II as an extension of programmable calculating machines. John von Neumann added the stored program concept in 1944, and this set off the explosive growth in computer hardware and software during the remainder of the twentieth century. As computing power increased, those using the computers began to think less about the underlying technology and more about what the computers allowed them to do with data. In this way, information, the organization of data in a way that facilitates decision making, became what was important, not the technology that was used to obtain the data. By 1984, organizations such as the Data Processing Management Association introduced a model curriculum defining the training required to produce professionals in information systems management, and information systems professionals began to manage the information of government and business.

In 1990, Computer scientist Tim Berners-Lee developed a Web browser, and the Web soon became the pervasive method of sharing information. In addition to businesses and governmental agencies, individuals became extensive users and organizers of information, using applications such as Google (search engine) and Facebook (social networking). In the early 2000's, it became clear that a new computer professional, specializing in information management, was needed to complement the existing information systems professionals. By 2008, the Association of Computing Machinery and the IEEE Computer Society released their recommendations for educating information technology professionals, authenticating the existence of this new computing field. The field of information technology has become one of the most active areas in computer science.

HOW IT WORKS

The principal activity of information technology is the creation, storage, manipulation, and communication of information. To accomplish these activities, information technology professionals must have a background from a number of fields and be able to use a wide variety of techniques.

Networking and Web Systems. Information is stored and processed on computers. In addition, information is shared by computers over networks. Information technology professionals need to have a good working knowledge of computer and network systems to assist them in acquiring, maintaining, and managing these systems. Of the many tasks performed by information technology professionals, none is more important than installing, updating, and training others to use applications software. The Web manages information by storing it on a server and distributing it to individuals using a browser, such as Internet Explorer or Foxfire. Building Web sites and applications is a big part of information technology and promises to increase in importance as more mobile devices access information provided by Web services.

Component Integration and Programming. Writing programs in a traditional language such as Java will continue to be an important task for information technology professionals, as it had been for other computer professionals in the past. However, building new applications from components using several Web services appears to be poised to replace traditional programming. For all types of custom applications, the creation of a user-friendly interface is important. This includes the careful design of the screen elements so that applications are easy to use and have well thought-out input techniques, such as allowing digital camera and scanner input at the appropriate place in a program, as well as making sure the application is accessible to visually and hearing impaired people.

Databases. Data storage is an important component of information technology. In the early days of computers, most information was stored in files. The difficulty of updating information derived from a file led to the first database systems, such as the information management system created by IBM in the 1960's. In the 1970's, relational databases became the dominant method of storing data, although a number of competing technologies are also in use. For example, many corporate data are stored in large spreadsheets, many personal data are kept in word-processing documents, and many Web data are stored directly in Web pages.

Information Security, Privacy, and Assurance. Regardless of how information is collected, stored, processed, or transferred, it needs to be secure. Information security techniques include developing security plans and policies, encrypting sensitive data during storage and transit, having good incident response and recovery teams, installing adequate security

Fascinating Facts About Information Technology

- In 1993, Intel introduced the Pentium processor as a Cray (supercomputer) on a chip. It contained 3.1 million transistors and was capable of performing 112 million instructions per second. The Pentium was the first of a number of low-cost but powerful processors that changed information technology.
- The iPhone, introduced in 2007, is a powerful and popular Internet-enabled smartphone. It allows users to access e-mail, the Web, and many cloud applications in a touch-friendly environment and promises to change the way people access Internet content.
- In 1981, IBM introduced a personal computer, which was copied by many other companies. Soon businesses, government agencies, and individuals were able to use computers to improve their productivity.
- Microsoft introduced Windows 3.1 in 1992, and its immediate popularity made Microsoft Windows the most popular operating system for personal computers.
- In 1980, Edgar Codd published "A Relational Model of Data for Large Shared Data Banks," a paper that defined the relational data model and led to the development of relational databases such as Oracle and DB2, which became the dominant storage technology for information technology.
- In 1989, Tim Berners-Lee introduced HTML (HyperText Markup Language) as a way to represent data on the Web. Later, Java and JavaScript were added to make programming on the Web almost as rich as programming on the desktop.
- Cloud computing, or providing software as a set of services to a computing device, is predicted to be the primary software model for the future. The origins of cloud computing go back to the early days of distributed computing, but exactly who should get credit for this exciting new area is not clear.

Increasingly, organizations want information handled in such a way that the organization can be assured that the data are accurate and have not been compromised.

Professionalism. Information technology professionals need to inform themselves of the ethical principles of their field (such as the Association of Computing Machinery' s code of ethical and professional conduct) and conscientiously follow this code. Increasingly, laws are being passed about how to handle information, and information technology professionals need to be aware of these laws and to follow them.

APPLICATIONS AND PRODUCTS

There are more useful applications of information technology than can be enumerated, but they all involve making information accessible and usable.

End-User Support. One of the most important aspects of information technology is its emphasis on providing support for a wide variety of computer users. In industrial, business, and educational environments, user support often starts immediately after someone is given access to a computer. An information technology professional assists the user with the login procedure, shows him or her how to use e-mail and various applications. After this initial introduction, technology professionals at a help desk answer the user's questions. Some companies also provide training courses. When hardware and software problems develop, users contact the information technology professionals in the computer support department for assistance in correcting their problems.

Information technology educational programs usually cover the theory and practice of end-user support in their general courses, and some offer courses dedicated to end-user support. This attention to end-user support is quite different from most programs in computer science, information systems, or software engineering and is one of the most important differences between information technology programs and other areas of computer science.

The Electronic Medical Record. A major use of information technology is in improving the operation of hospitals in providing medical services, maintaining flexible schedules, ensuring more accurate billing, and reducing the overall cost of operation. In much the same way, information technology helps doctors in clinics and individual practices improve the quality of their care, scheduling, billing,

controls, and providing security training for all levels of an organization. In addition to making sure that an organization's data are secure, it is also important to operate the data management functions of an organization in such a way that each individual's personal data are released only to authorized parties.

and operations. One of the early successes in using computers in medicine was the implementation of e-prescribing. Doctors can easily use the Internet to determine the availability of a drug needed by a patient, electronically send the prescription to the correct pharmacy, and bill the patient, or their insurance company, for the prescription. Another success story for medical information technology is in the area of digital imaging. Most medical images are created digitally, and virtually all images are stored in a digital format. The Digital Imaging and Communications in Medicine (DICOM) standard for storage and exchange of digital images makes it possible for medical images to be routinely exchanged between different medical facilities. Although the process of filling prescriptions and storing images is largely automated, the rest of medicine is still in the process of becoming fully automated. However, computers are likely to become as much a part of medical facilities as they are of financial institutions.

One of the major goals of information technology is the development of a true electronic medical record (EMR). The plan is to digitize and store all the medical information created for an individual by such activities as filling out forms, taking tests, and receiving care. This electronic medical record would be given to individuals in its entirety and to hospital and insurance companies as needed. Electronic medical records would be available to the government for data mining to determine health policies. Many professionals in information technology are likely to be involved in the collection, storage, and securitization of electronic medical records.

Geographic Information Management. Maps have been used by governments since the beginning of recorded history to help with the process of governing. Geographic information systems (GIS) are computerized systems for the storage, retrieval, manipulation, analysis, and display of geographic data and maps. In the 1950's, the first GIS were used by government agencies to assist in activities such as planning, zoning, real estate sales, and highway construction. They were developed on mainframes and accessed by government employees on behalf of those desiring the information contained in the systems. The early GIS required substantial computing resources because they used large numbers of binary maps but were relatively simple as information retrieval systems.

Geographic information systems still serve the government, but they are also used by industry and educational institutions and have gained many applications, including the study of disease, flood control, census estimates, and oil discovery. GIS information is easily accessed over the Internet. For example, zoning information about Dallas, Texas, is readily available from the city's Web site. Many general portals also provide GIS information to the public. For example, Google Maps allows travelers to print out their route and often provides a curbside view of the destination. Modern geographic information systems are complex, layered software systems that require expertise to create and maintain.

Network Management. Computer networking developed almost as quickly as the computing field itself. With the rapid acceptance of the Internet in the 1990's, computer connectivity through networking became as common as the telephone system. As computer networks developed, a need also developed for professionals to manage these networks. The first network managers were generally hardware specialists who were good at running cable through a building, adding hardware to a computer, and configuring network software so that it would work. Modern network specialists need all these early capabilities but also must be network designers, managers, and security experts.

Database Management. One of the major functions of information technology is the storage of information. Information consists of data organized for meaningful usage. Both data and information regarding how the data are related can be stored in many ways. For example, many corporate data are stored in word-processing documents, spreadsheets, and e-mails. Even more corporate data are stored in relational databases such as Oracle. Most businesses, educational institutions, and government agencies have database management specialists who spend most of their time determining the best logical and physical models for the storage of data.

Mobile Computing and the Cloud. Mobile phones have become so powerful that they rival small computers, and Web services and applications for computers or mobile devices are being developed at a breathtaking rate. These two technological developments are working together to provide many Web applications for smartphones. For example, any cell phone can play MP3 music. Literally dozens of Web services have been developed that will automatically download songs to cell phones and bill the owner a small fee. In addition, many smartphone applications will download and play songs from Web services.

Cloud computing and storage is beginning to change how people use computers in their homes. Rather than purchasing computers and software, some home users are paying for computing as a service and storing their data in an online repository. Microsoft's Live Office and Google's Apps are two of the early entrants in the software and storage as a service market.

Computer Integrated Manufacturing. Computer integrated manufacturing (CIM) provides complete support for the business analysis, engineering, manufacturing, and marketing of a manufactured item, such as a car. CIM has a number of key areas including computer-aided design (CAD), supply chain management, inventory control, and robotics.

Many of the areas of CIM require considerable use of information technology. For example, CAD programs require a good data management program to keep track of the design changes, complex algorithms to display the complicated graphical images, fast computer hardware to display the images quickly, and good project management tools to assist in completing the project on time. Robots, intelligent machines used to build products, require much innovative computer and machine hardware, very complex artificial intelligence algorithms simulating human operation of machine tools, and sophisticated computer networks for connecting the robotic components on the factory floor.

Computer Security Management. One of the most active areas of information technology is that of computer security management. As theoretical computer scientists develop new techniques to protect computers, such as new encryption algorithms, information technology specialists work on better ways to implement these techniques to protect computers and computer networks. Learning how to acquire the proper hardware and software, to do a complete risk analysis, to write a good security policy, to test a computer network for vulnerabilities, and to accredit a business's securing of its computers requires all the talents of an information technologist.

Web Site Development. The World Wide Web was first introduced in 1990, and since that time, it has become one of the most important information distribution technologies available. Information technology professionals are the backbone of most Web site development teams, providing support for the setup and maintenance of Web servers, developing the HTML pages and graphics that provide the Web content, and assisting others in a company in getting their content on the Web.

IMPACT ON INDUSTRY

The use of computers became pervasive in companies, government agencies, and universities in the second half of the twentieth century. Information technology gradually developed as a separate discipline in the late 1970's. By 1980, companies, government agencies, and universities recognized that it was the information provided by the computer that was important, not the computers themselves. Although access to and the ability to use the information was what people wanted, getting the information was difficult and required a number of skills. These included the building and maintenance of computers and networks, the development of computer software, providing end-user support, and using management information and decision support systems to add value to the information. Government and industry made their information needs known; universities and some computer manufacturers did basic and applied research developing hardware and software to meet these needs, and information technology professionals provided support to allow those in government and industry to actually use the information.

Although estimates of the total value of the goods and services provided by information technology in a year are impossible to determine, it is estimated to be close to $1 trillion per year.

Government, Industry, and University Research. The United States government exerts a great influence on the development of information technology because it purchases a large amount of computer hardware and software and makes substantial use of the information derived from its computers and networks. Companies such as IBM, SAP, and Oracle have geared much of their research and development toward meeting the needs of their government customers. For example, IBM's emphasis on COBOL for its mainframes was a result of a government requirement that all computers purchased by the government support COBOL compilers. SAP's largest customer base in the United States is federal, state, and local government agencies. One of the main reasons for Oracle's acquisition of PeopleSoft was to enhance its position in the educational enterprise resource planning market.

The National Science Foundation, the National Institutes of Health, and the U.S. Department of Defense have always provided support for university research for information technology. For example, government agencies supporting cancer research gave a great deal of support to Oracle and to university researchers in the development of a clinical research data management system to better track research on cancer. The government also has been a leader in developing educational standards for government information technology professionals, and these standards have greatly influenced university programs in information technology. For example, after the September 11, 2001, terrorist attacks on the United States, the National Security Agency was charged with coming up with security standards that all government employees needed to meet. The agency created an educational branch, the Committee on National Security Systems, which defined the standards and developed a set of security certifications to implement the standards. Its certifications have had a significant influence on the teaching of computer security.

The computer industry has been very active in developing better hardware and software to support information technology. For example, one of the original relational database implementations, System R and SQL/Data System, was developed at the IBM Research Laboratory in San Jose, California, in the 1970's, and this marked the beginning of the dominance of relational databases for data storage. There have also been a number of successful collaborations between industry and universities to improve information technology. For example, the Xerox PARC (later PARC) laboratories, established as a cooperative effort between Xerox and several universities, developed the Ethernet, the graphical user interface, and the laser printer. Microsoft has one of the largest industry research programs, which supports information technology development centers in Redmond, Washington; Cambridge, England; and six other locations. In 2010, Microsoft announced Azure, its operating system to support cloud computing. Azure will allow individuals to access information from the cloud from almost anywhere on a wide variety of devices, providing information on demand.

Information Technology Companies. Many successful large information technology companies are operating in the United States. Dell and HP (Hewlett Packard) concentrate on developing the computing hardware needed to store and process information. IBM produces both the hardware and software needed to store and process information. In the early 2010's, Microsoft produced more information technology software than any other company, but SAP was reported to have the greatest worldwide sales volume. Oracle, the largest and most successful database company, was the leader in providing on-site storage capacity for information, but Microsoft and Google had begun offering storage as a service over the Internet, a trend predicted to continue. Smaller but important information technology companies include Adobe Systems, which developed portable document format (PDF) files for transporting information and several programs to create PDF files.

CAREERS AND COURSE WORK

The most common way to prepare for a career in information technology is to obtain a degree in information systems. Students begin with courses in mathematics and business and then take about thirty hours of courses on information systems development. Those getting degrees in information systems often take information technology jobs as systems analysts, data modelers, or system managers, performing tasks such as helping implement a new database management system for a local bank.

A degree in computer science or software engineering is another way to prepare for an information technology career. The courses in ethics, mathematics, programming, and software management that are part of these majors provide background for becoming an information technology professional. Those getting degrees in these areas often take information technology jobs as programmers or system managers, charged with tasks such as helping write a new program to calculate the ratings of a stock fund.

Degree programs in information technology are available at some schools. Students begin with courses in problem solving, ethics, communications, and management, then take about thirty-six hours of courses in programming, networking, human-computer communications, databases, and Web systems. Those getting an information technology degree often take positions in network management, end-user support, database management, or data modeling. A possible position would be a network manager for a regional real estate brokerage.

In addition to obtaining a degree, many information technology professionals attend one of the many professional training and certification programs. One of the original certification programs was the Novell Certified Engineer, followed almost immediately by a number of Microsoft certification programs. These programs produced many information technology professionals for network management. Cisco has created a very successful certification program for preparing network specialists in internetworking (connecting individual local-area networks to create wide-area networks).

SOCIAL CONTEXT AND FUTURE PROSPECTS

The future of information technology is bright, with good jobs available in end-user support, network management, programming, and database management. These areas, and the other traditional areas of information technology, are likely to remain important. Network management and end-user support both appear to be poised for tremendous growth over the next few years. The growth in the use of the Internet and mobile devices requires the support of robust networks, and this, in turn, will require a large number of information technology professionals to install, repair, update, and manage networks. The greater use of the Internet and mobile devices also means that there will be a large number of new, less technically aware people trying to use the Web and needing help provided by information technology end-user specialists.

A large number of new applications for information technology are being developed. One of the areas of development is in medical informatics. This includes the fine-tuning of existing hospital software systems, development of better clinical systems, and the integration of all of these systems. The United States and many other nations are committed to the development of a portable electronic health record for each person. The creation of these electronic health records is a massive information technology project and will require a large workforce of highly specialized information technology professionals. For example, to classify all of the world's medical information, a new language, Health Level 7, has

been created, and literally thousands of specialists have been encoding medical information into this language.

The use of mobile devices to access computing from Web services is another important area of information technology development. Many experts believe the use of Web-based software and storage, or cloud computing, may be the dominant form of computing in the future, and it is likely to employ many information technology professionals.

Another important area of information technology is managing information in an ethical, legal, and secure way, while ensuring the privacy of the owners of the information. Security management specialists must work with network and database managers to ensure that the information being processed, transferred, and stored by organizations is properly handled.

George M. Whitson III, B.S., M.S., Ph.D.

FURTHER READING

Miller, Michael. *Cloud Computing: Web-Based Applications That Change the Way You Work and Collaborate.* Indianapolis, Ind.: Que, 2009. Provides an excellent overview of the new method of using computing and storage on demand over the cloud.

Reynolds, George. *Information Technology for Managers.* Boston: Course Technology, 2010. A short but very complete introduction to information technology by one of the leading authors of information technology books.

Senn, James. *Information Technology: Principles, Practices, and Opportunities.* 3d ed. Upper Saddle River, N.J.: Prentice Hall, 2004. An early information technology book that emphasizes the way information technology is used and applied to problem solving.

Shneiderman, Ben, and Catherine Plaisant. *Designing the User Interface.* Boston: Addison-Wesley, 2010. An excellent book about developing good user interfaces. In addition to good coverage on developing interfaces for desktop applications, it covers traditional Web applications and applications for mobile devices.

Stair, Ralph, and George Reynolds. *Principles of Information Systems.* 9th ed. Boston: Course Technology, 2010. Contains not only a complete introduction to systems analysis and design but also a very good description of security, professionalism, Web technologies, and management information systems.

WEB SITES

Association for Computing Machinery
http://www.acm.org

CompTIA
http://www.comptia.org/home.aspx

IEEE Computer Society
http://sites.computer.org/ccse

Information Technology Industry Council
http://www.itic.org

TechAmerica
http://www.itaa.org

See also: Communication; Computer-Aided Design and Manufacturing; Computer Networks; Computer Science; Human-Computer Interaction; Internet and Web Engineering; Software Engineering; Telephone Technology and Networks; Wireless Technologies and Communication.

INTEGRATED-CIRCUIT DESIGN

FIELDS OF STUDY

Mathematics; physics; electronics; quantum mechanics; mechanical engineering; electronic engineering; graphics; electronic materials science.

SUMMARY

The integrated circuit (IC) is the essential building block of modern electronics. Each IC consists of a chip of silicon upon which has been constructed a series of transistor structures, typically MOSFETs, or metal oxide semiconductor field effect transistors. The chip is encased in a protective outer package whose size facilitates use by humans and by automated machinery. Each chip is designed to perform specific electronic functions, and the package design allows electronics designers to work at the system level rather than with each individual circuit. The manufacture of silicon chips is a specialized industry, requiring the utmost care and quality control.

KEY TERMS AND CONCEPTS

- **Clock:** A multitransistor structure that generates square-wave pulses at a fixed frequency; used to regulate the switching cycles of electronic circuitry.
- **Fan-Out:** The number of transistor inputs that can be served by the output from a single transistor.
- **Flip-Flop:** A multitransistor structure that stores and maintains a single logic state while power is applied or until it is instructed to change.
- **Vibrator:** A multitransistor structure that outputs a constant sinusoidal pulse by switching between two unstable states, rather than by modifying an input signal.

DEFINITION AND BASIC PRINCIPLES

An integrated circuit, or IC, is an interconnected series of transistor structures assembled on the surface of a silicon chip. The purpose of the transistor assemblages is to perform specific operations on electrical signals that are provided as inputs. All IC devices can be produced from a structure of four transistors that function to invert the value of the input signal, called NOT gates or inverters. All digital electronic circuits are constructed from a small number of different transistor assemblages, called gates, that are built into the circuitry of particular ICs. The individual ICs are used as the building blocks of the digital electronic circuitry that is the functional heart of modern digital electronic technology.

The earliest ICs were constructed from bipolar transistors that function as two-state systems, a system high and a system low. This is not the same as "on" and "off," but is subject to the same logic. All digital systems function according to Boolean logic and binary mathematics.

Modern ICs are constructed using metal oxide semiconductor field effect transistor (MOSFET) technology, which has allowed for a reduction in size of the transistor structures to the point in which, at about 65 nanometers (nm) in size, literally millions of them can be constructed per square centimeter (cm^2) of silicon chip surface. The transistors function by conducting electrical current when they are biased by an applied voltage. Current ICs, such as the central processing units (CPUs) of personal computers, can operate at gigahertz frequencies, changing that state of the transistors on the IC billions of times per second.

BACKGROUND AND HISTORY

Digital electronics got its start in 1906, when American inventor Lee de Forest constructed the triode vacuum tube. Large, slow, and power-hungry as they were, vacuum tubes were nevertheless used to construct the first analogue and digital electronic computers. In 1947, American physicist William Shockley and colleagues constructed the first semiconductor transistor junction, which quickly developed into more advanced silicon-germanium junction transistors.

Through various chemical and physical processes, methods were developed to construct small transistor structures on a substrate of pure crystalline silicon. In 1958, American physicist and Nobel laureate Jack Kilby first demonstrated the method by constructing germanium-based transistors as an IC chip, and American physicist Robert Noyce constructed the first silicon-based transistors as an IC chip. The transistor structures were planar bipolar in nature, until the CMOS (complementary metal oxide semi-conductor) transistor was invented in 1963. Methods for producing CMOS chips efficiently were not developed for another twenty years.

Transistor structure took another developmental leap with the invention of the field effect transistor (FET), which was both a more efficient design than that of semiconductor junction transistors and easier to effectively manufacture. The MOSFET structure also is amenable to miniaturization and has allowed designers to engineer ICs that have one million or more transistor structures per centimeters squared.

How It Works

The electronic material silicon is the basis of all transistor structures. It is classed as a pure semiconductor. It is not a good conductor of electrical current or insulate well against electrical current. By adding a small amount of some impurity to the silicon, its electrical properties can be manipulated such that the application of a biasing voltage to the material allows it to conduct electrical current. When the biasing voltage is removed, the electrical properties of the material revert to their normal semiconductive state.

Silicon Manufacture. Integrated-circuit design begins with the growth of single crystals of pure silicon. A high-purity form of the material, known as polysilicon, is loaded into a furnace and heated to melt the material. At the proper stage of melt, a seed crystal is attached to a slowly turning rod and introduced to the melt. The single crystal begins to form around the seed crystal and the rotating crystal is "pulled" from the melt as it grows. This produces a relatively long, cylindrical, single crystal that is then allowed to cool and set.

Wafers. From this cylinder, thin wafers or slices are cut using a continuous wire saw that produces several uniform slices at the same time. The slices are subjected to numerous stages of polishing, cleaning, and quality checking, the end result of which is a consistent set of silicon wafers suitable for use as substrates for integrated circuits.

Circuitry. The integrated circuit itself begins as a complex design of electronic circuitry to be constructed from transistor structures and "wiring" on the surface of the silicon wafer. The circuits can be no more than a series of simple transistor gates (such as invertors, AND-gates, and OR-gates), up to and

Raymond Armstrong, an engineering program manager, holds a microprocessor integrated circuit in the Colorado Springs, Colorado offices of TAEUS, a company that specializes in reverse engineering and patent research. (AP Photos)

including the extremely complex transistor circuitry of advanced CPUs for computers.

Graphics technology is extremely important in this stage of the design and production of integrated circuits, because the entire layout of the required transistor circuitry must be imaged. The design software also is used to conduct virtual tests of the circuitry before any ICs are made. When the theoretical design is complete and imaged, the process of constructing ICs can begin.

Because the circuitry is so small, a great many copies can be produced on a single silicon wafer. The actual chips that are housed within the final polymer or ceramic package range in size from two to five cm^2. The actual dimensions of the circa 1986 Samsung KS74AHCT240 chip, for example, are just 1 cm x 2 cm. The transistor gate sizes used in this chip are 2 micrometers (um) (2×10^{-6} meters [m]), and each chip contains the circuitry for two octal buffers, constructed from hundreds of transistor gate structures. Transistor construction methods have become much more efficient, and transistor gate sizes are now measured in nanometers (10^{-9} m) rather than um, so that actual chip sizes have also become much smaller, in accord with Moore's law. The gate structures are connected through the formation of aluminum "wires" using the same chemical vapor deposition methodology used to form the silicon oxide and other layers needed.

Photochemical Etching. The transistor structures of the chip are built up on the silicon wafer substrate through a series of steps in which the substrate is photochemically etched and layers of the necessary materials are deposited. Absolute precision and an ultraclean environment are required at each step. The processes are so sensitive that any errant speck of dust or other contaminant that finds its way to the wafer's surface renders that part of the structure useless.

Accounting for losses of functional chips at each stage of the multistep process, it is commonly the case that as little as 10 percent of the chips produced from any particular wafer will prove viable at the end of their construction. If, for example, the procedure requires one hundred individual steps, not including quality-testing steps, to produce the final product, and a mere 2 percent of the chips are lost at each step, then the number of viable chips at the end of the procedure will be 0.98^{100}, or 13.26 percent of the number of chips that could ideally have been produced.

Each step in the formation of the IC chip must be tested to identify the functionality of the circuitry as it is formed. Each such step and test procedure adds significantly to the final cost of an IC. When the ICs are completed, the viable ones are identified, cut from the wafer, and enclosed within a protective casing of resin or ceramic material. A series of leads are also built into this "package" so that the circuitry of the IC chip can be connected into an electronic system.

APPLICATIONS AND PRODUCTS

Bipolar Transistors and MOSFETS. Transistors are commonly pictured as functioning as electronic on-off switches. This view is not entirely correct. Transistors function by switching between states according to the biasing voltages that are applied. Bipolar switching transistors have a cut-off state in which the applied biasing voltage is too low to make the transistor function. The normal operating condition of the transistor is called the linear state. The saturation state is achieved when the biasing voltage is applied to both poles of the transistor, preventing them from functioning. MOSFET transistors use a somewhat different means, relying on the extent of an electric field within the transistor substrate, but the resulting functions are essentially the same.

The transistor structures that form the electronic circuitry of an IC chip are designed to perform specific functions when an electrical signal is introduced. For simple IC circuits, each chip is packaged to perform just one function. An inverter chip, for example, contains only transistor circuitry that inverts the input signal from high to low or from low to high. Typically, six inverter circuits are provided in each package through twelve contact points. Two more contact points are provided for connection to the biasing voltage and ground of the external circuit. It is possible to construct all other transistor logic gates using just inverter gates. All ICs use this same general package format, varying only in their size and the number of contact points that must be provided.

MOSFETS have typical state switching times of something less than 100 nanoseconds, and are the transistor structures of choice in designing ICs, even though bipolar transistors can switch states faster. Unlike bipolars, however, MOSFETS can be constructed and wired to function as resistors and can be made to a much smaller scale than true resistors in the normal production process. MOSEFTS are easier to manufacture than are bipolars as they can be made much smaller in VLSI (very large scale integration) designs. MOSFETS also cost much less to produce.

NOTs, ANDs, ORs, and Other Gates. All digital electronic devices comprise just a few basic types of circuitry—called logic elements—of which there are two basic types: decision-making elements and storage elements. All logic elements function according to the Boolean logic of a two-state (binary) system. The only two states that are allowed are "high" and "low," representing an applied biasing voltage that either does or does not drive the transistor circuitry to function. All input to the circuitry is binary, as is all output from the circuitry.

Decision-making functions are carried out by logic gates (the AND, OR, NOT gates and their constructs) and memory functions are carried out by combination circuitry (flip-flops) that maintains certain input and output states until it is required to change states. All gates are made up of a circuit of interconnected transistors that produces a specific output according to the input that it receives.

A NOT gate, or inverter, outputs a signal that is the opposite of the input signal. A high input produces a low output, and vice versa. The AND gate outputs a high signal only when all input signals are high, and a low output signal only if any of the input signals are low. The OR gate functions in the opposite sense, producing a high output signal if any of the input signals are high and a low output signal only when all of the input signals are low. The NAND gate, which can be constructed from either four transistors and one diode or five diodes, is a combination of the AND and NOT gates. The NOR gate is a combination of the OR and NOT gates. These, and other gates, can have any number of inputs, limited only by the fan-out limits of the transistor structures.

Sequential logic circuits are used for timing, sequencing, and storage functions. The flip-flops are the main elements of these circuits, and memory functions are their primary uses. Counters consist of a series of flip-flops and are used to count the number of applied input pulses. They can be constructed to count up or down, as adders, subtracters, or both. Another set of devices called shift registers maintains a memory of the order of the applied input pulses, shifting over one place with each input pulse. These can be constructed to function as series devices, accepting one pulse (one data bit) at a time in a linear fashion, or as parallel devices, accepting several data bits, as bytes and words, with each input pulse. The devices also provide the corresponding output pulses.

Operational amplifiers, or OP-AMPs, represent a special class of ICs. Each OP-AMP IC contains a self-contained transistor-based amplifying circuit that provides a high voltage gain (typically 100,000 times or more), a very high input impedance and low output impedance, and good rejection of common-mode signals (that is, the presence of the same signal on both input leads of the OP-AMP).

Combinations of all of the gates and other devices constructed from them provide all of the computing power of all ICs, up to and including the most cutting-edge CPU chips. Their manufacturing process begins with the theoretical design of the desired functions of the circuitry. When this has been determined, the designer minimizes the detailed transistor circuitry that will be required and develops the corresponding mask and circuitry images that will be required for the IC production process. The resulting ICs can then be used to build the electronic circuitry of all electronic devices.

IMPACT ON INDUSTRY

Digital-electronic control mechanisms and appliances are the mainstays of modern industry, mainly because of the precision with which processes and

---

Fascinating Facts About Integrated-Circuit Design

- The ENIAC computer, built in 1947, weighed more than 30 tons; used 19,000 vacuum tubes, 1,500 relays, about one-half-million resistors, capacitors, and inductors; and consumed almost 200 kilowatts of electricity per hour.
- In 1997, the ENIAC computer was reproduced on a single triple-layer CMOS chip that was 7.44 millimeters (mm) x 5.29 mm in size and contained 174,569 transistors.
- In 1954, RAND scientists predicted that by the year 2000, some homes in the United States could have their own computer room, including a teletype terminal for FORTRAN programming, a large television set for graphic display, rows of blinking lights, and a double "steering wheel."
- The point transistor was invented in 1947, and the first integrated circuits were demonstrated in 1950 by two different researchers, leading directly to the first patent infringement litigation involving transistor technology.
- The flow of electricity in a bipolar transistor, and hence its function, is controlled by a bias voltage applied to one or the other pole.
- The flow of electricity, and hence its function, of a MOSFET is controlled by the extent of the electrical field formed within the substrate materials.
- Silicon wafers sliced from cylindrical single crystals of silicon undergo dozens of production steps before they are suitable for use as substrates for ICs.
- The production of ICs from polished silicon wafers can take hundreds of steps, after which as few as 10 percent of the ICs will be usable.

---

quality can be monitored through digital means. Virtually all electronic systems now operate using digital electronics constructed from ICs.

Digital electronics has far-reaching consequences. For example, in September, 2011, the telecommunications industry in North America ceased using analogue systems for broadcasting and receiving commercial signals. This development can be considered a continuation of a process that began with the invention of the Internet, with its entirely digital protocols and access methods. As digital transmission methods and infrastructure have developed to facilitate the transmission of digital communications between computer systems, the feasibility of supplanting traditional analogue communications systems with their digital counterparts became apparent. Electronic technology and the digital format enable the transmission of much greater amounts of information within the same bandwidth, with much lower energy requirements.

Modern industry without the electronic capabilities provided by IC design is almost inconceivable, from both the producer and the consumer points of view. The production of ICs themselves is an expensive proposition because of the demands for ultraclean facilities, special production methods, and the precision of the tools required for producing the transistor structures.

The transistors are produced on the surface of silicon wafers through photoetching and chemical-vapor-deposition methodologies. Photoetching requires the use of specially designed masks to provide the shape and dimensions of the transistors' component structures. As transistor dimensions have decreased from 35 um to 65 nm, the cost of the masks has risen exponentially, from about $100,000 to more than $3 million. At the same time, however, semiconductor industry revenues have increased from about $20 million in 1955 to more than $200 billion in the early twenty-first century, and they continue to increase at a rate of about 6 percent per year. This is mirrored by the amount of investment in research and development. In 2006, for example, such investment amounted to some $19.9 billion.

The consumer aspect of IC design is demonstrated by the market proliferation of consumer electronic devices and appliances. It seems as though everything, from compact fluorescent light bulbs and motion-sensing air fresheners to high-definition television monitors, personal computers, cellular telephones, and even automobiles depends on IC technology. This is a quantity whose value rivals the gross domestic product of the United States.

CAREERS AND COURSEWORK

IC design and manufacturing is a high-precision field. Students who are interested in pursuing a career working with devices constructed with ICs will require a sound basic education in mathematics and physics, as the basis for the study of electronics technology and engineering. As specialists in these fields, the focus will be on using ICs as the building blocks

of electronic circuits for numerous devices and appliances. A recognized college degree or the equivalent is the minimum requirement for qualification as an electronics technician or technologist. The minimum qualification for a career in electronics engineering is a bachelor's degree in that field from a recognized university. In addition, membership in professional associations following graduation, if maintained, will require regular upgrading of qualifications to meet standards and keep abreast of changing technology.

Electronics technology is an integral component of many technical fields, especially for automotive mechanics and transportation technicians, as electronic systems have been integrated into transportation and vehicle designs of all kinds.

For those pursuing a career in designing integrated circuitry and the manufacture of IC chips, a significantly more advanced level of training is required. Studies at this level include advanced mathematics and physics, quantum mechanics, electronic theory, design principles, and graphics. Integrated circuitry is cutting-edge technology, and the nature of ICs is likely to undergo rapid changes that will, in turn, require the IC designer to acquire knowledge of entirely new concepts of transistors and electronic materials. A graduate degree (master's or doctorate) from a recognized university will be the minimum qualification in this field.

SOCIAL CONTEXT AND FUTURE PROSPECTS

IC technology is on the verge of extreme change, as new electronic materials such as graphene and carbon nanotubes are developed. Research with these materials indicates that they will be the basic materials of molecular-scale transistor structures, which will be thousands of times smaller and more energy-efficient than VLSI technology based on MOSFETs. The computational capabilities of computers and other electronic devices are expected to become correspondingly greater as well.

Such devices will utilize what will be an entirely new type of IC technology, in which the structural features are actual molecules and atoms, rather than what are by comparison mass quantities of semiconductor materials and metals. As such, future IC designers will require a comprehensive understanding of both the chemical nature of the materials and of quantum physics to make the most effective use of the new concepts.

The scale of the material structures as well will have extraordinary application in society. It is possible, given the molecular scale of the components, that the technology could even be used to print ultra-high resolution displays and computer circuitry that would make even the lightest and thinnest of present-day appliances look like the ENIAC (Electronic Numerical Integrator and Computer) of 1947, which was the first electronic computer.

The social implications of such miniaturized technology are far-reaching. The RFID (radio-frequency identification) tag is now becoming an important means of embedding identification markers directly into materials. RFID tags are tiny enough to be included as a component of paints, fuels, explosives, and other materials, allowing identification of the exact source of the material, a useful implication for forensic investigations and other purposes. Even the RFID tag, however, would be immense compared with the molecular scale of graphene and nanotube-based devices that could carry much more information on each tiny particle.

The ultimate goal of electronic development, in current thought, is the quantum computer, a device that would use single electrons, or their absence, as data bits. The speed of such a computer would be unfathomable, taking seconds to carry out calculations that would take present-day supercomputers billions of years to complete. The ICs used for such a device would bear little resemblance, if any, to the MOSFET-based ICs used now.

Richard M. Renneboog, M.Sc.

FURTHER READING

Brown, Julian R. *Minds, Machines, and the Multiverse: The Quest for the Quantum Computer.* New York: Simon & Schuster, 2000. Provides historical insight and speculates about the future of computer science and technology.

Kurzweil, Ray. *The Age of Spiritual Machine: When Computers Exceed Human Intelligence.* New York: Penguin, 2000. Offers some speculations about computers and their uses through the twenty-first century.

Marks, Myles H. *Basic Integrated Circuits.* Blue Ridge Summit, Pa.: Tab Books, 1986. An introduction to functions and uses of integrated circuits.

Zheng, Y., et al. "Graphene Field Effect Transistors with Ferroelectric Gating." *Physical Review Letters* 105 (2010). Reports on the experimental development and successful testing of a graphene-based field-effect transistor system using gold and graphene electrodes with SiO_2 gate structures on a silicon substrate.

WEB SITES

U.S. Department of Commerce
http://www.bis.doc.gov/defenseindustrialbaseprograms

See also: Applied Mathematics; Applied Physics; Computer Engineering; Computer Science; Electrochemistry; Electronic Materials Production; Electronics and Electronic Engineering; Graphics Technologies; Liquid Crystal Technology; Nanotechnology; Quality Control; Surface and Interface Science; Transistor Technologies.

INTERNATIONAL SYSTEM OF UNITS

FIELDS OF STUDY

Metrology; agriculture; geography; history; sociology; ballistics; surveying; military applications and service; navigation; construction; civil engineering; mechanical engineering; chemical engineering; physics; biochemical analysis.

SUMMARY

The development of the metric system, which served as the basis of the International System of Units (Le Système International d'Unités; known as SI), occurred during the French Revolution in the mid-eighteenth century. This coincided with the beginning of the age of modern science, especially chemistry and physics, as the value of physical measurements in the conduct of those pursuits became apparent. As scientific activities became more precise and founded on sound theory, the common nature of science demanded an equally consistent system of units and measurements. The units in the SI have been defined by international accord to provide consistency in all fields of endeavor. The basic units are defined for only seven fundamental properties of matter. All other consistent units are derived as functions of these seven fundamental units.

KEY TERMS AND CONCEPTS

- **Base Unit:** Unit of measurement that is used for one of the seven fundamental properties; also known as defined unit.
- **Conversion Factor:** Ratio of one unit of measure in a particular system to the unit of measure of the same property in a different system of measurements.
- **Derived Unit:** Unit of measurement that is derived from the relationship of base units.
- **Dimensional Analysis:** Consideration and manipulation of the units involved in a measurement and their relationship to one another.
- **Fundamental Property:** Property ascribed to a phenomenon, substance, or body that can be quantified: length, mass, time, electric current, thermodynamic temperature, the amount of a substance, and luminous intensity.

- **International Standard:** Unit of measurement whose consistent value is recognized as a common unit of measurement by international agreement.
- **Measurement:** Ascribing a value to the determination of the quantity of a physical property.
- **Standard Temperature and Pressure (STP):** Value of the mean atmospheric pressure and mean annual temperature at sea level, chosen by international agreement to be designated as 1 standard atmosphere or 760 torr and 0 degrees Celsius/32 degrees Fahrenheit or 273.15 degrees Kelvin.
- **Unit:** Single definitive basic magnitude of portions used to measure various properties.

DEFINITION AND BASIC PRINCIPLES

The International System of Units (SI) is the internationally accepted standard system of measurement in use throughout the world. The units of the SI are ascribed to seven fundamental physical properties and two supplementary properties. These are length, mass, time, electric current, thermodynamic temperature, the amount of a substance, luminous intensity, and the magnitude of plane and solid angles.

Length is the extent of some physical structure or boundary in two dimensions, such as the distance from one point to another, how tall someone is, and the distance between nodes of a sinusoidal wave. The SI base unit associated with length is the meter.

Mass refers to the amount of material in a bulk quantity. The term is used interchangeably with weight, although the mass of any object remains constant while its weight varies according to the strength of the gravitational field to which it is subject. The SI base unit for mass is the gram.

Time is a rather more difficult property to define outside of itself, as it relates to the continuous progression of existence of some state through past, present, and future stages. The SI base unit of time is the second.

Electric current is the movement of electronic charge from one point to another, and it is ascribed by the SI base unit of the ampere. One ampere is defined as the movement of one coulomb of charge, as one mole of electrons, for a period of one second.

The amount of a substance is defined by an SI base unit called the mole. The term is essentially never used outside of the context of atoms, molecules, and certain subatomic particles such as electrons. A mole of any substance is the quantity of that substance containing the number of atoms or molecules as there are atoms in 12 grams of carbon-12; this number is referred to as Avogadro's number (or the Avogadro constant), which is 6.02214×10^{23} of atoms, molecules, or particles.

The term luminous intensity refers to the brightness, or the quantity, of light or other electromagnetic energy being emitted from a source. Initially luminous intensity was referred to by comparison to the light of a candle flame, but the variability of such a source is not conducive to standardization. The SI base unit for luminous intensity is the candela. This is about the amount of light emitted by a candle flame, but it has been standardized to mean an electromagnetic field strength of 0.00146 watts.

The derived unit known as the plane angle refers to the angular separation of two lines from a common point in a two-dimensional plane. The SI derived unit for plane angles is the radian. A complete rotation about a point origin is an angular displacement of 2π radians. The extension of this into three dimensions is known as the solid angle and is measured by the SI derived unit called the steradian.

Measurement of any property or quantity is accomplished by comparing the particular amount of the property or quantity to the amount represented by the standard unit. For example, an object that is 3.62 times as long as the distance defined as 1 meter is said to be 3.62 meters long. Similarly, an object that is proportionately 6,486 times as massive as the quantity defined as 1 gram is said to weigh or to have a mass of 6,486 grams. The use of standard reference units such as the SI units ensures that measurements and quantities have the same meaning in all places where that system is used.

BACKGROUND AND HISTORY

Traditionally, systems of measurement have employed units that related to various parts or proportions of the human body. Ready and convenient measuring tools in earlier times included the hand, the foot, the thumb, and the pace. Because these human measures tend to vary somewhat from person to person, no work that depended on measurements could be repeated exactly, even if the same person carried it out. Specialization in tasks eventually led to

the realization that standard units of measurement would be beneficial, and determination of certain units were established by royal decree as long ago as in the signing of the Magna Carta.

The Industrial Revolution in Europe and the growth of science as a common international pursuit drove the need for a unified system of measurement that would be independent of human variability and consistent from place to place. The International System of Units was developed and first put to use in about 1799. It represents the first real standardized system of measurements. Prior to this time, a broad variety of measurement systems were in use because many countries had developed their own measuring standards for use internally and in any territories that it held.

A need for a standardized system of measurement had been recognized by various luminaries and was proposed in 1670 by French scientist Gabriel Mouton. The incompatibility and variability of the many measurement systems in use at that time often resulted in unfair trade practices and power struggles. In 1790, the French Academy of Sciences was charged with developing a system of measurement that would be fixed and independent of any inconsistencies arising from human intervention. On April 7, 1795, the French National Assembly decreed the use of the meter as the standard unit of length. It was defined as being equal to one ten-millionth part of one quarter of the terrestrial meridian, specified by measurements undertaken between Dunkirk and Barcelona. The liter was later defined as being a unit of volume equal to the volume of a cube that is one-tenth of a meter on a side, and the kilogram as being equal to the mass of one liter of pure water. Larger and smaller quantities than these are indicated by the use of increasing and decreasing prefixes, all indicating units that were some power of ten times greater or lesser than the base units.

Standard reference models of the metric units were made and kept at the Palais-Royale, and it was then required that measuring devices being used for trade and commerce be regularly checked for accuracy against the official standard versions.

HOW IT WORKS

All measurement practices relate the actual properties and proportions of something to a defined standard that is relevant to that property or proportion.

For example, two-dimensional linear quantities are related to a standard of length, and the mass of an object is proportionately related to a standard unit of mass. In the measurement of length, for example, the standard unit of the IS is the meter. An object that is determined to be, for example, 6.3 times longer than the base unit of one meter is therefore said to be 6.3 meters long. Similarly, an object that is, say, 5.5 times heavier than the base unit of one kilogram is said to weigh 5.5 kilograms, and something that has a volume of 22.4 times that of the base unit of one liter is said to have a volume of 22.4 liters.

The modern metric system is simple to use, especially when compared with its most common predecessor systems, the British Imperial and the U.S. customary systems. Whereas these predecessor systems use units such as inches, feet, yards, ounces, and pounds that relate to one another irregularly, the metric system has always used the simple relation of factors of ten for all of its units. For example, the decameter is equal to ten meters (10×1 meter), while the decimeter is equal to one-tenth of a meter (0.1×1 meter). A complete series of prefixes indicates the size of the smaller and larger units that are being used in a measurement. Thus, a decimeter is ten times smaller than a meter, a centimeter is a hundred times smaller than a meter, a millimeter is a thousand times smaller, and so on. Correspondingly, there are ten millimeters in a centimeter, ten centimeters in a decimeter, and ten decimeters in a meter. There are one thousand milliliters in a liter, one hundred liters in a hectoliter, one thousand grams in a kilogram, and one thousand milligrams in a gram. The uniformity of the system readily allows one to visualize and estimate relative sizes and quantities.

The basic units of the SI have always been defined with the intention of relating to some universal and unalterable standard. In some cases, the definitions have been changed to relate the unit to a more permanent universal feature or a more accurately known property. For example, with the high accuracy of time measurement that has become available, the definition of the meter has been changed from being a specific fraction of the distance between two fixed terrestrial points to "the length of the path traveled by light in vacuum during a time interval of $1/299792458$ of a second."

The units of measurement of other properties and quantities reflect the fact that those properties and quantities can be viewed as combinations of the seven fundamental properties. That is, for any dimensional property Q, the dimensions of Q are derived from the expression:

$$\dim Q = L^{\alpha} M^{\beta} T^{\gamma} I^{\delta} \theta^{\varepsilon} N^{\zeta} J^{\eta}$$

The exponents a, b, c, d, e, f, and g are integers representing the degree of involvement of the corresponding property of length (L), mass (M), time (T), electric current (I), thermodynamic temperature (θ), amount of material (N), and luminous intensity (J). (Remember that any quantity raised to the power 0 has the numerical value of 1.)

APPLICATIONS AND PRODUCTS

Applications and products related to the metric system essentially fall into two categories: educational devices and training to promote familiarity with the system and metric versions of existing products and the tools required for their maintenance.

Educational Devices and Training. The long history of independent measurement systems has served to entrench those systems in common usage. Therefore, a very large body of materials, products, and devices have been constructed using nonmetric measurements. More significantly, those independent systems have been so deeply entrenched in the education systems of many nations that several generations of people have grown up using no other system. The familiarity gained through a lifetime of using a particular system generally gives individuals the ability to visualize and estimate quantities in terms of that measurement system. Making the change to a different measurement system such as the metric system represents a paradigm shift that leaves many individuals unable to associate the new measurement units with even the most familiar dimensions. Such a change, however, can be accommodated in a number of ways as the world continues to adopt the metric system of measurement as its universal standard.

The most basic method of replacing one system with another is to incorporate the new system into the basic education system, teaching it in such a manner that it becomes the entrenched measurement system as children progress through school. It is, therefore, important not only that teachers are educated in the use of the metric system but also that they actively replace their own reliance on any former system that they have used. This guarantees that, in time, the older system is completely displaced from the public lexicon and the metric system becomes the primary measurement system over a period of essentially one generation.

Those who have already left the school system can learn the metric system through training programs. Training in the metric system can be incorporated into professional development programs at the workplace or take the shape of formal training programs offered by local educational institutions or third-party providers.

Metric Versions of Existing Devices. The vast majority of goods and devices produced in countries that have not adopted the metric standard must be maintained using their original component dimensions because metric values and nonmetric values are not generally interchangeable. The fundamental difference between the two bases of measurement ensures that any coincidence of size from one system to the other is exactly that: coincidental. Therefore, switching to metric goods and devices requires that complete new lines of products be made with metric rather than nonmetric dimensions. This requirement places an odd sort of constraint on the situation because nonmetric parts and tools must still be produced in order to maintain existing nonmetric devices. At the same time, new devices to replace those that fail are produced in metric dimensions. This means that tradespeople, such as automobile mechanics, maintenance workers, and engineers, must obtain double sets of tools, and supply stores must maintain double sets of components and supplies. In addition, tools for taking measurements must be capable of using both metric and nonmetric dimensions, although the incorporation of electronic capabilities into many tools has greatly minimized the difficulties that arise from this dual requirement.

IMPACT ON INDUSTRY

In some countries that have officially adopted the metric system to replace a system such as the Imperial system, the change has been sudden and dramatic. For example, when Canada adopted the metric system in the 1970's, the changeover was quick and, at times, dramatic. The Imperial system of measurement was phased out and replaced by the metric system in a relatively short time, with the key feature being that after a certain date, Imperial units would no longer be used for any quantities that were subject to government regulation. On that date, gas stations

---

Fascinating Facts About International System Of Units

- On September 23, 1999, the Mars Climate Orbiter crashed on the planet Mars because engineers had neglected to take into account that some design data had been provided in metric units and some in U.S. customary units.
- The International System of Units is the first system of measurement based solely on scientific observation and physical characteristics that are not subject to change.
- Only three countries, the United States, Myanmar, and Liberia, still have not adopted the metric system.
- In the United States, the metric system is voluntary but preferred for trade and commerce.
- Demetrication (abandonment of the metric system) is taking place in some states and Canada. In 2005, the government of Ontario changed the secondary math curriculum to include Imperial measurements.

---

changed from price per gallon to price per liter, automobile speedometers were read in kilometers per hour, and Canadian schools taught using only the metric system. Average citizens were basically left on their own to learn and adapt to the new system. In many respects, it posed few hardships, because many goods were purchased by dollar amounts rather than unit quantities. Regulatory control of how goods were measured and delivered ensured that the consumer received the correct amount of what was purchased.

In other countries, such as the United States, the American democratic process has stalled conversion to the metric system. In 1800, the United States could have become the second nation in the world to adopt the SI because of its strong ties with France, where the SI was developed. It chose instead to adopt the U.S. customary system of measurement, which used units similar to the British Imperial system but sometimes varied in terms of quantities (an Imperial gallon held 160 fluid ounces, for example, while an American gallon held only 128 fluid ounces). The debate over adoption of the metric system in the United States continues, as many Americans oppose any official move to replace the customary system. As a result, the United States remains essentially the last major country in the world to maintain a system of measurement other than the metric system. In practical terms, the failure to completely adopt the metric system, even in the scientific and technical fields, has cost the country billions of dollars in lost trade and costly errors. For example, in 1999, a multimillion-dollar space probe crashed on Mars because the technical contractors had failed to notice that some of the data being used were in metric units while other data were in U.S. customary units.

CAREERS AND COURSE WORK

The International System of Units is not broad enough to provide a field of study or advanced course work and, by itself, is unlikely to form the basis of a career. However, during the process of converting from a nonmetric measurement system to a metric system, opportunities will arise for those familiar with SI to create and provide training to facilitate understanding and use of the metric system.

Anyone entering a technical or scientific field must become familiar with the metric system, as this has been the standard system of measurement in those areas since the SI was developed. As the metric system becomes more and more widely adopted and accepted, students and others should expect to carry out measurements and calculations using the appropriate metric system units.

Careers that depend specifically on measurement include quality-control engineering and most branches of scientific research and physical engineering. Specific examples of measurement-based careers include civil engineering, medical and biochemical analysis, analytical chemistry, metrology, mechanical engineering, and industrial chemical engineering.

SOCIAL CONTEXT AND FUTURE PROSPECTS

It seems likely that the metric system will ultimately become the standard system of measurement in the United States, as it already is in the rest of the world. As international trade and offshore manufacturing increase, the necessity for American industry to adopt the SI increases. In addition, educational systems have increased their focus on science and technology, and metric systems will become more familiar to children in American schools.

In the meantime, development of other measurement systems and hybrids of measuring systems progresses. For example, the manufacturing community has adopted a standard unit called the metric inch, which is the standard inch divided into hundredths, to be used on visual measuring devices. There is some logic to this adoption, as this measurement apparently corresponds with the limits of differentiation of which the human eye is capable to a better degree than does the millimeter division of the metric system. However, such developments are likely to delay and otherwise interfere with the adoption of the metric system.

Richard M. J. Renneboog, M.Sc.

FURTHER READING

Butcher, Kenneth S., Linda D. Crown, and Elizabeth J. Gentry. *The International System of Units (SI): Conversion Factors for General Use.* Gaithersburg, Md.: National Institute of Standards and Technology, 2006. A governmental guide to the International System of Units.

Cardarelli, Francois. *Encyclopedia of Scientific Units, Weights and Measures: Their SI Equivalences and Origins.* New York: Springer, 2003. A systematic review of the many incompatible systems of measurement that have been developed throughout history. It clearly relates those units to their modern SI equivalents and provides conversion tables for more than 19,000 units of measurement.

Finucane, Edward W. *Concise Guide to Environmental Definitions, Conversions, and Formulae.* Boca Raton, Fla.: Lewis, 1999. A comprehensive compilation of defined and derived units that clearly describes how each unit relates to the physical property that it quantifies.

Himbert, M. E. "A Brief History of Measurement." *European Physical Journal Special Topics* 172 (2009): 25-35. Describes measurement in a historical context.

Shoemaker, Robert W. *Metric For Me! A Layperson's Guide to the Metric System for Everyday Use with Exercises, Problems and Estimations.* 2d ed. South Beloit, Ill.: Blackhawk Metric Supply, 1998. A basic guide for the average person wanting to learn and understand the use of metric units.

WEB SITES

Bureau International des Poids et Mesures
http://www.bipm.org

National Institute of Standards and Technology
Weights and Measures
http://www.nist.gov/ts/wmd/index.cfm

See also: Maps and Mapping; Measurement and Units; Quality Control; Temperature Measurement; Time Measurement; Weight and Mass Measurement.

INTERNET AND WEB ENGINEERING

FIELDS OF STUDY

Computer science; computer programming; computer information systems; computer information technology; software engineering; Web design.

SUMMARY

The term "Internet" is often used to describe the web of computer networks connected to each other digitally that is accessible to the public around the world. The World Wide Web was developed by Oxford University graduate Tim Berners-Lee in 1989 while working at the European Council for Nuclear Research. The World Wide Web makes it possible for users to browse through various documents on different Web sites by clicking on hyperlinks located on Web pages. Numerous hardware and software advances have made the Internet an indispensable tool for business transactions and personal communication.

KEY TERMS AND CONCEPTS

- **Browser:** Software program that is used to view Web pages on the Internet.
- **Firewall:** Hardware device placed between the private network and Internet connection that prevents unauthorized users from gaining access to data on the network.
- **Hacker:** Person who tries to breach the security of other computers by using programming skills.
- **Hypertext Transfer Protocol (HTTP):** Set of standard rules that allow HTML data to be transferred between computers.
- **Operating System:** Computer program designed to manage the resources (including input and output devices) used by the central processing unit and that functions as an interface between a computer user and the hardware that runs the computer.
- **Portability:** Ability of a program to be downloaded from a remote location and executed on a variety of computers with different operating systems.
- **Static Web Page:** Web page that contains content that cannot be changed by the user.

- **Web Application:** Collection of Web pages that work together.
- **Wide Area Network (WAN):** Computer network that connects central processing units (CPUs), servers, and other computing devices at more than one physical site.

DEFINITION AND BASIC PRINCIPLES

The Internet is a type of wide area network (WAN) because digital devices of many different types, such as a personal computer (PC), the Apple Macintosh computer, or even cell phones, can all connect to the Internet from locations all over the world. Although sharing information between computers via the network connections that comprise the Internet is free, connecting to the Internet itself is not free in North America and requires the use of a private company that functions as an Internet service provider (ISP). Many of these ISPs were phone companies because during the 1990's most connections to the Internet were made via analogue dial-up connections through the phone lines made possible because of a device called a modem. A modem functions by translating the digital signal from a sending computer to the analogue signals transmitted by phone lines and then back into the digital signal required by the receiving computer.

New hardware devices and software have continuously been developed to make sharing information via the Internet easier and faster. A language called hypertext markup language (HTML) was created to design and format the user interface to Web page content for transmission over the Internet, based on hypertext transfer protocol (HTTP). As of 2011, there are many Web applications available to computer and cell phone users all over the world.

BACKGROUND AND HISTORY

The precursor to the Internet was created in 1969 by the Advanced Research Projects Agency (now the Defense Advanced Research Projects Agency) of the United States Department of Defense. This project proposed a method to link together the computers at several universities (University of California, Los Angeles and Santa Barbara, Stanford, and the University of Utah) to share computational data via networks.

This network became known as ARPANET. In the 1980's the International Organization for Standardization (ISO) implemented its Open Systems Interconnection—Reference Model (OSI—RM) to facilitate the interoperability of different hardware components. The creation in the 1990's of new software and programming languages, such as browsers and HTML, made the explosive growth of Internet activity and Web sites possible.

Along with the increasing use of microcomputers, there has also been an increase in the need for trained workers because companies need to be able to share information. This need for sharing information was one of the driving forces in the development of the Internet and the client-server model common to Web architecture that now allows workers to share files and access centralized databases.

How It Works

To transmit data across the Internet, there must be a level of communication possible between devices, analogous to two people shaking hands. For computers, this ability to communicate is called interoperability, and it can be classified as connection-oriented or connectionless. The connection-oriented mode of communication is somewhat similar to the process of a phone call because it requires that the sender of a message wait for the intended recipient to answer before any data is actually sent. It is more secure, but slower, for transmission of data over the Internet than the connectionless mode of data transmission. The connectionless mode has more of a broadcast nature, with data being transmitted before any secure connection is established. The connectionless mode is analogous to sending a letter or postcard through the regular mail. Different protocols, developed according to the Institute of Electrical and Electronics Engineers (IEEE), are used for communication across the Internet, depending on whether the connection or connectionless mode is best for a given situation.

Logical and Physical Topology Classifications. The different protocols function via alternative topologies that can be classified as logical or physical. Logical topologies describe the theoretical, but not visible, pathways for transmission of data signals throughout the physical topologies, which describe the tangible hardware connections of the actual pieces of equipment on a network. The two logical topologies are bus and ring, and the five physical topologies are bus, star, ring, mesh, and cellular.

The Seven Layers of the OSI Web Architecture Model. The OSI model was developed to be the Web architectural model for designing components that allow communication across the Internet. It continues to be useful as a theoretical construct to facilitate the development of different pieces of software and hardware from different manufacturers according widely accepted standards. If one single piece of hardware equipment within the network for one home or business fails, then that one piece can be purchased, even from a different manufacturer, and successfully replace the faulty equipment. This network architecture model was developed by the ISO in 1977, primarily to provide standardization among different manufacturers and vendors to allow data communication to flow uninterrupted across the Internet through different nations with complete interoperability of the software and hardware components. Therefore, the OSI model also helps with the diagnosis of connection problems by dividing the communication between two computers connected by a network into seven different layers.

The lowest level is called the physical layer because it consists of the most fundamental hardware devices that connect computers and transmit bits, including coaxial, twisted-pair, and fiber-optic cabling and connectors. Next is the data link Layer, which manages timing, flow control, error control, and encryption. It also recognizes the physical device addresses, which consist of 48 bits provided by manufacturers. Most bridges and switches are used to repeat data across different networks according to protocols such as Ethernet through this layer. The network layer uses the Internet protocol for five classes of addresses for devices, based on a 32-bit logical (not physical) address. These addresses are typically assigned by a network administrator for an organization. Routers consist of both hardware and software programs that can make decisions regarding the choice of routes for data transmission across different networks. The transport layer uses the transmission control protocol, which is a reliable, connection-oriented protocol that transmits data between two points at the same time. The session layer coordinates individual connection sessions and uses Structured Query Language (SQL) to retrieve information from databases. The presentation layer uses image, sound, and video formats such as graphics interchange format (GIF), musical instrument data interface (MIDI), and motion picture experts group (MPEG). It also encrypts data and compresses data for secure transmission. The top layer, the application layer, has the most direct interaction with a computer user. It contains the protocols for e-mail, Web access, operating systems, and file transfer protocol. These seven layers work together through encapsulation to make the transmission of data from one device to another device seamless.

APPLICATIONS AND PRODUCTS

In 1990, the World Wide Web began as a public web of networks linking digital content described using HTML at different digital locations defined by a uniform resource locator (URL). Initially the digital content was static and slow to display. To display forms that could interact with the user, Brendan Eich of Netscape developed the language initially called LiveScript in 1995, which later became widely known as JavaScript.

Web Applications. HTML was useful for allowing static Web pages to be displayed but not dynamic Web pages. Technologies were developed to provide dynamic Web page access, including ASP, Perl/CGI, JavaScript, and Cold Fusion. ColdFusion and ASP are examples of server-side technologies, which means that the computer of an individual user (client) is not required to have any special software or hardware to access the information from a remote geographical location other than any type of browser that allows access to the Internet. JavaScript, on the other hand is a client-side technology.

ColdFusion MX 6.1 is one example of a Web application server. Other examples are CGI and ASP.NET. These examples of server-side technologies are more secure because individual clients do not have any direct access to the code. Java libraries can be imported into Web pages by using ColdFusion. ColdFusion was created by Macromedia in 2003, and is supported by the operating systems Windows Server 2003, Red Hat Linux 8 and 9, and Solaris 9. ColdFusion's compiler can take Java byte code and directly compile it into ColdFusion code, which speeds up the Web page interactions.

Linus Torvalds created Linux in 1991 while he was a student at the University of Helsinki. Linux has continued to grow in popularity as an operating

The creation of the web by British computer software genius Tim Berners-Lee and other scientists at the European particle physics laboratory (CERN) paved the way for the Internet explosion that has changed our daily lives. (AFP/ Getty Images)

system ideal for Internet applications because it is free, open source (without any hidden or proprietary interfaces), and compatible with many devices and Web applications, including Oracle and IBM databases, and many browsers. Moreover, it is powerful enough to be able to handle many of the applications that were originally handled by the UNIX operating system. Some of the applications that use Linux include digital camera programs, sound cards, printers, modems, and electronic pocket organizers such as the PalmPilot.

To decrease the time for Web pages to be displayed, frames were introduced in the middle 1990's to divide a document into pieces and display them gradually. Navigator 2.0 was the browser created by Netscape and the first to use both frames and JavaScript. Microsoft's Explorer was the next browser to implement frames, and the creation of Dynamic HTML (DHTML) along with Cascading Style Sheets (CSS) and the Document Object Model (DOM) in the late 1990's revolutionized Web pages by allowing parts of Web pages to be modified and also provided structures to Web pages. Hidden frames, Extensible Markup Language (XML) and ActiveX controls such as XMLHttp became widespread after 2001.

These technologies were taken to the next level to make Web pages more dynamic in 2005 by Jesse James Garrett, cofounder of Adaptive Path. He called this new technology Ajax, an acronym for asynchronous JavaScript and XML. The XML parser is software that allows one to use XML to create one's own markup that can read and interpret XML data. Google was one of the first companies to incorporate this technology into Google Maps, Google Suggest, and Gmail Web applications still in use. Many additional companies have adopted Ajax as well, including Amazon, Microsoft, and Yahoo. Ajax increases interactivity with Web pages for users over the traditional "click and wait" process. JavaScript embedded within a Web page sends a message to a Web server that allows data to be retrieved and transmitted to the user without having to wait for the entire Web page to be displayed.

Flash is a programming language introduced by Macromedia to create dynamic images on a Web page by using vector graphic images instead of the bitmap images (common in digital cameras) that are smaller and slow down the dynamic animations and graphics of a Web page. This language has the byproduct stand-alone applications of Flash Player and Shockwave Flash files used to display movies on the Internet. It uses a JavaScript application programming interface and CSS. CSS are used by HTML and Flash as a template to keep consistent appearance among Web pages located on different Web sites.

Common Gateway Interface. Common Gateway Interface (CGI) is used to write programs that interact with Web pages to process information a person types into a form for registration, credit card purchase, or search on a Web site. CGI is a tool that functions as an interface between code written in Perl (Practical Extraction and Report Language) and a Web server. These CGI applications are referred to

as scripts instead of programs. Because CGI is able to interact dynamically to Web users, it is designed to locate URLs instantly using HTTP and HTML. Embedded with CGI scripts there are tags to create cache information, so that a user can quickly return to Web sites. Because the general use of CGI is to process dynamic data for forms and to assist with searches through forms, it also interacts with Perl.

Active Server Pages, ColdFusion, and PHP. The primary problem with CGI is that it requires a new and separate execution of the CGI script each time updated input is necessary. Therefore, several alternatives have been developed. One is called the Active Server Pages (ASP) engine, which is included as part of a Web server, so the extra step of accessing the Web server is not needed. In addition, ASP can be used with several programming languages, including JavaScript and Visual Basic, making ASP more versatile than CGI. ColdFusion allows functions to be programmed and then called from within a Web page, allowing more flexibility. Both ColdFusion and PHP are integrated into Web servers. PHP was developed in 1994 as the acronym for personal home page because it was used as a scripting language to facilitate the development of Web pages. PHP can be embedded within the HTML code of a Web page to make it more dynamic because the PHP code is interpreted rather than compiled.

IMPACT ON INDUSTRY

The growth of the Internet has led to the creation of many successful companies that are constantly leading the way in terms of innovative technologies and therefore new jobs. Among these are many companies originally founded in the United States that have since expanded worldwide, including Apple, Cisco, Google, Microsoft, Oracle, Sun Microsystems, Sybase, and Yahoo.

Cisco began in 1984 in San Francisco, primarily as a producer of some of the physical hardware devices such as routers and switches needed for the joining together of networks. It has since emerged as one of the largest technology companies in the world and has evolved into a company producing innovative software for Internet communication, such as its WebEx for video conferencing. Because it has become such an industry leader in the manufacture of networking devices, it has its own certifications including Cisco Certified Network Associate (CCNA) and Cisco Certified Design Associate (CCDA). These certifications are just a few of the most common certifications designed to enhance the skills for current and future employees.

Google was started by Larry Page and Sergey Brin, Stanford doctoral students, in 1996. The company has developed many new innovative software applications to accompany its search engine, including Gmail and Google Maps. As a result, its stock price has been more than $500 per share for many years.

Research in Motion (RIM) introduced its first smart phone to customers in 2002. These early smart phones allowed customers to send e-mail and text messages and access Web pages. RIM even developed its own software, including a new operating system, Internet service, database management system, and the innovative concept of push e-mail, whereby messages are sent instantly. Like Cisco, RIM has developed several certifications designed to help current and future employees attain the necessary job skills. Among the RIM certifications are BlackBerry Certified Enterprise Server Consultant and BlackBerry Certified Server Support Specialist. Because these

certifications, like those offered by Cisco, depend primarily on successfully passing various tests, there have been many opportunities for well-paying jobs within the technology sector for individuals possessing enough knowledge to pass these exams, but who may not necessarily have spent years to obtain a college degree.

CAREERS AND COURSE WORK

According to the Bureau of Labor Statistics, this career field is expected to grow by 23 percent during the period from 2008 to 2018, which is much faster than average. This career field refers to network administrators, network engineers, systems engineers, database administrators, help desk engineers and technicians, Web designers, security analysts, programmers, and project managers. There are ample opportunities for employment as help desk engineers and technicians, Web designers, database administrators, and network administrators for individuals who have an associate's degree in an appropriate field or work experience and certifications. In addition to some of the vendor-specific certifications offered by Cisco, RIM, Oracle, Sun Microsystems, and Microsoft, there are several certifications that are vendor neutral such as CompTIA's Network+, Security+, and iNet+. Network administrators design networks for organizations and maintain the security of these networks on a daily basis, as well as install shared software, assign logical Internet protocol (IP) addresses, and back up data. Web designers use Web page languages such as HTML, Perl, Java, and VB.NET. Certifications in these languages indicate useful skills as well. The bachelor's degree in computer science or a computer-related field is especially helpful to secure employment as a security analyst or project manager.

SOCIAL CONTEXT AND FUTURE PROSPECTS

The economy can be described as a digital one because the Internet is used as a tool to facilitate all kinds of online financial transactions involving the common use of bank accounts, credit cards, airline and hotel reservations, government benefits, and the stock market. These financial transactions can be completed using both wired and wireless computer internetworking. The rapid growth of Internet applications that facilitate communication and financial transactions has also been accompanied by a growth in identity theft and other cyber crimes.

Fascinating Facts About Internet and Web Engineering

- Employment of network systems and data communications analysts is estimated to increase by 53 percent from 2008 to 2018, according to the Bureau of Labor Statistics.
- Employment of database administrators, who are needed by all types of private companies and public organizations to manage online database access, is projected to grow by 20 percent from 2008 to 2018, according to the Bureau of Labor Statistics.
- Google currently has several thousand job openings at almost a dozen sites located throughout the United States. It has several more thousand job openings worldwide.
- Facebook was created in 2004 by Mark Zuckerberg, who was a college student at Harvard. By the end of 2010 it had become the largest social networking site in the world.
- At least 68 percent of the victims of cyber crime in 2005 suffered monetary losses of at least $10,000.

Web architecture has developed both in terms of providing convenience for consumer transactions and social networking sites, such as Facebook, and also in terms of providing security through the use of firewalls and new security software to try to prevent cyber crimes. Despite the security risks, the usage of the Internet for personal, financial, educational, and business-related activities is expected to increase, especially in many of the newly developing countries that have been lagging behind the United States and other developed nations.

Jeanne L. Kuhler, B.S., M.S., Ph.D.

FURTHER READING

Castro, Elizabeth. *HTML, XHTML, and CSS.* 6th ed. Berkeley, Calif.: Peachpit Press, 2007. This introductory text describes how a novice can set up an individual Web site and provides ample helpful screen shots.

Longino, Carlo. "Your Wireless Future." *Business 2.0.* May 22, 2006. http://money.cnn.com/2006/05/18/technology/business2_wirelessfuture_intro This article discusses the future of wireless technology in all sectors—business, entertainment, and communications.

Love, Chris. *ASP.NET 3.5 Website Programming: Problem-Design-Solution.* Indianapolis: Wiley Publishing, 2010. Describes the process for the design and construction of a Web site using ASP.NET 3.5.

Prayaga, Lakshmi, and Hamsa Suri. *Programming the Web with ColdFusion MX 6.1 Using XHTML.* Boston: McGraw-Hill, 2004. Explains the use of the ColdFusion language, with representative code examples and screen shots.

Spaanjaars, Imar. *Beginning ASP.NET 3.5 in C# and VB.* Indianapolis: Wiley Publishing, 2008. This text, although for beginners, assumes that readers have had some type of introduction to Web technology and programming.

Umar, Amjad. "IT Infrastructure to Enable Next Generation Enterprise." *Information Systems Frontiers* 7, no. 3 (July, 2005): 1201-1205. Looks at advances in network protocols and design.

Underdahl, Brian. *Macromedia Flash MX: The Complete Reference.* Berkeley, Calif.: McGraw-Hill/Osborne, 2003. This comprehensive volume covers all aspects of Flash.

WEB SITES

Association of Internet Researchers
http://aoir.org

Institute of Electronics and Electrical Engineers
http://www.ieee.org

WebProfessionals
http://webprofessionals.org

World Wide Web Consortium (W3C)
http://www.w3.org

See also: Computer Graphics; Computer Networks; Computer Science; Software Engineering; Wireless Technologies and Communication.

ION PROPULSION

FIELDS OF STUDY

Inorganic chemistry; general physics; plasma physics; astrophysics; mechanical engineering; electronics; electrical engineering; systems engineering; mathematics; statistics; material science.

SUMMARY

Ion propulsion is the ejection of charged particles from a space ship, causing the ship to move in the opposite direction from the ejected particles. Chemical rocket engines move ships in much the same way, but the speed of their exhaust is limited by the energy released in the chemical reaction. Ions, however, can be accelerated by electric and magnetic fields to far higher speeds, so ion engines require a far smaller mass of fuel than chemical engines.

KEY TERMS AND CONCEPTS

- **Burn Time:** Time it takes to burn up an amount of rocket fuel, and by extension, the time it takes an ion engine to eject a given amount of mass.
- **Geostationary Earth Orbit (GEO):** Geosynchronous orbit, 35,785 kilometers above the equator, so that the object in orbit appears to stay at a fixed point relative to the Earth's surface at all times.
- **Hall Effect:** Generation of a difference in potential across a conductor carrying an electric current that is exposed to an external magnetic field applied at a right angle.
- **Hall-Effect Ion Engine:** Ion engine that uses magnetic fields to accelerate ions.
- **Ion Engine:** Device that ionizes reaction mass, accelerates it with either electric or magnetic fields, and expels it from the rear of the engine, thereby driving the engine forward; generally refers to an engine that uses electric fields to accelerate ions.
- **Low Earth Orbit (LEO):** Orbit 160 to 2,000 kilometers above the Earth's surface.
- **Reaction Mass:** Mass ejected from the rear of a rocket that drives the rocket forward.
- **Space Charge:** Concentration of charge in a region that is large enough to significantly affect other charges moving toward that region.
- **Thrust:** Force exerted on a rocket by its motor; equals the mass ejected from the motor per unit time multiplied by the exhaust velocity.

DEFINITION AND BASIC PRINCIPLES

The fundamental principle of rocket propulsion is Sir Isaac Newton's third law of motion: For every action, there is an equal and opposite reaction. The mass of fuel ejected rearward from a rocket results in the rocket being propelled forward. More precisely, the mass of fuel multiplied by the exhaust velocity equals the force pushing the rocket forward multiplied by the burn time. It follows then that the greatest amount of mass possible must be ejected each second at the highest possible speed to provide maximum acceleration to a rocket. The maximum exhaust speed for a chemical rocket is about 3 to 5 kilometers per second, but the exhaust speed for an ion engine can be 30 to 50 kilometers per second or greater.

The best thrust of an ion engine is only a small fraction of its weight, so ion engines cannot be used to lift a space ship into orbit. Chemical engines must perform that task. However, once the space ship is in orbit, ion engines can be very useful. Short bursts from an ion engine are ideal for station keeping (maintaining position and orientation with respect to another craft or object). Because ion engines use so little mass, they may be fired for days or months instead of the minutes that a chemical engine can burn. The low thrust of an ion engine fired for a long time can make major changes in the orbit of a space ship. For example, an ion engine took the National Aeronautics and Space Administration's Deep Space 1 from low Earth orbit to fly by asteroid 9969 Braille and Comet Borrelly.

BACKGROUND AND HISTORY

Anyone working with electric fields and charged particles soon discovers that it is easy to accelerate charged particles to high speeds. Rocket pioneers Robert Goddard (in 1906) and Konstantin Tsiolkovsky (in 1911) discussed the idea of an ion engine. In 1916, Goddard built an ion engine and demonstrated that it produced thrust. In 1964, National Aeronautics and Space Administration

(NASA) scientist Harold R. Kaufman built and successfully tested an ion engine that used mercury as reaction mass in the suborbital flight of the Space Electric Rocket Test 1 (SERT 1).

During the 1950's and 1960's, both the United States and the Soviet Union (now Russia) worked on a Hall-effect ion engine that used magnetic fields to accelerate ions. The United States continued to work on the electric field ion engine but dropped out of the competition for the Hall-effect engine. Russia eventually developed Hall-effect engines and began using them in space in 1972.

In 1992, the West adopted some of the Russian technology and started using Hall-effect engines as well as the electric field ion engine. In 1998, NASA launched Deep Space 1, which used an ion engine and flew by asteroid 9969 Braille and Comet Borrelly. The Deep Space 1 used an ion engine as its main source of propulsion, and its success paved the way for future missions powered by ion engines. In the 2000's, NASA, the European Space Agency (ESA), and the Japan Aerospace Exploration Agency launched spacecraft that used versions of ion engines.

How It Works

Heavy atoms are usually selected as the reaction mass for an ion engine because heavy ions increase the mass ejected per second. It is also desirable to use a substance that requires relatively little energy to ionize it. Mercury, cesium, xenon, bismuth, and argon have all been used for reaction mass. However, mercury and cesium are poisonous and require special handling, so the substance most commonly used for reaction mass is xenon. Xenon can be stored under pressure as a liquid for long periods of time. As a gas, it is easy to transfer from storage to the rocket motor, the molecules are relatively heavy and easily ionized, and it is chemically inert, so it will not corrode the engine. Ion engine bodies are often made of boron nitride, a good insulator that can withstand the operating conditions of the engine.

A small amount of the reaction mass is leaked at a controlled rate into a plasma chamber, where it is bombarded with electrons to ionize it. Fields from permanent magnets are used to confine and control the resulting plasma.

The walls of the plasma chamber are maintained at a high positive voltage (such as +1,100 volts) so that free electrons migrate to the walls and are absorbed.

Two grids or wire screens are at the rear of the plasma chamber. The first grid is at a lower voltage, such as +1,065 volts, so that positive ions from the chamber are attracted to it. A second grid is placed 2 or 3 millimeters beyond the first grid and electrified at −180 volts. Positive ions will be accelerated by the electric field in the gap between the two grids and will gain 1,245 electron volts of energy. This means that xenon ions will exit the engine at an impressive 43 kilometers per second.

There are three areas that have to be addressed for ion propulsion to work. First, because the craft is ejecting positive charge into space, it is becoming more negative. If nothing were done about this, the craft would eventually become so negative that the positive ion exhaust would be attracted back to the ship, and there would be no propulsion. To deal with this, a special electrode sprays negative charge onto the ion plume behind the craft to keep the craft's charge neutral. Second, a positive charge being accelerated between the grids constitutes a space charge density. This space charge repels approaching positive charge, and this limits the ion current to fairly small values, which in turn limits the thrust of an ion engine to small values. These limits on thrust must be carefully calculated and considered. Third, high-energy ions strike electrodes and erode them. Careful attention to the electrode design and the materials used can extend electrode lives to 10,000 hours.

Hall-Effect Thrusters. The Hall effect (named for Edwin Hall, who discovered it in 1879) states that if a conductor carrying a current is placed in a magnetic field perpendicular to the current, the magnetic field will cause a new current to flow perpendicular to both the magnetic field and the original current.

As an example, consider a horizontal tube lying along the bottom of a page. Let the tube contain a vertical electric field pointing from the bottom to the top of the page. Suppose further that there is a horizontal magnetic field pointing out of the paper. If electrons and ionized xenon atoms are introduced at the left end of the tube, electrons will be propelled by the electric field to sink to the bottom side of the tube along the lower edge of the page. Because of the greater mass of the xenon ions, their progress toward the upper side of the tube will be far slower. A charged particle moving in a magnetic field experiences a force perpendicular both to its velocity and to the magnetic field. This causes the electrons that were

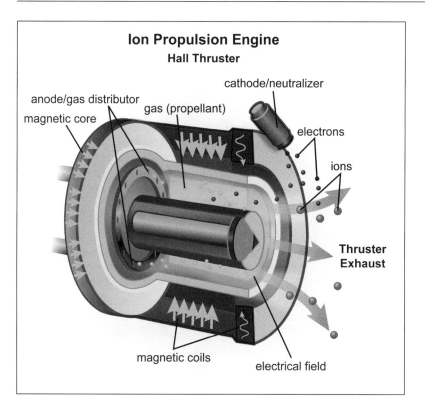

Ion Propulsion Engine
Hall Thruster

cathode/neutralizer

anode/gas distributor

gas (propellant)

magnetic core

electrons

ions

Thruster
Exhaust

magnetic coils

electrical field

microwaves heating the plasma, by changing the rate at which reaction mass is delivered to the engine, or by changing the size of the magnetic field nozzle. Engines using 100 and 200 kilowatts have been built and ground-tested. One day it may be possible to build a 200-megawatt engine powered by a small nuclear reactor. A craft powered by such an engine could make the trip from Earth to Mars in only 39 days, a considerable improvement over the 255 days required by conventional rockets.

APPLICATIONS AND PRODUCTS

Numerous ion engines have been used for station keeping, and an increasing number of missions are using them as main engines.

SERT 1. Launched by NASA in 1964, SERT 1 showed that an ion engine actually functioned as expected in space. In 1970, SERT 2 was sent into space to test two ion engines using mercury as reaction mass. The engines operated 2,011 hours and 3,781 hours respectively and were restarted about three hundred times.

Deep Space 1. Launched by NASA in 1998, Deep Space 1 tested twelve high-risk technologies, including the NSTAR ion engine. The engine drew up to 2.5 kilowatts from solar cells and produced 0.092 newtons of thrust. It carried 81.5 kilograms of xenon reaction mass, enough to last 678 days. Each day, the engine added 24 to 32 kilometers per hour to the craft's speed. It flew by asteroid 9969 Braille and Comet Borrelly. Because of a software crash, the pictures of Braille were not as good as had been hoped, but the pictures of Comet Borrelly were outstanding. A camera had to be reprogramed to take over tracking duties when the star tracker failed, but most of the other instruments worked well. The mission ended with engine shutdown on December 18, 2001.

Artemis. The European Space Agency launched the Artemis in 2001, but the telecommunications satellite failed to reach its intended orbit. It used its remaining chemical fuel to transfer to a higher orbit, then used its xenon ion engine for eighteen months to raise it to the intended geostationary Earth orbit.

sinking to the bottom of the tube to move from left to right down the axis of the tube. These electrons will strike other electrons and ions, forcing them to move down the tube as well. This plasma now moving along the tube's axis will become the exhaust jet that propels the spacecraft. The mass exhausted is not limited by space charge and can be as large as the device can handle.

VASIMR. The VASIMR (Variable Specific Impulse Magnetoplasma Rocket) project of the Ad Astra Rocket Company in Webster, Texas, seems to hold the key to more powerful ion engines. The VASIMR engine uses microwaves to ionize and heat the propulsive gas. The temperature of the gas can be controlled by increasing or decreasing microwave intensity. The absence of electrodes means they cannot be eroded by hot plasma and suffer degraded performance as happens in a normal ion engine. Magnetic fields confine, direct, and accelerate the plasma to exhaust speeds up to 50 kilometers per second. Magnetic fields form a rocket nozzle inside the metal nozzle. Gas introduced between the metal nozzle and the plasma is heated by the plasma and adds to the thrust. This thrust can be modified by changing the

Hayabusa. The Japan Aerospace Exploration Agency launched Hayabusa in 2003. It reached the asteroid Itokawa in 2005, but the lander module MINERVA flew by the asteroid instead of landing as planned. Astronomers have long suspected that some asteroids are rubble piles held together by only self-gravitation. Images sent from Hayabusa seem to show exactly that. Hayabusa successfully landed and activated a collection capsule. The craft was propelled by four xenon ion engines that amassed a combined total of 31,400 operating hours. These were the first engines in space to use microwaves to form and heat the xenon plasma. Unfortunately, the engines failed one by one. None was operational to bring Hayabusa home, but operators were able to use the ion generator from one engine with the electron gun neutralizer from another engine to make one working engine. The sample capsule returned to Earth on June 13, 2010. It had several grains of dust inside, and if they prove to be from the asteroid, it will be the first such sample ever obtained.

SMART-1. The European Space Agency launched SMART-1 (Small Missions for Advanced Research in Technology 1) in 2003. A Hall-effect ion engine, using xenon as the reaction mass, allowed SMART-1 to travel from low Earth orbit to lunar orbit. SMART-1 surveyed the chemical elements on the lunar surface, and then was driven into the Moon in a controlled crash in 2006. Although the Russians had used Hall-effect engines for station keeping for many years, this was the first use of a Hall-effect engine as the main engine.

Dawn and GOCE. In 2007, NASA launched Dawn to explore the asteroid Vesta in 2011 and the dwarf planet Ceres in 2015. The spacecraft is propelled by three xenon ion engines but typically uses only one at a time. The European Space Agency launched GOCE (Gravity Field and Steady-State Ocean Circulation Explorer) in 2009. GOCE orbits only 260 kilometers above the Earth's surface so that it can map small changes in the Earth's field. It uses a xenon ion engine to make up for loses caused by air drag.

GSAT-4. The Indian Space Research Organization launched the GSAT-4, a geostationary satellite for navigation and communications, in April, 2010. It had an ion engine for station keeping, but the third stage failed to ignite, and the satellite was lost. The ion engine would have extended GSAT-4's normal ten-year life to fifteen years.

LISA. Although gravity waves have never been detected directly, theory predicts their existence, and they best explain the orbital decay of binary neutron stars. LISA (Laser Interferometer Space Antenna) is a very sensitive device that can detect the waves if they exist. The device will consist of three satellites containing special masses, mirrors, and lasers. The satellites will be 5×10^6 kilometers apart, each at a different corner of an equilateral triangle. The lasers should be able to measure small changes in the positions of the satellites caused by a gravity wave. LISA and LISA Pathfinder are joint projects of NASA and the European Space Agency. LISA Pathfinder will test several techniques to be used on LISA, including colloid thrusters. These thrusters are ion engines that use charged liquid drops for reaction mass. Their thrust is very small, perhaps only 20×10^6 newtons. Such small forces are useful in making tiny adjustments to the speed or position of a satellite. LISA Pathfinder is set for launch in 2011.

IMPACT ON INDUSTRY

Ever since the Chinese developed gunpowder rockets in the eleventh century, the word "rocket" has usually referred to a device powered by explosive chemical reactions. It soon became apparent to those trying to improve rockets that increasing the exhaust speed of the reaction mass was the key to faster rockets. Therefore, in 1906, Robert Goddard conceived of an ion engine as a solution. One of the best chemical rocket fuel combinations is liquid oxygen and liquid hydrogen. When burned in an efficient engine, the resultant steam has an exhaust velocity of 4.5 kilometers per second, but singly ionized xenon atoms accelerated through 1,000 volts in an ion engine have an exhaust velocity of 38 kilometers per second. Modern rocket designers have increasingly turned toward using the ion engine.

Government Development. Because ion engines are roughly ten times as efficient as chemical engines, all the major players in space exploration have adopted ion engines wherever possible. The Soviet Union pioneered Hall-effect engines in the 1960's, and they were gradually adopted for station keeping and later for powering spacecraft. Missions using ion engines have been launched by the Russian Federal Space Agency (Roscosmos), NASA, the Indian Space Research Organization, the Japan Aerospace Exploration Agency, and the European Space Agency (its member nations are Austria, Belgium, the Czech Republic, Denmark, Finland, France, Germany, Greece, Ireland, Italy, Luxembourg, the Netherlands, Norway, Portugal, Spain, Sweden, Switzerland and the United Kingdom).

Industrial Development. Private companies are increasingly involved in the creation of engines to power spacecraft and satellites as well as the satellites themselves. The Ad Astra Rocket Company in Webster, Texas, is developing the VASIMR engine, and the British company QinetiQ is preparing engines to power the European Space Agency's BepiColombo to the planet Mercury. The Satellite Business Development Office of NEC built the four ion engines for

Japan's Hayabusa. The Boeing Satellite Development Center bills itself as the world's leading manufacturer of commercial communications satellites, a major supplier of spacecraft and equipment, and as the supplier for weather satellites for the United States and Japan.

CAREERS AND COURSE WORK

Any job in design and development is likely to be exciting and interesting to those who like to solve problems and see ideas made into reality. A strong background in the physical sciences—at least a bachelor's degree in physics, engineering, electrical engineering, mechanical engineering, materials science, or chemistry—is required to design and develop ion engines or other space hardware. A high school student should take all available courses in physics, chemistry, and mathematics. The same is true for college students, who will need at least one year of basic chemistry, at least one year of calculus-based physics with laboratory practice, and mathematics through differential equations and matrices. Classes in statistics and computer programing may be helpful.

Other useful classes include writing and speech (for reports and presentations) and a simple business course. Those wishing to work in research and development need a feel for how things work and some level of creativity. Those working in quality control may need to design tests that demonstrate that a device works and that it will continue to work reliably.

SOCIAL CONTEXT AND FUTURE PROSPECTS

As Hall-effect ion engines and VASIMR engines become more powerful, rocket fuel will account for less of the spacecraft's total weight. This will allow for larger payloads and enable robot missions to conduct more scientific experiments. If high-powered engines can be developed, the only source capable of providing enough power is a small nuclear reactor. Fission reactors have been used in space by Russia, which has used more than thirty to power satellites and by the United States, which flew a test reactor, the SNAP-10A, in 1965. With a nuclear reactor for power and with a large ion engine, a spacecraft could travel from Earth to Mars in 39 days instead of the 255 days that it would take a conventionally powered vessel. This may make a manned mission possible because of the reduction in days of exposure to cosmic rays. Such an engine might make it possible to fly to

---

Fascinating Facts About Ion Propulsion

- If space travel is ever to become much more common, it seems likely that it will be powered by ion engines.

- The bluish color of the exhaust from an ion engine comes from the recombination of electrons with xenon ions.

- NASA's Evolutionary Xenon Thruster (NEXT) generates 2.5 times as much thrust as the NSTAR engine used on the Dawn mission. It provides a record-breaking 10 million Newton-seconds of total impulse (the overall acceleration available to a spacecraft).

- Microwaves will heat the plasma in the VASIMR engine to millions of Kelvins. Magnetic fields hold the plasma away from the rocket's walls so they do not melt.

- The VASIMR VX-200 ion engine produces a thrust of 5 newtons at 5,000 seconds specific impulse. In comparison, the NEXT engine produces 0.327 newtons at 4,300 seconds specific impulse and the NSTAR engine 0.092 newtons at 3,300 seconds specific impulse.

- The Australian National University designed and built the dual-stage four-grid (DS4G) ion thruster for the European Space Agency. An exhaust speed of 210 kilometers per second was recorded during testing in January, 2006. This engine, however, remains in the prototype stage.

- To reduce the risk of radioactive contamination, nuclear reactors used to power ion engines will not be turned on until they are in space.

an asteroid and exploit its minerals. (A single metallic asteroid would most likely contain more nickel than has ever been mined on Earth.) An additional possible benefit is that people's apprehensions about using nuclear reactors for power on Earth might lessen if nuclear reactors in space built up a good safety record.

Charles W. Rogers, B.A., M.S., Ph.D.

FURTHER READING

Bruno, Claudio, ed. *Nuclear Space Power and Propulsion Systems.* Reston, Va.: American Institute of Aeronautics and Astronautics, 2008. Essays examine the use of nuclear power on spacecraft. One chapter specifically examines ion thrusters and nuclear power.

Doody, Dave. *Deep Space Craft: An Overview of Interplanetary Flight.* New York: Springer, 2009. Contains a chapter on propulsion and discusses ion propulsion in the greater context of spacecraft.

Gilster, Paul. *Centauri Dreams: Imagining and Planning Interstellar Exploration.* New York: Copernicus Books, 2004. Includes a good discussion of various means of propulsion, including ion engines.

Goebel, Dan M., and Ira Katz. *Fundamentals of Electric Propulsion: Ion and Hall Thrusters.* Hoboken, N.J.: John Wiley & Sons, 2008. Provides a guide to the technology and physics of Hall-effect and ion thrusters. Contains photographs of successful engines and figures and tables that show thruster and cathode schematics.

Turner, Martin J. L. *Rocket and Spacecraft Propulsion: Principles, Practice and New Developments.* 2d ed. Chichester, England: Praxis, 2006. Contains general discussions of propulsion methods, including ion engines. Excellent diagrams and pictures.

WEB SITES

Ad Astra Rocket Company
http://www.adastrarocket.com

European Space Agency
Propulsion, Advanced Concepts Team
http://www.esa.int/gsp/ACT/pro/pp/DS4G/background.htm

National Aeronautics and Space Administration
Glenn Research Center
http://www.nasa.gov/centers/glenn/home/index.html

NEC
Hayabusa's Seven-Year Journey
http://www.nec.com/global/ad/hayabusa

Propulsion Engineering Research Center
http://www.psu.edu/dept/PERC

Propulsion Research Center
http://prc.uah.edu/Propulsion_Research_Center/Welcome.html

See also: Aeronautics and Aviation; Jet Engine Technology; Propulsion Technologies; Spacecraft Engineering; Space Science.

J

JET ENGINE TECHNOLOGY

FIELDS OF STUDY

Physics; chemistry; thermodynamics; gas dynamics; aerodynamics; heat transfer; materials science; metallurgy.

SUMMARY

Jet engines are machines that add energy to a fluid stream and generate thrust from the increase of momentum and pressure of the fluid. Jet engines, which usually include turbomachines to raise the pressure, vary greatly in size. Applications include microelectromechanical gas turbines for insect-sized devices; mid-sized engines for helicopters, ships, and cruise missiles; giant turbofans for airliners; and scramjets for hypersonic vehicles. Jet engine development pushes technology frontiers in materials, chemical kinetics, measurement techniques, and control systems. Born in the desperation of World War II, jet engines have come to power most modern aircraft.

KEY TERMS AND CONCEPTS

- **Cycle Efficiency:** Ratio of the useful work delivered by a thermodynamic system to the heat put into the system.
- **Equivalent Exhaust Velocity:** Thrust divided by the mass flow rate of propellant through the exhaust.
- **Fuel-to-Air Ratio:** Ratio of the mass of fuel burned per unit of time to the mass of air passing through the burner per unit of time.
- **Overall Pressure Ratio:** Ratio between the highest pressure in the propulsion system and the lowest pressure, typically the atmospheric pressure at the altitude at which the vehicle is flying.
- **Propulsive Efficiency:** Ratio of work done by the propulsion system on the vehicle to the kinetic energy imparted by the system to the working fluid.

- **Specific Impulse:** Equivalent exhaust velocity divided by the standard value of acceleration due to gravity; expressed in seconds.
- **Stage Pressure Ratio:** Ratio of the stagnation pressure at the end of a single stage of a turbomachine to the stagnation pressure ahead of the stage.
- **Thermal Efficiency:** Ratio of work done by the propulsion system to the heat put into the system.
- **Thrust:** Force exerted by the propulsion system on the vehicle.
- **Thrust-Specific Fuel Consumption:** Weight of fuel consumed per unit thrust force, per hour.
- **Turbine Inlet Temperature:** Highest temperature of the working fluid, typically reached at the end of heat addition and before work is taken out in the turbine.

DEFINITION AND BASIC PRINCIPLES

The term "jet engine" is typically used to denote an engine in which the working fluid is mostly atmospheric air, so that the only propellant carried on the vehicle is the fuel used to release heat. Typically, the mass of fuel used is only about 2 to 4 percent of the mass of air that is accelerated by the vehicle.

BACKGROUND AND HISTORY

British engineer Frank Whittle and German engineer Hans Pabst von Ohain independently invented the jet engine, earning patents in Britain in 1932 and in Germany in 1936, respectively. Whittle's engine used a centrifugal compressor and turbine. The gas came in near the axis, was flung out to the periphery, and returned to the axis through ducts. Ohain's engine used a combination of centrifugal and axial-flow turbomachines. The gas direction inside the engine was mostly aligned with the axis of the engine, but it underwent small changes as it passed through the compressor and turbine. The axial-flow engine had higher thrust per unit weight of the engine and smaller frontal area than the centrifugal machine.

Initially, the axial-flow machine, which had a large number of stages and blades, was much more prone to failure than the centrifugal machine, which had sturdier blades and fewer moving parts. However, these problems were resolved, and most modern jet engines use purely axial flow.

Several early experiments with jet engines ended in explosions. Ohain's engine was successfully used to power a Heinkel He 178 aircraft on August 27, 1939. Ohain's engine led to the Junkers Jumo 004 axial turbojet engine, which was mass-produced to power the Messerschmitt Me 262 fighter aircraft in 1944. The British Gloster E.28/39, which contained one of Whittle's engines, flew in 1941. In September, 1942, a General Electric engine powered the Bell XP-59 Airacomet, the first American jet fighter aircraft.

The de Havilland Comet jet airliner service began in 1952, powered by the Rolls-Royce Avon. In 1958, a de Havilland Comet jet operated by British Overseas Airways Corporation flew from London to New York, initiating transatlantic passenger jet service. In the mid-1950's, the Rolls-Royce Conway became the first turbofan in airliner service when it was used for the Boeing 707. The Rolls-Royce Olympus afterburning turbofan powered the supersonic Concorde in 1969. The majority of aircraft have come to be powered by jet engines, mostly turbofans, the largest, as of 2010, being the General Electric GE90-115, which produces nearly 128,000 pounds (rated at 115,000 pounds, or 570,000 Newtons) of thrust.

From 1903 to the 1940's, the power per unit weight of engines increased from 0.05 to about 0.8 horsepower per pound (hp/lb), with the largest engines producing 4,000 hp. With jet engines, the power per unit weight has risen to more than 20 hp/lb, with the largest engines producing more than 100,000 hp. The compressor pressure ratio has risen from 3:1 for initial engines to more than 40:1 for modern engines.

HOW IT WORKS

Jet engines operate by creating thrust through the Brayton thermodynamic cycle. In a process called isentropic compression, a mixture of working gases is compressed, with no losses in stagnation pressure. Heat is chemically released or externally added to the fluid, ideally at constant pressure. A turbine extracts work from the expanding gases. This work runs the compressor and other components such as a fan or propeller, depending on the engine type and application. The gas leaving the turbine expands further through a nozzle, exiting at a high speed.

Jet engine developers try to maximize the pressure and temperature that the engine can tolerate. The thermal efficiency of the Brayton cycle increases with the overall pressure ratio; therefore, designers try to get the highest possible pressure at the end of the compression. However, the temperature rise accompanying the pressure rise limits the amount of heat that can be added before the temperature limit of the engine is reached. Thrust is highest when the greatest net momentum increase is added to the flow. The propulsive efficiency of the engine is highest when the speed of the jet exhaust is close to the flight speed of the vehicle. These considerations drive jet engine design in different directions depending on the application. For very-high-speed applications (typically military engines), engine mass and frontal area must be kept low, so a smaller amount of air is ingested and accelerated through a large speed difference. For engines such as those used on airliners, a large amount of air is ingested using a large diameter intake and accelerated through a small speed difference for best propulsive efficiency. The major components of a jet engine are the inlet, diffuser, compressor, fan, propeller, combustor, turbine, afterburner, nozzles, and gearbox.

Inlet and Diffuser. The engine inlet is designed to capture the required airflow without causing flow separation. An aircraft flying at supersonic speeds has a supersonic inlet in which a series of shocks slows down the flow to subsonic speeds with minimal losses in pressure. Once the flow is subsonic, it is slowed further to the Mach number of about 0.4 needed at the face of the compressor or fan. A supersonic inlet and diffuser may lose 5 to 10 percent of the stagnation pressure of the incoming flow when used with aircraft flying in excess of Mach 2.

Compressor, Fan, and Propeller. In 1908, René Lorin patented a jet engine in which a reciprocating piston engine compressed the fluid. In 1913, he patented a supersonic ramjet engine in which enough compression would occur simply by slowing the flow down to subsonic speeds. In 1921, Maxime Guillaume patented a jet engine with a rotating axial-flow compressor and turbine. Most modern engines use the turbomachine in some form. The compressor is built in several stages, with the pressure ratio of each stage limited to prevent flow separation and stall. A

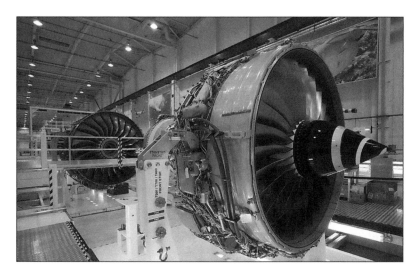

A jet engine is produced in a factory in Derbyshire, England. (Tim Graham/ Getty Images)

centrifugal compressor stage consists of a rotor that imparts a strong radial velocity to the flow, and the flow is flung out at the periphery of the blades with added kinetic energy. The flow is then turned and brought back near the axis by a diffuser stage in which the static pressure rises and the flow speed decreases. In an axial compressor, work is added to the flow in several stages, as many as fifteen in some engines. In each stage, a spinning rotor wheel with many blades that act as lifting wings, imparts a swirl velocity to the flow. This added energy is then converted to a pressure rise in stator blade passages, bringing the flow back to being axial, but with increased pressure. Some newer compressors have counter-rotating wheels in each stage instead of a rotor and a stator. Shock-in-rotor stages use blades moving at supersonic speed relative to the flow to create large pressure rise because of shocks. Supersonic through-flow compressor designs are being developed for future high-speed engines.

The fans of most engines are extensions of the first, or low-pressure, compressor stages. Fans have only one or two stages, and fewer, larger blades than the compressor stages.

Propellers are used in turboprop engines to produce a portion of the engine's thrust.

Combustor. The combustor is designed to mix and react the fuel with the air rapidly and to contain the reaction zone within an envelope of cooling air.

At the exit of the combustor, these flows are mixed to ensure the most uniform temperature distribution across the gases entering the turbine. Older combustors were either several cans connected by tubes, arranged around the turbomachine shaft, or an annular passage. Some modern combustors are arranged in a reverse-flow geometry to enable better mixing and reaction. Ideally, a jet engine combustor must add all the heat that can be released from the fuel, at a constant pressure, with minimal pressure losses because of flow separation and turbulence. Heat addition must also be done at the lowest flow Mach number possible.

Turbine. The critical limiting temperature in a jet engine is the temperature at the inlet to the first turbine stage. This is usually tied to the strength of the blade material at high temperatures. A turbine has only a few stages, typically four or fewer. The highest mass flow rate through the engine is limited to the flow rate at which the passages at the final turbine stage are choked, in other words, when the Mach number reaches 1 at these passages. Turbine stage disks and blades are integrated into blisks and can be made as a single piece, with cooling passages inside the blades, using powder metallurgy.

The turbine is directly attached to the compressor through a shaft. To enable starting the compressor and better matching the requirements of the different stages, a twin-spool or three-spool design is used, where the outer (low-pressure, lower rotation speed) stages of the compressor, fan, and turbine are connected through an inner shaft, and the inner (high-pressure) stages are connected through a concentric, outer shaft.

Afterburner. Older military engines use an afterburner (also known as reheat) duct attached downstream of the turbine, where more fuel is added and burned, with temperatures possibly exceeding the turbine inlet temperature. Afterburners are highly inefficient but produce a large increment of thrust for short durations. Therefore, afterburners are used at takeoff, in supersonic dashes, or in combat situations.

Nozzles. For jet engines without afterburners, the exit velocity is at most sonic and the nozzle is just a converging duct. If the exhaust is expanded to supersonic speeds, the nozzle has a convergent-divergent contour. Thrust-vectoring nozzles either rotate the whole nozzle, as in the case of the Harrier or the F-22, or use paddles in the exhaust, as in the case of the Sukhoi-30.

Gearbox. When the engine must drive a rotor, propeller, or counter-rotating fan, a gearbox is used to reduce the speed or change the direction of rotation. This usually adds considerable weight to the engine.

APPLICATIONS AND PRODUCTS

Ramjet and Scramjet. Ramjets are used to power vehicles at speeds from about Mach 0.8 to 4. The diffuser slows the flow down to subsonic speeds, increasing the pressure so much that thrust can be generated without a mechanical compressor or turbine. Beyond Mach 4, the pressure loss in slowing down the flow to below Mach 1 is greater than the loss due to adding heat to a supersonic flow. In addition, if such a flow were decelerated to subsonic

conditions, the pressure and temperature rise would be too high, either exceeding engine strength or leaving too little room for heat addition. In this regime, the supersonic combustion ramjet, or scramjet, becomes a better solution.

Turbojet. The turbojet is the purest jet engine, with a compressor and turbine added to the components of the ramjet. The turbojet can start from rest, which the pure ramjet cannot. However, since the turbojet converts all its net work into the kinetic energy of the jet exhaust, the exhaust speed is high. High propulsive efficiency requires a high flight speed, making the turbojet most suitable near Mach 2 to 3. Because jet noise scales as the fifth or sixth power of jet speed, the turbojet engine was unable to meet the noise regulations near airports in the 1970's and was rapidly superseded by the turbofan for airliner applications.

Turbofan. The turbine of the turbofan engine extracts more work than that required to run the compressor. The remaining work is used to drive a fan, which accelerates a large volume of air, albeit through

---

Fascinating Facts About Jet Engine Technology

- Hans Pabst von Ohain's first demonstrator jet engine used hydrogen as fuel, and the first Heinkel jet engine patent application was for a hydrogen combustion system.

- The Pratt & Whitney J-58 engines of the Lockheed SR-71 Blackbird aircraft started as afterburning turbojets but morphed into ramjets to allow the aircraft to reach more than three times the speed of sound.

- A turbojet engine built with microelectromechanical systems (MEMS) technology at the Lincoln Laboratories in Boston has a 4 millimeter-diameter turbine that spins at more than 1 million revolutions per minute.

- The General Electric GE90-115 turbofan engine for the Boeing 777, which produces 128,000 pounds of thrust, has a diameter of 3.43 meters, nearly as large as the cabin of the Boeing 737 aircraft.

- The blue light seen shooting out of jet engine nozzles indicates incomplete combustion, with opportunities for further gains in efficiency.

- In 1985, twin-engine airliners were allowed to fly transatlantic routes because jet engine reliability was determined to be so high that failures of other critical aircraft components were far more likely.

- Ramjet engines proposed for vehicles designed to explore the atmosphere of the planet Jupiter would use atmospheric hydrogen as the working fluid and add oxygen as fuel for combustion.

- The Bussard intersteller ramjet engine concept, first proposed in the 1960's, involves capturing hydrogen ions from interstellar gas using an electromagnetic field, compressing the flow in a diffuser, and heating it using an nuclear thermal reactor.

- The Pratt & Whitney F135 and the competing General Electric F136 engines for the F-35 joint strike fighter use fans and compressors that are attached to turbine stages that spin in opposite directions, saving the weight of the cooled turbine stator disks that would otherwise be needed.

- Modern jet engine turbine blades are grown as single crystals, or the entire blade-disk assembly of a stage is cast as a single piece, hollow inside.

- Supersonic combustion ramjet engines for hypersonic vehicles must provide enough distance for fuel to ignite. An ignition delay of 1 millisecond, for a flow moving at 2,000 meters per second, means an ignition distance of 2 meters.

a small pressure ratio. The air that goes through the fan may exit the engine through a separate fan nozzle or mix with the core exhaust that goes through the turbine before exiting. Because the overall exhaust speed of the turbofan is much lower than that of the turbojet, the propulsive efficiency is high in the transonic speed range where airliner flight is most efficient, yet airport noise levels are far lower than with a turbojet. Turbofan engines are used for most civilian airliner applications and even for fighter and business jet engines.

Turboprop. In the turboprop engine, a separate power turbine extracts work to run a propeller instead of a fan. The propeller typically has a larger diameter than a fan for an engine of comparable thrust. However, the rotating speed of a propeller, constrained by the Mach number at the tip, is only on the order of 3,000 to 5,000 revolutions per minute, as opposed to turbomachine speeds, which may be three to ten times higher. Therefore, a gearbox is required.

Turboshaft. Instead of a propeller, a helicopter rotor or other device may be driven by the power turbine. Automobile turbochargers, turbopumps for rocket propellants, and gas turbine electric power generators are all turboshaft engines.

Propfan. Propfans are turbofans in which the fan has no cowling, so that it resembles a propeller and has a larger capture area, but the blades are highly swept and wider than propeller blades.

Air Liquefaction. The high pressures encountered in high-speed flight make it possible to liquefy some of the captured and compressed air at lower altitudes, using heat transfer to cryogenic fuels such as hydrogen. The oxygen from this liquid can be separated out and stored for use as the vehicle reaches the edge of the atmosphere and beyond. Turboramjet engines using this technology could enable routine travel to and from space, with fully reusable, single-stage vehicles.

IMPACT ON INDUSTRY

The jet engine allowed airplanes to fly efficiently, well above the weather and the troposphere in the smooth air of the stratosphere. The smooth stratosphere allowed airplanes to travel at high speeds without exceeding structural strength limits posed by air turbulence. This in turn has enabled intercontinental flights in jet-smooth comfort and the construction of very large airliners. An immense expansion of commerce has resulted from these developments. Engine size and power have increased far beyond what was possible with the internal combustion engine. Outside the aerospace industry, gas turbines are used to power ships and in most thermal power plants.

CAREERS AND COURSE WORK

Jet engine development and manufacturing are parts of a highly specialized industry. Those interested in jet engine technology must have a very good basic understanding of thermodynamics and dynamics and can specialize in combustion, turbomachine aerodynamics, gas dynamics, or materials engineering. Airlines operate engine test cells and employ many technical workers to diagnose and repair problems and ensure proper maintenance and operating procedures. The National Aeronautics and Space Administration (NASA) and the U.S. Department of Defense offer many opportunities in all aspects of jet engine technology. Engine developers include the very large corporations that supply airliner engines and the smaller companies that develop engines for business jets, cruise missiles, and other applications. Companies doing research, development, and manufacture of jet engines are in the United States, Europe, Russia, Japan, and China. Many other nations also produce jet engines under license.

SOCIAL CONTEXT AND FUTURE PROSPECTS

Jet engines have developed rapidly since the 1940's and have come to dominate the propulsion market for atmospheric flight vehicles. Because the air that makes up more than 95 percent of the working fluid is free and does not have to be carried up from the ground, air-breathing propulsion offers a huge increase in specific impulse over rocket engines for flight in the atmosphere. As technology advances to enable rotating machinery to tolerate higher temperatures, pressures, and stresses, jet engines can become substantially lighter and more efficient per unit of thrust. Hydrogen-fueled engines can operate much more efficiently than hydrocarbon-fueled engines. Turbo-ramjet engines may one day enable swift and inexpensive access to space. Helicopter engines and the lift fans developed for vertical-landing fighter planes bring personal air vehicles closer to reality. Supersonic airline travel using hydrogen fuel is much closer to becoming routine. At the other extreme, micro jet engines are finding use in devices to power actuators for many applications, including surgical tools and devices to control stall on wings and larger engines. These advances create the prospect for revolutionary advances in human capabilities. Jet engine development promises to remain a leading-edge technological field for many years to come.

Narayanan M. Komerath, Ph.D.

FURTHER READING

Constant, Edward W. *The Origin of the Turbojet Revolution.* Baltimore: The Johns Hopkins University Press, 1980. Traces the history of the turbojet.

Conway, Erik M. *High-Speed Dreams: NASA and the Technopolitics of Supersonic Transportation, 1945-1999.* 2005. Reprint. Baltimore: The Johns Hopkins University Press, 2008. Written by a NASA historian, this book traces the early development of engines for military supersonic flight and then analyzes the economic and political factors that led to the failure of three attempts to develop an American supersonic airliner.

Golley, John, Frank Whittle, and Bill Gunston. *Whittle: The True Story.* Washington, D.C.: Smithsonian Institution Press, 1987. Biography of Whittle, giving the British perspective on the invention and development of early jet engines.

Hill, Philip Graham, and Carl R. Peterson. *Mechanics and Thermodynamics of Propulsion.* 2d ed. Reading, Mass.: Addison-Wesley, 2010. This popular textbook provides succinct presentations of the theory and applications of propulsion.

Hünecke, Klaus. *Jet Engines: Fundamentals of Theory, Design and Operation.* 1998. Reprint. London: Zenith Press, 2005. Begins with the history of jet engines and continues with the basics of how they operate and their structure.

Mattingly, Jack. *Elements of Gas Turbine Propulsion.* 1996. Reprint. Reston, Va.: American Institute of Aeronautics and Astronautics, 2005. Contains large sections on turbomachines and fighter engines. The authoritative foreword by Hans Pabst von Ohain provides historical insight and perspective.

WEB SITES

American Institute of Aeronautics and Astronautics
http://www.aiaa.org

General Electric Aviation
http://www.geae.com/engines/index.html

National Aeronautic Association
http://www.naa.aero

National Aeronautics and Space Administration
Gas Turbine Propulsion
http://www.grc.nasa.gov/WWW/K-12/airplane/turbine.html

See also: Aeronautics and Aviation; Propulsion Technologies; Spacecraft Engineering; Vertical Takeoff and Landing Aircraft.

K

KINESIOLOGY

FIELDS OF STUDY

Anatomy; physics; biomechanics; physiology; exercise physiology; chemistry; biochemistry; motor behavior; sport psychology; sport sociology; coaching; ergonomics.

SUMMARY

Kinesiology is a multidisciplinary field that specializes in the science of human movement. It can focus on improving health or performance or on preventing injuries. Traditionally, kinesiology concentrated on the structural anatomy and the mechanics of movement. Later, the field expanded to include the physiological and mental aspects of movement. Some common applications of kinesiology include proper running and jumping mechanics, correct weightlifting techniques, and perfecting the execution of sports skills. Kinesiology also can encompass the mechanics of work-related activities such as lifting, repetitive movements, and sitting at a desk.

KEY TERMS AND CONCEPTS

- **Action:** Type of movement made by a muscle contraction.
- **Adenosine Triphosphate:** Form of stored energy that can be used directly by muscle cells for contraction.
- **Aerobic Exercise:** Physical activity that is vigorous, continuous, and rhythmical.
- **Alveoli:** Lung structure in which oxygen is transferred from air to the blood.
- **Antagonist:** Muscle that has an opposite action from its paired muscle and limits its action.
- **Applied Kinesiology:** Diagnostic technique used in chiropractic medicine.
- **Hemoglobin:** Component in blood that attaches to and transfers oxygen.

- **Insertion:** Point of muscle attachment on the bone that moves.
- **Isokinetic Machine:** Instrument that measures the strength of muscles through the joint's full range of motion.
- **Metabolism:** Sum of all chemical reactions in a living organism.
- **Motor:** Relating to or involving movements of a muscle.
- **Origin:** Point of muscle attachment on the bone that remains stationary.

DEFINITION AND BASIC PRINCIPLES

Kinesiology is the study of human movement. Although it technically is not limited to humans, as an applied science, it almost always is. Kinesiology is primarily concerned with all kinds of physical activity, including competitive sports and activities designed to maintain and improve health. It is a major part of physical rehabilitation and injury prevention, and it can be used to design workstations that are ergonomic and minimize hazards.

A key component of kinesiology is to understand which muscles are involved in specific movements. Each major joint in the human body has identifiable planes of movement. Each movement in the plane is named according to its direction, and each movement has certain muscles that contribute to it. Muscles can be primary movers at the joint or can assist in the movement. Using this information, kinesiologists can observe movement patterns, identify which joints are involved, determine the movement at each joint, and ascertain which muscles contribute to the movement. Furthermore, they can develop training programs to work the appropriate muscles to improve the strength, movement, and muscle balance at the joint.

The field of kinesiology should not be confused with applied kinesiology, which is a diagnostic technique used in chiropractic medicine. The clinician applies a force to a muscle or muscle group, and the patient resists the force. Based on the patient's response to the force, the clinician makes a diagnosis. This is not a generally accepted practice in medicine and not related to the field of kinesiology.

BACKGROUND AND HISTORY

The field of kinesiology is believed to have begun in ancient Greece. The term "kinesiology" comes from the Greek words *kinein* meaning "to move" and *logos* meaning "discourse." Aristotle is considered the father of kinesiology for his work using geometry to describe the movement of humans. It was not until the fifteenth century that Leonardo da Vinci helped expand the knowledge of human movement by studying the mechanics of standing, walking, and jumping.

One of the greatest contributors to kinesiology, Sir Isaac Newton, did not actually study human movement. In the late 1600's and early 1700's, Newton developed three laws of rest and movement that laid the foundation for the analysis of human movement in the following years. The development of photography in the 1800's enabled researchers to study how animals moved by taking a number of pictures in rapid succession. They first studied horses, then humans. Cinematography further enabled researchers to understand human movement.

In 1990, the American Academy of Physical Education (later the American Academy of Kinesiology and Physical Education) recommended that programs of study involving human movement be called kinesiology. This idea gained wide acceptance, and kinesiology as a field came to include many other specialized fields beyond the traditional anatomy and biomechanics. Kinesiology can include any area that relates to human movement, such as history, sociology, psychology, physiology, philosophy, and motor behavior.

HOW IT WORKS

Movement Analysis. Traditionally, courses in kinesiology focused on the anatomy and mechanics of human movement. Although more advanced courses sometimes take the same approach, introductory courses tend to be an overview of the broader field of kinesiology.

A thorough understanding of muscle and skeletal anatomy is required to understand movement. This includes the names of the bones and the names, origins, insertions, and actions of the major muscles of the human body. The origin and insertion of a

muscle are the locations where it attaches to bones. These locations determine the action of the muscle, based on the angle of pull on the bones. This knowledge is important in understanding which muscles are involved in specific movements.

To describe movements of the body, planes and rotations are defined at the various joints. Some of these movements include flexion, extension, adduction, abduction, internal rotation, and external rotation. Kinesiologists watch a specific movement and determine the actions at the joint or joints being evaluated. They can also evaluate movements to determine if they are completed properly. Slow-motion cinematography is helpful when analyzing the very fast movements often found in sports.

After determining the movement at a joint, kinesiologists can use their knowledge of anatomy to determine which muscles are involved. Additionally, they can use this information to develop strength training programs to develop the specific muscles used in the activity. It is important to note that muscles that generate a movement (agonist muscles) have antagonist muscles that stop the movement. Therefore, strengthening an antagonist muscle is just as important as strengthening the muscle that initiates the movement.

Physiological Function. Human movement requires oxygen and energy beyond the levels needed to simply survive. Exercise physiology includes the study of how the body gets food and oxygen from the environment to the working muscles. Oxygen is very important for sustaining activity during intense exercise, during which the cardiovascular, respiratory, and muscular systems are primarily involved.

The respiratory system transfers oxygen from the atmosphere into the blood. Air, which is about 21 percent oxygen, is inhaled into the lungs, and much of it enters the alveoli at the ends of the airways. Oxygen diffuses from the alveoli into the blood, where it binds with the hemoglobin. The pumping action of the heart carries the blood to the muscles. When the oxygenated blood gets near muscle cells that are low in oxygen, the hemoglobin releases the oxygen, which goes into the cell. A strong, healthy heart and blood with normal amounts of hemoglobin are capable of delivering sufficient amounts of oxygen to support high levels of muscle movement and activity.

When oxygen enters the muscle cell, it is metabolized. It goes through a series of chemical reactions in which oxygen and energy are converted into adenosine triphosphate (ATP). Only ATP can be used by the muscle to make the fibers contract and the body move. When highly trained people exercise at a very high intensity, the amount of oxygen consumed can increase more than twenty times. It is the ability of the muscle cells to use oxygen to produce ATP that limits high-intensity human movement. The oxygen consumed and several other measures can be determined with a metabolic cart, which is an important type of testing equipment used in exercise physiology.

Behavioral Control. The areas of kinesiology that involve the brain and nervous system are motor development, sport psychology, and sport sociology. Motor development is concerned with skills that take stored movement patterns in the brain and communicates them to the muscles. Most skilled movements are an organized, synchronous set of smaller movements. Therefore, the series of muscle contractions needed to perform the skill must be stored and retrieved often. Practicing the movement patterns on a regular basis is required to refine the skills.

Sport psychology and sociology are the segments of kinesiology that relate to the mental aspects of human movement and performance. One of the largest fields is clinical sport psychology, in which psychologists assist athletes with aggression, stress, motivation, mood, adherence, and leadership as well as a number of other related issues. Sport sociology is a smaller element of kinesiology that focuses on social relationships in sport and how sports affect different segments of society and organizational structures.

APPLICATIONS AND PRODUCTS

Rehabilitation. Kinesiology techniques, especially those regarding the muscular and skeletal systems, are used in physical rehabilitation every day. Physicians, chiropractors, physical therapists, and athletic trainers use their knowledge of muscle origins, insertions, and actions to diagnose injuries and determine exercises that will help people recover from them. Isokinetic machines can be used to determine the muscular strength at any joint through the entire range of motion. Identifying points of weakness during movement helps determine which muscles need to be strengthened. A muscle and its antagonist can be tested to see if one is weaker than the other. Good muscle balance is needed to prevent injuries and reinjury. Limbs (arms and legs) can be tested to see if the left and right sides are equally strong or if one side, usually the injured side, is weaker. Kinesiology is a very important component of muscular and skeletal rehabilitation.

Cardiovascular rehabilitation focuses on the area of kinesiology that includes the cardiovascular and respiratory systems. In this area, exercise is used to strengthen the heart muscle. When a person has a heart attack, some of the heart tissue dies and is replaced by connective tissue. With some of the muscle gone and not able to pump blood, the heart is weaker. In cardiac rehabilitation, patients perform aerobic exercises to increase the strength of the heart muscle that is left. Through kinesiology applications, patients can strengthen their hearts and improve their ability to engage in daily living activities.

Health Promotion and Injury Prevention. The principles of kinesiology are used to maintain good health and prevent injuries. For many years, exercise has been recognized as an important component of health promotion. Kinesiology studies have involved developing and researching the best types of exercise to improve health. Exercise specialists such as physical education instructors, fitness trainers, and sports coaches rely on kinesiology to help people exercise safely and efficiently. They create exercise programs or develop exercise sessions based on kinesiology principles. Exercise can take the form of group classes or individual instruction (personal training).

Cardiovascular exercise is very important for health and fitness. Exercise specialists are often charged with helping a person develop the endurance to engage in physical activity for extended periods. Based on an assessment of the person's health and fitness, exercise specialists use exercise physiology principles to write a prescription for activity. The prescription, or exercise plan, is often based on heart rate so that the individual can use his or her pulse to gauge whether the right level of effort is being attained. Another important type of exercise for health and fitness is strength training. Exercise specialists use kinesiology principles to demonstrate proper lifting techniques and help clients get stronger. Attention is paid to balancing muscle strength across the body. Movement mechanics are also used to improve flexibility. A stretching program is designed to improve or maintain flexibility throughout the body and help clients move more freely. A good training program will help clients stay healthier throughout their lives and enable them to perform the activities of daily living more easily and longer.

An overriding factor in exercise for health is adherence to the program. Health benefits are obtained only with regular participation. The psychological area of kinesiology studies provides information about getting and keeping exercisers motivated.

Coaching. Coaches use kinesiology in many of their activities. Sports skill development requires regular evaluation of movement to determine if the sports skill is being performed properly, which maximizes performance and reduces the chances of injuries. Coaches must consider all the involved joints, the type of muscle contractions, and the planes of movements. Of great importance is the synchronization of the movements around the involved joints. Energy transfer from one joint to the next is critical for superior performance. Coaches use their knowledge of kinesiology to teach proper sport skills.

Coaches also must use kinesiology for strength and conditioning. Weightlifting and other exercises must be performed with proper mechanics. The variety of available equipment makes a basic understanding of kinesiology imperative for teaching athletes effectively. Additionally, coaches must determine which exercises should be performed and which muscles are to be strengthened.

Coaches also use sport psychology to motivate athletes to perform their best and to develop the leadership skills that are important for success. Athletic competition can be stressful, especially as most athletes must deal with demands on their time and concentration that stem from their social and academic or work-related obligation. Coaches often teach athletes stress management techniques to help them cope.

Ergonomics. Kinesiology can be applied at the workplace in the area of ergonomics, which science uses body mechanics concepts to design workstations that are more comfortable and minimize overuse injuries from repetitive movements. Any workstation can be analyzed and appropriately modified. Ergonomic solutions for people in desk jobs are relatively simple, but finding answers for people in jobs that require lifting and carrying require the use of more kinesiology principles. After the proper techniques for lifting and carrying have been determined, the worker must be trained to perform the movements properly.

IMPACT ON INDUSTRY

The health, fitness, and sports sector of the U.S. economy has been undergoing substantial growth, and this trend is also found in the international market. In 2009, gyms and health and fitness clubs (includes ice and roller rinks, tennis courts, and swimming facilities) generated $23.6 billion in revenues. Although economically developed countries spend the most on health and fitness, growth is found in most countries of the world. Kinesiology applications for the medical and business industries are also common.

Sports and Fitness. Traditionally, the field of kinesiology has been applied to sports skills and conditioning for performance. This is still the major

---

Fascinating Facts About Kinesiology

- In 1972, University of Oregon track coach Bill Bowerman invented the Nike Waffle Racer, a shoe with extra traction, when he poured a liquid rubber compound into his wife's waffle iron.

- Exercise is believed to boost mental function, increase energy, reduce stress, build relationships, prevent disease, strengthen the heart, and allow a person to eat more. However, on an average day, only about 16 percent of Americans over the age of fifteen participated in sports or exercised.

- In 2007, there were 9.9 million health club members over the age of fifty-five, more than four times the number in 1990.

- In 2010, the Washington, D.C., area was ranked as the fittest metropolitan area in the United States for the third time in a row.

- A leading professional organization in the kinesiology field, the American College of Sports Medicine, was founded in 1954.

- There are 639 skeletal muscles and 206 bones in the human body. Numbers vary slightly depending on how the muscles and bones are counted.

- Indiana University psychologist Norman Triplett is considered the first sport psychologist for his work that found exercisers performed better in a group than when exercising alone. He concluded that people became more competitive in a group setting.

- Exercise physiology in the United States is believed to have originated at the Harvard Fatigue Laboratory in 1927.

application. When working with athletes, coaches and trainers use their knowledge in areas such as anatomy, biomechanics, exercise physiology, and sports psychology to help athletes maximize their performance. World records continue to be broken in most sports, so it appears that athletic performance can still be improved. The level of competition in many sports, whether professional or amateur, has increased. Much of the credit for these improvements goes to coaches, trainers, and others who have had education in kinesiology.

Higher levels of training and more intense competition has resulted in an increase in injuries. Kinesiology principles have been used to develop programs to rehabilitate injured athletes and to decrease the chances of injury. As sports become more popular and generate higher revenues, sports managers want to ensure that their best players remain active during the season and, if injured, return to play as quickly as possible. Kinesiology training enables rehabilitation professionals to facilitate the athlete's return to competition.

Workplace. Since the 1970's, the numbers of organizations offering fitness centers at their work sites has increased. Employees can more easily fit exercise into their workday at a facility that is convenient and often competitively priced. By offering group classes and quality supervision in the weight rooms, a company can be assured that its employees can exercise safely and obtain the health benefits. Workplace health programs have been shown to reduce health care costs and absenteeism and to improve productivity.

Kinesiology concepts—in the form of ergonomics—have been applied to workstations. Movement analysis can lead to better workstations and fewer overuse and acute injuries. Ergonomics can save business and industry millions of dollars in medical costs and workers' compensation.

Health. Kinesiology programs have placed a greater emphasis on recreational sports and health since the 1970's, when interest in fitness began to grow among the general population. People began running, bicycling, and lifting weights. Cardiovascular exercise, including aerobics, gained popularity. Gyms and clubs began offering places to exercise safety and obtain information about health and fitness. Helping participants reach and maintain healthy weights is one of the major benefits of these programs.

During this same time period, exercise began to be used to rehabilitate individuals with chronic diseases. The major program has been cardiac rehabilitation, in which patients who have had heart surgery or experienced a heart attack or other problem are placed in a supervised and monitored exercise program. Patients who participate in cardiac rehabilitation typically recover faster and are more likely to resume an independent lifestyle. Cardiac rehabilitation programs may include individuals with other chronic diseases such as diabetes, lung disease, and cancer. Sometimes specialized programs are created for those with other chronic diseases, but the common goal is to design an effective program that improves the improved health and function of participants.

CAREERS AND COURSE WORK

Students who are interested in a career in kinesiology need a basic science background in anatomy, physics, chemistry, and math. Kinesiology courses apply these basic science principles to human movement. Although a bachelor's degree provides an overview of the major areas within kinesiology, a master's degree allows for specialization. An advanced degree is not required for most entry-level positions but it does provide more opportunities, particularly in management. Some graduates may prefer to use their kinesiology education to pursue physical therapy, chiropractic medicine, or other clinical degrees. Those who want to do research in kinesiology will need a doctorate.

Careers in the kinesiology field are many and varied. Students can coach and train athletes. With the increase in the number of youth competing in sports and the parents' willingness to pay, opportunities for coaching are expanding. The most common job is that of a health and fitness trainer in a private or corporate fitness club or a community recreation facility. An experienced health and fitness trainer can become a personal trainer and work with clients individually or in small groups to prescribe training programs and help them meet their exercise goals.

Clinical settings such as hospitals offer positions in cardiac rehabilitation and sports medicine. Exercise physiologists can work in cardiac rehabilitation, and athletic trainers can work in sports medicine. These positions are for those who like the clinical setting and working with individuals and smaller groups. Some opportunities exist for sport psychologists who use mental strategies with athletes to improve performance.

SOCIAL CONTEXT AND FUTURE PROSPECTS

Programs that use kinesiology professionals have traditionally been paid for by the participants. Kinesiology benefited those who could pay for good sport trainers, health and fitness clubs, and sports medicine. More programs, including youth sports programs, are emerging to provide access to exercise facilities and sports training to those who cannot afford to pay. Some insurance plans provide free or discounted gym membership, and many companies offer employee gyms or discounted gym memberships.

Kinesiology continues to be a growing field. Professional sports remain popular around the world, and the emergence of new competitive sports such as extreme sports results in more research and more need for people trained in kinesiology. Also, more people are interested in maintaining good health, and positions in fitness training are likely to increase to meet the demands of these exercisers. With more people exercising and participating in sports, the number of sports-related injuries is likely to grow. Kinesiology education will be needed to train rehabilitation professionals to research injuries, educate the public about injuries, and rehabilitate those injured.

Bradley R. A. Wilson, Ph.D.

FURTHER READING

Floyd, R. T. *Manual of Structural Kinesiology.* New York: McGraw-Hill, 2009. An introductory text that focuses on movements of the major joints.

Hoffman, Shirl J., ed. *Introduction to Kinesiology: Studying Physical Activity.* 3d ed. Champaign, Ill.: Human Kinetics, 2009. Includes an overview of the major areas of kinesiology and a discussion of professions.

Klavora, P. *Foundations of Kinesiology: Studying Human Movement and Health.* Toronto: Sport Books, 2007. An overview of basic anatomy and physiology, human performance, and the major components of kinesiology.

_____. *Introduction to Kinesiology: A Biophysical Perspective.* Toronto: Sport Books, 2009. A succinct text on basic kinesiology and human movement.

Kornspan, Alan S. *Fundamentals of Sport and Exercise Psychology.* Champaign, Ill.: Human Kinetics, 2009. Discusses basic opportunities and goals in the field.

Oatis, Carol A. *Kinesiology: The Mechanics of Human Movement.* Baltimore: Lippincott Williams & Wilkins, 2009. Covers biomechanics, movement at the major joints, and posture.

WEB SITES

American Academy of Kinesiology and Physical Education
http://www.aakpe.org

American Kinesiology Association
http://www.americankinesiology.org

Energy Kinesiology Association
http://www.energykinesiology.com

See also: Biomechanics; Ergonomics; Occupational Health; Rehabilitation Engineering; Sports Engineering.

L

LANDSCAPE ARCHITECTURE AND ENGINEERING

FIELDS OF STUDY

Architecture; ecology; forestry; landscape engineering; industrial design; landscape design; environmental science; ecological restoration; natural resource management; physical, natural, and social sciences; land-use planning; environmental psychology; botany; landscape ecology; geographic information systems (GIS); computer modeling.

SUMMARY

Landscape architecture is the development and design of landscapes that take into account environmental and aesthetic concerns while addressing practical human use of the space. Landscape architecture aims to incorporate the natural environment into the design and construction of a space in a minimally disruptive way that also conserves natural resources in the short and long term. This can pose challenges and create limitations in terms of techniques, applications, and judgments used in the design and construction processes. Landscape architects design and plan spaces such as gardens, parks, campuses, and commercial centers, and restore historic sites and degraded environments, such as landfills and mined land.

KEY TERMS AND CONCEPTS

- **Degraded Environment and Ecosystem:** Space that has been negatively affected by humans (through events such as oil spills or the clear cutting of forests) or natural disasters (such as earthquakes or tsunamis).
- **Greenfield Site:** Forested area, agricultural land, or any area outside of a city where development can occur without hindrance from previous structures and that has been identified for industrial or commercial development.
- **Hardscape:** Inanimate or nonliving elements of landscaping, including features such as water fountains, masonry, and woodwork.
- **Landscape Design:** Art of designing and altering landscape features for function or aesthetics; frequently divided into hardscape and softscape design.
- **Landscape Ecology:** Study of the relationship between spatial patterns and ecological processes on a large number of diverse landscapes.
- **Native Vegetation:** Plant or tree that is indigenous to the area.
- **Softscape:** Animate or living elements of landscaping, such as plants and trees.
- **Sustainability:** Degree of success in integration of a site's natural land, water, and energy resources with human needs and structures in development of an area.

DEFINITION AND BASIC PRINCIPLES

Landscape architecture attempts to unify the human-constructed environment with the existing natural environment to create a practical, sustainable, and aesthetically pleasing space. Landscape architecture tends to incorporate urban planning, ecological restoration, and green infrastructure in the design of open-air public spaces.

The title of landscape architect is often used interchangeably with landscape designer, and although both do undertake landscape design, the American Society of Landscape Architects states that the two professions are different because a landscape architect must have a higher level of training and skill and usually needs post-secondary education. Fundamentally, it takes a greater investment of time to study landscape architecture.

All visual, ecological, and cultural issues involved in the development of the natural landscape must be considered in landscape architecture. Essentially, landscape architecture involves designing open-air

American landscape architect Frederick Law Olmsted (1822-1903). (Getty Images)

public spaces that meet certain environmental, social, and aesthetic goals while making prudent use of finite land and natural resources. It follows the theories, principles, and practices of general architecture and also considers and assesses the social, ecological, and geological processes occurring within and affecting the landscape.

BACKGROUND AND HISTORY

Human society has long attempted to blend natural and artificial environments, as evidenced by the English theater gardens of the seventeenth and eighteenth centuries, the Zen gardens in China and Japan, and large urban parks in the United States. Although the design philosophies and approaches toward unifying architecture with the natural environment differ, the concept is shared by Western and Eastern cultures.

Landscape architecture, particularly as it relates to gardening and building design, is a discipline as old as human society and the human desire to control

and modify the natural landscape. The term "landscape architecture" was first coined by the Scottish gentleman Gilbert Laing Meason in 1828. The concept of landscape architecture reached the mainstream about thirty years later, when Frederick Law Olmsted and Calvert Vaux were commissioned to design Central Park in New York City. Olmsted is considered the father of American landscape architecture. The profession continued to gain significant credibility and growth at the beginning of the twentieth century with the founding of the American Society of Landscape Architects in 1899 and the introduction of a landscape architecture program at Harvard University in 1900. This program was instrumental in shifting the focus of landscape architecture from private gardens to open-air public spaces.

Landscape architecture has had a long history in the West, but countries in other parts of the world have also been active in the field. Since the mid-twentieth century, a significant amount of research in and application of landscape architecture has taken place in China. There, landscape architecture has advanced significantly, focusing on incorporating gardens into urban landscapes.

How IT WORKS

Landscape architecture involves the design and development of outdoor public areas that are not only functional but also aesthetically pleasing and environmentally friendly. It is the job of a landscape architect to blend design, function, beauty, and the natural environment to achieve desired social, environmental, and aesthetic outcomes.

Typically, landscape architecture involves overseeing project design, often of urban regeneration programs; site inspections of both ecological aspects and preexisting human-made structures; and environmental impact assessments. Landscape architects also design projects and present their proposals before government bodies charged with planning and conservation responsibilities. The design process usually involves a number of steps, including preplanning, project planning, preliminary and final design, report writing and approvals, and implementation and development. Landscape architects work in conjunction with architects, surveyors, hydrologists, environmental scientists, foresters, and engineers to determine the design and arrangement of structures and infrastructure and the most appropriate way to protect and use natural areas and resources. Landscape architects analyze the natural elements present in the area and produce detailed designs that incorporate environmental aspects of construction such as climate, soil, topography, vegetation, and habitat, with structures for human use, including walkways, fountains, and decorative features.

APPLICATIONS AND PRODUCTS

Landscape architecture is a multidisciplinary field of applied science and engineering, and its applications are vast and varied. Wherever humans and nature meet, landscape architecture theory and application is necessary. The applications of landscape architecture are diverse and range from small to large. They include the design and management of vast wilderness areas and national parks, small-scale urban gardens and public parks, resorts, residential estates, and open areas in business parks. Landscape architecture also deals with the reclamation of environmentally degraded areas, such as landfills, and the preservation of historic sites and districts. Landscape architects must be versatile, as their work can encompass urban and rural areas and involve both terrestrial and aquatic environments.

For the most part, however, landscape architecture is employed in urban design and planning, environmental restoration, ecotourism, parks, gardens, recreational area design and development, and green infrastructure implementation.

Urban Design. Urban design generally refers to the organizational and developmental stage of landscape architecture in towns and cities and often involves development of public spaces in urban areas. Landscape architects use the knowledge gained through preliminary assessments to determine the ideal site for development within a specific urban environmental space. For instance, when choosing a site for what is to be a widely used structure or area, a landscape architect would want the site to be easily accessible, near public transportation or a densely populated residential area. This would minimize travel time and expense, reduce car emissions and pollution, and conserve fossils fuels.

Environmental Restoration. The growing human population is placing significant pressure on the natural landscape. The impact that humans have on nature can be seen in polluted bodies of water, abandoned industrial areas, dilapidated waterfronts, neglected

railroad yards, and landfills. As humans continue to encroach into natural habitats and true greenfield sites become rarer, landscape architects and designers look toward degraded areas, either to restore them to their natural state or to permit renewed human use. The reclamation of derelict urban and rural areas for public use is vital for the sustainable development of human communities. It not only demonstrates a commitment to sustainable land use and protection of the environment but also improves the quality of life for members of the community.

Ecotourism, Parks, Gardens, and Recreational Areas. Many people living within urban areas use and visit parks and gardens. Landscape architects design and develop these areas for both recreational and aesthetic purposes, while minimizing their impact on the natural environment. Landscape architects also design resorts and ecotourism infrastructures. Tourism based on ecological principles has become increasingly popular. One of the applications of landscape architecture is the development of green infrastructure in wilderness areas that will accommodate tourists without adversely affecting the surrounding landscape and the native flora and fauna.

IMPACT ON INDUSTRY

Landscape architects, designers, and engineers have moved away from the manipulation of landscapes to the design of landscapes that are aligned with the environmental elements within a given area. Instead of fighting environmental conditions, modern landscape design works with existing conditions to provide for human use. This shift in approach requires that landscape architects not only know the visible conditions but also understand subsurface conditions (such as groundwater and soils), conservation and larger planning requirements and directives, and the laws that guide their efforts.

Major Organizations. The profession of landscape architecture is globally represented by the International Federation of Landscape Architects, founded at Cambridge, England, in 1948. The federation aims to cultivate a public understanding of landscape architecture as being environmentally, culturally, and socially responsible. Another major organization is the American Society of Landscape Architects, a professional association for landscape architects founded in 1899, with members in the United States and in more than sixty countries around the world. Despite the significant number of organizations involved in landscape architecture, the level of government legislation and regulation within the field varies significantly among countries.

Many countries possess professional landscape architecture institutes and agencies that are responsible for the promotion and regulation of the profession. Such institutes include the Council of Landscape Architectural Registration Boards in the United States and Canada, the Landscape Institute in Great Britain, the Canadian Society of Landscape Architects, and the Australian Institute of Landscape Architects.

Government Research and Regulation. Most governments apply landscape architecture principles and practices to urban design and town planning. Evaluating the environmental impact of development within urban areas is becoming ever more important. In particular, many governments are investigating the reclamation of degraded sites, including their restoration and potential use.

In 2009, Barack Obama's administration created the Partnership for Sustainable Communities to encourage the sustainable development and restoration of urban areas. In addition, the U.S. Congress has considered bills that relate to numerous aspects of landscape architecture, such as green infrastructure, historic restoration, and environmental reclamation. In other countries, including China, concepts such as underground landscape design and architecture are being considered in an effort to maximize use of the available space.

CAREERS AND COURSE WORK

For entry into the landscape architecture profession, a person should obtain a bachelor of landscape architecture or a bachelor of science in landscape architecture from an accredited university. Programs in landscape architecture have been offered at more than sixty universities in the United States, including Harvard University, Pennsylvania State University, and the University of Wisconsin, Madison. Students who seek a career in landscape architecture will usually have a talent for design and possess strong analytical skills. They should also have a strong interest in the environment. Landscape architecture courses include the history of landscape architecture, surveying, landscape design and construction, landscape ecology, plant and soil science, site design, and urban and regional planning.

Upon completion of their undergraduate studies, students will have gained knowledge and awareness of urban and landscape design, ecological planning, heritage conservation, regional landscape planning, land development, park and recreation planning, and environmental reclamation and restoration. In addition, students will be familiar with computer-aided design, model building, geographic information systems (GIS), and three-dimensional visualization and simulation.

In many countries, including the United States, the United Kingdom, Canada, and Australia, graduates must register to obtain a license to practice landscape architecture, and they may have to continue their education to maintain registration. Registration processes and procedures, however, differ from country to country. Many landscape architects are self-employed, but graduates can also embark on careers in the private sector, in specialized government organizations and agencies, and in universities undertaking teaching and research.

SOCIAL CONTEXT AND FUTURE PROSPECTS

As the public's environmental consciousness develops, greater opportunities in landscape architecture will emerge, as will the necessity for lasting and significant decisions regarding landscape planning and design. Novel projects such as green roofing to reduce the environmental and economic costs of heating and cooling and water conservation and pollution reduction through design and management of storm water are being investigated and implemented by landscape architects in the early twenty-first century.

In addition, as GIS and computer graphics technologies continue to advance, so will the ability of landscape architects to design projects using simulated three-dimensional visualizations and modeling. In particular, GIS-based data can assist in creating realistic representations of softscape design elements such as vegetation and improve terrain and landscape imagery. Advances in technology will allow landscape architects to produce highly detailed simulations of natural and urban environments, enabling them to provide much more accurate predictions of possible environmental impacts.

Christine Watts, Ph.D., B.App.Sc., B.Sc.

FURTHER READING

Benedict Mark A., and Edward T. McMahon. *Green Infrastructure: Linking Landscapes and Communities.* Washington, D.C.: Island Press, 2006. Examines the large-scale design and integrated action involved in the advancement of green infrastructure and focuses on smart land development.

Downton, Paul. *Ecopolis: Architecture and Cities for a Changing Climate.* Collingwood, Australia: CSIRO, 2009. Discusses architectural theories and practices, planning, and ecology, with a focus on the development of sustainable future cities through the integration of academic and practical knowledge.

Dramstad, Wenche, James D. Olson, and Richard T. T. Forman. *Landscape Ecology Principles in Landscape Architecture and Land-Use Planning.* Washington, D.C.: Island Press, 1996. An overview of landscape ecology theory and principles with a focus on how they are applied in design and planning.

Newman, Peter, and Isabella Jennings. *Cities as Sustainable Ecosystems: Principles and Practices.* Washington, D.C.: Island Press, 2008. Examines how living systems can be applied to sustainable urban design strategies through the exploration of flows of energy, materials, and information. Considers the relationship between human and nonhuman elements.

Register, Richard. *EcoCities: Rebuilding Cities in Balance with Nature.* Gabriola, B.C.: New Society Publishers, 2006. This book examines the concepts and ideas behind rebuilding urbanized areas based on the principles of ecology, with the aim of long-term sustainability, cultural vitality, and environmental health.

Swaffield, Simon. *Theory in Landscape Architecture: A Reader.* Philadelphia: University of Pennsylvania Press, 2002. A resource for seminal theory in the field, covering fifty years of study on the relationship between landscape and social, cultural, and political structures.

WEB SITES

American Society of Landscape Architects
http://www.asla.org

International Federation of Landscape Architects
http://www.iflaonline.org

See also: Civil Engineering; Ecological Engineering; Environmental Engineering; Landscape Ecology; Land-Use Management.

LANDSCAPE ECOLOGY

FIELDS OF STUDY

Ecology; spatial ecology; conservation; forest ecology; forestry; agronomy; disturbance ecology; ecosystem management; environmental planning; natural resources policy and management; international development; regional and urban planning; landscape geography; landscape architecture; human ecology; animal behavior; plant biology; spatial statistics; modeling; remote sensing; geographic information systems (GIS).

SUMMARY

Landscape ecology is a relatively new field of science that focuses on spatial heterogeneity and is the study of how ecosystems, including the built environment, are arranged and how these arrangements affect the wildlife and environmental conditions that form them. In other words, it is the study of the abundance, distribution, and origin of elements within landscapes, coupled with their impact on the ecology of an area. Landscape ecology brings a spatial perspective to the integration of people with natural ecosystems. Because of this emphasis, landscape ecology has the potential to be of significant use in ecological and urban planning and management and in solving global scale problems.

KEY TERMS AND CONCEPTS

- **Connectivity:** Continuity, in a spatial sense, of habitat across an area (such as a landscape).
- **Corridor:** Narrow area, including line corridor, strip corridor, and riparian/stream corridor, that differs from the adjacent habitat on both of its sides.
- **Edge:** Outer boundary area of a landscape, which will often exhibit environmental conditions different from those within the landscape.
- **Fragmentation:** Division, separation, or breakup of a landscape into patches and corridors.
- **Heterogeneity:** Condition of diversity or consisting of dissimilar elements.
- **Landscapes:** Diverse land areas made up of repeating and comparable patterns of interconnected and interacting ecosystems.

- **Matrix:** Extensive and highly connected background area of a landscape, which surrounds and influences patches and corridors. Examples include forest or moorland.
- **Patch:** Broad and homogeneous area that differs in some respect from the surrounding landscape.
- **Scale:** Spatial and temporal dimensions of a landscape measured by grain (finest level of measurement) and extent (total area sample).

DEFINITION AND BASIC PRINCIPLES

Landscape ecology, also referred to as landscape science, is an interdisciplinary science. Ecology is the study of the relationships and interconnections between organisms and the environment in which they live in an attempt to determine how living (biotic) and nonliving (abiotic) patterns and processes influence organism abundance and distribution. A landscape is a diverse land area made up of repeating and comparable patterns of interconnected and interacting ecosystems. According to the International Association for Landscape Ecology, landscape ecology is the study of "spatial variation in landscapes at a variety of scales" that "includes the biophysical and societal causes and consequences of landscape heterogeneity."

One of the most fundamental concepts of landscape ecology is that all landscapes are heterogeneous, no matter whether the environment is natural or modified or large or small. Landscape ecology most significantly deviates from classic ecological studies (and more closely resembles geography) in its focus on spatial patterns and its emphasis on broad-scale research on the ecological consequences of spatial patterning on ecosystems. Because landscape ecologists study such spatial heterogeneity, landscape ecology theory and concepts are often considered to be somewhat outside the rigid structure of traditional science.

BACKGROUND AND HISTORY

Although landscape ecology is considered to be a young scientific discipline, the study of landscape ecology has a rich history in central and Eastern Europe and, as a concept, can be traced back to the 1930's. The German geographer Carl Troll first coined the term in 1939 based on the emergence of aerial photography as a tool of geography. Although he believed landscape ecology to be more of a philosophy than a branch of science, he also stated that it was the "perfect marriage between geography and biology." Over the following decades, landscape ecology gained acknowledgment as a branch of ecology in its own right but primarily in central and Eastern Europe and in strong connection with land planning.

The study of landscape ecology gained a measure of global recognition in 1981 following the first international landscape ecology congress in the Netherlands. The primary problems that emerged from this congress on landscape ecology differed widely depending on the region involved, and these differences were instrumental in the development of two distinct approaches, the European and the North American.

The concept of landscape ecology in Europe had been based on environments that had undergone significant human manipulation, and research was often focused on cultural landscapes. The North American approach, however, saw a much greater integration of ecological principles, and research tended to concentrate on natural and forest landscapes, studying the relationship of disturbance, spatial patterns, and change. Research in the theory and application of landscape ecology has developed rapidly since the 1980's, as landscape ecology has moved from a regional to a global science and has been significant in ecosystem management and planning.

HOW IT WORKS

The concept of landscape ecology is perhaps best clarified by its fundamental themes related to spatial heterogeneity, patterns, and the relationships between natural science and human beings. Specifically, according to the International Association for Landscape Ecology, landscape ecology research tends to concentrate on "the spatial pattern or structure of landscapes, ranging from wilderness to cities; the relationship between pattern and process in landscapes; the relationship of human activity to landscape pattern, process and change; and the effect of scale and disturbance on the landscape." In other words, research in this area of science attempts to understand spatial heterogeneity and its influence on ecological processes, interactions within and between heterogeneous landscapes, and the

management of such landscapes and spatial heterogeneity.

Landscape ecology differs from classical ecology in that it focuses on spatial patterns and broadscale research on how spatial patterning affects ecosystems. Theories of landscape ecology concentrate on the interaction and impact of people on landscape structure, function, and change. Landscape structure is determined by the spatial relationships between the elements common to all landscapes: patches, edges, boundaries, corridors, and matrices. Landscape function is defined by interactions between spatial elements and patterns, such as quantity of habitat, size of landscape structures, and landscape connectivity. Landscape change is determined by modifications in the structure and function of landscape over time.

Landscape ecology aims to establish the influence of human activity on the structure and function of landscapes. It aims to highlight the interactions between people and landscapes and to provide a basis for managing landscapes, restoring degraded landscapes, recovering from landscape disturbances, and planning land uses.

APPLICATIONS AND PRODUCTS

Landscape ecology is a young branch of ecology, incorporating theories from social, biological, and geographical sciences. Researchers and scientists in the field have stated that significant research is still required and an improved theoretical and conceptual basis needs to be developed. Once a more unified concept of landscape ecology has been established, further applications and products can be developed. Because of its emphasis on the interaction between natural landscapes and people, landscape ecology has the potential to be of significant use in ecological planning and management, and in solving global-scale problems.

Although the field of landscape ecology has been hampered by a lack of unified methodologies and theories, it has the potential to provide a true joining of people with the natural world. The practical application of landscape ecology has experienced rapid growth, especially in solving conflicts between human development and conservation of the natural landscape. The most successful applications of landscape ecology in relation to environmental and spatial planning rely heavily on effective translation and communication of base data, which have been

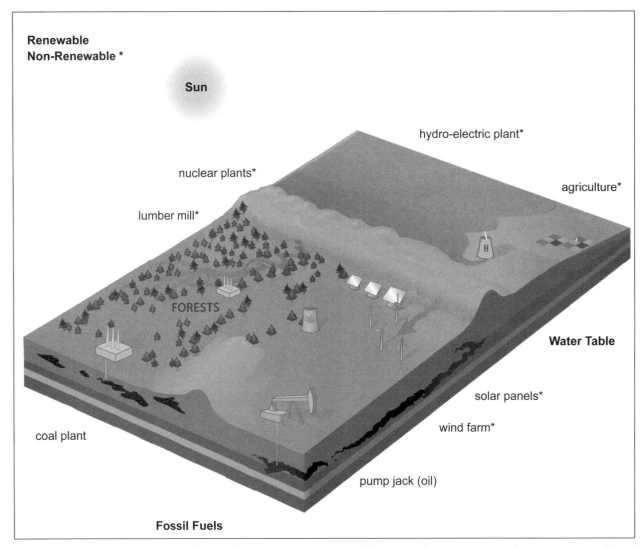

Renewable
Non-Renewable *

Sun

hydro-electric plant*

nuclear plants*

agriculture*

lumber mill*

FORESTS

Water Table

coal plant

solar panels*

wind farm*

pump jack (oil)

Fossil Fuels

Landscape ecology concepts are implemented by most governments in their approach to environmental management and land use, including the harnessing of natural resources.

garnered using such tools as modeling, remote sensing, and geographic information systems (GIS).

As human impact on the environment increases, the demand grows for solid scientific data that will permit more effective management of entire landscapes and will further the identification and understanding of the effects of spatial heterogeneity in relation to land and resource management. Pressure is mounting for resources to be managed as a part of the entire ecosystem rather than in and of themselves and for the temporal and spatial sensitivity of certain management and utilization activities to be

considered. Generally, therefore, applications of landscape ecology can be categorized as being in nature conservation or in land management, which includes sustainable land-use decisions, forest management, ecological/environmental impact assessments, and broadscale environmental monitoring.

Landscape Planning and Land-Use Decisions. Human activity, particularly agriculture, mining, and urbanization, can fundamentally alter landscape structure by changing the dominant land cover and use from natural habitats to human-use areas. This changes the spatial patterns of natural landscapes and influences biodiversity. The study of such modifications in the structures of natural landscapes can be applied to decisions regarding the feasibility and preferred location of a development and the suitability of a specific land use. In agriculture and mining, for example, landscape ecology research tries to establish better options for the management of the environmental risks and hazards that accompany modern farming and mining practices.

Forest Management. A fundamental element of landscape ecology is the impact of forestry activity and natural processes on the spatial patterns of the forest landscape. Landscape ecology research can be applied to forest management to measure growth and decline of the world's forests, understand and minimize habitat fragmentation in forest landscapes, and determine harvesting processes and impacts on both the spatial and the temporal scale.

Ecological/Environmental Risk Assessments. Ecological and environmental risk assessments are fundamental to ecosystem management and policy development. They provide a basis from which to judge, estimate, and clarify the ecological risks associated with environmental developments, hazards, or specific land-use practices. As some hazards, such as climate change, affect large areas and can be influenced by landscape patterns, landscape ecology has been particularly applicable in addressing issues of long-term resource management over a wide range of time periods and distances.

Broadscale Environmental Monitoring. Examining and supervising large-scale landscapes is a significant application of landscape ecology, made more effective by the use of remote sensing and satellite imagery to determine landscape modifications. This application is particularly important because of large-scale problems such as climate change and its subsequent effects on biodiversity.

IMPACT ON INDUSTRY

Although landscape ecology is not a fully mature science, it has certainly experienced rapid development and interest since the 1980's. After its shift from a regional (mostly European) science to a global one, many government agencies, universities, industries, corporations, and organizations worldwide began researching and working in this field of applied science.

Major Organizations. One of the most significant organizations in the field of landscape ecology is the International Association for Landscape Ecology. This organization provides the infrastructure and the means for a global community of landscape ecologists to interact, collaborate, and network across disciplines. Many organizations also use landscape ecology approaches and research in large-scale conservation planning. For example, the World Wide Fund for Nature uses landscape ecology to "develop cost-effective, spatially-explicit strategies that meet the ecological needs of wildlife and habitats while minimizing human-wildlife conflicts and maximizing benefits to resident populations."

Government Research and Regulation. Most governments attempt to use landscape ecology concepts

---

Fascinating Facts About Landscape Ecology

- Forest landscape simulation models are computer programs that predict changes in the forest landscape over time. Landscape ecologists use them to test various ecological assumptions or management options.

- Researchers in China used landscape modeling to analyze the effects of climate warming on the forest ecosystem in Changbai Natural Reserve. Among items of interest was the rate at which Korean pine, spruce, and fir would be displaced by broadleaf trees.

- The U.S. Geological Survey uses gap analysis to identify the distribution of plant and animal species over a large area. This landscape ecology approach uses GIS maps of vegetation and habitat types to look at the broader picture and find problem areas.

- Environmental resource maps are an important tool in landscape ecology. The California Digital Atlas, an online resource, allows users to create maps showing layers of natural resources, including wetlands, rivers, and soils, and other environmental information, such as fire history and urban areas and projected growth.

- The U.S. Environmental Protection Agency's Environmental Monitoring and Assessment Program conducted a five-year program to examine the streams in the western United States and provide a database for long-term monitoring. This program is one of a number of the agency's landscape ecology projects.

in their approach to environmental management. Such agencies and departments include environmental protection agencies and government or academic departments dealing with climate change, environmental issues, natural resource management, city planning, and the mining industry.

CAREERS AND COURSE WORK

Many universities offer science-based undergraduate and postgraduate degrees in landscape ecology. Most commonly, students who wish to pursue a career in landscape ecology will possess an ecological or geographical background. Because landscape ecology is widely recognized as an interdisciplinary field, students can also come from such varied backgrounds as the social sciences, architecture, animal behavior, plant biology, or conservation. Following graduation, students should have a solid understanding of spatial heterogeneity and its influences on natural processes, and the importance of landscape ecology to conservation management. In addition, because landscape ecologists provide the science-based data used for management and planning of areas at a variety of scales, students should gain knowledge and experience in the use of tools such as modeling, remote sensing, and geographic information systems. Students who study landscape ecology can pursue careers in environmental consulting, environmental advocacy, and other environmental fields in the private sector, nongovernmental organizations, and the government, particularly in the development of conservation proposals and environmental management plans.

SOCIAL CONTEXT AND FUTURE PROSPECTS

The world's human population is continuing to grow and the impact of human activity on the natural environment is increasing. To meet the challenge, landscape ecology must become a more holistic science, oriented toward solving problems on both the small and the large scale. It is important, particularly given the increasing global population, that concepts such as urban and natural sustainability are approached in an interdisciplinary and holistic way.

One of the most significant areas for future landscape ecology research is climate change. Models of climate change have predicted significant changes, particularly in relation to the frequency and timing of large-scale weather events. Climate change will affect the structure and function of landscapes and will alter land-use patterns and influence entire ecosystems. Scientific inquiry is playing an important role in the debate on climate change and in finding solutions. The strategies for combating global climate change must incorporate concepts and research in landscape ecology. Examples of how landscape concepts are influenced by climate change include land-use changes, such as modifying agricultural practices or using land to sequester carbon. The continued impact of greenhouse gases will fundamentally change landscapes for many years and perhaps forever. Landscape ecology can increase the scientific understanding of landscape resilience, recovery, and disturbance. Such information can provide an important framework for the sustainable management of landscapes and ecosystems.

Christine Watts, Ph.D., B.App.Sc., B.Sc.

FURTHER READING

Burel, Françoise, and Jacques Baudry. *Landscape Ecology: Concepts, Methods, and Applications.* Enfield, N.H.: Science Publishers, 2003. A comprehensive look into landscape structure, ecological processes, and dynamics, and how research can be applied to landscape management and planning.

Farina, Almo. *Principles and Methods in Landscape Ecology: Towards a Science of the Landscape.* Boston: Kluwer Academic, 2006. This holistic textbook is a useful tool for undergraduate and graduate students studying landscape ecology and coming from a variety of different disciplines.

Hong, Sun-Kee, et al., eds. *Landscape Ecological Applications in Man-Influenced Areas.* Dordrecht, the Netherlands: Springer, 2008. Concentrates on landscape ecology as it relates to urbanization, biodiversity, and land alteration. It provides a number of specific case studies for developing sustainable management through the examination of spatial analysis and landscape modeling.

Turner, Monica, Robert Gardner, and Robert O'Neill. *Landscape Ecology in Theory and Practice: Pattern and Process:* New York: Springer, 2001. Provides a fundamental framework regarding contemporary landscape ecology, while particularly focusing on the relationship and interaction between spatial patterns and ecological processes. Useful for both practicing ecologists and graduate students.

Wiens, John, et al., eds. *Foundation Papers in Landscape Ecology.* New York: Columbia University Press, 2007. Presents information on the origins and progress in the science of landscape ecology from a number of different standpoints, methodologies, and geographical areas.

Wu, Jianguo, and Richard Hobbs, eds. *Key Topics in Landscape Ecology*. New York: Cambridge University Press, 2007. Some of the most noted figures in landscape ecology discuss advances in the field and the concepts and methods used in understanding landscape patterns and in applying the science in a novel way.

WEB SITES

California Natural Resources Agency
Map Server
http://atlas.resources.ca.gov

International Association of Landscape Ecology
http://www.landscape-ecology.org

U.S. Environmental Protection Agency
Landscape Ecology
http://www.epa.gov/esd/land-sci/default.htm

See also: Climate Modeling; Ecological Engineering; Environmental Engineering; Flood-Control Technology; Forestry; Landscape Architecture and Engineering; Land-Use Management; Maps and Mapping; Remote Sensing; Soil Science; Water-Pollution Control; Wildlife Conservation.

LAND-USE MANAGEMENT

FIELDS OF STUDY

Civil engineering; landscape architecture; regional and urban planning; design; water resource planning; agriculture; agricultural engineering; agronomy; environmental science; forestry; geography; social science; soil science.

SUMMARY

Land-use management concerns allocation of the landscape and its natural resources to appropriate uses for the purposes of providing food and shelter, ensuring adequate social and economic life-support systems, and protecting public health and safety, while preserving and sustaining the affected environment and natural resources for future generations. Another way to define land-use management is as a process that alters nature to benefit humanity and also strives to diminish the impact of the built and cultivated environment on natural resources such as soil, water, and air. The ultimate goal is to manage the competition for land and natural resources through integrated planning based on clearly stated objectives.

KEY TERMS AND CONCEPTS

- **Biodiversity:** Number and variety of plant and animal species in a particular area.
- **Built Environment:** Human-made features, such as buildings and structures, found in the environment.
- **Geospatial:** Relating to the relative position of things on Earth's surface.
- **Hydrology:** Study of water on the earth, underground, and in the atmosphere.
- **Land-Use Planning:** Integration of social, economic, and environmental objectives to ensure the best use of land and natural resources.
- **Mitigation:** Reduction of the negative impact of the built environment on the natural environment.
- **Stakeholders:** All those who have an interest in the success of a project or business.
- **Subdivision Control:** Regulation of land division, road construction, and infrastructure installation to serve specific uses.
- **Zoning:** Governmental regulation of buildings, structures, and land uses.

DEFINITION AND BASIC PRINCIPLES

Land-use management is a multidisciplinary science that integrates natural land systems with the built environment. Land-use managers assess, evaluate, analyze, and study geospatial data in setting public policy and proposing land-use controls that mitigate environmental impacts caused by development. The collaboration of multiple stakeholders and a high level of communication are necessary components in land-use management.

The graphic arts are often employed to prepare maps or plans that are used in communicating land-use management proposals. Land-use management plans provide viable support for a variety of management purposes. Maps and plans may depict existing and future uses for land such as transportation corridors, hazardous and solid-waste disposal sites, and historical districts, or water management solutions such as aquifers and watersheds. Maps of limiting factors, such as wetlands, floodplains, carrying capacity assessments, and the results of land suitability studies, help determine how land should be used and zoned. Soil and erosion mapping can aid in planning agricultural uses, and forest studies protect the habitats of flora and fauna from development.

When a lack of planning and land-use management creates haphazard and unsustainable growth, governments are often forced to take on the burden of correcting the situation.

BACKGROUND AND HISTORY

Around 8000 B.C.E., people turned from hunting to agriculture, and eventually, city-states began to spring up along rivers and bodies of water. In the fourth century B.C.E., the Greek philosopher Aristotle espoused a holistic view of city design, and laws began to be written to manage water resources for irrigation. People began constructing walls around cities to protect them from invaders, and they created geometric land patterns, including gardens that served both practical and aesthetic purposes. Ancient societies began to strategically locate buildings and select sites for monuments and religious structures based on purposes such as uniting the heavens with the earth. As people discovered metals and other resources that could be bartered, they developed transportation networks and began to exploit land with little regard for the environmental impact.

Modern land-use management evolved from eighteenth century schools of thought such as Western classicism. In the United States, Thomas Jefferson is well known for his management of the landscape based on Renaissance traditions. Landscape architect Frederick Law Olmsted developed urban parks, and in partnership with architect Calvert Vaux, he planned communities such as Riverside Estate in Chicago. The depletion of natural resources and uncontrolled development that resulted from industrialization highlighted the need for planning and land management. Pioneers in land-use planning include

English town planner Sir Ebenezer Howard, the founder of the garden city concept, and American historian and philosopher Lewis Mumford, author of *The Story of Utopias* (1922) and cofounder in 1923 of the Regional Planning Association of America, a group devoted to planned development of urban areas.

In 1909, the first National Conference on City Planning and the Problems of Congestion was held in Washington, D.C., to address problems such as those uncovered by an extensive survey of Pittsburgh that had begun in 1907. Policymakers began to examine questions as to what American cities should provide for their citizens and how they could be economically and socially beneficial as well as aesthetically pleasing. This interest in land-use management resulted in the founding of organizations such as the American City Planning Institute (1917, later the

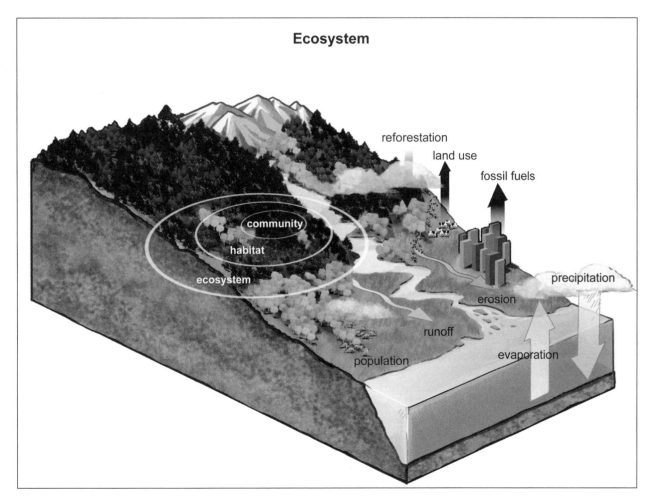

1105

American Planning Association), the creation of city planning boards (Boston, 1912), and the enacting of the first zoning ordinances (New York, 1916). Land use gradually became controlled by federal, state, and local government regulations and policies.

HOW IT WORKS

Land-use management begins with an inventory of past and existing land-use patterns for the purposes of proposing sustainable land uses and promulgating accompanying management policies. In preparing land-use plans, establishing public policy, and adopting comprehensive land-use laws and regulations, land-use managers make extensive use of technologies such as remote sensing and geographic information systems (GIS) to detect changes in land-use patterns and AutoCAD (two- or three-dimensional computer-aided design, or CAD, software) to depict existing and future features of selected areas.

Land-Use Controls. Those involved in managing land use must balance, evaluate, and integrate social, economic, and environmental concerns before arriving at a final management plan. Governmental regulators will usually seek public and private input as part of the decision-making process, and in many jurisdictions, land-use controls must be approved by legislative and administrative bodies as well as by the public. The types of land-use management controls are many and varied, but most include zoning and subdivision control regulations adopted by municipal and county governments based on comprehensive master plans. Other public land-use controls might include ordinances and bylaws that regulate water resources, earth removal, and building construction. Land use can also be managed through easements, both public and private, and proprietary regulations such as those found in real estate development covenants. Permitting and enforcement of land-use laws and regulations by local, state, and federal governmental entities also provide a form of land-use control, especially in the areas of environmental and transportation planning.

New Technologies. Many new technologies support modern land-use management. Private and public tracking systems are available to governments and others for the purposes of collecting and sharing geographic data. The U.S. Navy's electronic Land Use Control Information System (LUCIS) is a national digitized mapping system. Private companies also have created information gathering and tracking systems, such as Google Earth, which provides satellite images of the world as well as maps. Many of these new systems allow users not only to collect and track data on land-use controls and permitting but also to manage land-use activities by integrating with GIS and CAD systems.

Environmental modeling is another technology that has become popular as a land-use management tool, especially in the international community. Computer modeling allows experimentation that suggests an outcome when something occurs in nature. Modeling might be used to study the relationship between industrialization, air pollution, and environmental degradation or to improve agricultural production by suggesting the amount of surface-water runoff and erosion that can be expected based on soil and ground cover conditions.

Sustainable Land Use. Land-use management concerns the impact that the developed environment has on the natural environment and the strictures that the natural environment might place on land development. Climate, weather conditions, topography, soils, and hydrology are some of the physical features of the environment that land-use managers rely on in developing sustainable policies. Land-use managers must make difficult decisions such as whether developments should be allowed in areas subject to hurricanes or whether expansion into undeveloped areas should be limited so as to prevent urban overcrowding, sprawl, and unhealthy living conditions. The dilemma in achieving sustainable land-use management, therefore, is how to reach equilibrium between the built and natural environments and satisfy the multiple economic, social, and environmental interests.

APPLICATIONS AND PRODUCTS

Spatial Analysis. Geographic information systems (GIS) together with remote sensing technologies are used for spatial analysis of land uses. One application of spatial analysis is management of agricultural production. Remote sensing through agricultural satellites allows for monitoring of crop yields and distribution, atmospheric conditions that affect agricultural growth, moisture needs, pest and disease damage, and soil erosion. Planners also use spatial analysis to generate land-use maps and informational databases that promote sustainable development through evaluation of land-use change and adoption of regulatory schemes.

Land Cover Evaluation. In preparing land cover maps, land-use managers classify and quantify existing land uses into categories such as water resources, pavement, forest lands, agricultural lands, grasslands, and the built environment. This mapped information, together with database knowledge gleaned from new technologies and field experimentation, allows planners to gain an understanding of existing land-use conditions and monitor future changes throughout the world. Agricultural sustainability has benefited greatly from land cover analysis. Evaluation of land cover has also been used for predicting storm-water runoff to design drainage systems; urban planning, including integration of green space and recreational uses into the built environment; fire hazard mapping based on the amount of plant coverage that would provide fuel for a fire; and assessment of hydrology needs in arid and semi-arid regions of the world.

Land Surface Sensing. Remote-sensing technologies that use satellites to gather data enable land-use managers to study objects on the surface of the land, such as forest growth, buildings, pavement, and bodies of water, as well as the surrounding biosphere. Those involved in land-use management associated with natural resource conservation are likely to use land surface sensing together with field measurements to manage forests and ensure biodiversity by gathering data on plant growth, density of the canopy, and undercover species. Other applications for land surface sensing include meteorological studies, evaluation of soils and topography, collection of data concerning radiation from natural and non-natural sources, monitoring of ground temperature and snow and ice accumulation and decline, and water resources management.

Growth Management. Governmental regulation of growth allows for economic development in ways that make better use of existing land and natural resources. Land uses are managed by clustering the built environment around transportation networks, water resources, and other essential services, integrating open space into the developed landscape, and preserving natural land features. Smart growth is an example of growth management that promotes high-density, compact projects in areas best suited for development based on topography, geographic features,

---

Fascinating Facts About Land-Use Management

- Without implementation of land-use policies that preserve cultivable land and promote sustainable agriculture, global agricultural production is unlikely to be able to keep pace with the world population, which is increasing by more than 90 million people per year.
- The world's deserts are expanding. For example, the Sahara Desert is expanding at a rate of about 30 miles each year.
- The increase in the worldwide destruction of forests results in depletion of an energy source for more than three-quarters of the world's population, changes in rainfall patterns, increased river flooding, and extinction of flora and fauna.
- Every day, between 150 and 200 species of plants and animals become extinct because their habitats have been destroyed by the overdevelopment of land.
- Urban expansion has a major impact on land use: Each day, a city with a population of 1 million people uses about 625,000 metric tons of water and generates about 500,000 metric tons of wastewater and 2,000 metric tons of solid waste.
- More than 70 percent of the global population lacks access to clean water, and as a result, thousands of people die each day.

---

water resources, and infrastructure. Smart growth consists of mixed-use developments, including residential, commercial, industrial, and recreational uses that support a self-sustaining community.

Environment Issues and Hazard Mitigation. Land-use managers can address environmental pollution issues by using modeling to predict the impact of various land uses on the environment and develop mitigation strategies. For example, planners use modeling to propose regulations to mitigate possible impacts of fertilizers and pesticides used in agriculture on water resources. In addition, GIS and related technologies enable land-use managers to locate and monitor hazards such as mine and hazardous-waste sites.

Natural Resources Protection. Those involved with the economic side of land-use management do not always provide the best protection for natural resources including water, air, soil, and plant and animal habitats. Ensuring that natural resources are preserved and open space is integrated into the built environment are important applications of land-use management. In addition, land-use managers are involved in land conservation through outright purchase of land or easements for conservation or through restricting land development in ways that preserve fragile ecosystems. Modern land-use management technologies provide the means for determining which land should be protected.

Habitat Modeling. The use of habitat modeling as a land-use management tool aids in protecting plant and animal habitats, including those that are rare; conserving biodiversity; monitoring ecosystems and their functionality; and predicting changes in habitats caused by nonnatural and natural disturbances. Often habitat modeling supports the preservation of specific species. Habitat modeling is used in conjunction with GIS, remote sensing, and fieldwork, including soil classification and plant coverage measurements. By using data from multiple sources, land-use managers are able to compare variables to make inferences concerning habitat health. Variables used in habitat modeling include tree types, soil classes, topography, and hydrology.

Open Land Management. Globally, as development continues to encroach on open land, governments have begun to regulate land uses to preserve rangeland, grazing land, grassland habitats, and agricultural areas. Land-use managers play a key role in monitoring land-use changes through new technologies and advising governments about which nonagricultural uses are compatible with open land uses. The preservation of open land uses is important for sustaining the future of life on the planet.

IMPACT ON INDUSTRY

The connections involving land use and development, climate change, the quantity and quality of water resources, and preservation of plant and animal habitats are some of the most important issues of the twenty-first century. The United Nations and international organizations such as the Earthwatch Institute, a nonprofit entity with offices in the United States, England, Australia, and Japan, are promoting sustainable development. Through research and education, Earthwatch and similar organizations are attempting to balance the social and economic needs of society with protection of the environment.

Governmental and Nongovernmental Organizations. Many federal and state agencies are involved in land-use management. At the federal level, the U.S. Department of Agriculture addresses agricultural land uses and rural and community development. Within the department is the U.S. Forest Service (USFS), which manages federal forests and grasslands with the goal of preserving flora and fauna habitats, enhancing water resources, and responding to climate change, while allowing multiple uses of public lands (for recreation, as rangeland, and as a source of forest products). To manage the forest ecosystem, the agency evaluates human activity such as cutting and harvesting trees as well as natural systems including disease and insect infestations that affect the forest. The USFS employs several new technologies and tools, including integrated geographic information systems, simulators, and scale models to analyze how land-use management practices affect forest growth and watersheds. The U.S. Geological Survey provides scientific information for managing land-use resources including water and energy, and the Bureau of Land Management (in the U.S. Department of the Interior) manages public lands, especially those related to renewable energy resources, and is also responsible for habitat maintenance under the national wild horse and burro program. The Environmental Protection Agency (EPA) is the watchdog agency over all land uses that negatively affect the environment. The agency's mandate is accomplished through regulation, permitting, fines, and civil and criminal legal actions.

Colleges and Universities. Academic institutions, often in conjunction with the federal government, conduct research on land-use management. For example, the University of Washington and Yale University have worked with the USFS in analyzing forest ecosystems and generating database inventories of forest resources. Purdue University and the EPA have conducted hydrology research, including estimating changes in air and water resources as a result of existing and proposed land-use changes. This research is especially useful for localities that depend on groundwater for their public drinking water supply. Many universities with landscape architecture programs, often state universities that derived from land-grant colleges, conduct research and participate in public and private planning related to land-use management. For example, the University of Massachusetts brings together public and private stakeholders through its Citizen Planner Training Collaborative.

Scientific and Professional Organizations. The Interstate Technology Regulatory Council, whose members come from state and federal governmental agencies and private industries within the United States, conducts research and provides education on environmental public policy issues. The organization considers many of the land uses that affect the environment, including landfills, brownfields, and mining. Another scientific organization involved in land-use analysis is the Institute for Applied Ecology. Working with agencies and organizations including the U.S. Forest Service, Bureau of Land Management, and the Nature Conservancy, the institute's mission is preservation by evaluating the impact of land uses, such as transportation systems, energy generators, and landfills, on native plant and animal habitats. Internationally, an important scientific organization is the Global Observation of Forest and Land Cover Dynamics, which uses information on forest change obtained from satellites, aerial photography, and ground measurements to analyze the impact of multiple land uses on the environment.

Professional organizations include the International Federation of Landscape Architects, the American Planning Association, and the Urban Land Institute. The International Federation of Landscape Architects promotes not only design of the landscape but also worldwide land-use management and conservation through the profession of landscape architecture, while members of the American Planning Association are involved in land-use planning in all settings from urban to rural. The Urban Land Institute assists global leaders in the development of sustainable land use.

CAREERS AND COURSE WORK

An undergraduate or advanced degree in one of several planning disciplines, such as urban, regional, spatial, environmental, or transportation planning, is an excellent background for launching a land-use management career. Students also should take classes in planning theory, graphic arts and design, geographic information systems, tracking and notification computer databases, economics, sociology, political science, public administration, botany, statistics, law, and natural sciences such as biology. Additional

course work might include hydrology for a focus on water resource management or soil science and geology for students interested in soil management, especially as it relates to agricultural land use.

Job opportunities related to land-use management are best for students with advanced degrees that include training in modern technologies for geographical data collection, tracking, and sharing. Most advanced degrees are in one of the planning disciplines, and most planners also seek membership in professional organizations such as the American Institute of Certified Planners. Although doctorate degrees in the area of land-use management are becoming common, the typical advanced degree program culminates with a master's degree.

The majority of land-use management careers are in local, state, and federal governmental settings, often as a planner or public policy administrator. Opportunities also exist with international bodies, nongovernmental agencies, and nonprofit organizations in diverse fields associated with agriculture or environmental and natural resource management and conservation. Private companies that provide consultation services to governmental agencies also employ land-use managers. Some specialty careers in land-use management are related to oil and gas exploration, mining (including reclamation), and historic preservation.

SOCIAL CONTEXT AND FUTURE PROSPECTS

Although the ability to study and manage land use has been enhanced by geographic information systems, soil degradation, lack of quality water, and loss of biodiversity continue to be significant land-use management issues. Gaining global support to tackle these matters is not always viable, as less-developed nations do not have the economic resources or the social desire to develop policies that enable sustainable land use. In addition, some stakeholders in the most developed nations are unwilling to forgo immediate economic gain in exchange for future sustainability.

Individual nations and the international community, however, are working together to solve some of the most pressing land-use management concerns. The greater availability of geographical data, including those collected by satellites, has led to the implementation of better agricultural practices in areas where soil degradation is obvious, the evaluation of water resources to solve issues regarding the quality and quantity of water, and better management of waste disposal. These data have also clarified land-use patterns and encroachment on the natural environment, including deforestation and loss of habitat for plants and animals. Refinements of existing technology and future advances are likely to add to the knowledge base that planners and policy makers can use and share in making informed land-use management decisions that affect the future of global life.

Carol A. Rolf, B.S.L.A., M.Ed., M.B.A., J.D.

FURTHER READING

Aspinall, Richard J., and Michael J. Hill, eds. *Land Use Change: Science, Policy, and Management.* Boca Raton, Fla.: CRC Press, 2007. Discusses land-use data analysis through remote-sensing technologies.

Birnbaum, Charles A., and Stephanie S. Foell, eds. *Shaping the American Landscape: New Profiles from the Pioneers of American Landscape Design Project.* Charlottesville: University of Virginia Press, 2009. Features articles authored by historic pioneers involved in diverse fields related to land-use management.

Butterfield, Jody, Sam Bingham, and Allan Savory. *Holistic Management Handbook: Healthy Land, Healthy Profits.* Chicago: Island Press, 2006. Provides step-by-step instructions for land management decision making using holistic practices.

Jellicoe, Geoffrey, and Susan Jellicoe. *The Landscape of Man: Shaping the Environment from Prehistory to the Present Day.* 3d ed. London: Thames and Hudson, 1995. Comprehensively covers the history of land-use management and the many schools of landscape design and planning.

Peterson, Gretchen N. *GIS Cartography: A Guide to Effective Map Design.* Boca Raton, Fla.: CRC Press, 2009. Practical guidebook for using GIS to create clear and informative maps that will be effective communication tools.

Randolph, John. *Environmental Land Use Planning and Management.* Chicago: Island Press, 2003. Considers advances in geographical data collection to mitigate environmental impacts, especially on water resources.

WEB SITES
American Planning Association
http://www.planning.org

Institute for Applied Ecology
http://appliedeco.org

International Federation of Landscape Architects
http://www.iflaonline.org

U.S. Department of Agriculture
U.S. Forest Service
http://www.fs.fed.us

U.S. Department of the Interior
Bureau of Land Management
http://www.blm.gov/wo/st/en.html

U.S. Geological Survey
Land Use History of North America
http://biology.usgs.gov/luhna/contents.html

See also: Agricultural Science; Agroforestry; Environmental Engineering; Erosion Control; Forestry; Hazardous-Waste Disposal; Hydrology and Hydrogeology; Landscape Ecology; Soil Science.

LASER INTERFEROMETRY

FIELDS OF STUDY

Control engineering; electrical engineering; engineering metrology; interferometry; laser science; manufacturing engineering; materials science; mechanical engineering; nanometrology; optical engineering; optics; physics.

SUMMARY

Laser interferometry includes many different measurement methods that are all based on the unique interference properties of laser lights. These techniques are used to measure distance, velocity, vibration, and surface roughness in industry, military, and scientific research.

KEY TERMS AND CONCEPTS

- **Beam Splitter:** Partially reflecting and partially transmitting mirror.
- **Constructive Interference:** Addition of two or more waves that are in phase, leading to a larger overall wave.
- **Destructive Interference:** Addition of two or more waves that are out of phase, leading to a smaller overall wave.
- **Heterodyne Detection:** Mixing of two different frequencies of light to create a detectable difference in their interference pattern.
- **Homodyne Detection:** Mixing of two beams of light at the same frequency, but different relative phase, to create a detectable difference in their interference pattern.
- **Monochromatic:** Containing a single wavelength.
- **Spatial Coherence:** Measure of the phase of light over a defined space.
- **Temporal Coherence:** Measure of the phase of light as a function of time.

DEFINITION AND BASIC PRINCIPLES

Laser interferometry is a technique that is used to make extremely precise difference measurements between two beams of light by measuring their interference pattern. One beam is reflected off a reference surface and the other either reflects from or passes through a surface to be measured. When the beams are recombined, they either add (constructive interference) or subtract (destructive interference) from each other to yield dark and light patterns that can be read by a photosensitive detector. This interference pattern changes as the relative path length changes or if the relative wavelength or frequency of the two beams changes. For instance, the path lengths might vary because one object is moving, yielding a measurement of vibration or velocity. If the path lengths vary because of the roughness of one surface, a "map" of surface smoothness can be recorded. If the two beams travel through different media, then the resulting phase shift of the beams can be used to characterize the media.

Lasers are not required for interferometric measurements, but they are often used because laser light is monochromatic and coherent. It is principally these characteristics that make lasers ideal for interferometric measurements. The resulting interference pattern is stable over time and can be easily measured, and the precision is on the order of the wavelength of the laser light.

BACKGROUND AND HISTORY

The interference of light was first demonstrated in the early 1800's by English physicist Thomas Young in his double-slit experiment, in which he showed that two beams of light can interact like waves to produce alternating dark and light bands. Many scientists believed that if light were composed of waves, it would require a medium to travel through, and this medium (termed "ether") had never been detected. In the late 1800's, American physicist Albert Michelson designed an interferometer to measure the effect of the ether on the speed of light. His experiment was considered a failure in that he was not able to provide proof of the existence of the ether. However, the utility of the interferometer for measuring a precise distance was soon exploited. One of Michelson's first uses of his interferometer was to measure the international unit of a meter using a platinum-iridium metal bar, paving the way for modern interferometric methods of measurement. Up until the mid-twentieth century, atomic sources were used in interferometers, but their use for measurement was limited to their coherence length, which was less than a meter. When lasers were first developed in the 1960's, they quickly replaced the spectral line sources used for interferometric measurements because of their long coherence length, and the modern field of laser interferometry was born.

HOW IT WORKS

The most common interferometer is the Michelson interferometer, in which a laser beam is divided in two by use of a beam splitter. The split beams travel at right angles from each other to different surfaces, where they are reflected back to the beam splitter and redirected into a common path. The interference between the recombined beams is recorded on a photosensitive detector and is directly correlated with the differences in the two paths that the light has traveled.

In the visible region, one of the most commonly available lasers is the helium-neon laser, which produces interference patterns that can be visually observed, but it is also possible to use invisible light lasers, such as those in the X-ray, infrared, or ultraviolet regions. Digital cameras and photodiodes are routinely used to capture interference patterns, and these can be recorded as a function of time to create a movie of an interference pattern that changes with time. Mathematical methods, such as Fourier analysis, are often used to help resolve the wavelength composition of the interference patterns. In heterodyne detection, one of the beams is purposefully phase shifted a small amount relative to the other, and this gives rise to a beat frequency, which can be measured to even higher precision than in standard homodyne detection. Fiber optics can be used to direct the light beams, and these are especially useful to control the environment through which the light travels. In this case, the reflection from the ends of the fiber optics have to be taken into account or used in place of reflecting mirrors. Polarizers and waveretarding lenses can be inserted in the beam path to control the polarization or the phase of one beam relative to the other.

While Michelson interferometers are typically used to measure distance differences between the two reflecting surfaces, there are many other configurations. Some examples are the Mach-Zehnder and Jamin interferometers, in which two beams are reflected off of identical mirrors but travel through different media. For instance, if one beam travels through a gas, and the other beam travels through vacuum, the beams will be phase shifted relative to the other, causing an interference pattern that can be interpreted to give the index of refraction of the gas. In a Fabry-Perot interferometer, light is directed into a cavity consisting of two highly reflecting surfaces. The light bounces between the surfaces multiple times before exiting to a detector, creating an interference pattern that is much more highly resolved than in a standard Michelson interferometer. Several other types of interferometers are described below in relation to specific applications.

APPLICATIONS AND PRODUCTS

Measures of Standards and Calibration. Because of the accuracy possible with laser interferometry, it is widely used for calibration of length measurements. The National Institute of Standards and Technology (NIST), for example, offers measurements of gauge blocks and line scales for a fee using a modified Michelson-type interferometer. Many commercial companies also offer measurement services based on laser interferometer technology. Typical services are for precise measurement of mechanical devices, such as bearings, as well as for linear, angular, and flatness calibration of other tools such as calipers, micrometers, and machine tools. Interferometers are also used to measure wavelength, coherence, and spectral purity of other laser systems.

Dimensional Measurements. Many commercial laser interferometers are available for purchase and can be used for measurements of length, distance, and angle. Industries that require noncontact measurements of complex parts use laser interferometers to test whether a part is good or to maintain precise positioning of parts during fabrication. Laser interferometers are widely used for these purposes in the automotive, semiconductor, machine tool, and medical- and scientific-parts industries.

Vibrational Measurements. Laser vibrometers make use of the Doppler shift, which occurs when one laser beam experiences a frequency shift relative to the other because of the motion of the sample. These interferometers are used in many industries to measure vibration of moving parts, such as in airline or automotive parts, or parts under stress, such as those in bridges.

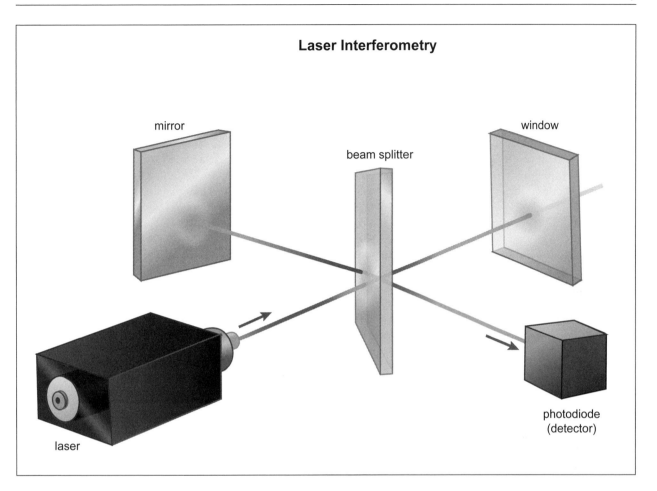

Laser Interferometry

mirror

window

beam splitter

laser

photodiode
(detector)

Optical Metrology. Mirrors and lenses used in astronomy require high-quality surfaces. The Twyman-Green and Fizeau interferometers are variations on the Michelson interferometer, in which an optical lens or mirror to be tested is inserted into the path of one of beams, and the measured interference pattern is a result of the optical deviations between the two surfaces. Other industrial optical testing applications include the quality control of lenses in glasses or microscopes, the testing of DVD reader optical components, and the testing of masks used in lithography in the semiconductor industry.

Ring Lasers and Gyroscopes. In the last few decades, laser interferometers have started to replace mechanical gyroscopes in many aircraft navigation systems. In these interferometers, the laser light is reflected off of mirrors such that the two beams travel in opposite directions to each other in a ring, recombining to produce an interference pattern at the starting point. If the entire interferometer is rotated, the path that the light travels in one direction is longer than the path length in the other direction, and this results in the Sagnac effect: an interference pattern that changes with the angular velocity of the apparatus. These ring interferometers are widely available from both civilian and military suppliers such as Honeywell or Northrop Grumman.

Ophthalmology. A laser interferometry technique using infrared light for measuring intraocular distances (eye length) is widely used in ophthalmology. This technique has been developed and marketed primarily by Zeiss, which sells an instrument called the IOL Master. The technique is also referred to as partial coherence interferometry (PCI) and laser Doppler interferometry (LDI) and is an area of active research for other biological applications.

Sensors. Technology based on fiber-optic acoustic sensors to detect sound waves in water have been developed by the Navy and are commercially available from manufacturers such as Northrop Grumman.

Gravitational Wave Detection. General relativity predicts that large astronomical events, such as black hole formation or supernova, will cause "ripples" of gravitational waves that spread out from their source. Several interferometers have been built to try to measure these tiny disturbances in the local gravitational fields around the earth. These interferometers typically have arm lengths on the order of miles and require a huge engineering effort to achieve the necessary mechanical and vibrational stability of the lasers and the mirrors. There are currently efforts to build space-based gravity-wave-detecting interferometers, which would not be subject to the same seismic instability as Earth-based systems.

Research Applications. Laser interferometers are used in diverse form in many scientific experiments. In many optical physics applications, laser interferometers are used to align mirrors and other experimental components precisely. Interferometers are also used for materials characterization in many basic research applications, while ultrasonic laser interferometers are used to characterize velocity distributions and structures in solids and liquids. A more recent technology development is interferometric sensors, which are used to monitor chemical reactions in real time by comparing laser light directed through waveguides. The interference pattern from the two beams changes as the chemical reaction progresses. This technique is often referred to as dual-polarization interferometry.

IMPACT ON INDUSTRY

The invention of the laser in 1960 opened up the field of interferometry to a huge range of applications, allowing measurements both at long distances and with extremely high precision. The following decades saw the initial application of laser interferometry to vibration and dimensional measurements in industrial processes. The invention of cheap semiconductor diode lasers in the last decade has further decreased the price of laser interferometers and has widened the range of accessible wavelengths. Many laser interferometer systems are now available commercially, and they are used in a large segment of the semiconductor, automotive, and measurement industries. Systems now sold by many companies give compact, reliable ways to measure surface roughness,

calibrate mechanical components, align or position parts during manufacturing, or for precision machining. Lasers are a multibillion-dollar industry, and laser interferometers are a constantly growing segment of this market.

Laser interferometers have hugely affected the quality control of lenses in the optical-manufacturing industries. Interferometers are the method of choice to measure curvature and smoothness of lenses used in microscopes, telescopes, and eyeglasses. Companies specializing in optical measurements are available for on- or off-site testing of optical components. Astronomy, in particular, has benefited from the precision testing of polished surfaces made possible by laser interferometers; mirrors such as the ones used in the Hubble Space Telescope would not be possible without a laser interferometry testing system.

The National Science Foundation (NSF) has invested substantially in ground-based interferometers for the measurement of gravity waves to test the predictions of general relativity. In the 1990's, the NSF funded the building of two very large-scale Michelson interferometers in the United States–one in Hanford, Washington, and the other in Livingston, Louisiana. Together they make up the Laser Interferometer Gravitational-Wave Observatory (LIGO) and consist of interferometer arms that are miles long. Though gravity waves have not yet been detected, government-funded efforts around the world continue with large-scale interferometers currently operating in a half-dozen countries.

Sensor technology is a growing field in laser interferometry applications. Sensors based on interference of laser light using fiber optics have been developed to detect environmental changes, such as temperature, moisture, pressure, strain on components, or chemical composition of an environment. These types of sensors are not yet widely commercially available but are in development by industry and by the military for applications such as chemical-agent detection in the field or for use in harsh or normally inaccessible environments. Combined with wireless technology, they could be used, for instance, to monitor environmental conditions far underground or within the walls of buildings.

CAREERS AND COURSE WORK

Basic research on laser interferometry and its applications is conducted in academia, in many

Fascinating Facts About Laser Interferometry

- When the first laser was demonstrated in 1960, it was called a "solution looking for a problem."
- The Laser Interferometer Gravitational-Wave Observatory (LIGO) is sensitive to disturbances in the local gravitational fields that are caused by astronomical events as far back in time as 70 million years.
- Sensors based on laser interferometers are being developed to detect acoustic signals for surveillance. A buried sensor can detect the sound of footsteps from as far away as 30 feet.
- The Laser Interferometer Space Antenna (LISA) will consist of three spacecraft orbiting at 5 million kilometers apart and will provide information on the growth and formation of black holes and other events never before seen.
- Laser interferometers will be used to measure the optical quality of the mirror used in the James Webb Space Telescope, scheduled to launch in 2014. The surface smoothness must be unprecedented: If the mirror were scaled up to a size of 3,000 miles across, the surface height would be allowed to vary only by a foot at most.
- Quantum interferometers have been built to recreate the classic double-slit interference experiment, using single atoms instead of a light beam.

metrology industries, and in government laboratories and agencies. For research careers in new and emerging interferometry methods in academia or as a primary investigator in industry, a doctorate degree is generally required. Graduate work should be in the area of physics or engineering. The undergraduate program leading into the graduate program should include classes in mathematics, engineering, computer, and materials science.

For careers in industries that provide or use commercial laser interferometers but do not conduct basic research, a master's or bachelor's degree would be sufficient, depending on the career path. Senior careers in these industries involve leading a team of engineers in new designs and applications or guiding a new application into the manufacturing field. In this case, the focus of course work should be in engineering.

Mechanical, electrical, optical, or laser engineering will provide a solid background and an understanding of the basic theory of interferometer science. Additional courses should include physics, mathematics, and materials science. A bachelor's degree would also be required for a marketing position in laser interferometry industries. In this case, focus should be on business, but a strong background in engineering or physics will make a candidate much more competitive. Technical jobs that do not require a bachelor's degree could involve maintenance, servicing, or calibration of laser interferometers for measurements in industry. They could involve assembly of precision optomechanical systems or machining of precision parts.

SOCIAL CONTEXT AND FUTURE PROSPECTS

The development of increasingly precise interferometers in the last few decades, such as for gravitational-wave measurement, has spurred corresponding leaps in mechanical and materials engineering, since these systems require unprecedented mechanical and vibrational stability. Laser interferometers are beginning to be used in characterization of nanomaterials, and this will push the limits of resolution of laser interferometers even further. As the cost of lasers and optical components continues to decrease, the use of laser interferometers in many industrial manufacturing applications will likely increase. They are an ideal measurement system in that they do not contain moving parts, so there is no wear on parts, and they do not mechanically contact the sample being measured.

Active research is conducted in the field of laser interferometric sensors, with potential applications in military and manufacturing industries. Oil and gas companies may also drive development of sensors for leak and gas detection during drilling. In addition, commercial applications for laser sensors will open up in areas of surveillance as acoustic laser interferometry technology is developed.

Research continues in academic labs in areas that may someday become real-world applications, such as using laser interferometers for the detection of seismic waves, or in the area of quantum interferometry in which single photons are manipulated to interfere with each other in a highly controlled manner.

Corie Ralston, B.S., Ph.D.

FURTHER READING

Beers, John S., and William B. Penzes. "The NIST Length Scale Interferometer." *Journal of Research of the National Institute of Standards and Technology* 104, no. 3 (May/June, 1999): 225-252. Detailed description of the original NIST Michelson interferometer, with historical context.

Halsey, David, and William Raynor, eds. *Handbook of Interferometers: Research, Technology and Applications.* New York: Nova Science Publishers, 2009. A comprehensive text on current interferometry technologies. Designed for readers with a strong physics background.

Hariharan, P. *Basics of Interferometry.* 2d ed. Burlington, Mass.: Academic Press, 2007. Covers most of the current interferometer configurations and includes sections on applications.

_____. *Optical Interferometry.* 2d ed. San Diego: Academic Press, 2003. Overview of the basic theory of interferometry methods, including sections on laser interferometry.

Malacara, Daniel, ed. *Optical Shop Testing.* 3d ed. Hoboken, N.J.: John Wiley & Sons, 2007. An introduction to interferometers used in testing optical components.

Sirohi, Rajpal S. *Optical Methods of Measurement: Whole-field Techniques.* 2d ed. Boca Raton, Fla.: CRC Press, 2009. Covers the basics of wave equations and interference phenomena. Includes multiwavelength techniques of interferometry.

Tolansky, Samuel. *An Introduction to Interferometry.* 2d ed. London: Longman, 1973. Includes many interferometry methods that do not use lasers.

WEB SITES

Laser Interferometer Gravitational-Wave Observatory
http://www.ligo.caltech.edu

National Aeronautics and Space Administration
James Webb Space Telescope
http://www.jwst.nasa.gov

National Institute of Standards and Technology
http://www.nist.gov

See also: Computer Science; Electrical Engineering; Engineering; Laser Technologies; Mechanical Engineering; Optics.

LASER TECHNOLOGIES

FIELDS OF STUDY

Cosmetic surgery; dermatology; fiber optics; laparoscopic surgery; medicine; ophthalmology; optical disk manufacturing; physics; printer manufacturing; spectroscopy; surgery.

SUMMARY

The term "laser" is an acronym for light amplification by stimulated emission of radiation. A laser device emits light via a process of optical amplification; the process is based on the stimulated emission of photons (particles of electromagnetic energy). Laser beams exhibit a high degree of spatial and temporal coherence, which means that the beam does not widen or deteriorate over a distance. Laser applications are numerous and include surgery, skin treatment, eye treatment, kidney-stone treatment, light displays, optical disks, bar-code scanners, welding, and printers.

KEY TERMS AND CONCEPTS

- **Fiber Optics:** Transmission of light through extremely fine, flexible glass or plastic fibers.
- **Laparoscopic Surgery:** Minimally invasive surgery that is accomplished with a laparoscope and the use of specialized instruments, including laser devices, through small incisions.
- **Laser Guidance:** Guidance via a laser beam, which continuously illuminates a target; missiles, bombs, or projectiles (for example, bullets) home on the beam to strike the target.
- **Optical Disk:** Plastic-coated disk that stores digital data, such as music images, or text; tiny pits in the disk are read by a laser beam.
- **Photon:** Electromagnetic energy, which is regarded to be a discrete particle that has zero mass, no electrical charge, and an indefinitely long lifetime.
- **Spectroscopy:** Analysis of matter via a light beam.

DEFINITION AND BASIC PRINCIPLES

A laser consists of a highly reflective optical cavity, which contains mirrors and a gain medium. The gain medium is a substance with light-amplification properties. Energy is applied to the gain medium via an electrical discharge or an optical source such as another laser or a flash lamp. The process of applying energy to the gain medium is known as pumping. Light of a specific wavelength is amplified in the optical cavity. The mirrors ensure that the light bounces repeatedly back and forth in the chamber. With each bounce, the light is further amplified. One mirror in the chamber is partially transparent; amplified light escapes through this mirror as a light beam.

Many types of gain media are employed in lasers, they include: gases (carbon dioxide, carbon monoxide, nitrogen, argon, and helium/neon mixtures); silicate or phosphate glasses; certain crystals (yttrium aluminum garnet); and semiconductors (gallium arsenide and indium gallium arsenide). A basic concept of laser technology is "population inversion." Normally, most of the particles comprising a gain medium lack energy and are in the ground state. Pumping the medium places most or all the particles in an excited state. This results in a powerful, focused laser beam. Lasers are classified as operating in either continuous or pulsed mode. In continuous mode, the power output is continuous and constant; in pulse mode, the laser output takes on the form of intermittent pulses of light.

BACKGROUND AND HISTORY

In 1917, Albert Einstein theorized about the process of stimulated emission, which makes lasers possible. The precursor of the laser was the maser (microwave amplification by stimulated emission of radiation). A patent for the maser was granted to American physicists Charles Hard Townes and Arthur L. Schawlow on March 24, 1959. The maser did not emit light; rather it amplified radio signals for space research. In 1958, Townes and Schawlow published scientific papers, which theorized about visible lasers. Also in 1958, Gordon Gould, a doctoral student at Columbia University under Townes, began building an optical laser. He did not obtain a patent until 1977. In 1960, the first gas laser (helium-neon) was invented by Iranian-American physicist Ali Javan. The device converted electrical energy to light, and this type of laser has many practical applications such as laser surgery. In 1962, American engineer Robert Hall invented the semiconductor laser, which has many applications for communications systems and electronic appliances. In 1969, Gary Starkweather, a Xerox researcher, demonstrated the use of a laser beam to print. Laser printers were a marked improvement over the dot-matrix printer. The print quality was much better and it could print on a single sheet of paper rather than a continuous pile of fan-folded paper. In September, 1976, Sony first demonstrated an optical audio disk. This was the precursor to compact discs (CDs) and digital versatile discs (DVDs). In 1977, telephone companies began trials using fiber-optic cables to carry telephone traffic. In 1987, New York City ophthalmologist Stephen Trokel performed the first laser surgery on a patient's eyes.

HOW IT WORKS

Laser Ablation. Laser ablation involves removing material from a surface (usually a solid but occasionally a liquid). Pulsed laser is most commonly used; however, at a high intensity, continuous laser can ablate material. At lower levels of laser energy, the material is heated and evaporates. At higher levels, the material is converted to plasma. Plasma is similar to gas, but it differs from gas in that some of the particles are ionized (a loss of electrons). Because of this ionization, plasma is electrically conductive.

Laser Cutting. Laser cutting uses laser energy to cut materials either by melting, burning, or vaporizing. Laser cutting is extremely focusable to about 25 microns (one-quarter the width of a human hair); thus, a minimal amount of material is removed. The three common types of lasers used for cutting are: carbon dioxide (CO_2), neodymium (Nd), and neodymium-yttrium aluminum garnet (Nd-YAG).

Laser Guidance. Laser guidance involves the use of a laser beam to guide a projectile (a bomb or a bullet) to a target. In its simplest form, such as a beam emitted from a rifle, the shooter points the laser beam so that the bullet will hit the target. A much more complex process is the guidance of a missile or a bomb. In some cases, the missile contains a laser homing device in which the projectile "rides" the laser beam to the target. More commonly, a technique referred to as semi-active laser homing (SALH) is employed. With SALH, the laser beam is kept pointed at the target after the projectile is launched. Laser energy is scattered from the target, and as the missile

Historic image of a physicist at Litton Industries in San Carlos, California, experimenting with lasers. (Joyce R. Wilson/Photo Researchers, Inc.)

approaches the target, heat sensors home in on this energy. If a target does not reflect laser energy well, the beam is aimed at a reflective source near the target.

Laser Lighting Displays. The focused beam emitted by a laser makes it useful for light shows. The bright, narrow beam is highly visible in the night sky. The beam can also be used to draw images on a variety of surfaces, such as walls or ceilings, or even theatrical smoke. The image can be reflected from mirrors to produce laser sculptures. The beam can be moved at any speed in different directions by the use of a galvanometer, which deflects the beam via an electrical current. The variety of vivid colors available with lasers enhances the visual effects.

Laser Printing. Laser printing involves projecting an image onto an electrically charged, rotating drum that is coated with a photoconductor. When exposed to light, the electrical conductivity of the photoconductor increases. The drum then picks up particles of dry ink (toner). These particles are picked up by varying degrees depending on the amount of charge. The toner is then applied to a sheet of paper. The process involves rapidly "painting" the image line by line. Fast (and more expensive) laser printers can print up to 200 pages per minute. Printers are both monochrome (one color–usually black) or color. Color printers contain toners in four colors: cyan, magenta, yellow, and black. A separate drum is used for each toner. The mixture of the colors on the toners produces a crisp, multicolored image. Duplex laser printers are available, which print on both sides of a sheet of paper. Some duplex printers are manual devices, which require the operator to flip one or more pages manually when indicated. Automatic duplexers mechanically turn each sheet of paper and feed it past the drum twice.

Optical Disks. Optical disks are flat, circular disks that contain binary data in the form of microscopic pits, which are non-reflective and form a binary value of 0. Smooth areas are reflective and form a binary value of 1. Optical disks are both created and read with a laser beam. The disks are encoded in a continuous spiral running from the center of the disk to the perimeter. Some disks are dual layer; with these disks, after reaching the perimeter, a second spiral track is etched back to the center. The amount of data storage is dependent on the wavelength of the laser beam. The shorter the wavelength, the greater the storage capacity (shorter-wavelength lasers can read a smaller pit on the disk surface). For example, the high-capacity Blu-ray Disc uses short-wavelength blue light. Laser can be used to create a master disk from which duplicates can be made by a stamping process.

Laser 3-D Scanners. Laser 3-D scanners analyze an object via a laser beam. The collected data is used to construct a digital, three-dimensional object of the model. In addition to shape, some scanners can replicate color.

APPLICATIONS AND PRODUCTS

Laser technology includes a vast number of applications for business, entertainment, industrial, medical, and military use. The reflective ability of a laser beam has one major drawback: It can inadvertently strike an unintended target. For example, a reflected laser beam could damage an eye, so protective goggles are worn by laser operators. Products range from inexpensive laser pointers and CDs to surgical, industrial, and military devices costing hundreds of thousands of dollars. The following applications and products are a representative sample and are by no means comprehensive.

Optical Disks. Most optical disks are read-only; however, some are rewritable. They are used for the storage of data, computer programs, music, graphic images, and video games. Since the first CD was introduced in 1982, this technology has evolved markedly.

Optical data storage has in a large part supplanted storage on magnetic tape. Although optical storage media can degrade over time from environmental factors, they are much more durable than magnetic tape, which loses its magnetic charge over time. Magnetic tape is also subject to wear as it passes through the rollers and recording head. This is not the case for optical media in which the only contact with the recording surface is the laser beam. CDs are primarily used for the storage of music: A five-inch CD

Fascinating Facts About Laser Technologies

- In early 1942, long before laser guidance systems appeared, psychologist B. F. Skinner at the University of Minnesota investigated weapons guidance by trained pigeons. The birds were trained with slides of aerial photographs of the target. They were placed in a harness inside the guidance system and kept the target in the crosshairs by pecking. Skinner never overcame official skepticism, and the project was abandoned.

- The radial keratotomy procedure originated in Soviet Russia with physician Svyatoslav Fyodorov in the 1970's. Fyodorov treated a boy whose glasses had been broken in a fight and cut his cornea. After the boy recovered, his eyesight had improved—his myopia was significantly lessened.

- Physician Stephen Trokel, who patented the excimer laser for vision correction, adapted the device from its original use: etching silicone computer chips.

- In 1960, American inventor Hildreth Walker, Jr., developed a ruby laser, which precisely measured the distance from the Earth to the Moon during the Apollo 11 mission.

- The original Hewlett Packard (HP) LaserJet printer retailed for more than $3,000 and could print eight pages per minute at a resolution of 300 dots per inch. The only font available was Courier, and it could print only a quarter-page of graphics at that resolution. As of 2011, much smaller laser printers with a multitude of features, including higher speed and print quality, retail for around $100.

- Laboratory lasers can measure distances accurate to less than a thousandth of the wavelength of light, which is less than the size of an atom.

can hold an entire recorded album, replacing the vinyl record, which was subject to wear and degradation. A limitation of the CD is its storage capacity: 700 megabytes of data (eighty minutes of music). Actually, three years before the introduction of the CD, the larger LaserDisc appeared for home video use. However, it never attained much popularity in the United States. The DVD, which appeared in 1996, rapidly gained popularity and soon outpaced VHS tape for the storage of feature-length movies. The DVD can store 4.7 gigabytes of data in single-layer format and 8.5 gigabytes in dual-layer format. The development of high-definition (HD) television fueled the development of higher-capacity storage media. After a format war of several year's duration, the Blu-ray DVD won out over the HD disc. The Blu-ray Disc can store about six times the amount of data as a standard DVD: 25 gigabytes of data in single-layer format and 50 gigabytes of data in dual-layer format.

Medical Uses. Medical applications of laser technology exist for abdominal surgery, ophthalmic surgery, vascular surgery, and dermatology. The narrow laser beam can cut through tissue and cauterize small blood vessels at the same time. Laser can be used with an open surgical incision as well as laparoscopic procedures in which the surgery is performed through small incisions for passage of the laparoscope and surgical instruments. The surgeon and his or her assistants can view images on a video monitor. Mirrors can be used to deflect the laser beam in the desired direction. However, the surgeon must be extremely careful when directing a laser beam at an internal structure–inadvertent reflection of the beam can damage tissue (for example, puncture the bowel).

A common ophthalmic procedure is the radial keratotomy. With this procedure, fine laser cuts are made in a radial fashion around the cornea. These precision cuts can correct myopia (nearsightedness) and hyperopia (farsightedness). Laser is also used to treat a number of eye problems such as a detached retina (separation of the imaging surface of the retina from the back of the eye), glaucoma (increased pressure within the eyeball), and cataracts (clouding of the lens).

Atherosclerosis (a narrowing of the blood vessels due to plaque formation) is a common disease. Laser can be used to vaporize the plaque. It also can be used to bore small holes within the heart to improve circulation within the heart muscle.

Dermatology procedures can be performed with the laser–many of which are cosmetic procedures done to improve appearance. Some laser dermatologic applications are: acne treatment and acne scar removal, removal of age spots, skin resurfacing, scar removal, tattoo removal, spider vein removal, and hair removal.

Innovative uses of laser for medical purposes are being reported frequently. For example, in December, 2010, French researchers reported a noninvasive technique to diagnose cystic fibrosis prenatally. Cystic fibrosis is a genetic disease in which thick mucus secretions form in the lungs and glands such as the pancreas, which result in progressive deterioration of the affected organs. The technology involves the identification of affected fetuses by laser microdissection of a fetal cell, which was circulating in the mother's bloodstream. The procedure avoids the risk of a miscarriage when either chorionic villus sampling or amniocentesis is used. Another innovative use of laser was also reported in December, 2010. Medical devices, which are made of plastic or metal, are surgically placed within the body. Examples of medical devices are stents, which are inserted to improve blood flow to the heart, and hip-replacement prosthetics. These devices can become coated with biofilm, which is of bacterial origin. This biofilm is resistant to antibiotics; however, a laser technique was reported to be successful in removing biofilm.

Industrial Uses. Lasers can make precision cuts on metal or other materials. A minimal amount of material is removed, leaving a smooth, polished surface on both sides of the cut. It also can weld and heat-treat materials. Other industrial uses include marking and measuring parts.

Business Uses. The supermarket bar-code scanner was one of the earliest laser applications–it appeared in 1974. Laser printers are ubiquitous in all but the smallest business offices. They range in price from less than $100 to more than $10,000. The higher-priced models have features such as duplexing capability, color, high-speed printing, and collating.

Military Uses. In addition to marking targets and weapons guidance, laser is used by the military for defensive countermeasures. It also can produce temporary blindness, which can temporarily impair an enemy's ability to fire a weapon or engage in another harmful activity. The military also uses laser guidance systems. Defensive countermeasure applications include small infrared lasers, which confuse heat-seeking missiles to intercept lasers, and power boost-phase intercept laser systems, which contain a complex system of lasers that can locate, track, and destroy intercontinental ballistic missiles (ICBMs). Intercept systems are powered by chemical lasers. When deployed, a chemical reaction results in the quick release of large amounts of energy.

Laser Lighting Displays. Laser light displays are popular worldwide. The displays range from simple to complex and are often choreographed to music. A popular laser multimedia display is Hong Kong's Symphony of Lights, which is presented nightly. The display is accentuated during Tet, the Vietnamese New Year. The exteriors of 44 buildings on either side of Victoria Harbour are illuminated with a synchronized, multicolored laser display, which is accompanied by music. More than four million visitors and locals have viewed the display.

IMPACT ON INDUSTRY

Laser technology is a major component of many industries, including electronics, medical devices, data storage, data transmission, research, and entertainment. Laser products enjoy significant repeat business. Over time the devices fail and need to be repaired or replaced. Often, before device failure occurs, the consumer will be attracted to an improved rendition of the product. For example the original LaserDisc, which never achieved popularity in the United States, has evolved to the Blu-ray Disc, which boasts vastly improved storage capacity as well as image and sound quality. In addition, many laser devices come with a hefty price tag–ranging from several thousand dollars to well in excess of 1 million dollars.

Industry and Business. The average home contains laser devices such as CDs (discs and players) and DVDs (discs and players); and home computers, televisions, and telephones that are often connected to a fiber-optic cable network. Many homes and businesses use one or more laser printers. Virtually every retail establishment has a laser bar-code scanner. The military uses laser devices ranging from inexpensive pointers to guidance systems, which can cost many thousands of dollars. These expensive devices offer a continuous revenue stream for a vendor as they are only used once. The United States automobile industry has been using laser to cut the steel for its vehicles since the 1980's. As of 2011, many manufacturing industries use laser for the cutting and welding of metal parts. The medical-device industry is responsible for many high-end products. Laparoscopic surgery is currently popular, and laser instruments are well-adapted to operating by remote control through small incisions. Ophthalmologists are heavy users of laser instruments. Dermatologic procedures, particularly in the realm of cosmetic surgery, often employ expensive laser devices. Scientific procedures such as spectroscopy are commonplace in all industrialized nations.

Government and University Research. One of the biggest sources of funding for laser research in the United States is the Defense Advanced Research Projects Agency (DARPA). It is also the biggest client for certain kinds of laser applications. The agency is primarily concerned with lasers with a military focus, such as guidance systems, missile countermeasures, and laser-based weapons. Another source of funding for laser research is the United States Department of Energy (DOE).The DOE is currently focused on energy efficiency and renewable energy.

DARPA and the DOE supply funds to many universities in the United States for laser research and development. For example, the two major universities in the Los Angeles area, University of California, Los Angeles (UCLA), and the University of Southern California (USC), are actively engaged in highly technical laser research. UCLA is the home to the Plasma-Accelerator Group, which publishes papers such as "Highest Power CO_2 Laser in the World" (2010). The USC Center for Laser Studies contains a number of research groups: Optical Devices Research Group, Research at High Speed Technology, Semiconductor Device Technology, Solid State Lasers, and Theoretical Research.

CAREERS AND COURSE WORK

The laser technology industry offers careers ranging from entry-level positions, such as lighting display operators, to high-level and highly technical positions, which require a scientific or engineering degree. The high-level positions require at least a bachelor's degree with course work in several fields related to laser technology: engineering, physics, computer science, mathematics, and robotics. Many of the positions require a master's or doctorate degree.

Positions are available for individuals with a degree in laser engineering in both the government and private sectors. The ability to be a team player is often of value for these positions because ongoing research is often a collaborative effort. University research positions are also available. In this arena, the employee is expected to divide his or her time between research and teaching.

Laser technicians are needed in a variety of fields including medical, business, and entertainment. Many of these positions require some training beyond high school at a community college or trade school. Dermatologists and other medical specialists often employ technicians to perform laser procedures. If a company employs a number of laser technicians, supervisory positions may be available.

SOCIAL CONTEXT AND FUTURE PROSPECTS

As of 2011, laser applications have become innumerable and ubiquitous. Many aspects of daily living that are taken for granted, such as ringing up groceries, making a phone call, or playing a video, are dependent on this technology. Virtually every branch of laser technology is continually evolving. Furthermore, existing technologies are being implemented in increasing numbers. For example, use of the CD for data storage is being replaced with the much higher capacity DVD. Research in the realm of military applications is particularly vigorous. The science-fiction weapon, the laser blaster, will soon become a reality. A primary obstacle to developing lasers as weapons has been the generation of enough power to produce a laser blast with a sufficient level of destruction. That power level is rapidly approaching. In March, 2009, Northrop Grumman first fired up their 100 kilowatt (100,000 watts) weapons-grade laser. (One hundred kilowatts is enough energy to power about six U.S. homes for a month). Tests are ongoing for the device. Grumman announced in late 2010 that the device would soon be transferred to the White Sands Missile Range in New Mexico for further testing. Medical applications for laser are expanding rapidly. The consumer, however, must be aware that adding the "laser" adjective to a procedure does not necessarily make it superior to previous techniques. The laser scalpel, for example, is basically just another means of cutting tissue.

Robin L. Wulffson, M.D., F.A.C.O.G.

FURTHER READING

Bone, Jan. *Opportunities in Laser Technology Careers.* New York: McGraw-Hill, 2008. Offers a comprehensive overview of career opportunities in laser technology; includes salary figures as well as experience required to enter the field.

Hecht, Jeff. *Understanding Lasers: An Entry-Level Guide.* 3d ed. Piscataway, N.J.: IEEE Press, 2008. This introductory text is suitable for students at the advanced high school level.

Lele, Ajey. *Strategic Technologies for the Military: Breaking New Frontiers.* Thousand Oaks, Calif.: Sage, 2009. Describes the nuances of technological development in a purely scientific manner and provides a social perspective to their relevance for future warfare and for issues such as disarmament and arms control, as well as their impact on the environment.

Sarnoff, Deborah S., and Joan Swirsky. *Beauty and the Beam: Your Complete Guide to Cosmetic Laser Surgery.* New York: St. Martin's, 1998. Provides information for the consumer on laser dermatology procedures and aids the reader in selecting the safest and most experienced practitioner. It also describes what each surgical procedure entails and details costs in different regions.

Silfvast, William Thomas. *Laser Fundamentals.* 2d ed. New York: Cambridge University Press, 2004. Covers topics from laser basics to advanced laser physics and engineering.

Townes, Charles H. *How the Laser Happened: Adventures of a Scientist.* New York: Oxford University Press, 1999. A personal account by a Nobel laureate that describes some of the leading events in twentieth-century physics.

WEB SITES

American Society for Laser Medicine and Surgery
http://www.aslms.org

Defense Advanced Research Projects Agency
http://www.darpa.mil

USC Center for Laser Studies
http://www.usc.edu/dept/CLS/page.html

See also: Dermatology and Dermatopathology; Ophthalmology; Photonics; Printing; Surgery.

LIGHT-EMITTING DIODES

FIELDS OF STUDY

Electrical engineering; materials science; semiconductor technology; semiconductor manufacturing; electronics; physics; chemistry; mathematics; optics; lighting; environmental studies; physics.

SUMMARY

Light-emitting diodes (LEDs) are diodes, semiconductor devices that pass current easily in only one direction, that emit light when current is passing through them in the proper direction. LEDs are small and are easier to install in limited spaces or where small light sources are preferred, such as indicator lights in devices. LEDs are also generally much more efficient at producing visible light than other light sources. As solid-state devices, when used properly, LEDs also have very few failure modes and have longer operational lives than many other light sources. For these reasons, LEDs are gaining popularity as light sources in many applications, despite their higher cost compared with other more traditional light sources.

KEY TERMS AND CONCEPTS

- **Anode:** More positive side of the diode, through which current can easily flow into the device when forward biased.
- **Cathode:** More negative side of the diode, through which current can flow out of the device when forward biased.
- **Color Temperature:** Temperature of a blackbody radiating thermal energy having the same color as the light emitted by the LED.
- **Forward Bias:** Orientation of the diode in which current most easily flows through the device.
- **Photon:** Quantum mechanical particle of light.
- **P-N Junction:** Junction between positive type (p-type) and negative type (n-type) doped semiconductors on which all diodes are based.
- **Radiant Efficiency:** Ratio of optical power output to the electrical power input of the device.
- **Reverse Bias:** Orientation of the diode in which current does not easily flow through the device.
- **Reverse Breakdown Voltage:** Maximum reverse biased voltage that can be applied to the device before it begins to conduct electricity, often in an uncontrolled manner; sometimes simply called the breakdown voltage.
- **Thermal Power Dissipation:** Rate of energy per unit time dissipated in the device in the form of heat.

DEFINITION AND BASIC PRINCIPLES

Diodes act as one-way valves for electrical current. Current flows through a diode easily in one direction, and the ideal diode blocks current flow in the other direction. The very name diode comes from the Greek meaning two pathways. The diode-like behavior comes from joining two types of semiconductors, one that conducts electricity using electrons (n-type semiconductor) and one that conducts electrons using holes, or the lack of electrons (p-type semiconductor). The electrons will try to fill the holes, but applying voltage in the proper direction ensures a constant supply of holes and electrons to conduct electricity through the diode. The electrons and holes have different energies, so when the electrons combine with holes, they release energy. For most diodes this energy heats the diode. However, by adjusting the types and properties of the semiconductors making up the diodes, the energy difference between holes and electrons can be made larger or smaller. If the energy difference corresponds to the energy of a photon of light, then the energy is given off in the form of light. This is the basis of how LEDs work.

LEDs are not 100 percent efficient, and some energy is lost in current passing through the device, but the majority of energy consumed by LEDs goes into the production of light. The color of light is determined by the semiconductors making up the device, so LEDs can be fabricated to make light only in the range of wavelengths desired. This makes LEDs among the most energy-efficient sources of light.

BACKGROUND AND HISTORY

In 1907, H. J. Round reported that light could be emitted by passing current through a crystal rectifier junction under the right circumstances. This was the ancestor of the modern LED, though the term diode had not yet been invented. Though research continued on these crystal lamps, as they were called, they were seen as impractical alternatives to incandescent and other far less expensive means of producing light. By 1955, Rubin Braunstein, working at RCA, had shown that certain semiconductor junctions produced infrared light when current passed through them. Scientists Robert Biard and Gary Pittman, however, managed to produce a usable infrared LED, receiving a patent for their device. Nick Holonyak, Jr., a scientist at General Electric, then created a red LED–the viable and useful visual spectrum LED–in 1961. Though these early LEDs were usable, they were far too expensive for widespread adoption. By the 1970's, Fairchild Semiconductor had developed inexpensive red LEDs. These LEDs were soon incorporated into seven segment numeric indicators for calculators produced by Hewlett Packard and Texas Instruments. Red LEDs were also used in digital watch displays and as red indicator lights on various pieces of equipment.

Early LEDs were limited in brightness, and only the red ones could be fabricated inexpensively. Eventually, other color LEDs and LEDs capable of higher light output were developed. As the capabilities of LEDs expanded, they began to see more uses. By the early twenty-first century, LEDs began to compete with other forms of artificial lighting based on their energy efficiency.

How It Works

An LED is a specific type of solid-state diode, but it still retains the other properties typical of diodes. Solid-state diodes are formed at the junction of two semiconductors of different properties. Semiconductors are materials that are inherently neither good conductors nor good insulators. The electrical properties of the materials making up semiconductors can be altered by the addition of impurities into the crystal structure of the material as it is fabricated. Adding impurities to semiconductors to achieve the proper electrical nature is called doping. If the added impurity has one more electron in its outermost electron shell compared with the semiconductor material, then extra electrons are available to conduct electricity. This is a negative doped, or n-type, semiconductor. However, if the impurity has one fewer electron in its outermost electron shell compared with the semiconductor material, then there are too few electrons in the crystal structure, and an electron

can move from atom to atom to fill the void. This results in a missing electron moving from place to place and acts like a positive charge moving through the semiconductor. Engineers call this missing electron a hole, and semiconductors in which holes dominate are called positive doped, or p-type, semiconductors.

The P-N Junction. To make a diode, a device is fabricated in which a p-type semiconductor is placed in contact with a n-type semiconductor. The shared boundary between the two types of semiconductors is called a p-n junction. In the vicinity of the junction, the extra electrons in the n-type region combine with the holes of the p-type region. The results in the removal of charge carriers in the vicinity of the p-n junction and the area of few charge carriers is called the depletion region.

When a voltage is applied across the p-n junction, with the p-type region having the higher voltage, then electrons are pulled from the n-type region and holes are pulled from the p-type region into the depletion region. Additionally, electrons are pulled into cathode (the exterior terminal connecting to the n-type region) replenishing the supply of electrons in the n-type region, and electrons are pulled from the anode (the exterior terminal connecting to the p-type region) replenishing the holes in the p-type region. This is the forward-bias orientation of the diode, and current flows through the diode when voltage is applied in this manner. However, when voltage is applied in the reverse direction, electrons are pulled from the n-type region and into the

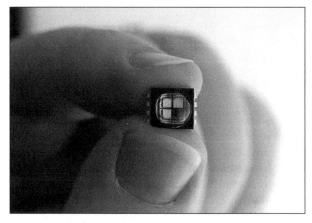

LEDs are small and generally much more efficient at producing visible light than other light sources. (Bloomberg via Getty Images)

p-type region, resulting in a larger depletion region and fewer available charge carriers. Electric current does not flow through the diode in this reverse-bias orientation.

Electroluminescence. Electrons and holes have different energy levels. When the electrons and holes combine in the depletion region, therefore, they release energy. For most diodes, the energy difference between the p-type holes and the n-type electrons is fairly small, so the energy released is correspondingly small. However, if the energy difference is sufficiently large, then when an electron and hole combine, the amount of energy released is the same as that of a photon of light, and the energy is released in the form of light. This is called electroluminescence. Different wavelengths or colors of light have different energies, with infrared light having less energy than visual light, and red light having less energy than other forms of visual light. Blue light has more energy than other forms of visual light. The color of light emitted by the recombination of electrons and holes is determined by the energy difference between the electrons and holes. Different semiconductors have different energies of electrons and holes, so p-n junctions made of different kinds of semiconductors with different doping result in different colors of light emitted by the LED.

Efficiency. Most light sources emit light over a wide range of wavelengths, often including both visual and nonvisual light as well as heat. Therefore, only a portion of the energy used goes into the form of light desired. For an idealized LED, all of the light goes into one color of light, and that color is determined by the composition of the semiconductors making the p-n junction. For real LEDs, not all of the light makes it out of the material. Some of it is internally reflected and absorbed. Furthermore, there is some electrical resistance to the device, so there is some energy lost in heat in the LED–but nowhere near as much as with many other light sources. This makes LEDs very efficient as light sources. However, LED efficiency is temperature dependent, and they are most efficient at lower temperatures. High temperatures tend to reduce LED efficiency and shorten the lifetime of the devices.

APPLICATIONS AND PRODUCTS

LEDs produce light, like any other light source, and they can be used in applications where other light sources would have been used. LEDs have certain properties, however, that sometimes make their use preferable to other artificial light sources.

Indicator Lights. Among the first widespread commercial use of LEDs for public consumption was as indicator lights. The early red LEDs were used as small lights on instruments in place of small incandescent lights. The LEDs were smaller and less likely to burn out. LEDs are still used in a similar way, though not with only the red LEDs. They are used as the indicator lights in automobile dashboards and in aircraft instrument panels. They are also used in many other applications where a light is needed and there is little room for an incandescent bulb.

Another early widespread commercial use of LEDs was the seven-segment numeric displays used to show digits in calculators and timepieces. However, LEDs require electrical current to operate, and calculators and watches would rather quickly discharge the batteries of these devices. Often the display on the watches was visible only when a button was pressed to light up the display. However, the advent of liquid crystal displays (LCDs) has rendered these uses mostly obsolete since they require far less energy to operate, and LEDs are needed to light the display at night only.

Replacements for Colored Incandescent Lights. Red LEDs have become bright enough to be used as brake lights in automobiles. Red, green, and yellow LEDs are sometimes used for traffic lights and for runway lights at airports. LEDs are even used in Christmas-decoration lighting. They are also used in message boards and signs. LEDs are sometimes used for backlighting LCD screens on televisions and laptops. Colored LEDs are also frequently used in decorative or accent lighting, such as lighting in aquariums to accentuate the colors of coral or fish. Some aircraft use LED lighting in their cabins because of energy efficiency. Red LEDs are also used in pulse oximeters used in a medical setting to measure the oxygen saturation in a patient's blood.

The biggest obstacle to replacing incandescent lights with LEDs for room lighting or building lighting is that they produce light of only one color. Several strategies are in development for producing white light using LEDs. One strategy is to use multiple-colored LEDs to simulate the broad spectrum of light produced by incandescent lights or fluorescent lights. However, arrays of LEDs produce a set of

Fascinating Facts About Light-Emitting Diodes

- Most of the device often called a light-emitting diode is really just the packaging for the LED. The actual diode is typically very tiny and embedded deep inside the packaging.
- The fist inexpensive digital watches marketed to the public used red LEDs for displays.
- Most remote controls for televisions, DVD players, and similar devices use an infrared LED to communicate between the remote control and the devices.
- LEDs operate more efficiently when they are cold than when they are hot.
- Many automobile manufacturers use LEDs in brake lights, particularly the center third brake light.
- An early term for the LED was "crystal lamp." This term eventually gave way to the term "light-emitting diode," later commonly abbreviated LED.
- LEDs, if properly cared for and operated under specified conditions, can last for upward of 40,000 hours of operation.
- The first LEDs produced were infrared-emitting diodes. The first mass-produced visible-light LEDs emitted red light.
- LEDs switch on and off far more quickly than most other light sources, often being able to go from off to fully on in a fraction of a microsecond.

LEDs are very energy efficient, but they tend to be less efficient at high power and high light output. LEDs are slightly more efficient, and far more expensive, than high-efficiency fluorescent lights, but research continues.

Nonvisual Uses for LEDs. Infrared LEDs are often used as door sensors or for communication by remote controls for electronic devices. They can also be used in fiber optics. The rapid switching capabilities of LEDs makes them well suited for high-speed communication purposes. Ultraviolet LEDs are being investigated as replacements for black lights for purposes of sterilization, since many bacteria are killed by ultraviolet light.

IMPACT ON INDUSTRY

Government and University Research. Research continues to produce less expensive and more capable LEDs. The nature of LEDs makes them potentially among the most energy-efficient light sources. The United States Department of Energy is a significant driving force behind development of energy-efficient solid-state lighting systems, specifically the development of LED technology for widespread use. University researchers, particularly faculty in engineering and materials science, receive government grants to study improving LED performance.

Industry and Business. There are two different ways that the private sector is impacted by advances in LEDs and LED uses. Major corporations, particularly those in semiconductor manufacturing, are doing research alongside government and university researchers to develop more efficient LEDs with greater capabilities. The increased demand for LED lighting from energy-conscious consumers spurs companies to compete to develop new LEDs for consumer use.

However, no matter what types of new LEDs are developed, they must be used in products that consumers want to purchase in order to be commercially viable. In the early twenty-first century, environmentally conscious consumers began looking for more energy-efficient lights. A great many companies began to make products that used LED lights in place of incandescent lights for these energy-conscious consumers who were willing to pay more for a product that was perceived as being more environmentally friendly and energy efficient. A prime example of this sort of development was the advent

discrete colors of light rather than all colors of the rainbow, thus distorting colors of objects illuminated by the LED arrays. This is aesthetically unpleasing to most people. Another strategy for producing white light from LEDs is to include a phosphorescent coating in the casing around the LED. This coating would provide the different colors of light that would mimic the light of fluorescent bulbs; however, such a strategy removes much of the efficiency of LEDs. Research continues to produce a pleasing white light from LEDs.

Despite the color problems and the high initial cost of LEDs, LEDs have many properties that make them attractive replacements for incandescent or fluorescent lights. LEDs typically have no breakable parts and being solid-state devices are very durable and have low susceptibility to vibrational damage.

of LED Christmas lights; however, many other applications have been developed. The long operational life of LEDs is also attractive to consumers, since they are less likely to need a replacement compared with other light sources. The continued development of increasingly inexpensive LEDs with more different colors of light possible is spurring more companies to develop more products using LEDs.

CAREERS AND COURSE WORK

LEDs are used in many industries, not just in electronics, which means that there are many different degree and course-work pathways to working with LEDs.

The development of new types of LEDs requires detailed understanding of semiconductor physics, chemistry, and materials science. Typically, such research requires an advanced degree in physics, materials science, or electrical engineering. Such degrees require courses in physics, mathematics, chemistry, and electronics. The different degrees will have different proportions of those courses.

Utilization of LEDs in circuits, however, requires a quite different background. Technicians and assembly workers need only basic electronics and circuits courses to incorporate the LEDs into circuits or devices.

Lighting technicians and lighting engineers also work with LEDs in new applications. Such careers could require bachelor's degrees in their field. New LED lamps are being developed and LEDs are seen as a possible energy-efficient alternative to other types of lighting. They also have long operational lives, so there is continual development to include LEDs in any type of application where light sources of any sort are used.

SOCIAL CONTEXT AND FUTURE PROSPECTS

At first, LEDs were a niche field, with limited uses. However, as LEDs with greater capabilities and different colors of emitted light were produced, uses began to grow. LEDs have evolved past the point of simply being indicator lights or alphanumeric displays. Developments in semiconductor manufacturing have driven down the cost of many semiconductor devices, including LEDs. The reducing cost combined with the energy efficiency of LEDs has led these devices to become more prominent, particularly where colored lights are desired. Research

continues to produce newer LEDs with different colors, different power requirements, and different intensities. Newer techniques are being developed to produce white light using LEDs. These technological developments will make LEDs even more practical replacements for current light sources, despite their higher initial up-front costs.

Research continues on LEDs to make them more commercially and aesthetically viable as alternatives to more traditional light sources. However, research is also continuing on other alternative light sources. The highest-efficiency fluorescent lights have similar efficiencies to standard LEDs, but they cost less and are able to produce pleasing white light that LEDs do not yet produce. LEDs will continue to play an increasing role in their current uses, but it is unclear if they will eventually become wide-scale replacements for incandescent or fluorescent lights.

Raymond D. Benge, Jr., B.S., M.S.

FURTHER READING

Held, Gilbert. *Introduction to Light Emitting Diode Technology and Applications.* Boca Raton, Fla.: Auerbach, 2009. A comprehensive overview of light-emitting diode technology and applications of LEDs.

Mottier, Patrick, ed. *LEDs for Lighting Applications.* Hoboken, N.J.: John Wiley & Sons, 2009. A detailed book about LEDs, the manufacture of LEDs, and use of the devices in artificial lighting.

Paynter, Robert T. *Introductory Electronic Devices and Circuits: Conventional Flow Version.* 7th ed. Upper Saddle River, N.J.: Prentice Hall, 2006. An excellent and frequently used introductory electronics textbook, containing several sections on diodes, with a very good description of light-emitting diodes.

Razeghi, Manijeh. *Fundamentals of Solid State Engineering.* 3d ed. New York: Springer, 2009. An advanced undergraduate textbook on the physics of semiconductors, with a very detailed explanation of the physics of the p-n junction, which is the heart of diode technology.

Schubert, E. Fred. *Light-Emitting Diodes.* 2d ed. New York: Cambridge University Press, 2006. A very good and thorough overview of light-emitting diodes and their uses.

Žukauskas, Artūras, Michael S. Shur, and Remis Gaska. *Introduction to Solid-State Lighting.* New York: John Wiley & Sons, 2002. A fairly advanced and very thorough treatise on artificial lighting technologies, particularly solid-state lighting, such as LEDs.

WEB SITES
LED Magazine
http://www.ledsmagazine.com

The Photonics Society
http://photonicssociety.org

Schottkey Diode Flash Tutorial
http://cleanroom.byu.edu/schottky_animation.phtml

University of Cambridge
Interactive Explanation of Semiconductor Diode
http://www-G.eng.cam.ac.uk/mmg/teaching/linearcircuits/diode.html

U.S. Department of Energy
Solid-State Lighting
http://www1.eere.energy.gov/buildings/ssl/about.html

See also: Diode Technology; Electrical Engineering; Electronics and Electronic Engineering; Optics.

LIQUID CRYSTAL TECHNOLOGY

FIELDS OF STUDY

Electronics; chemistry; organic chemistry; physics; optics; materials science; mathematics

SUMMARY

Liquid crystal devices are the energy efficient, low-cost displays used in a variety of applications in which information or images are presented. The operation of the devices is based on the unique electrical and optical properties of liquid crystal materials.

KEY TERMS AND CONCEPTS

- **Active Matrix Liquid Crystal Display:** A display with transistors to control voltage at each pixel, which allows for a sharper picture.
- **Anisotropic:** Molecules that are not similar in all directions but have a longer axis in certain planes, so certain measurements may depend on direction.
- **Calamitic Liquid Crystal:** An anisotropic liquid crystal with a longer axis in one direction, giving it a rodlike shape (also referred to as prolate).
- **Discotic Liquid Crystal:** An anisotropic liquid crystal with one axis shorter than the other two, giving it a disclike shape (also referred to as oblate).
- **Lyotropic Liquid Crystal:** Liquid crystals achieved when the molecule is combined to a high enough concentration with a solvent (rather than through temperature change). Soap is one example.
- **Nematic State:** Liquid crystals that are not distinctly layered but can be oriented and aligned by a charge.
- **Smectic State:** Liquid crystals that are somewhat disordered but maintain distinct layers.
- **Thermotropic Liquid Crystals:** Liquid crystals in which the specific properties are maintained within a specific temperature range.
- **Thin Film Transistor (TFT):** Transistors in a liquid crystal display (LCD) that help control the voltage, and therefore, the coloring, at individual pixels.
- **Twisted-Nematic Effect (TNE):** The effect in which, in a turned-on LCD, light travels through the display changing from being polarized in one direction to being twisted 90 degrees and polarized in that direction.
- **Twisted-Nematic Liquid Crystal Display (TN LCD):** The dominant type of LCD display, which utilizes the TNE.

DEFINITION AND BASIC PRINCIPLES

Liquid crystal technology is the use of a unique property of matter to create visual displays that have become the standard for modern technology.

Originally discovered as a state existing between a solid and a liquid, liquid crystals were later found to have applications for visual display. While liquid crystals are less rigid than something in a solid state of matter, they also are ordered in a manner not found in liquids. As anisotropic molecules, liquid crystals can be polarized to a specific orientation to achieve the desired lighting effects in display technologies.

Liquid crystals themselves can exist in several states. These range from a well-ordered crystal state to a disordered liquid state. In-between states are known as the smectic phase, which have layering, and the nematic phase, in which the separate layers no longer exist but the molecules can still be ordered.

Liquid crystal displays (LCDs) at this point typically use crystals in the nematic state. They also use calamitic liquid crystals, whose rodlike shape and orientation along one axis allow the display to lighten and darken.

Over time, researchers have made advances in the materials used for LCDs and the route of power for manipulating the crystals, allowing for the low-cost, high-resolution, energy-efficient displays that have become the dominant technology for displays such as computers and television sets. Future research should allow for improvements in the response time of the displays and for better viewing from different angles.

BACKGROUND AND HISTORY

Liquid crystals were discovered by Austrian botanist and chemist Friedrich Reinitzer in 1888. While working with cholesterol, he discovered what appeared to be a phase of matter between the solid (crystal) state and the liquid state. While attempting to find the melting point, Reinitzer observed that within a certain temperature range he had a cloudy mixture, and only at a higher temperature did that mixture become a liquid. Reinitzer wrote of his discovery to his friend, German physicist Otto Lehmann, who not only confirmed Reinitzer's discovery—that the liquid crystal state was unique and not simply a mixture of solid and liquid states—but also noted some distinct visual properties, namely that light can travel in one of two different ways through the crystals, a property known as birefringence.

After the discovery of liquid crystals, the field saw a lengthy period of dormancy. Modern display applications have their roots in the early 1960s, in part because of the work of French physicist and Nobel laureate Pierre-Gilles de Gennes, who connected research in liquid crystals with that in superconductors. He found that applying voltages to liquid crystals allowed for control of their orientation, thus allowing for control of the passage of light through them.

In the early 1970s, researchers, including Swiss physicist and inventor Martin Schadt at the Swiss company Hoffman-LaRoche, discovered the twisted-nematic effect—a central idea in LCD technology. (The year of invention is typically said to be 1971, although patents were awarded later.) The idea was patented in the United States at the same time by the International Liquid Xtal Company (now LXD), which was founded by American physicist and inventor James Fergason in Kent, Ohio. (Fergason was part of the Liquid Crystal Institute at Kent State University. The institute was founded by American chemist Glenn H. Brown in 1965.) Licensing the patents to outside manufacturers allowed for the production of simple LCDs in products such as calculators and wristwatches.

In the 1980s, LCD technology expanded into computers. LCDs became critical components of laptop computers and smaller television sets. With research continuing on liquid crystals into the twenty-first century, LCD televisions overtook cathode ray tubes (CRTs) as the dominant technology for television sets.

HOW IT WORKS

LCDs have a similar structure, whether in a digital watch or in a 40-inch television. The liquid crystals are held between two layers of glass. A layer of transparent conductors on the liquid crystal side of the

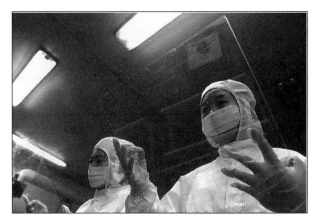

Quality inspectors of a LCD company inspect liquid crystal display panels on the product line in China. (Getty Images)

glass allows the liquid crystal layer to be manipulated. Polarized film layers are placed on the outside ends of the glass, one of which will face the viewer and the other will remain at the back of the display.

Polarizers alter the course of light. Typically, light travels outward in random directions. Polarizers present a barrier, blocking light from traveling in certain directions and preventing glare. The polarizers in an LCD are oriented at 90-degree angles from each other. With the polarizers in place alone, all light would be blocked from traveling through an LCD, but the workings of liquid crystals allow that light to come through.

The electrical current running through the liquid crystals controls their orientation. The rodlike crystals, without voltage, are oriented perpendicular to the glass of the screens. In this state, the crystals do not alter the direction of the light passing through. As voltage is applied, the crystals turn parallel to the direction of the screen. Like the polarized films, the conductors are oriented at 90-degree angles to each other, as are the crystals next to each screen (that is, crystals on one end are at a 90-degree angle from those on the other end). Between, however, the crystals orient in a twisting pattern, so light polarized in one direction will be redirected and turned 90 degrees when it emerges at the other end of the display. This is known as the twisted-nematic effect.

Thus, the voltage applied to the crystals controls the light coming through the LCD screen. At lower voltage levels, some, but not all, light is allowed through the display. By manipulating the intensity of the incoming light, the LCD can display in a gray scale.

Because liquid crystals are a state of matter, they exist only at a certain temperature: between the melting and freezing points of the material. Thus, LCD displays may have trouble working in extreme heat or extreme cold. One of the primary challenges of LCD display is finding materials that remain in liquid crystal forms at the temperatures in which the devices are likely to be used. Some of the challenge also lies in finding materials that may display better color or allow for lower energy consumption. However, materials that do one of these things better may make other features of a display worse. In some cases, a mixture of compounds for the liquid crystals in a device may be used.

Simple LCDs. The simplest LCD displays, such as calculators and watches, typically do not have their own light sources. Instead, they have what is known as passive display. In back of the LCD display is a reflective surface. Light enters the display and then bounces off the reflective surface to allow for the screen display. Simple LCDs are monochromatic and have specific areas (typically bars or dots) that become light or dark. While these devices are lower-powered, some do still use a light source of their own. Alarm clocks, for example, have light-emitting diodes (LEDs) as part of their display so that they can be seen in the dark.

Personal Computers and Televisions. For larger monitors that display complex images in color, the setup for an LCD becomes more complicated. Multicolor LCDs need a significant light source at the back of the display.

The glass used for more sophisticated LCD displays will have microscopic etchings on the glass plates at the front and back of the display. As with the polarizing filters and the conductors, the etchings are at 90-degree angles from each other, vertical on one plate and horizontal on the other. This alignment forms a matrix of points in each location where the horizontal and vertical etchings cross, resulting in what are known as pixels. Each pixel has a unique "address" for the electronic workings of the display. Many television sets are marketed as having 1080p, referring to 1,080 horizontal lines of pixels.

An active matrix (AM) display will have individual thin film transistors (TFTs) added at each pixel to allow for control of those sites. Three transistors are actually present at each pixel, each accompanied by an additional filter of red, green, or blue. Each of those transistors has 256 power levels. The blending of the different levels of those three colors (256^3, or 16,777,216 possible combinations) and the number of pixels allows for the full-color LCD displays.

While light is displayed as a combination of red, green, and blue on screens, printing is typically done on a scale that uses cyan, magenta, yellow, and black as base colors. This accounts for some discrepancy between colors that appear on screen and those that show up on paper.

APPLICATIONS AND PRODUCTS

Liquid crystals are used in displays for a number of products. Early uses included digital thermometers, digital wristwatches, electronic games, and calculators. As the power needed for an LCD display and resolution improved, LCDs came to be used in computer monitors, television sets, car dashboards, and cellphones.

Calculators and Digital Watches. Watches and calculators use what is known as a seven-segment display, wherein each of the seven segments that make up a number are "lit" or "unlit" to represent the ten digits. Looking closely at an LCD will reveal that most numbers come from seven segments, which can be lit to display the ten different digits. Without the polarizing layer, the display would not work. Placing the polarized layers in parallel on the surface would, for example, cause the outlines of all the numbers and other areas on the display to illuminate (appearing as 8s) and leave as blanks the rest of the display.

Early electronic games also used a segment display. Fixed places on the display would be either lit or unlit, allowing game characters to appear to move across the screen.

Temperature Monitors. Because of their sensitivity to heat, liquid crystals have been studied for their use as temperature monitors. Molecules in the smectic liquid crystal state rotate around their axes, and the angle at which they rotate (the pitch) can be temperature sensitive. At different temperatures, the wavelength of light given off will change. Some liquid crystal mixtures are fairly temperature sensitive, and so the mixture of the colors will change with relatively small changes in the temperature. Because of this, they can be used for displays such as infrared or surface temperatures.

Computer Monitors. LCDs have been, and will likely remain, the standard for laptop computers. They have been used for monitors since the notebook computer was introduced. Because of the low power consumption and thinness of the monitor, their use is likely to continue.

Television Sets. The workings of LCDs in televisions have been outlined in the foregoing section. As will be discussed further in the next section, LCDs have had a marked impact on television displays, both in the quality of displays in the home and in industry overall.

There are several developments that could affect LCD technology in the near future. One example is the development of LEDs for use as backlighting for LCDs. By using LEDs rather than a fluorescent bulb, as LCD technology now uses, LCDs can manifest greater contrast in different areas of the screen. Other areas of LCD development include photoalignment and supertwisted nematic (STN) LCDs.

Grooves are made in glass used for LCDs, but this has raised some concern about possible electric charges, reducing the picture quality. Additionally, photoalignment—a focus on the materials used to align the liquid crystals in the display—should ultimately allow for liquid display screens that are flexible or curved, rather than rigid (as are glass panels).

STN LCDs are modified versions of TN LCDs. Rather than twisting the crystals between the layers a total of 90 degrees, STN LCDs rotate the crystals by 270 degrees within the display. This greater level of twisting allows for a much greater degree of change in the levels of brightness in a display. At the same time, it presents a challenge because the response time for the screen is significantly slower.

IMPACT ON INDUSTRY

The introduction of LCDs allowed for the replacement of CRTs, which had been the standard in the industry. LCDs require less energy than CRTs, and because they do not need the space for the cathode ray, the physical sets are lighter and take up less space for a similar-sized screen.

In larger displays, LCDs compete in the consumer market with plasma televisions, which utilize small compartments of gas for the light on their screens.

Fascinating Facts About Liquid Crystal Technology

- Although it was the dominant display technology in televisions for more than sixty years, cathode ray tubes were discovered close to twenty years after liquid crystals.
- Partly for his work with liquid crystals, Pierre-Gilles de Gennes was awarded the Nobel Prize in Physics in 1991.
- The world's largest LCDs are in Cowboys Stadium in Arlington, Texas. Each screen of the four-sided display is 72 feet high and 160 feet wide. The screens are backlit by a total of 10.5 million LEDs, and altogether, the display weighs 1.2 million pounds.
- The first LCD television, the black-and-white Casio TV-10, came to market in 1983.
- Introduced in 1982, the Epson HX-20, generally considered the world's first standard-sized laptop, utilized an LCD.
- The discovery of liquid crystals was made by Friedrich Reinitzer while working with the compound cholesteryl benzoate. In the time since its discovery, thousands of other organic compounds have been found to have a liquid crystal state.
- In addition to use in displays, liquid crystals have some application in polymers. Because they can be oriented, they can produce stronger fibers. As a result, they are used in the production of Kevlar.

There are still a number of reasons a consumer might choose plasma over LCD, or vice versa.

Because of their polarized light, LCDs do not have the problems with glare that plasma displays may have. However, because LCDs have polarized films on the screen, LCD images vary greatly when they are not viewed straight on. While plasma televisions may have a slightly faster response rate than LCD televisions, LCD televisions consume less energy. While there is some competition between LCDs and plasma screens in the large television market, plasmas are made only in larger sizes, so smaller television sets and other applications such as computer monitors continue to use LCDs.

Political Change. With the proliferation of high-definition television sets, whether in plasma or LCD form, came a corresponding need for higher quality signals for broadcast. Cable and satellite television providers have offered packages of channels that are distributed at higher quality.

Additionally, the presence of these televisions in homes was partly responsible for the industry's decision to move to a different broadcast format for television and to free up parts of the broadcast spectrum. Starting in 2005 and 2006 (depending on the size of the set), television sets were required to have digital receivers. As of June 12, 2009, all television signals were broadcast at a high definition (HD) frequency.

CAREERS AND COURSEWORK

Much of liquid crystal technology is oriented toward the production of display screens, but the complexity of the subject leaves a number of career path options. Master's degrees and doctorates are available in the area of liquid crystal research. Programs typically involve interdisciplinary study in chemistry and physics, and in other potentially relevant areas. Liquid crystal technology builds off basic knowledge of physics, chemistry, and organic chemistry.

Some research in the area of liquid crystals focuses on the material of the crystals themselves and in the development of crystals that improve upon current LCD displays. A background in chemical analysis and optics is important. Design of the products themselves involves knowledge about the design of circuits and backlighting.

There are a number of areas for prospective research and product development. These include the design of LCDs themselves, design of the manufacturing process, and the process of creating the molecules used in the displays.

SOCIAL CONTEXT AND FUTURE PROSPECTS

Some of the concerns and problems with LCDs are being confronted by society as a whole. One concern is the high energy consumption of fluorescent lamps used by LCDs. In contrast, LED lights, which use less energy, are being used more and more in LCDs. There is concern, however, about the environmental hazards LEDs may create when they are disposed of in landfills. Another possibility is the use of carbon nanotubes, which would provide LCD backlighting but would use even less energy than LEDs.

Durability concerns may also come to play a role. The grooves in the glass necessary for high definition LCDs also lead to physical wear and tear on the product. Refining the technology further may produce more durable sets while also alleviating some of the concerns about electronics disposal. Future work on LCDs also will involve altering components to overcome picture quality and durability concerns. Given the prominence of the products that utilize liquid crystals, the technology is likely to be important for development for the foreseeable future.

Joseph I. Brownstein, M.S.

FURTHER READING

Chandrasekar, Sivaramakrishna. *Liquid Crystals.* 2d ed. New York: Cambridge University Press, 1992. Originally written in 1977, this work is considered one of the classic textbooks in the field and provides early history and an overview of work in liquid crystals.

Chigrinov, Vladimir G., Vladimir M. Kozenkov, and Hoi-Sing Kwok. *Photoalignment of Liquid Crystalline Materials.* Chichester, England: John Wiley & Sons, 2008. This book covers some areas of development in improving screens for LCD devices.

Collings, Peter J., and Michael Hird. *Introduction to Liquid Crystals.* New York: Taylor & Francis, 1997. As an introduction to the field, this book goes through the basics of liquid crystals and then some of the applications, using less technical language than many other texts on the subject.

Delepierre, Gabriel, et al. "Green Backlighting for TV Liquid Crystal Display Using Carbon Nanotubes." *Journal of Applied Physics* 108, no. 4 (September 2010). This article examines the possibility of using carbon nanotubes to backlight LCDs, potentially reducing both production and energy costs.

WEB SITES

Kent State University, Liquid Crystal Institute
http://www.lcinet.kent.edu

Nobel Prize Foundation
.http://www.nobelprize.org

See also: Applied Physics; Computer Engineering; Computer Graphics; Computer Science; Diode Technology; Electrical Engineering; Electronic Materials Production; Electronics and Electronic Engineering; Graphics Technologies; Information Technology; Light-Emitting Diodes; Optics; Surface and Interface Science; Television Technologies.

LITHOGRAPHY

FIELDS OF STUDY

Printing; photolithography; photography; physics; chemistry; mathematics; calculus; optics; mechanical engineering; material science; graphic design; electromagnetics; microfabrication; semiconductor manufacturing; laser imaging.

SUMMARY

Lithography is an ink-based printing process that was first used in Europe at the end of the eighteenth century. Unlike an older printing press, in which individual pieces of raised type were pressed down onto sheets of paper, lithography uses a flat plate to transfer an image to a sheet of paper. Nearly all books, newspapers, and magazines being published are printed using lithography, as are posters and packing materials. A specialized subfield of lithography known as photolithography is also used in the making of semiconductors for computers. Career opportunities in lithography are growing in specialized areas but overall are neither increasing or decreasing because of the rise of electronic publishing and marketing.

KEY TERMS AND CONCEPTS

- **Emulsion:** Mixture of two chemicals; in lithography, often used on plate surfaces.
- **Hydrophilic:** Chemical property on a plate's surface that attracts and holds a water-based ink; the opposite is hydrophobic.
- **Image:** Words, pictures, or both on a printing plate.
- **Imagesetter:** Device that transfers an image from a computer directly to a plate without the use of photographic negatives.
- **Offset:** Transfer of an image from a plate to a secondary surface, often a rubber mat, that reverses it before final printing.
- **Photolithography:** Process that uses high-precision equipment and light-sensitive chemicals to make products such as semiconductors.
- **Photomask:** Flat surface into which holes have been cut to allow light to pass through; used in photolithography.
- **Plate:** Printing surface, made of metal or stone, on which areas have been chemically treated to attract or repel ink.

DEFINITION AND BASIC PRINCIPLES

Lithography is the process of making an image on a flat stone or metal plate and using ink to print the image onto another surface. Areas of the plate are etched or treated chemically in order to attract or repel ink. The ink is then transferred, directly or indirectly, to the surface where the final image appears.

Unlike a process such as letterpress, where raised letters or blocks of type are coated with ink and pressed against a surface such as paper, lithographic printing yields a result that is smooth to the touch. Lithography differs from photocopying in that plates must be created and ink applied before prints can be made. Photocopying uses a process known as xerography, in which a tube-shaped drum charged with light-sensitive material picks up an image directly from a source. Laser printing is another application of xerography and is not the same as lithography.

Photolithography is a process that imitates traditional lithography in several ways but is not identical. Its high level of precision–a photolithographic image can be accurate down to the level of a micrometer or smaller–is useful in applications such as the manufacturing of computer components.

BACKGROUND AND HISTORY

Lithography was invented in 1798 by Alois Senefelder, a German playwright. Senefelder, who was looking for a way to publish his plays cheaply, discovered that printing plates could be made by writing on a flat stone block with grease pencil and etching away the stone surface around the writing. Eventually Senefelder developed a process by which ink adhered only to the parts of a flat surface not covered by grease. He later expanded the process to include multiple ink colors and predicted that lithography would one day be advanced enough to reproduce works of fine art.

German and French printers in the early 1800's made additional innovations. A patent was issued in 1837 to artist Godefroy Engelmann in France for a process he called chromalithography, in which colors were layered to create book illustrations. Interest in lithography and color printing also spread to North America, where printers in Boston invented new technologies that made the mass production of lithographic prints both high quality and economical. The process quickly spread from books to greeting cards, personal and business cards, posters, advertisements, and packaging labels. Lithography is still the leading process by which mass-produced reading material and packaging are printed.

How It Works

Lithography in the context of printing follows a different set of steps than photolithography as used to make microprocessors.

Offset Lithography. While there are many ways to print on paper or packaging using lithographic techniques, most items involve a process known as offset lithography. The term "offset" refers to the fact that the printing plate does not touch the paper or item itself.

In offset lithography, a plate is first created with the image to be printed. The plate may be made of metal, paper, or a composite such as polyester or Mylar. Lithographic printing plates were flat at one time, but modern printing presses use plates shaped like cylinders, with the image on the outside. To transfer the image to the plate, the surface of the plate is roughened slightly and covered with a light-sensitive chemical emulsion. A sheet of photo film with a reverse, or negative, of the image is laid over the emulsion. When an ultraviolet light is shone on the negative, the light filters through the image only in the areas where the negative is translucent. The result is a positive image–essentially, a negative of the negative–left on the printing plate.

The plate is treated again with a series of chemicals that make the darker areas of the image more likely to pick up ink, which is oil based. The lighter areas of the image are made to be hydrophilic, or water loving. Because oil and water do not mix, water blocks ink from being absorbed by these areas. A water-based mixture called fountain solution is applied to the surface of the plate and is picked up by the hydrophilic areas of the image. Rollers then coat the plate with ink, which adheres only to the hydrophobic (water-fearing) areas that will appear darker on the final image. Once the plate is inked, the press rolls it against a rubber-covered cylinder known as a blanket. The ink from the plate is transferred to the blanket in the form of a negative image. Excess water from the ink as well as fountain solution is removed in the process. The blanket is rolled against the sheet of paper or other item that will receive the final image. Finally, the paper carrying the newly inked image passes through an oven, followed by a set of water-chilled metal rollers, to set the image and prevent the ink from smudging.

Photolithography. Like lithography, the process of photolithography depends on the making of a plate coated with a light-sensitive substance. The plate is known as the substrate, while the light-sensitive chemical is known as the photoresist. Instead of a photo negative with the image to be printed, a photomask is used to shield the photoresist from light in some areas and expose it in others.

The similarities to traditional lithography end here, however. In photolithography, the substrate—rather than a sheet of paper or packaging material—is the final product. Once the image is transferred through the photo mask onto the photoresist, the substrate is treated with a series of chemicals that engraves the image into the surface. In lithographic printing, the image is never engraved directly onto the plate. Unlike printing plates and blankets, which are cylindrical, substrates are always flat. The result is a thin sheet of silicon, glass, quartz, or a composite etched precisely enough to be used as a microprocessing component.

Applications and Products

Lithography as a printing technology has developed in multiple, almost opposing, directions throughout its history. Because lithographic plates can be used to make large numbers of impressions, the development of lithography allowed for printing of images and type on a mass scale that was commercially viable, a major change from the letterpress and intaglio methods of printing that came earlier. Over time, lithography came to be associated with lower-cost editions of books and other printed matter intended to be short-lived, such as newspapers, magazines, and catalogs. Lithography has also evolved as a method of artistic printmaking that can produce works of great beauty and high value. On the photolithography side, the technology has kept pace with the needs of generations of computers.

Web-Fed Offset Printing. Large numbers of copies of a printed work–in the range of 50,000 copies and up–require printing processes that can run quickly and efficiently. Web-fed offset printing takes its name from the way in which paper is fed into the press. A web press uses a roll of paper, known as a web, which is printed and later cut into individual sheets. The largest web presses stand nearly three stories tall, print images on both sides of a sheet at once, and can print at a rate of 20,000 copies per hour. Major newspapers and magazines as well as best-selling books with high print runs are printed on web presses. One of the disadvantages of using a web press is that post-print options, such as folding and binding, are limited. Page sizes are highly standardized and cannot be changed easily to meet the needs of an individual print run. Image quality also is not as high as that offered by other types of lithographic presses.

Sheet-Fed Offset Printing. As its name suggests, sheet-fed offset printing uses a paper supply of individually cut sheets rather than a paper roll. Each press has a mechanism that feeds paper sheets into the machine, one at a time. This process is less efficient than web-fed printing and can lead to a higher rate of mechanical problems, such as damage to the rubber blanket when more than one sheet is fed into the press in error. However, sheet-fed printing allows for a greater degree of customization for each printing job. The size and type of the paper can be changed, as can the area of the page on which each image is to be printed. A paper of heavier, higher-quality grade may be used in a sheet-fed printer. A wider range of post-print options are also available. Sheet-fed print runs can be bound using a number of different methods, including lamination and glue. These features make it more suitable for products such as sales brochures, corporate annual reports, coffee-table books, and posters.

Lithography in Art. When it first appeared in the United States in the mid-1800's, lithography was associated with high-quality printing, particularly reproductions of works of art. The later introduction of technologies such as photogravure printing eventually made lithographic illustrations in mass-produced printed matter obsolete. At the same time, a number of artists on both sides of the Atlantic Ocean were making advances in lithographic printing as an art form. Henri Toulouse-Lautrec depended on lithography to achieve the bold lines and fields of

Alois Senefelder is famous for the invention in 1796 of the printing method of lithography. This was an improvement on the chemical etching that was then used to form a template for printing. (Science Source)

color in his iconic posters for the Moulin Rouge and other French cabarets in the late 1800's. Another surge of interest in lithography came in the 1920's with works from painters Wassily Kandinsky, Georges Braque, and Pablo Picasso. In some cases, such as Toulouse-Lautrec's posters, these works were originally commercial in nature and intended to be reproduced in large print runs. Artists who experimented with lithography in the twentieth century were more likely to be drawn to the medium for its visual characteristics and possibilities for expression, not for its ability to generate copies. Paris was a major center of lithographic art until World War II, at which point many artists relocated to New York. A revival of the technique emerged in the 1950's with new prints from artists such as Sam Francis, Jasper Johns, and Robert Rauschenberg. Lithography is taught in many fine-arts schools. Some artists prefer to work directly with the stone or metal printing plates, while others draw or paint images and rely on third parties to transfer the work from the page to the plate.

Semiconductor Manufacturing. Photolithography has been used to manufacture semiconductors and microprocessing components for about fifty years. When it was first developed, photolithography depended on the use of photomasks that came into direct contact with the photoresist. This contact often damaged the photomasks and made the manufacturing process costly. Next, a system was developed in which photomasks were suspended a few microns above the photoresist without touching it. This strategy reduced damage, but also lowered the precision with which a photomask could project an image. Since the 1970's manufacturing plants have used a system known as projection printing, in which an image is reflected through an ultrahigh-precision lens onto a photoresist. This technology has allowed manufacturers to fit increasingly higher numbers of integrated circuits onto a single microchip. In 1965, Gordon Moore, a technology executive who would go on to cofound Intel, predicted that the number of transistors that could be placed on a microchip would double about every two years. The prediction has been so accurate that the principle is now known as Moore's law.

IMPACT ON INDUSTRY

Commercial Printing. The market for commercial lithographic printing is a mature one and is not expected to see significant growth in the next several years. Unlike many other areas of technology, lithography in printing does not receive government research funding or have programs at academic institutions devoted to its study.

The commercial printing industry in North America is divided into tiers by company size and niche. As a market, lithographic printing on a large scale is led by RR Donnelley followed by Quad/Graphics. These firms and their competitors dominate market segments such as books, magazines, directories, catalogs, and direct-mail marketing pieces. Beyond corporations such as these, however, the commercial printing industry is made up primarily of small businesses with local clientele. According to the U.S. Bureau of Labor Statistics, seven out of ten companies offering lithographic printing services have fewer than ten employees. Taken as a whole, commercial printers earned about $100 billion in revenue in 2009. The roughly 1,600 daily newspapers in the United States make up another major market segment in lithographic printing. Because most newspapers own and operate their own printing facilities, they are seen as belonging to a related but separate industry, and their revenues are not included in most printing-industry estimates.

Most of the innovations in large-scale lithographic printing that have occurred since the 1990's involve the use of digital technology. Rather than using film and photo negatives to create reverse images, computers allow typesetters and printers to transfer an image directly onto the surface of a printing plate. However, many developments in commercial printing methods largely involve technologies that do not rely on lithography. When this trend is taken into account, along with the migration of many books and news sources from print formats to electronic ones, the future of commercial lithographic printing seems very limited.

Photolithography. Prospects for photolithography and semiconductor manufacturing are much brighter. The semiconductor industry reported sales worldwide of $298.3 billion in 2010, reflecting a 32 percent increase over 2009. Much of the growth was due to a rise in microchip purchases by customers in the Asia Pacific region, which makes up slightly more than half of the global market by volume. Microchip buyers in this context are not consumers, but rather companies that manufacture computers, mobile phones, and other types of hardware. The leading semiconductor manufacturers also reflect the global nature of this industry. Intel tops the list by volume. Its competitors include Analog Devices, Texas Instruments, and Micron Technology in the United States; STMicroelectronics and NXP Semiconductors in Europe; and, Samsung, Toshiba, NEC Electronics, and Taiwan Semiconductor Manufacturing in Asia.

For many years, industry sources have predicted that photolithography would be replaced by other technologies because of an increasing need for precision in the making of microchips. Innovations such as excimer laser technology have allowed lithography to become so precise that features smaller than a single wavelength of light can be printed accurately. The vast expense of microtechnology on this scale prevents many nonprofit institutions such as universities from devoting significant resources to its study. Instead, advances are most likely to come directly from manufacturing companies themselves, which reinvest about 15 percent of sales into research and development each year. Research funding is also supplied by government sources such as the National Science Foundation, the National Institute of Standards and Technology, the U.S. Department of Energy, and the U.S. Department of Defense.

CAREERS AND COURSE WORK

The course work required for a career in lithography varies widely with the nature of the product and the stage in the printing process.

In traditional lithography, one major professional area is media printing. Books, magazines, and newspapers must be designed and laid out page by page before lithographic plates can be created. Many of the professionals who hold these jobs have earned bachelor's or master's degrees in academic areas such as art, graphic design, industrial and product design, and journalism. A background of this type could include course work in typography, color theory, digital imaging, or consumer marketing. Students seeking opportunities in media design also pursue internships with publishers and other companies in their fields of interest.

The mechanical process of lithography has become highly automated. Fewer employees are needed in printing plants than before. Most lithographic press operators receive their training on the job and through apprenticeships. Formal education is offered through postsecondary programs at community colleges, vocational and technical schools, and some universities. Students take courses in mechanical engineering and in the maintenance and repair of heavy equipment. Additional course work may include mathematics, chemistry, physics, and color theory.

Lithography as an artistic printing technique is taught in many college and university art departments. While it is considered too specialized by most institutions for a degree, artists may choose to use lithographic printing to create visual works on paper and other materials.

Photolithography is a highly specialized area of technological manufacturing. Its course work and career track are notably different from those in traditional lithography. Professionals working in photolithography have undergraduate or graduate degrees in fields such as engineering, physics, mathematics, and chemistry. An extensive knowledge of microtechnology and the properties of light-sensitive materials is needed. Because most of the world's semiconductor manufacturing takes place outside North America, careers in photolithography can involve frequent travel to areas such as Asia.

SOCIAL CONTEXT AND FUTURE PROSPECTS

As a broad category, lithographic printing offers very limited job growth. Consumers are increasingly concerned about the environmental impact of paper use in catalogs and other sources of bulk mail. In an effort to respond to these concerns, many companies have reduced their use of paper-based marketing

---

Fascinating Facts About Lithography

- Commercial lithography changed the way in which companies marketed products to consumers, starting in the late 1800's. Inexpensive, mass-produced pictures launched the catalog industry. Product advertisements and labels became more vivid and colorful.

- Christmas cards printed by L. Prang and Company with images of Santa Claus and evergreen trees first appeared in England in 1873 and in the United States in 1874. Affordable and attractive, the cards were popular instantly and launched the tradition of sending holiday cards to friends and family.

- Alois Senefelder, the inventor of lithography, taught himself to write backwards quickly in order to make his first printing plates.

- The idea of using a mixture of wax and ink written on stone came to Senefelder by accident. He was interrupted in his work one day by his mother, who needed him to write a bill for a washerwoman waiting at the door. Senefelder scrawled the information onto a new stone printing plate. The result inspired him to develop the process that became lithography.

- Senefelder's first play was called *Die Maedchenkenner* (*The Connoisseur of Girls*). It is believed to have been his only commercial success as a playwright.

- In 1846 in Boston, inventor Richard M. Hoe redesigned a lithographic flatbed press by putting the plates onto a rotary drum. The new machine could print six times faster, earning it the nickname "the lightning press."

campaigns. This change has lowered the demand for commercial offset lithography. A similar trend has affected the printing of checks and invoices, which are being replaced by electronic systems and online banking.

The growth of electronic media, from the Internet to handheld e-book readers, has also lowered the need for lithography in the publishing industry. Newspapers are reducing the circulation and length of their paper editions and shifting their publishing efforts to Web sites and news feeds. While the demand for paper-based books is not likely to disappear in the near future, sales of new books in electronic formats are growing at a more rapid pace than their print counterparts. In the print segment, new technology is boosting the use of print-on-demand systems for books, which use digital printing techniques rather than lithography.

The prospects for growth in photolithography are more optimistic. Photolithography continues to be one of the most effective and precise ways to make semiconductors. Until it is replaced by a new technology, the field is expected to keep growing with new demand for smaller, faster computers.

Julia A. Rosenthal, B.A., M.S.

FURTHER READING

Devon, Marjorie. *Tamarind Techniques for Fine Art Lithography*. New York: Abrams, 2009. A hands-on manual of techniques for artists seeking to produce lithographic prints, written by the director of the Tamarind Institute of Lithography at the University of New Mexico in Albuquerque.

Landis, Stefan, ed. *Lithography*. Hoboken, N.J.: Wiley-I STE, 2010. A new detailed textbook on lithography as applied to the design and manufacturing of microtechnology.

Meggs, Philip B., and Alston W. Purvis. *Meggs' History of Graphic Design*. 4th ed. Hoboken, N.J.: John Wiley & Sons, 2006. A broad history of graphic design and printing processes throughout the world, including the role of lithography.

Senefelder, Alois. *Senefelder on Lithography: The Classic 1819 Treatise*. Mineola, N.Y.: Dover Publications, 2005. A reproduction of an essay published nearly two centuries ago by the founder of lithography.

Suzuki, Kazuaki, and Bruce W. Smith, eds. *Microlithography: Science and Technology*. 2d ed. Boca Raton, Fla.: CRC Press, 2007. A series of essays by several contributors on the processes behind microlithography, one of the manufacturing techniques used to make semiconductors.

Wilson, Daniel G. *Lithography Primer*. 3d ed. Pittsburgh: GATF Press, 2005. An illustrated overview of each step in the lithographic printing process. Includes chapters on topics such as plate imaging, inks, and papers.

WEB SITES

American Institute of Graphic Arts (AIGA)
http://www.aiga.org

National Association of Litho Clubs
http://www.graphicarts.org

Tamarind Institute
http://tamarind.unm.edu/index.html

See also: Calculus; Mechanical Engineering; Optics; Photography; Printing; Typography.

LONG-RANGE ARTILLERY

FIELDS OF STUDY

Physics; ballistics; chemistry; mathematics; geometry; trigonometry; military science; mechanical engineering; materials science; political science; military history.

SUMMARY

Artillery is any weapon that projects power from afar, striking an enemy over distance with decisive destruction at great rates of fire. Historically, artillery has been the technological advancement driving the evolution of warfare. Over the centuries, artillery in one form or another has been the weapons system either deciding a battle's outcome or offering the necessary support required to win a victory.

KEY TERMS AND CONCEPTS

- **Battery:** Three to six artillery pieces grouped according to caliber and type.
- **Breechloader:** Gun in which shells are loaded through the breech at the back of the weapon.
- **Corning:** Compressive process of formulating granular gunpowder.
- **Gun:** Long-range, long-rifled barrel artillery cannon.
- **Howitzer:** Cannon capable of both high-angle and low-angle firing.
- **Mortar:** Muzzle-loading, high-angle cannon capable of delivering ordnance at very short ranges.
- **Muzzle Loader:** Gun in which the propellant and round are loaded down the barrel muzzle.

DEFINITION AND BASIC PRINCIPLES

In war, the ability to strike an enemy from afar is paramount: If an army can project power from a distance, its soldiers can stay farther from harm's way. Artillery evolved as a means to this end, and until the beginning of the twentieth century, the role of explosive-force artillery–as field, siege, naval, or fortress guns–was largely unchanged. In World War I, additional specialized artillery was developed. Explosive propellant artillery pieces are classified according to their projectile trajectories: Mortars lob objects in high arc parabolas; guns tend to have straight, high-velocity trajectories; and howitzers are a compromise, propelling large shells at slower speeds, over moderate distances.

The history of artillery is the quest for precision and accuracy. Ballistic science is concerned with the properties of classical physical mechanics governing the motions of bodies under force. With artillery, these motions involve the mechanics of gun machinery, the dynamics of propellants, and the trajectory of discharged projectiles. The basic dynamics of artillery fire–whether bow and arrow, catapult, howitzer, or railroad gun–are based on Newton's second law of motion: Net force is the product of the mass times the acceleration. Traditionally, for artillery this has meant that the amount of destruction was equal to weight of the projectile times how fast it could be propelled. In modern warfare, this destructive force is multiplied by adding explosives and submunitions to the projectile.

Newton's third law of motion–for every action there is an equal and opposite reaction–also plays a role in the history and development of artillery. Explosions propelling projectiles out of barrels result in recoil, moving the entire gun, carriage and all, backward. Until barrel recoil could be countered, guns had to be aimed after each shot, disrupting their accuracy and consistency to target. Accuracy requires the ability to calculate the trajectory, path geometry, and position of an object over time. Knowledge of projectile weight, the force of acceleration, the forces of gravity, the Earth's curvature, atmospheric conditions, and the effects of friction are necessary to accurately predict the parabolic arc and subsequent impact point of a projectile. The result is the proper angle of elevation required to propel an object to a specified location at a certain distance.

BACKGROUND AND HISTORY

Before the invention of gunpowder and its practical application in launching projectiles, four types of mechanical advantage were used to increase the range of missiles. During the Middle Ages, volley-fired archery dominated battlefields and great siege engines were required to attack large fortified cities and castles. The first form of practical artillery used in battle was the bow and arrow. The bow is a tension device capable of launching an arrow farther than a person could throw a dart. Arrows became the fastest, farthest flying, and most accurate of historical artilleries, delivering great destructive power, but they were incapable of bringing down fortifications.

Large mechanical devices were used to overcome the limitations of archery. Although these machines could not match the distance and accuracy of the bow and arrow, they brought massive destructive power to the battlefield. Counterweight lever weapons had a short range and slow velocity, but could propel massive missiles. The best-known counterweight lever weapon is the trebuchet, the most powerful weapon of the Middle Ages. Spring-powered artillery used either an oversized double-stave bow or a series of laminated leaf springs to store energy that was released when the missile was shot. The spring's power was limited to the elastic strength of the materials used to build it. The most well-known spring artilleries are the giant crossbow, the spring engine catapult, and the spring engine strike-hurler. Torsion artillery used the elasticity of a twisted bundle of fiber to store energy until it was released to hurl an object. Torsion artillery–such as engine catapults, ballistas, and onagers–throw objects with great accuracy and high velocity.

Using the explosive force of gunpowder to launch projectiles was the greatest step forward in weapons technology. Although gunpowder was invented in China during the eleventh century, it was not exploited as a weapons propellant until the early thirteenth century in Europe. Early gunpowder mixtures burned slowly, resulting in less explosive force. The development of grained gunpowder by corning produced a fast-burning powder with more explosive force, which equated to greater velocity, more range, and more smashing power. Since the inception of propellant-based artillery, engineers have sought to improve it by maximizing the rate of fire, lengthening the flight distance, perfecting accuracy, and increasing the lethality of munitions. They have also striven to minimize discharge signatures, barrel fouling and overheating, recoil, and concussion.

The first explosive-force artillery were muzzle-loading guns. These required a charge of powder to be pushed to the bottom of the gun barrel, followed by a wad, then a projectile. An ignition source applied to a vent would explode the powder and propel the missile. Muzzle-loaded cannons had a relative short range and were aimed directly at their targets. The smooth bore of muzzle-loading guns required the shot to be round, causing the shot to be unstable and tumble in flight, resulting in relative inaccuracy. However, sustained bombardment of massed ranks of infantry and cavalry by multiple muzzle-loading guns produced devastating effects. Maintaining accuracy with muzzle-loading guns requires readjusting them after each firing to compensate for recoil. In the 1860's, the rifling of cannon barrels allowed muzzle-loading guns to put spin on projectiles, giving them gyroscopic stability. Point-first ammunition was designed to take advantage of this stability, providing for larger, heavier shells that could fly greater distances with high accuracy.

The next step forward in artillery technology was the breech-loading mechanism: a system of sealing blocks, screws, or interrupters designed to allow loading from the rear of a gun and prevent the escape of propellant gases on firing. When combined with self-contained ammunition (in which a metal casing containing explosive propellant, fuse, and projectile was loaded into the gun's breech), it allowed for rapid firing. As quickly as one round was fired and its casing ejected, another round could be placed into the gun and discharged. To exploit the advantages afforded by breech-loading guns, recoil had to be countered. The development of a reliable recoil-suppression mechanism in combination with breech-loading technology resulted in artillery capable of sustained, accurate firing beyond the line of sight. In the 1890's, French engineers designed a 75-millimeter field gun with a hydropneumatic recoil-suppression system: Rates of fire for artillery doubled from ten rounds per minute to twenty. Furthermore, recoil suppression meant the gun did not have to be readjusted after each round to stay on target.

During World War I, the application of indirect artillery fire on static enemy positions transformed war forever: Mass bombardments and indirect artillery barrages became the driving force in ground warfare until the end of the twentieth century. World War I also saw the development of three new types of artillery designed to counter advances in military technology and doctrine: antiaircraft guns, mechanized armor (tanks and tracked artillery), and antitank guns.

A self-propelled Howitzer stands on display at the Kansas National Guard Museum in Topeka, Kansas. (AP Photo)

In World War II, artillery's role changed again. Fast-moving operations equipped with mobile infantry and armored targets became difficult to destroy with indirect bombardment. Artillery technology changed to reflect a need for direct-fire guns to take out designated targets. The United States began to establish centers to coordinate artillery operations, allowing multiple gun batteries to accurately bombard targets designated by forward observers either on the ground or in the air. Since World War II, automatic data processing, digital communications, lasers, radar, Global Positioning Systems (GPS), and satellite technology have made it possible for a targeting sensor to directly communicate with artillery units and pinpoint fire. These technological applications reduce the time needed for targeting and increase accuracy.

How It Works

Artillery pieces are large, unwieldy, and complex, with multiple components, making their operation a team effort. Specialized training, organization, and team cooperation are required to effectively deploy, target, and fire artillery. Modern artillery bombardments are mostly indirect fire support of ground operations. For gunnery teams to strike their target requires a series of coordinated actions organized and timed for maximum efficient use of valuable ammunition and manpower. Forward observers on the ground or in the air, usually senior artillery officers, determine

targets and communicate them to a fire direction center. The center sets priorities, selects ordnance type, directs fire control, and selects the battery units to complete the mission against priority targets. The gun battery calculates firing data needed to hit the target with the proper gun or rocket launcher. The forward observer gives the order to fire to the battery and can communicate needed corrections to register the target. The battery continues to fire until a cease-fire order is given.

Applications and Products

Throughout history, artillery guns have been designed to accomplish specific tasks. Field guns, such as howitzers, are designed to accompany a military force on campaign. They were towed by horses and then by trucks and later became self-propelled gun platforms. In World War I, arms manufacturer Friedrich Krupp designed and built the Big Bertha (*Dicke Bertha*), a massive howitzer, so that the German army could destroy forts along the Belgian frontier. Krupp also built the massive traversing railroad cannon, the Paris Gun (*Paris-Geschütz*), which bombarded Paris from 75 miles away. The gun had poor accuracy but was never meant to destroy Paris but rather to terrorize the populace. After World War I, antiaircraft artillery became a highly specialized weapons system with the sole purpose of destroying aircraft, and this type of artillery has been largely replaced by guided missiles.

Traversing, stabilized, high-velocity, quick-firing rifled cannons were developed for armored fighting vehicles, specifically tanks, and specialized oversized high-velocity rifled guns were developed to counter tanks. Large-caliber naval rifles were designed to defeat the armor of battle fleets before airpower rendered their use obsolete. Multiple launch rocket systems were developed as a cost-saving means of providing indirect artillery fire. The ballistic missile is self-propelled artillery that can reach any place on the planet. As of the early twenty-first century, the most advanced artillery gun was the U.S. M777 howitzer, which uses a digital fire-control system to shoot a 155-millimeter GPS-guided M982 Excalibur fire-and-forget projectile. When fired from a distance of 24 miles, the round will land within 30 feet of its target.

IMPACT ON INDUSTRY

Military technology reacts to existing needs, and military doctrine dictates what weapons systems are needed. Warfare became the driving force of the Industrial Revolution as the need for mass-produced weapons inspired manufacturers to discover ways to meet the demands of warring nations. The quicker the military could be armed with the most powerful and reliable weapons, the better the chance of victory and the greater the arms manufacturer's profits. The need for bigger, stronger cannons to hold and direct the explosive forces necessary to launch larger projectiles drove the creation of new metal casting techniques and new designs to perform precision machine turning of rifled cannon barrels. The need for advanced artillery design resulted in the development of high-tensile steel alloys to withstand high compression and heat from propellants used in larger rifle-barreled long-range field guns.

Through World War II, one of the most prolific and successful artillery makers was Friedrich Krupp (which became part of Thyssen Krupp in 1999), a German company that produced advanced alloy steels, ammunition, and armaments, especially high-quality artillery pieces.

CAREERS AND COURSE WORK

Careers in artillery are limited to the military and weapons research, development, and testing. The U.S. Army Field Artillery School teaches members of the military to use cannons, rockets, and missiles in coordination with combined arms operations. Nonmilitary careers in artillery design and testing require degrees in engineering, physics, or materials sciences. Most jobs are with large defense contractors such as Lockheed-Martin, BAE Systems Bofors, or Raytheon, or governmental agencies such as the Defense Advanced Research Projects Agency (DARPA), part of the U.S. Department of Defense.

SOCIAL CONTEXT AND FUTURE PROSPECTS

The modern design and application of artillery is a direct by-product of the interdependence of political and military doctrine and advances in technology. Substantial technological advances since the end of World War II have improved the accuracy and lethality of artillery fire and increased the mobility of

---

Fascinating Facts About Long-Range Artillery

- From the mid-eighteenth to mid-nineteenth centuries, half of all battlefield casualties during war resulted from artillery bombardment.
- Artillery accounted for 70 percent of all deaths in World War I.
- The British fired 1 million artillery shells in the first week of the Battle of the Somme, which began on July 1, 1916, and lasted until November 18.
- The German Big Bertha howitzer was named after Alfred Krupp's daughter, Bertha Krupp von Bohlen und Halbach, heiress to the Krupp steel and weapons manufacturing empire.
- German traversing railroad guns of both world wars required gun crews of eighty men to operate efficiently.
- In World War II, artillery was so valued as a means to achieving victory on the battlefield that between 30 and 40 percent of Allied armies were made up of artillery gunnery teams.

- In 1953, the United States built and test fired Atomic Annie, an artillery gun intended to shoot a tactical 280-millimeter shell with a fission warhead to yield a 15-kiloton explosion.
- Modern mortars are the last surviving muzzle-loading artillery, and along with direct-fire recoilless guns, are no longer classified as field artillery but as infantry weapons.
- Since the 1990's, technological advances, such as laser, radar, and Global Positioning Systems, have been used to guide smart bombs and fire-and-forget missile systems, which have lessened the importance of field artillery.
- Crews manning large-scale artillery guns usually fire only three rounds before being replaced by a new crew because extensive, prolonged, or continuous exposure to the gun's percussive forces can cause internal organ damage.

both guns and support crews. Stronger and lighter alloys have been adopted in the manufacture of modern artillery, reducing weight and allowing for greater mobility. Advances in the chemical composition of propellant charges and the development of self-propelled rounds has increased missile velocities, equating to longer range. These chemical changes, as well as the introduction of caseless ammunition, result in less-corrosive barrel wear, increasing barrel life. Innovations in recoil suppression reduce fatigue to gun parts and allow for increased rates of fire with minimal aiming corrections. The use of laser systems to measure range has created nearly pinpoint artillery accuracy. Laser, electro-optical, infrared, GPS, radio-beam, and radar target acquisition and designation systems eliminate nearly all error in target allocation, increasing the likelihood of first-round accuracy. The use of computer fire-control systems allows modern artillery to be electronically aimed, with computers making any necessary corrections in range, elevation, azimuth, and depression. This increases accuracy and reduces the time between target acquisition and firing.

A large portion of modern artillery consists of mounted, self-propelled weapons, which reflects the modern military's preference for fast, mobile fighting forces. Modern field artillery can advance, stop, set up, target, fire, and move on in minutes, all before it can be located and counterattacked. The United States military is testing the non-line-of-sight (NLOS) cannon, a lightweight, self-loading, highly computerized, self-propelled gun that requires only two people to operate it. The gun can fire four shells in sequence at differing trajectories and land them all on target simultaneously. Future advances in artillery technology will most likely be linked to changes in mobile rocket launchers, electronic targeting systems, projectiles, propellants, self-propelled gun carriages, and tactical changes on the battlefield. Many long-range artillery missions are being replaced by precision-guided aerial munitions, or smart bombs. As the accuracy of aerial attack munitions grows, and mechanized armor and armored personal carriers increase their speeds over the battlefield, mobile artillery will need to keep pace technologically if it wishes to maintain a place in the modern arsenal.

Randall L. Milstein, B.S., M.S., Ph.D.

FURTHER READING

Bailey, B. A. *Field Artillery and Firepower.* Rev. ed. Annapolis, Md.: U.S. Naval Institute Press, 2003. Documents the changing technology, tactics, and strategy of artillery usage and then analyzes artillery effectiveness under multiple combat conditions.

Bidwell, Shelford, Brian Blunt, and Tolley Taylor, eds. *Brassey's Artillery of the World: Guns, Howitzers, Mortars, Guided Weapons, Rockets, and Ancillary Equipment in Service with the Regular and Reserve Forces of All Nations.* 2d ed. Elmsford, N.Y.: Pergamon Press, 1981. A classic work on the characteristics of modern field artillery pieces.

Foss, Christopher F., ed. *Jane's Armour and Artillery, 2010-2011.* Alexandria, Va.: Jane's Information Group, 2010. An authoritative, essential recognition guide to tanks and artillery for armies of all nations.

Grice, Michael D. *On Gunnery: The Art and Science of Field Artillery from the American Civil War to the Dawn of the Twenty-first Century.* North Charleston, S.C.: Booksurge, 2009. Documents American artillery doctrine from the Civil War through Operation Iraqi Freedom.

Grossman, Dave. *On Killing: The Psychological Cost of Learning to Kill in War and Society.* New York: Little, Brown, 2009. Although it focuses on the Vietnam War, the book reflects on the devastating effect prolonged exposure to intense artillery bombardment had on the psychological health of soldiers in World War I and World War II, as well as the artillery men whose guns killed from great distances.

Hogg, Ian. *Twentieth-Century Artillery: Three Hundred of the World's Greatest Artillery Pieces.* New York: Friedman-Fairfax, 2001. Hogg, one of the world's leading experts on guns of all types, documents some of the most famous and widely used artillery pieces.

Manchester, William. *The Arms of Krupp: The Rise and Fall of the Industrial Dynasty That Armed Germany at War.* 1968. Reprint. Boston: Little Brown, 2003. A biography of the Krupp family, innovators and manufacturers of weapons that shaped modern warfare.

Manson, M. P. *Guns, Mortars, and Rockets.* London: Brassey's United Kingdom, 1997. Documents the historical development of artillery, emphasizing three types of missile launch platforms–guns, mortars, and rockets.

Miller, Henry W. *The Paris Gun: The Bombardment of Paris by the German Long-Range Guns and the Great*

German Offensives of 1918. 1930. Reprint. East Sussex, England: Naval and Military Press, 2003. A detailed account of the construction and use of the Paris gun; richly illustrated with photographs and maps.

WEB SITES
American Artillery Association
http://www.americanartillery.org

U.S. Army Field Artillery School
http://sill-www.army.mil/USAFAS

U.S. Field Artillery Association
http://fieldartillery.org

See also: Antiballistic Missile Defense Systems; Military Sciences and Combat Engineering.

M

MAGNETIC RESONANCE IMAGING

FIELDS OF STUDY

Radiology; diagnostics; radiofrequency; pathology; physiology; radiation physics; instrumentation; anatomy; microbiology; imaging; angiography; nuclear physics.

SUMMARY

Magnetic resonance imaging (MRI) is a noninvasive form of diagnostic radiography that produces images of slices or planes from tissues and organs inside the body. An MRI scan is painless and does not expose the patient to radiation, as an X ray does. The images produced are detailed and can be used to detect tiny changes of structures within the body, which are extremely valuable clues to physicians in the diagnosis and treatment of their patients. A strong magnetic field is created around the patient, causing the protons of hydrogen atoms in body tissues to absorb and release energy. This energy, when exposed to a radiofrequency, produces a faint signal that is detected by the receiver portion of the MRI scanner, which transforms it into an image.

KEY TERMS AND CONCEPTS

- **Artifact:** Feature in a diagnostic image, usually a complication of the imaging process, that results in an inaccurate representation of the tissue being studied.
- **Axial Slice:** Horizontal imaging plane that corresponds with right to left and front to back.
- **Claustrophobia:** Abnormal and persistent fear of closed spaces, of being closed in or being shut in.
- **Computed Tomography (CT) Scan:** Image of structures within the body created by a computer from multiple X-ray images.
- **Contrast Agent:** Dye used to provide contrast, for example, between blood vessels and other tissue.
- **Functional Magnetic Resonance Imaging (fMRI):** Use of MRI to study physiological processes rather than just anatomy.
- **Gradient Coils:** Coils of wire used to generate the magnetic field gradients that are used in MRI.
- **Magnetic Resonance Angiogram (MRA):** Noninvasive complement to MRI to observe anatomy of blood vessels of certain size in the head and neck.
- **Sagittal Slice:** Vertical imaging plane that corresponds with front-to-back and top-to-bottom.
- **Scan:** Single, continuous collection of images.
- **Session:** Time that a single subject is in the magnetic resonance scanner; can be two hours for fMRI.
- **Tesla:** Unit of magnetic field strength; named for Nikola Tesla who discovered the rotating magnetic field in 1882.

DEFINITION AND BASIC PRINCIPLES

Magnetic resonance imaging (MRI), sometimes called magnetic resonance tomography, is a noninvasive medical imaging method used to visualize the internal structures and some functions of the body. MRI provides much greater detail and contrast between the different tissues in the body than are available from X rays or computed tomography (CT) without using ionizing radiation. MRI uses a powerful magnetic field to align the nuclear magnetization of protons of hydrogen atoms in the body. A radio frequency alters the alignment of the protons, creating a signal that is detectable by the scanner. The signals are processed through a mathematical algorithm to produce a series of cross-sectional images of the desired area. Image resolution and accuracy can be further refined through the use of contrast agents.

The detailed images produced are extremely valuable in detection and diagnosing of medical conditions and disease. The need for exploratory surgery has been greatly reduced, and surgical procedures and treatments can be more accurately directed by the ability to visualize structures and changes within the body.

BACKGROUND AND HISTORY

Magnetic resonance imaging is a relatively new scientific discovery, and its application to human diagnostics was first published in 1977. Two American scientists, Felix Bloch at Stanford University and Edward Mills Purcell from Harvard University, were both independently successful with their nuclear magnetic resonance (NMR) experiments in 1946. Their work was based on the Larmor relationship, named for Irish physicist Joseph Larmor, which stated that the strength of the magnetic field matched the radiofrequency. Bloch and Purcell found that when certain nuclei were in the presence of a magnetic field, they absorbed energy in the radiofrequency range of the electromagnetic spectrum and emitted this energy when the nuclei returned to their original state. They termed their discovery nuclear magnetic resonance: "nuclear" because only the nuclei of certain atoms reacted, "magnetic" because a magnetic field was required, and "resonance" because of the direct frequency dependence of the magnetic and radiofrequency fields. Bloch and Purcell were awarded the Nobel Prize in Physics in 1952.

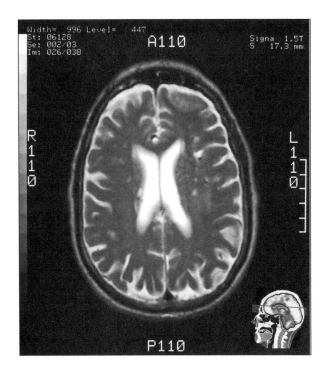

MRI of a human brain. (BSIP/Photo Researchers, Inc.)

NMR technology was used for the next few decades as a spectroscopy method to determine the composition of chemical compounds. In the late 1960's and early 1970's, Raymond Damadian, a State University of New York physician, found that when NMR techniques were applied to tumor samples, the results were distinguishable from normal tissue. His results were published in the journal Science in 1971. Damadian filed patents in 1972 and 1978 for a NMR system large enough to accommodate a human being that would emit a signal if tumor tissue was detected but did not produce an image. Paul Lauterbur, a physicist from State University of New York, devised technology that could run the signals produced by NMR through a computed back projection algorithm, which produced an image. Peter Mansfield, a British physicist from the University of Nottingham, further refined the mathematical analysis, improving the image. He also discovered echo-planar imaging, which is a fast imaging protocol for MRI and the basis for functional MRI. Lauterbur and Mansfield shared the 2003 Nobel Prize in Physiology or Medicine. Some controversy still exists regarding Damadian's exclusion from this honor.

HOW IT WORKS

The human body is made up of about 70 percent water. Water molecules are made up of two hydrogen atoms and one oxygen atom. When exposed to a powerful magnetic field, some of the protons in the nuclei of the hydrogen atoms align with the direction of the field. When a radio frequency transmitter is added, creating an electromagnetic field, a resonance frequency provides the energy required to flip the alignment of the affected protons. Once the field is turned off, the protons return to their original state. The difference in energy between the two states is called a photon, which is a frequency signal detected by the scanner. The photon frequency is determined by the strength of the magnetic field. The detected signals are run through a computerized algorithm to deliver an image. The contrast of the image is produced by differences in proton density and magnetic resonance relaxation time, referred to as T1 or T2.

An MRI scanning machine is a tube surrounded by a giant circular magnet. The patient is placed on an examination table that is inserted through the tube space. Some individuals experience claustrophobia when lying in the closed space of the scanning tube and may be given a mild sedative to reduce anxiety. Children are often sedated or receive anesthesia for an MRI. Patients are required to remain very still during the scan, which normally takes between thirty to ninety minutes to complete. During the scan, patients are usually provided with a hand buzzer or communication device so that they may interact with technicians. The magnetic field is created by passing electric current through a series of gradient coils. The strong magnetic fields are normally safe for patients, with the exception of people with metal implants such as pacemakers, surgical clips or plates, or cochlear implants, making them ineligible for MRI. During the scanning procedure, patients will hear a loud humming, beeping, or knocking noise, which can reach up to 120 decibels. (Patients are often provided with ear protection.) The noise is caused by the interaction of the gradient magnetic fields with the static magnetic field. The gradient coils are subject to a twisting force each time they are switched on and off, and this creates a loud mechanical vibration in the cylinder supporting the coils and surrounding mountings.

To enhance the images, contrast agents can be injected intravenously or directly into a joint. MRI is being used to visualize all parts of the body by producing a series of two-dimensional images that appears as cross sections or slices. These slices can also be reconstructed to create three-dimensional views of the entire body or specific parts.

APPLICATIONS AND PRODUCTS

Research into the applications of magnetic resonance imaging technology beyond basic image generation has been progressing at a tremendous rate. Although the basic images are immensely valuable to physicians and scientists, the application of the scientific principles in the development of specialized scans is reaching far beyond original expectations and benefiting health care delivery and patient care.

Functional MRI (fMRI) is based on the changes in blood flow to the parts of the brain that accompany neural activity, and it provides visualization of these changes. This has been critical in detecting the brain areas involved in specific tasks, processes, or emotions. fMRI does not detect absolute activity of areas of the brain but it detects differences in activity. During the scan, the patient is asked to perform tasks

---

Fascinating Facts About Magnetic Resonance Imaging

- Magnetic resonance imaging (MRI) is based on nuclear magnetic resonance techniques, but it was named "magnetic" rather than "nuclear" resonance imaging because of the negative connotations associated with the word "nuclear" in the 1970's.
- An overwhelming majority of American physicians identified computed tomography (CT) and MRI as the most important medical innovations for improving patient care during the 1990's.
- In 1977, Raymond Damadian and colleagues built their first magnetic resonance scanner and named it "Indomitable." It is housed in the Smithsonian.
- MRI scanners were primarily developed for use in medicine but are also used to study fossils and historical artifacts.
- Functional MRI can create a video of blood flow in the brain and has been used to study monks serving under the Dalai Llama and the control they exert over mental processes through meditation.
- As of 2009, there were 7,950 magnetic resonance imaging systems in the United States, compared to only 266 in Canada.
- Tattoos received before 1990 may contain small amounts of metal in the ink, which may interfere with or be painful during MRI.
- While using functional magnetic resonance imaging (fMRI) to research children with attention deficit disorder, Pennsylvania psychiatrist Daniel Langleben discovered that deception activates regions in the prefrontal cortex, showing that fMRI could be used as a lie detector.

---

or is presented with stimuli to trigger thoughts or emotions. The detection of the brain areas that are used is based on the blood oxygenation level dependent (BOLD) effect, which creates a variation signal, linked with the concentration of oxy-/deoxy-hemoglobine in each area. These scans are performed every two to three seconds over a period of minutes at a low resolution and do not often require additional contrast media to be used.

Diffusion MRI can measure the diffusion of water molecules in biological tissues. This is incredibly useful in detecting the movement of molecules in neural fiber, which can enable brain mapping, illustrating connectivity of different regions in the brain, and examination of areas of the brain affected by neural degeneration and demyelination, as in multiple sclerosis. Diffusion MRI, when applied to diffusion-weighted imaging, can detect swelling in brain cells within ten minutes of the onset of ischemic stroke symptoms, allowing physicians to direct reperfusion therapy to specific regions in the brain. Previously, computed tomography would take up to four hours to detect similar findings, delaying cerebral perfusion therapy to salvageable areas.

Interventional magnetic resonance imaging is used to guide medical practitioners during minimally invasive procedures that do not involve any potentially magnetic instruments. A subset of this is intraoperative MRI, which is used during surgical procedures; however, most often images are taken during a break from the procedure in order to track progress and success and further guide ongoing surgery.

Magnetic resonance angiography (MRA) and venography (MRV) provide visualization of arteries and veins. The images produced can help physicians evaluate potential health problems such as narrowing of the vessels or vessel walls at risk of rupture as in an aneurysm. The most common arteries and veins examined are the major vessels in the head, neck, abdomen, kidneys, and legs.

Magnetic resonance spectroscopy (MRS) measures the levels of different metabolites in body tissues, usually in the evaluation of nervous system disorders. Concentrations of metabolites such as N-acetyl aspartate, choline, creatine, and lactate in brain tissue can be examined. Information on levels of metabolites is useful in determining and diagnosing specific metabolic disorders such as Canavan's disease, creatine deficiency, and untreated bacterial brain abscess. MRS has also been useful in the differentiation of high-grade from low-grade brain tumors.

Precise treatment of diseased or cancerous tissue within the body is a tremendous advance in health care delivery. Radiation therapy simulation uses MRI technology to locate tumors within the body and determine their exact location, size, shape, and orientation. The patient is carefully marked with points corresponding to this information, and precise radiation therapy can be delivered to the tumor mass. This drastically reduces excess radiation therapy and

limits damage to healthy tissues surrounding the tumor. Similarly, magnetic resonance guided focused ultrasound (MRgFUS) allows ultrasound beams to achieve more precise and complete treatment and the ablation of diseased tissues is guided and controlled by magnetic resonance thermal imaging.

IMPACT ON INDUSTRY

In 1983, the Food and Drug Administration approved MRI scanners for sale in the United States. Magnetic resonance imaging is experiencing rapid growth on the global market and is expected to sustain this trend in the future as more clinics and health care centers use the imaging technique. The aging of the American population has increased the demand for efficient, effective, and noninvasive diagnosis, especially for neurological and cardiovascular diseases. Although a basic MRI system can cost more than $1 million, and the cost of construction of the suite to accommodate it can exceed $500,000, installation of these imaging systems in medical institutions still promises a positive return on investment for the health care providers. Some companies have begun offering refurbished MRI systems at reduced cost. MRI technology is an asset in challenging economic times because it provides advanced and quick diagnoses, leading to faster patient care and greater patient turnover, leading to greater revenues for the institution. In the United States, health care insurers and the federal government provide very good reimbursement for the scan itself and a professional fee for a review of the resulting images by a radiologist.

Commercial development and industrial growth depends on advancements in research, which is being funded and conducted around the world. Some projects that are moving from experimental to commercial and driving the industry are integrated systems combining MRI with another modality, more portable MRI systems, improved contrast agents, advanced image-processing techniques, and magnetic coil technology.

Positron emission tomography (PET) is used in the diagnosis of cancer, but because it lacks anatomical detail, it is used in combination with CT scans. MRI scans provide superior contrast to CT for determining tumor structure and integrating PET and MRI into a single modality would be highly desirable and is in development.

Research is being done to improve the enhancement of the contrast agents used with MRI. Specifically, research involves targeted contrast agents that will bind only to desired tissue at the molecular and cellular level. Visualizing changes at a cellular level would make it possible to detect and diagnose disease sooner and provide opportunities for more focused treatment.

A growing niche market consists of MRI systems that allow patients to remain upright during the scan and open systems that can accommodate patients who previously were unable to have a scan because of their size or their claustrophobia.

Portable and handheld MRI technology has entered the international market. These small devices promise to deliver quality images at low cost, without the need for a special room to accommodate the device.

CAREERS AND COURSE WORK

The most common career choice in the field of magnetic resonance imaging is the MRI technician or technologist, individuals who operate the MRI system to effectively produce the desired images for diagnostic purposes while adhering to radiation safety measures and government regulations. Researchers and government agencies are exploring potential occupational hazards to personnel because of prolonged and frequent exposure to magnetic fields. Technicians first explain the procedure to patients. Then, they ensure that patients do not have any metal present on their person or in their body and position them correctly on the examination table. Some technicians also administer intravenous sedation or contrast media to the patients. During the scan, the technologist observes the patient as well as the display of the area being scanned and makes any needed adjustment to density or contrast to improve picture quality. When the scan is complete, the technologist will evaluate the images to ensure that they are satisfactory for diagnostic purposes. The MRI training program may result in an associate's degree or certificate; some programs require prior completion of radiology or sonography programs and core competencies in writing, math, anatomy, physiology, and psychology. Once admitted to an accredited program, students receive training that includes patient care, magnetic resonance physics, and anatomy and physiology. The American Registry of Magnetic Resonance Imaging Technologists requires that students complete one thousand hours of clinical training. To satisfy this clinical training requirement, students are assigned to a specific hospital or are rotated through different hospitals. Becoming an MRI technician can take two to three years.

Diagnostic radiologists are physicians who have specialized in obtaining and interpreting medical images such as those produced by MRI. Becoming a radiologist in the United States requires the completion of four years of college or university, four years of medical school, and four to five years of additional specialized training.

MRI physicists are specialized scientists with a diverse background covering nuclear magnetic resonance (NMR) physics, biophysics, and medical physics, in combination with basic medical sciences, including human anatomy, physiology, and pathology. They also have a good understanding of engineering issues involving advanced hardware, such as large superconducting magnets, high-power radio frequencies, fast digital data processing, and remote sensing and control. Industrial MRI physicists often work in research and development for biotechnology companies or they implement new applications and provide support for equipment already installed in health care centers. Academic MRI physicists work in a university laboratory or in cooperation with a medical center involved in clinical research and training. Academic research may involve basic science in MRI spectroscopy, functional imaging, contrast media, or echo-planar imaging.

SOCIAL CONTEXT AND FUTURE PROSPECTS

Magnetic resonance imaging provides physicians with the ability to see detailed images of the inside of their patients to more easily diagnose and guide treatment. It provides researchers with valuable insight into the metabolism and physiology of the body. Still, there are drawbacks that make this scientific advance unavailable to some patients because of their economic circumstances or body shape. For patients who do not have health care insurance, the price of an MRI scan, which can range from $700 to $2500 depending on body part and type of examination, may be beyond what they can afford. Also, MRI systems have weight and circumference restrictions that make many people unsuitable candidates, limiting the quality of acute care for them. Typically, the weight limit is 350 to 500 pounds for the examination table, and the size of the patients is additionally limited by the diameter of the magnetic tube. As people tend to be larger in size, biotechnology companies are working on scanning systems, such as the upright or open concept scanner, that can accommodate larger people.

April D. Ingram, B.Sc.

FURTHER READING

Blamire, A. M. "The Technology of MRI—The Next Ten Years?" *British Journal of Radiology* 81 (2008): 601-617. Looks at the clinical status and future of MRI.

Filler, Aaron. "The History, Development and Impact of Computed Imaging in Neurological Diagnosis and Neurosurgery: CT, MRI, and DTI." *The Internet Journal of Neurosurgery* 7, no. 1 (2010). Provides a good history of the development of diagnostic imaging techniques.

Haacke, Mark, et al. *Magnetic Resonance Imaging: Physical Principles and Sequence Design.* New York: John Wiley & Sons, 1999. Explains the key fundamental and operational principles of MRI from a physics and mathematical viewpoint.

Simon, Merrill, and James Mattson. *The Pioneers of NMR and Magnetic Resonance in Medicine: The Story of MRI.* Ramat Gan, Israel: Bar-Ilan University Press, 1996. Describes the history of MRI, from its development of scientific principles to application in health care.

Weishaupt, Dominik, Vitor Koechli, and Borut Marincek. *How Does MRI Work? An Introduction to the Physics and Function of Magnetic Resonance Imaging.* New York: Springer, 2006. A good resource for those who wish to familiarize themselves with the workings of magnetic resonance imaging and have some of the challenging concepts explained. It uses conceptual rather than mathematical methods to clarify the physics of MRI.

WEB SITES

Clinical Magnetic Resonance Industry
http://www.cmrs.com

International Society for Magnetic Resonance in Medicine
http://www.ismrm.org

MedlinePlus
Magnetic Resonance Imaging
http://www.nlm.nih.gov/medlineplus/mriscans.html

See also: Cardiology; Computed Tomography; Radiology and Medical Imaging; Ultrasonic Imaging.

MAGNETIC STORAGE

FIELDS OF STUDY

Computer engineering; computer networking systems; electrical engineering; information technology; materials engineering.

SUMMARY

Magnetic storage is a durable and non-volatile way of recording analogue, digital, and alphanumerical data. In most applications, an electrical current is used to generate a variable magnetic field over a specially prepared tape or disk that imprints the tape or disk with patterns that, when "read" by an electromagnetic drive "head," duplicates the wavelengths of the original signal. Magnetic storage has been a particularly enduring technology, as the original conceptual designs were published well over a century ago.

KEY TERMS AND CONCEPTS

- **Biasing:** Process that pre-magnetizes a magnetic medium to reproduce the magnetic flux from the recording head more exactly.
- **Bit:** Storage required of one binary digit.
- **Byte:** Basic unit, composed of eight bits, of measuring storage capacity and the basis of larger units.
- **Ferromagnetic Media:** Iron-based media sometimes used as a source of tape, hard, or flexible disks.
- **Magnetic Anisotrophy:** Tendency, in certain substances, to hold onto a magnetically induced pattern even without the presence of electrical current.
- **Magnetization:** Property of an object determining whether it can be affected by magnetism.
- **Random Access Memory (RAM):** Property of certain kinds of storage media where all elements of a sequence may be accessed in the same length of time; also called direct access.
- **Sequential Access Memory:** Property of certain kinds of storage media where remote elements require a longer access time than elements located in the immediate vicinity.

DEFINITION AND BASIC PRINCIPLES

Magnetic storage is a term describing one method in which recorded information is stored for later access. A magnetized medium can be one or a combination of several different substances: iron wire, steel bands, strips of paper or cotton string coated with powdered iron filings, cellulose or polyester tape coated with iron oxide or chromium oxide particles, or aluminum or ceramic disks coated with multiple layers of nonmetallic alloys overlaid with a thin layer of a magnetic (typically a ferrite) alloy. The varying magnetic structures are encoded with alphanumerical data and become a temporary or permanent nonvolatile repository of that data. Typical uses of magnetic storage media range from magnetic recording tape and hard and floppy computer disks to the striping material on the backs of credit, debit, and identification cards as well as certain kinds of bank checks.

BACKGROUND AND HISTORY

American engineer Oberlin Smith's 1878 trip to Thomas Edison's laboratory in Menlo Park, New Jersey, was the source for Smith's earliest prototypes of a form of magnetic storage. Disappointed by the poor recording quality of Edison's wax cylinder phonograph, Smith imagined a different method for recording and replaying sound. In the early 1820's, electrical pioneers such as Hans Ørsted had demonstrated basic electromagnetic principles: Electrical current, when run through a iron wire, could generate a magnetic field, and electrically charged wires affected each other magnetically. Smith toyed with the idea, but did not file a patent—possibly because he never found the time to construct a complete, working model. On September 8, 1888, he finally published a description of his conceptual design, involving a cotton cord woven with iron filings passing through a coil of electrically charged wire, in *Electrical World* magazine. The concept in the article, "Some Possible Forms of Phonograph," though theoretically possible, was never tested.

The first actual magnetic audio recording was Danish inventor Valdemar Poulsen's telegraphone, developed in 1896 and demonstrated at the Exposition Universelle in Paris in 1900. The telegraphone was composed of a cylinder, cut with grooves along its surface, wrapped in steel wire. The electromagnetic head, as it passed over the tightly wrapped iron wire, operated both in recording sound and in playing back the recorded audio. Poulsen, trying to reduce distortion in his recordings, had also made early attempts at biasing (increasing the fidelity of a recording by including a DC current in his phonograph model) but, like Oberlin Smith's earlier model, his recorders, based on wire, steel tape, and steel disks, could not easily be heard and lacked a method of amplification.

Austrian inventor Fritz Pfleumer was the originator of magnetic tape recording. Since Pfleumer was accustomed to working with paper (his business was cigarette-paper manufacturing), he created the original magnetic tape by gluing pulverized iron particles (ferrous oxide) onto strips of paper that could be wound into rolls. Pfleumer also constructed a tape recorder to use his tape. On January 31, 1928, Pfleumer received German patent DE 500900 for his sound record carrier (lautschriftträger), unaware that an American inventor, Joseph O'Neill, had filed a patent—the first—for a device that magnetically recorded sound in December, 1927.

How It Works

The theory underlying magnetic storage and magnetic recording is simple: An electrical or magnetic current imprints patterns on the magnetic-storage medium. Magnetic tape, magnetic hard and floppy disks, and other forms of magnetic media operate in a very similar way: An electric current is generated and applied to a demagnetized surface to vary the substratum and form a pattern based on variations in the electrical current. The biggest differences between the three dominant types of magnetic-storage media (tape, rigid or hard disks, and flexible or floppy disks) are the varying speeds at which stored data can be recovered.

Magnetic Tape. Magnetic tape used to be employed extensively for archival computer data storage as well as analogue sound or video recording. The ferrous- or chromium-impregnated plastic tape, initially demagnetized, passes at a constant rate over a recording head, which generates a weak magnetic field proportional to the audio or video impulses being recorded and selectively magnetizes the surface of the tape. Although fairly durable, given the correct storage conditions, magnetic tape has the significant disadvantage of being consecutively ordered—the

recovery of stored information depends on how quickly the spooling mechanism within the recorder can operate. Sometimes the demand for high-density, cheap data storage outweighs the slower rate of data access. Large computer systems commonly archive information on magnetic tape cassettes or cartridges. Despite the archaic form, advances in tape density allow magnetic tape cassettes to store up to five terabytes (TB) of data in uncompressed formats.

For audio or video applications, sequential retrieval of information (watching a movie or listening to a piece of music) is the most common method, so a delay in locating a particular part is regarded with greater tolerance. Analogue tape was an industry standard for recording music, film, and television until the advent of optical storage, which uses a laser

The magnetic strip on the back of a credit card encodes the information to contact the cardholder's financial institution and connect with their account. (GIPhotoStock/Photo Researchers, Inc.)

to encode data streams into a recordable media disk and is less affected by temperature and humidity.

Magnetic Disks. Two other types of recordable magnetic media are the hard and floppy diskettes—both of which involve the imprinting of data onto a circular disk or platter. The ease and speed of access to recorded information encouraged the development of a new magnetic storage media form for the computer industry. The initial push to develop a nonlinear system resulted, in 1956, with the unveiling of IBM's 350 Disk Storage Unit—an early example of what is currently known as a hard drive. Circular, ferrous-impregnated aluminum disks were designed to spin at a high rate of speed and were written upon or read by magnetic heads moving radially over the disk's surface.

Hard and floppy disks differ only in the range of components available within a standard unit. Hard disks, composed of a spindle of disks and a magnetic read-write apparatus, are typically located inside a metal case. Floppy disks, on the other hand, were packaged as a single or dual-density magnetic disk (separate from the read-write apparatus that encodes them) inside a plastic cover. Floppy disks, because they do not contain recording hardware, were intended to be more portable (and less fragile) than hard disks—a trait that made them extremely popular for home-computer users.

Another variant of the disk-based magnetic storage technology is magneto-optical recording. Like an optical drive, magneto-optical recording operates by burning encoded information with a laser and accesses the stored information through optical means. Unlike an optical-storage medium, a magneto-optical drive directs its laser at the layer of magnetic material. In 1992, Sony released the MiniDisc, an unsuccessful magneto-optical storage medium.

APPLICATIONS AND PRODUCTS

Applications for magnetic storage range from industrial or institutional uses to private-sector applications, but the technology underlying each of these formats is functionally the same. The technology that created so many different inventions based on electrical current and magnetic imprinting also had a big impact on children's toys. The Magna Doodle, a toy developed in 1974, demonstrates a simple application of the concept behind magnetic storage that can shed light on how the more complex applications of

the technology also work. In this toy, a dense, opaque fluid encapsulates fine iron filings between two thin sheets of plastic. The upper layer of plastic is transparent and thin enough that the weak magnetic current generated by a small magnet encased in a cylinder of plastic (a magnetic pen) can make the iron filings float to the surface of the opaque fluid and form a visible dark line. Any images produced by the pen are, like analogue audio signals encoded into magnetic tape, nonvolatile and remain visible until manually erased by a strip of magnets passing over the plastic and drawing the filings back under the opaque fluid.

Magnetic Tape Drives. It is this basic principle of nonvolatile storage that underlies the usage of the three basic types of magnetic storage media—magnetic tape, hard disks, and floppy disks. All three have been used for a wide variety of purposes. Magnetic tape, whether in the form of steel bands, paper, or any one of a number of plastic formulations, was the original magnetic media and was extensively used by the technologically developed nations to capture audio and, eventually, video signals. It remains the medium of choice for archival mainframe data storage because large computer systems intended for mass data archival require a system of data storage that is both capable of recording vast amounts of information in a minimum of space (high-density) and is extremely inexpensive—two qualities inherent to magnetic tape.

Early versions of home computers also had magnetic tape drives as a secondary method of data storage. In the 1970's, IBM offered their own version of a magnetic cassette tape recorder (compatible with its desktop computer) that used the widely available cassette tape. By 1985, however, hard disks and floppy disks had dominated the market for computer systems designed to access smaller amounts of data frequently and quickly, and cassette tapes became obsolete for home-computer data storage.

Hard Disk Drives. In 1955, IBM's 350 Disk Storage Unit, one of the computer industry's earliest hard drives, had only a five megabyte (MB) capacity despite its massive size (it contained a spindle of fifty twenty-four-inch disks in a casing the size of a large refrigerator). However, the 350 was just the first of a long series of hard drives with ever-increasing storage capacity. Between the years of 1950 and 2010, the average area density of a hard drive has doubled every few years, starting from about three megabytes to the current high-end availability of three terabytes. Higher-capacity drives are in development, as computer companies such as Microsoft are redefining the basic unit of storage capacity on a hard drive from 512 bytes (IBM's standard unit, established during the 1980's) to a far larger 4 kilobytes (KB). The size of the typical hard drive made it difficult to transport and caused the development, in 1971, of another, similar form of magnetic media—the floppy disk.

Early hard drives such as IBM's 350 were huge (88 cubic feet) and prohibitively expensive (costing about $15,000 per megabyte of data capacity). Given that commercial sales of IBM computers to nongovernmental customers were increasing rapidly, IBM wanted some way to be able to deliver software updates to clients cheaply and efficiently. Consequently, engineers conceived of a way of separating a hard disk's components into two units—the recording mechanism (the drive) and the recording medium (the floppy disk).

Floppy Disk Drives. The floppy disk itself, even in its initial eight-inch diameter, was a fraction of the weight and size needed for a contemporary hard disk. Because of the rapidly increasing storage needs of the most popular computer programs, smaller disk size and higher disk density became the end goal of the major producers of magnetic media—Memorex, Shugart, and Mitsumi, among others. As with the hard drive, floppy disk size and storage capacity respectively decreased and increased over time until the 3.5-inch floppy disk became the industry standard. Similarly, the physical dimensions of hard drives also shrunk (from 88 cubic feet to 2.5 cubic inches), allowing the introduction of portable external hard drives into the market. Floppy disks were made functionally obsolete when another small, cheap, portable recording device came on the market—the thumb or flash drive. Sony, the last manufacturer of floppy disks, announced that, as of March, 2011, it would stop producing floppy disks.

Magnetic Striping. Magnetic storage, apart from tape, hard and floppy disk, is also widely used for the frequent transmission of small amounts of exclusively personal data—namely, the strip of magnetic tape located on the back of credit, debit, and identification cards as well as the ferrous-impregnated inks that are used to print numbers along the bottom of paper checks. Since durability over time is a key factor,

encasing the magnetic stripe into a durable, plastic card has become the industry standard for banks and other lending institutions.

IMPACT ON INDUSTRY

At the base of any industry is the need to store information. Written languages, ideographic art, and even musical notation were developed for the purpose of passing knowledge on to others. At the most basic, any individual, company, or government division that uses a computer system relies on some form of magnetic storage to encode and store data.

Government Use. Few government entities in even moderately technologically developed countries can operate without using some form of computer system and the requisite magnetic storage of either magnetic tape, internal or external hard drive, or both.

In 1942, the Armour Institute of Technology sold literally thousands of wire sound recorders to the American military, since the Army intelligence service was experimenting with technology that intercepted foreign radio transmissions and wished to record those transmissions for the purpose of decoding them.

In June, 1949, IBM envisioned a new kind of magnetic storage device that could act as a data repository for another new invention—the computer. On May 21, 1952, the IBM 726 Tape Unit with the IBM 701 Defense Calculator was unveiled. On September 13, 1956, a IBM announced their creation of the 305 RAMAC (Random Access Memory Accounting System) computer. These devices (and other, similar inventions) allowed both the government and the private sector to start phasing out punch-card-based computer storage systems. In the following decades, governmental agencies were able to switch from the massive boxes of punch cards to the more efficient database system based on external magnetic tape cassettes and internal magnetic disks. In the United States, for example, the Internal Revenue Service, which records and processes financial data on every working individual in the country identified by a Social Security Number, is one of many governmental agencies that must store massive amounts of sensitive information. As a result, the federal government purchases roughly a million reels every year.

Business Use. Besides the magnetic media used in computers, magnetic storage media in the form of magnetic tape has had a dramatic impact on the

Fascinating Facts About Magnetic Storage

- One of the earliest magnetic recording devices, Valdemar Poulsen's telegraphone, was not only intended for the recording of telephone calls but was actually constructed out of telephone components.
- In 2011, the highest-capacity desktop hard drive was the three-terabyte Hitachi Deskstar 7K1000. It contains the highest number of hard disk platters—five—and rotates them at 7,200 rpm.
- In 2011, the highest-capacity floppy disk was 3M's 3.5-inch LS-240 SuperDisk, which could store up to 240 megabytes of data.
- The oldest form of magnetic storage media, magnetic tape, still has the highest capacity of all of the current forms of magnetic data storage, at 5 terabytes per cassette.
- Forrest Parry, under contract to the United States government, invented the first security card in 1960—a plastic card with a strip of magnetic tape affixed to the back. "Magstripes" (typically three tracks per stripe) are universally used to record identifying data for a variety of security and financial purposes: credit cards, banking identification cards, gift cards, and government benefit cards.
- Twenty-one states in the United States and six provinces in Canada use driver licenses that have magnetic striping to carry identifying data about the individual user depicted on the card. Some passports also have a magnetic stripe for easier handling by immigration officials.

industrial and business sectors of the developed countries of the world. Miniature cassette recorders, introduced in 1964 by the Philips Corporation, became a common sight in college classrooms, executive offices, and journalists' hands for recording lectures, memos, or interviews because of their small size and adjustable tape speed. These recorders suggested the importance of knowing exactly what another individual has said for reasons of accuracy and truth in reporting.

Entertainment Use. The invention of magnetic storage allowed for a more precise system of audio and video recording. Dramatic shifts in popular entertainment and the rise of the major music labels

such as Victor or RCA occurred precisely because there were two alternative methods of audio recording that, ultimately, introduced music into vastly different venues and allowed individuals to take an interest in music that was as lucrative to the music company as it was pleasurable to the consumer. Phonograph recordings made on wax, lacquered wood, or vinyl brought music into a family's living room and millions of teenagers' bedrooms, while electromagnetic recordings on tape allowed music to be played as background music in vehicles and workplaces and to be carried from place to place in portable devices. Competition between the rival manufacturers of records (such as Columbia and EMI) and tapes (such as BASF and TDK) was fierce at times, even inspiring copyright-infringement campaigns by such organizations as the British Phonographic Industry (BPI) against the manufacturers and users of blank cassette tapes in the 1980's. Regardless, the portability and shock resistance of magnetic tape-driven music players ensured that magnetic storage was unlikely to be made obsolete by either phonograph technology or music-industry politics.

Consumer Use. Magnetic-storage media probably would not have the impact it did on Western society if it had not been so enthusiastically embraced by the willingness of the private individual to purchase and use the technology. For example, although magnetic media gained a reputation for illicit use because of music- and film-industry complaints about some consumers' misuse of blank magnetic media to record illegal (bootleg) copies of phonograph recordings, television programs, and movies broadcast over cable or satellite systems, magnetic-storage media has also given consumers the ability to use blank media as a vehicle for the recording of their own audio and video creations. The technology was robust enough to make a durable recording but flexible enough to allow repeated erasures and rerecordings on a single tape.

CAREERS AND COURSE WORK

Individuals interested in further developing the technology of magnetic media should study computer engineering, electrical engineering, or materials engineering depending on whether they wish to pursue increasing storage capacity of the existing electromagnetic technology or finding alternative methods of using electromagnetic theory on the problems of data storage. Some universities are particularly sensitive to the related nature of computer-component design and electrical-system design and allow for a dual degree in both disciplines. Supplemental course work in information technology would also be helpful, as an understanding of how networks transmit data will more precisely define where future storage needs may be anticipated and remedied.

SOCIAL CONTEXT AND FUTURE PROSPECTS

The current social trend, both locally and globally, is to seek to collect and store vast amounts of data that, nevertheless, must be readily accessible. An example of this is in meteorology, which has been attempting over the past few decades to formulate complex computer models to predict weather trends based on previous trends observed and recorded globally during the past century. Weather and temperature records are increasingly detailed, so computer storage for the ideal weather-predicting computer will need to be able to keep up with the storage requirements of a weather-system archive and meet reasonable deadlines for the access and evaluation of a century's worth of atmospheric data.

Other storage developments will probably center on making more biometric data immediately available in identification-card striping. With surveillance, rather than incarceration, of petty criminals being considered as a method by which state and federal government might cut expenses, an improvement in the design and implementation of identification cards might be one magnetic-storage trend for the future. Another archival-related goal might be the increasing need for quick comparisons of DNA evidence in various sectors of law enforcement.

Julia M. Meyers, M.A., Ph.D.

FURTHER READING

Bertram, H. Neal. *Theory of Magnetic Recording*. Cambridge, England: Cambridge University Press, 1994. A complete and thorough handbook on magnetic recording theory; goes well beyond the typical introductory-style discussion of magnetic and strives for a more advanced discussion of magnetic recording and playback theory.

Hadjipanayis, George C, ed. *Magnetic Storage Systems Beyond 2000*. Dordrecht, The Netherlands: Kluwer Academic, 2001. This book is the collected papers presented at the "Magnetic Storage Systems Beyond 2000" Conference held by the NATO Advanced Study Institute (ASI) in Rhodes, Greece. Although the papers are technical discussions of the limitations of magnetic technology (such as magnetic heads, particulate media, and systems still in development), the speculative emphasis often describes new areas of possible development. Each paper has a tutorial, which provides an introduction to the ideas discussed.

Mee, C. Denis, and Eric D. Daniel. *Magnetic Recording Technology*. 2d ed. New York: McGraw-Hill, 1996. An older handbook, but one with particularly clear description and analysis of many of the emergent trends in the last few decades in the field of data recording and storage.

National Research Council of the National Academies. *Innovation in Information Technology*. Washington, D.C.: National Academies Press, 2003. A good handbook for the study of storage media and other issues in the field of information technology.

Prince, Betty. *Emerging Memories: Technologies and Trends*. Norwell, Mass.: Kluwer Academic, 2002. A good primer for understanding the development of memory in computing systems. Has application both to data storage and volatile memory forms.

Wang, Shan X., and Alexander Markovich Taratorin. *Magnetic Information Storage Technology*. San Diego: Academic Press, 1999. Provides the basic principles of magnetic storage and digital information and describes the technological need for data recording and the resulting push for increased capacity, faster access rates, and greater durability of media.

WEB SITES

IEEE Magnetics Society
http://www.ieeemagnetics.org

Information Technology Association of America
http://www.itaa.org

International Disk Drive Equipment and Materials Association
http://www.idema.org

See also: Computer Engineering; Computer Networks; Computer Science; Electrical Engineering.

MAPS AND MAPPING

FIELDS OF STUDY

Cartography; geography; geology; graphic arts; mathematics; statistics; geographic information systems; remote sensing.

SUMMARY

Mapping encompasses methods for representing geospatial information in paper and digital forms as well as newer digital technologies such as remote sensing and geographic information systems. Maps are needed in areas ranging from civil engineering to regional planning. Computer technologies and the Internet have brought profound changes to the ways maps are created and used, and a growing number of mapping applications have been developed for portable devices such as car navigation systems.

KEY TERMS AND CONCEPTS

- **Cartography:** Art and science of making maps.
- **Digital Orthophoto Quadrangle:** Aerial photograph that has been rectified to remove camera distortion and tied to a system of coordinates such as latitude and longitude.
- **Digitizing:** Process of converting analogue information into a digital or machine-readable format.
- **Geographic Information System (GIS):** Computer-based system for storing, analyzing, and visualizing geographic information.
- **Global Positioning System (GPS):** Satellite-based positioning system developed and operated by the U.S. Department of Defense.
- **Map Projection:** Mathematical method for transferring information from the spherical Earth to a flat surface.
- **Scale:** Relationship in measurement units between a map and the real-world location it represents.
- **Thematic Map:** Map representing data corresponding to a single theme such as rainfall or wheat production.
- **Virtual Map:** Temporary map displayed on a computer screen.

DEFINITION AND BASIC PRINCIPLES

Maps assume many forms, including spherical globes, folded paper charts, wall-sized murals, and images displayed on tiny electronic screens. As scale models of reality, maps show a selection of information. In addition to describing features on the surface of the Earth, maps can represent underwater areas, the interiors of caves, and celestial bodies such as planets. The term "cartography" refers to the process of creating maps and related geographic products. According to the International Cartographic Association, cartography is the art, science, and technology of making maps.

Maps can be divided into two categories: general-purpose and thematic maps. General-purpose maps show a variety of information, including lakes, rivers, roads, cities, and administrative or political boundaries. Most general-purpose maps, such as road and topographic maps, are designed to serve as reference tools. In contrast, the purpose of thematic maps is to show patterns and distributions corresponding to a single topic or theme such as rainfall or corn production.

BACKGROUND AND HISTORY

For more than 4,000 years, maps have been used to record and communicate information about the Earth. The ancient Greeks created maps to characterize the Earth's spherical shape, while legions of Roman soldiers used them as tools in the conquest of new territories. Drawn by hand, early maps showed landscape features using pictorial symbols to represent mountains, rivers, and other physical features. Later technologies such as printing eliminated the need for maps to be painstakingly copied and increased the speed at which maps could be reproduced and disseminated. Exploration, especially during the fifteenth and sixteenth centuries, created a demand for maps to assist in ocean navigation and to document newly discovered locations.

Modern maps play many roles in society, from guiding aircraft to providing reference information within school textbooks. Under most circumstances, maps are superior to charts and graphs in their ability to represent distances, directions, and the relative sizes of objects over space. In addition to showing locations of features, maps are able to efficiently represent geographic patterns and the spatial extent of physical entities such as rivers and mountains. They are also useful for showing cultural features such as property boundaries and political or administrative areas. As archives of spatial information, maps are used extensively for recording the locations of historical events as well as spatially referenced scientific data. In addition to their use in storing information, maps serve as important research tools for visualizing geographic data needed for evaluating hypotheses about spatial distributions.

HOW IT WORKS

Scale and Projection. All maps of the Earth's surface require some level of reduction. Map scale is an important concept to maps and mapping because it determines the level of map reduction necessary. Scale is the ratio of a distance measured on a map to the distance it represents in the real world, using the same units of measurement. For example, 1 inch measured on a map with a scale of 1:24,000 represents 24,000 inches (or 2,000 feet) in the real world. The scale used on any given map determines the size of the area that can be represented as well as the level of detail that can be depicted. In addition to the representative fraction (for example, 1:100,000), scale can be shown using a verbal statement such as "1 inch equals 1 mile" or a graphic or bar scale printed directly on the map.

Cartographers draw a distinction between small- and large-scale maps. Small-scale maps represent very large areas of the world and are useful for representing entire continents. In contrast, large-scale maps are capable of showing very small areas at higher levels of detail. City road maps and topographic maps are examples of large-scale maps. Given that mapmaking involves reductions, the cartographer must make choices about the features to be shown and the symbols used to represent those features. This process, called selection, must be closely tied to the purpose of the map. In addition, complex objects such as coastlines must be simplified to reduce unnecessary detail.

Another important consideration is the choice of projection for displaying mapped information. Projection refers to a mathematical transformation of geographic information from the spherical Earth to a flat or developable surface. Because spherical surfaces cannot be flattened without distortion, the cartographer must select the most important properties to maintain during this transformation. For example, on a world map, it may be desirable to maintain the relative sizes of countries for making comparisons. Although maintaining the property of equivalence, this type of projection distorts the shapes of countries near the poles. The well-known Mercator projection is useful for ocean navigation because it preserves shapes as well as lines of constant geographic direction as straight lines. However, a disadvantage of the Mercator projection is that it severely distorts the relative sizes of polar areas, making the island of Greenland in the North Atlantic look larger than the continent of South America, which is actually fifteen times larger than Greenland.

Geographic Grids. Cartographers use geographic grids, or coordinate systems, for the placement of points, lines, and areas to systematically organize geographic information. Most maps are created using either the latitude-longitude coordinate system or a rectangular coordinate system. Latitude-longitude and rectangular grids use Cartesian coordinates (x,y). The point where the prime meridian (0 degrees longitude) and the equator (0 degrees latitude) cross serves as an origin against which all other locations are referenced. Smaller geographic areas are often mapped with rectangular coordinates that use an origin located outside and to the southwest of the area of interest to make all coordinate values positive.

Map Design. Map design is a systematic process involving the selection and arrangement of map elements in a way that facilitates a user's correct interpretation of ideas and concepts. Cartographic design has benefited from developments in both graphic design and cognitive psychology. Because the principal objective of mapmaking is communication, it is important for cartographers to select visual elements and arrangements that unambiguously present concepts and spatial relationships. The design process can require several stages as alternatives are tested and modified. Basic graphic elements manipulated in the design process include points, lines, and area symbols. In addition, the selection and placement of text is an important design element. As suggested by French cartographer Jacques Bertin, graphic elements used in cartography can be made more or less prominent through changes in hue (color), value (darkness or lightness), texture, orientation, size, and shape. Design is affected not only by the choice and design of symbols but also by the arrangement of map elements. In the design process, cartographers must be careful to use space efficiently and to avoid symbols that create unwanted attention or noise. Other important concepts within map design include achieving balance in the arrangement of map elements and creating a visual hierarchy that draws the map reader's attention to features most important to the map's purpose.

Map Output. Developments in map reproduction have followed innovations in printing, beginning with Johann Gutenberg's invention of the printing press in 1440. Early printing technology employed a process called letterpress, in which ink was applied to raised portions of wooden blocks that were then pressed against paper. Introduced in the early 1700's, engraving involved the application of ink to depressions in a metal printing plate. Engraving improved map reproduction because it enabled the use of finer line widths and improved gray tones and patterns. Lithographic reproduction, a technique based on the incompatibility of grease and water, was introduced in the 1800's. Most modern maps are reproduced using lithographic plates, which are applied to a rotating drum to transfer an image to a moving sheet of paper. Four-color maps use lithographic plates corresponding to the subtractive primary colors of cyan, magenta, and yellow plus black. A wide variety of colors can be created by combining percentages of each primary color. Modern printing presses use computer files to create plates needed to produce color maps, with sheets of paper passing through the press once for each plate.

Maps can also be reproduced in smaller quantities using low-cost output devices such as laser printers and ink-jet plotters. Electrostatic (or xerographic) copying is a common method for reproduction. Laser and ink-jet printers have also become a popular method for reproducing small numbers of maps. The readability of map lettering and the sharpness of lines and grey tones is a function of the printer's resolution as represented in dots per inch (dpi). Low-end printers typically offer 300-dpi output, meaning that each square inch has 90,000 dots. Higher-end printers and plotters offer resolutions exceeding 2,000 dpi.

The introduction of personal communications devices with high-resolution screens has led to an

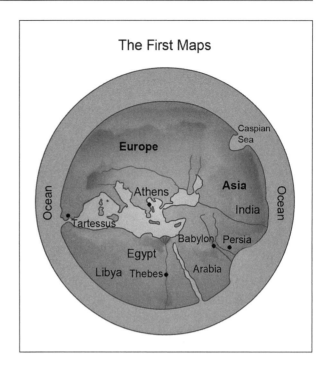

The First Maps

increasing number of maps that are never printed. These virtual maps may be used for checking weather conditions or finding a street address. Internet-based Google Maps has been an innovator in making virtual maps, aerial photographs, and satellite images accessible to the public. Google Maps and Google Earth enable the simultaneous viewing of aerial photographs and map information in the form of roads, contour lines, and political boundaries. Mapping tools enable the user to zoom in or out, pan in any direction, and select specific map features and points of interest to be displayed.

Computers and Mapmaking. The single most significant change to cartography during the twentieth century was the introduction of computers that enabled tasks previously requiring hours to be carried out in seconds. Computer cartography, or the use of computers in map production, has been particularly beneficial in enabling cartographers to experiment. Using mapping software, the cartographer can resize and reposition map elements such as text labels, scale bars, and legend boxes. Computers also facilitate mathematical and statistical transformations of tabular data needed for making thematic maps. Some programs enable map images to be combined with other spatial data such as aerial photographs or satellite images. Increasingly, digital databases are

1161

replacing paper maps as the principal method for storing and retrieving geographic information.

Before geographic information can be manipulated by a computer, it must be converted to a digital form. The process of capturing points, lines, and areas shown on a paper map in a digital format is called digitizing. Analogue information can be digitized by tracing features of interest individually using a tablet or table-sized digitizer. Digitizing can also be accomplished by scanning an entire map sheet. Before scanned data are useful, they must be transformed into a system of coordinates such as latitude and longitude. This process is called geocoding or georeferencing.

APPLICATIONS AND PRODUCTS

Since 1985, a significant amount of data used in computer mapping has been converted into digital format. For example, the U.S. GeoData digital line graphs are computer files containing transportation and other line information that has been captured from topographic maps maintained by the United States Geological Survey (USGS). The set of digital files called digital raster graphics includes information scanned from USGS topographic maps. USGS digital elevation model data files contain grids of elevation values captured every 30 meters and are useful for constructing contour maps, three-dimensional terrain diagrams, and computer simulations such as aerial fly-bys.

Computer technology also makes it possible to create new types of cartographic visualizations such as animated maps. Maps that use a sequence of digital images are called temporal animations. This type of map is useful for representing changes over time, such as the movement of military fronts during a battle.

Computer technologies have influenced maps and mapping through developments in closely related fields such as aerial photography, satellite remote sensing, and geographic information systems (GIS). Remote sensing involves the capture of information about the Earth using sensors and is commonly divided into photographic methods such as aerial photography and nonphotographic imaging that include sensors looking outside visible portions of the electromagnetic spectrum.

Digital orthophoto quadrangles (DOQs) have emerged as an important source of information for constructing maps. DOQs are aerial photographs in

***

Fascinating Facts About Maps and Mapping

- Faculty in the Department of Geography at the University of California, Santa Barbara, designed a mapping system for blind persons that uses audio signals coupled with GPS guidance.
- Geographic information systems and remote sensing were used extensively after Hurricane Katrina to assist in directing emergency responders and organizations involved with recovery efforts. Geographic information systems were also important in the recovery effort following the September 11, 2001, attack on the World Trade Center in New York City.
- The U.S. military is using geographic information systems to provide battlefield commanders with tactical views of personnel and weapons systems to improve situational awareness.
- Assisted by French cartographer Louis-Alexandre Berthier, George Washington used hinged overlay maps during the Battle of Yorktown to study troop movements.
- Most modern cell phones contain GPS receivers that can guide emergency responders to the telephone's location.
- The destruction wrought by the 1906 earthquake and resulting fire in San Francisco was first documented in aerial photographs taken by cameras carried by kites.

***

digital form that have been rectified (corrected) for camera distortion. In addition to serving as a source from which to extract the location of physical and cultural features, they can be used as background images for maps. Multispectral scanners used by Landsat and other satellites capture large amounts of data using scanners that are sensitive to narrow wavelengths of energy emitted from the Earth's surface. Such data can be useful for revealing information that cannot be seen in photographic images, including stress or disease in vegetation.

Radio detection and ranging (radar) is another nonphotographic imaging system that can be used for mapping surface features, underground areas, and atmospheric conditions. Radar uses transmitted energy that is reflected by objects and then captured and converted into an image. Imaging radar used on space shuttle missions has become an important source of elevation data used in a variety of mapping applications. Light detection and ranging (lidar) is a innovation for mapping the Earth by measuring the properties of scattered light as it comes into contact with distant objects. Lidar images are created by measuring the time needed for laser pulses to reach the Earth's surface. Some applications for lidar include three-dimensional mapping of urban areas and assessments following natural disasters

IMPACT ON INDUSTRY

Computer technology has not only affected the way maps are created but also introduced new ways of accessing, analyzing, visualizing, and sharing geographic information. Among the most important developments has been the growth of geographic information systems (GISs). GIS has been defined as a system for storing, analyzing, and visualizing spatial information. Although there is no well-defined boundary between computer cartography and GIS, the former is concerned with the production of maps, while the later focuses on the analysis and visualization of geographic data. Examples of GIS applications include surface-water modeling, tracking the spread of invasive plant and animal species, determining optimal locations for bus stops, and developing strategies to improve emergency response following natural and human-caused disasters. An important area of GIS development has been precision agriculture, in which GIS maps of soil properties and other site-specific factors are used to guide tractors moving through fields as they apply fertilizers, insecticides, or herbicides.

Another innovation that has greatly influenced mapping is the Global Positioning System (GPS). GPS uses Earth-orbiting satellites to provide a stream of data that can be used by portable receivers for pinpointing Earth locations within less than a meter. GPS receivers have introduced possibilities for the rapid collection of geographic coordinates corresponding to points, lines, and areas. GPS receivers determine latitude, longitude, and elevation above sea level by measuring the time needed for a radio signal to be received from four or more satellites. Data captured using GPS receivers can be used to update cartographic data sets or develop real-time mapping applications that track the location and speed of cars, trucks, boats, or aircraft.

CAREERS AND COURSE WORK

College and university courses in cartography are most often taught within departments of geography. Students seeking careers in the mapping sciences should complete college course work in geography, mathematics, statistics, and technical writing. Depending on the type of maps they will be involved in producing, they may also take course work in graphic arts, computer programming, physics, remote sensing, computer-aided drafting, database management, planning, or geology. A wide variety of agencies and organizations employ cartographers, GIS analysts, and remote-sensing specialists. The U.S. Geological Survey employs cartographers to maintain both paper and digital map products. Its National Imagery and Mapping Agency supports mapping and imagery needs for the military and policy making. The U.S. Department of Agriculture's Natural Resources Conservation Service maintains the Soil Survey Geographic Database, an important spatial data set defining soil boundaries within the United States. The Central Intelligence Agency provides mapping products used internally and by other intelligence agencies. State and local government agencies and private organizations also employ geospatial experts. State agencies produce road maps used for promoting tourism, and local government departments maintain property and planning maps. Maps and proprietary data used in mapping are also created by private companies.

SOCIAL CONTEXT AND FUTURE PROSPECTS

Maps and other cartographic products are important within areas ranging from civil engineering and planning to the humanities. For example, maps are used in weekly magazines, in newspapers, and as part of Web sites. Journalistic maps typically show the countries, cities, or accident sites that are the focus of news events.

Transportation maps such as nautical and aviation charts are of critical importance to the safe movement of ships and aircraft. Within the United States, the National Ocean Service of the National Oceanic and Atmospheric Administration (NOAA) publishes four types of nautical charts: general, sailing, coast, and harbor charts. The Federal Aviation Administration publishes several aviation charts, including the world aeronautical charts at a scale of 1:1,000,000 and sectional charts that represent smaller areas in greater detail at a scale of 1:500,000. Nautical and aviation charts must be updated frequently to reflect changes that can affect safety. Other types of transportation maps include street and highway maps, road atlases, maps showing airline or train routes, and bicycle or walking trail maps.

Topographic maps show rivers, lakes, mountains, and other surface features, as well as cultural features such as roads, cities, levees, and towers. Elevations on topographic maps are shown using contours defined as lines that join points of equal elevation. Bathymetric or hydrographic charts are a related type of map that uses lines to represent underwater depths. The U.S. Geological Survey maintains topographic maps as well as maps showing geological cross sections, groundwater areas, earthquake faults, and seismically active areas. Geologic maps have a wide range of uses, from mineral exploration to civil engineering. Most geologic maps use colors, gray tones, patterns, and sometimes letter codes to represent different rock types. In addition to government agencies, energy and mining companies also maintain maps showing fossil fuels, mineral reserves, and areas with the potential to generate power using solar or wind energy.

Maps are important for recording the location of property boundaries and for planning applications. Cadastral maps are maintained by local government agencies to show ownership boundaries as a basis for property taxes. This type of map is also important to insurance companies. Many local government planning agencies maintain maps showing the city or county infrastructure, such as water resources, sewers, electric lines, and gas lines. City and regional planners use maps for zoning changes and to maintain 911 emergency response systems.

The NOAA's National Weather Service is the principal federal agency responsible for disseminating information about weather conditions. Weather maps are used by forecasters and others to monitor or predict atmospheric conditions including temperature, precipitation, wind direction and speed, humidity, cloud cover, and anomalies such as thunderstorms, hurricanes, and tornadoes. Weather maps may show a single parameter such as surface pressure or multiple types of information. Climate maps include some of the same information shown on weather maps but for longer timeframes. Examples include maps depicting monthly temperature highs and lows or average yearly rainfall.

Widespread availability of the Internet has created possibilities for the publication of new types of mapping products such as electronic atlases that allow users to generate customized maps by selecting the data to be shown. Electronic atlases enable users to search on place-names and easily locate cities and find other information. An added benefit of the Internet is its ability to incorporate updates as soon as new information becomes available.

Thomas A. Wikle, Ph.D.

FURTHER READING

Cartwright, William, Michael P. Peterson, and Georg F. Garner, eds. *Multimedia Cartography.* 2d ed. New York: Springer, 2007. Examines the growing field of interactive mapping that has gained interest as a result of the Internet's expansion.

Dodge, Martin, Rob Kitchin, and Chris Perkins, eds. *Rethinking Maps: New Frontiers in Cartographic Theory.* New York: Routledge, 2009. Discusses the changing meaning of maps and how their construction and use has evolved. Chapters are written by theorists who examine topics ranging from sustainable mapping to the role of maps in fields such as anthropology.

Kraak, Menno-Jan, and Ferjan Ormeling. *Cartography: Visualization of Spatial Data.* 3d ed. New York: Prentice Hall, 2010. Examines map design, statistical mapping, and data acquisition.

Okada, Alexandra, Simon J. Buckingham Shum, and Tony Sherborne, eds. *Knowledge Cartography: Software Tools and Mapping Techniques.* London: Springer-Verlag London, 2008. Knowledge cartography refers to the mapping of intellectual landscapes. Topics include visual languages, the role of hypertext maps, and methods for evaluating maps.

Robinson, Arthur H., et al. *Elements of Cartography.* 6th ed. New York: John Wiley & Sons, 1995. A comprehensive, well-illustrated text on cartography.

Wilford, John N. *The Mapmakers: The Story of the Great Pioneers in Cartography from Antiquity to the Space Age.* London: Pimlico, 2002. Looks at the origins of mapmaking, beginning with Ptolemy's crude maps and extending through technological achievements that enable scientists to map distant planets.

WEB SITES

Federal Aviation Administration
World Aeronautical Charts
http://www.naco.faa.gov/index.asp?xml=naco/catalog/charts/vfr/world

Global Positioning System
http://gps.gov

Google Maps
http://maps.google.com

National Oceanic and Atmospheric Administration
National Ocean Service, Office of Coastal Survey

http://www.nauticalcharts.noaa.gov/staff/chartspubs.html

National Resources Conservation Service
Soil Survey Geographic Database
http://soils.usda.gov/survey/geography/ssurgo

U.S. Geological Survey
Maps, Imagery, and Publications
http://www.usgs.gov/pubprod

See also: Aeronautics and Aviation; Air Traffic Control; Geoinformatics; Meteorology; Navigation; Oceanography; Plane Surveying; Remote Sensing.

MARINE MINING

Mining engineering; petroleum engineering; geological engineering; ocean engineering; environmental engineering; geological oceanography; economic geology; applied science; environmental science; materials science; mineralogy; metallurgy, gemology.

SUMMARY

Marine mining is the mining of minerals and other natural resources from the ocean. Valuable resources in the near-shore zone include sand and gravel for construction, salt, phosphate deposits needed for fertilizers, coral for marine aquariums, carbonate sands, and diamonds, gold, tin, and other minerals in placers. Farther out are oil and natural gas deposits, sulfur for industry, and gas hydrates. On the seafloor are nodules and crusts containing strategically important metals such as manganese and cobalt and sulfide deposits around the black smokers in the rift zones, which contain gold, silver, and other useful metals.

KEY TERMS AND CONCEPTS

- **Black Smoker:** Seafloor opening that emits jets of superheated, mineral-rich water.
- **Dynamic Positioning:** Computer-controlled way to keep a ship directly over a drilled hole.
- **Gas Hydrate:** Gas-rich layer of ice found in ocean sediments.
- **Manganese Nodule:** Metal-rich lump found on the deep seafloor.
- **Placer:** Accumulation of mineral grains washed by rivers into the sea.
- **Salt Dome:** Salt mass that rises through sediments because of salt's lesser density.
- **Semisubmersible Rig:** Deep-water drilling rig supported by pontoons.
- **Sulfide Deposit:** Metal-rich crust found surrounding a back smoker.

DEFINITION AND BASIC PRINCIPLES

Marine mining deals with the extraction of economically valuable substances from the ocean. These may be gases, liquids, minerals, rocks, base and precious metals, or ornamental stones such as pearls, diamonds, and coral. Near-shore mining has been done for years, and drilling for oil and gas at sea began in the 1890's, but the mineral deposits on the seafloor have been largely ignored because of the technological difficulty in obtaining them. The substances obtained by marine mining fall into three categories: rocks and minerals found in shallow, near-shore waters; oil and gas deposits found in waters up to thousands of feet deep; and metal-rich deposits found on the seafloor, two miles or more down. Near-shore sand and gravel and the minerals in placers can be scooped or suctioned up by giant dredges, salt can be evaporated in shallow coastal ponds, and pearls and ornamental coral for aquariums can be obtained by divers. Oil wells have been drilled in the ocean for more than one hundred years, and during that time, marine oil and gas production has grown so rapidly that by the twenty-first century, thirty-four percent of the world's oil and twenty-eight percent of the world's natural gas come from below the seafloor. Someday the gas hydrates, manganese nodules, and sulfide deposits may be tapped as well.

BACKGROUND AND HISTORY

The ocean has provided salt for human dietary purposes since ancient times, so evaporation of salt from seawater probably represents one of the earliest examples of marine mining. Another early example was the collecting of minerals washed into the sea by rivers to form placers. Shoreline tin was mined in Indonesia in the early 1900's using dredges, and diamonds carried into the sea by the Orange River were mined along the coast of South Africa at about the same time. Oil has been produced on land using dug or drilled wells for hundreds, if not thousands, of years, but the first offshore oil wells were not drilled until the early 1890's, using piers or pilings driven into shallow water.

Not until the 1930's did oil drillers venture more than a mile from shore, and the immense floating platforms used in the twenty-first century were not introduced until the 1960's. The first manganese nodules were discovered on the seafloor by the HMS *Challenger* expedition in the 1870's, but the techniques for mining them and the related sulfide deposits are still being developed.

HOW IT WORKS

Near-Shore Resources. Humans must have found early on that the salt crystals in dried-up lagoons near the ocean were useful as a seasoning and for the curing and preserving of foods. The construction of evaporating ponds was a logical next step, and these are common in countries that have access to the ocean and lack underground salt deposits. The mining of placers uses large, floating dredges that either suck the minerals up in long tubes or scoop them up in a series of circulating buckets. Other valuable substances found in shallow water include phosphate for fertilizer, pearls for jewelry, live coral for aquariums, and pure carbonate sand. Although phosphate deposits are widely present in the ocean, they are not mined because phosphate deposits on land can be tapped at a lower cost. The pearls are grown from oysters in pearl farms, which consist of floating wooden rafts or posts driven into the bottom of the sea. A few pearls are still brought up by divers. Divers also bring up the live coral so prized by marine aquarium hobbyists. In the Bahamas, shallow-water shoals of pure carbonate sand are mined as a source of lime for industry.

Oil and Natural Gas. To extract oil and natural gas, a fixed or floating platform is built so that a hole may be drilled into the seabed. The first wells used fixed platforms, either built on pilings at the ends of piers or on artificial islands. The first true offshore well in the Gulf of Mexico was spudded in 1937, one mile from shore in the then-intimidating water depth of 14 feet. The first well located out of sight of land in the Gulf of Mexico was drilled in 1947; it was 10 miles from shore and in 18 feet of water. In the 2010's, companies drill wells in the Gulf of Mexico that are 50 miles out and in nearly 10,000 feet of water. For wells in moderate water depths, the drilling platform rests on a concrete structure or a steel tower. Deeper waters necessitate the use of semisubmersible rigs in which the platform rests on underwater pontoons, moored to the bottom by steel cables or kept in place by dynamic positioning.

Related Products. Sulfur is commonly associated with oil deposits, especially when salt domes are present. As the salt moves upward because of its lesser density, it arches up the sediments, and chemical reactions between the salt and the sediments may form a cap rock that is rich in sulfur. The sulfur is mined by drilling into the cap rock, melting the sulfur with high-pressure steam, and bringing it to the surface as a liquid for processing. The methane gas hydrates in the ocean, which contains a large amount of natural gas, and has not been mined yet. In these hydrates, the gas is bonded with water in the form of ice instead of being associated with oil.

Deep-Sea Resources. Technology for mining the manganese nodules and sulfide deposits found on the seafloor is still in the developmental stages. The nodules are lumps about the size of hamburgers and are scattered about. Someday it may be possible to vacuum them up with suction hoses or scoop them up with a series of buckets. The sulfide deposits around the black smokers present a greater challenge. They are welded to the seafloor and would have to be broken loose.

APPLICATIONS AND PRODUCTS

Sand and Gravel. Sand and gravel for construction purposes are in short supply in many parts of the world. When available deposits on land have been exhausted, the only other supplies are in the ocean. This situation is particularly true for some of the older countries in Europe, such as England, where construction has gone on for hundreds of years and real estate values are so high that the few remaining deposits on land are prohibitively expensive. Deposits of sand and gravel in the ocean have the advantage of being well sorted, with grains all of the same size, so they are suitable for building purposes once the salt has been washed out of them. Sand and gravel are mined along the coast of the United States to rebuild beaches that have lost their sand because of erosion. The sand is pumped directly onto the beach from the offshore zone in a process known as beach nourishment. The cost of nourishing a beach may exceed several million dollars for each mile of beach treated, and a significant problem is that if the sand is fine grained, it may wash away quickly in storms. In the Bahamas, shallow-water shoals of carbonate sand are mined as a source of lime for making cement and for separating iron from its ore in blast furnaces.

Placers. Gold, tin, and diamonds are the most sought-after substances found in placers, but platinum, iron minerals, and other gemstones may also be found in them. The gold, platinum, and gems are used in the manufacture of jewelry, with the gold also being needed for dentistry, coinage, and monetary purposes. Both gold and platinum have electrical properties that are useful for industry, and platinum is required in the manufacture of catalytic converters for automobiles. Tin was highly prized in ancient times because it was needed for the manufacture of bronze and pewter. In modern times, it is used in solder and for the tin plating of steel to prevent corrosion. The familiar so-called tin can used to package food is simply a steel container with a thin, protective coating of tin.

Salt. Because of its dietary importance and usefulness for food preservation, salt has been valued by humans throughout history. It is also widely used in the chemical industry as a water softener, for snow and ice removal, and by anyone who makes homemade ice cream. However, only 86 percent of the dissolved material in the ocean is table salt. The remainder includes bromine, which is used in medicines, chemistry, and antiknock gasoline; magnesium, an important light-weight metal used in fireworks and flares because it is flammable; potassium, which is used in fertilizers; and calcium sulfate (gypsum), an important component of wallboard.

Oil and Natural Gas. By far the most valuable of the resources obtained from the sea are oil and natural gas. In addition to being used for fuels, such as gasoline, diesel fuel, kerosene, fuel oil for heating homes, and jet fuel, oil is the raw material that is the basis of grease, motor oil and other lubricants, asphalt and tar, and a variety of chemical products, including nylon fibers for stockings. Even the paraffin wax used to seal jars of homemade jelly comes from oil. The natural gas found with oil is used domestically for cooking and heating, in the power industry to generate electricity, and in the production of ammonia for making fertilizer. The methane gas hydrates deposited on the seafloor, which have not yet been exploited commercially, may someday have similar uses.

Sulfur. Although sulfur is not derived from the seafloor in the twenty-first century, it has been in the past. Sulfuric acid, which is widely used in chemistry and industry, is its most important product. Because of its toxicity to bacteria, sulfur is an ingredient in medicines, insecticides, fungicides, and pesticides. Sulfur is also flammable and is used in the manufacture of matches and gunpowder.

Manganese Nodules. Although the manganese nodules and crusts found on the seafloor are not being mined commercially, the United States has an interest in them because of the metals they contain. Three of the metals—iron, needed for making steel; copper, used in wiring; and nickel, important in the manufacture of stainless steel—are already available on land. The other two metals, strategically important in wartime, are not available in the United States and must be imported. They are manganese, used in making armor plate, and cobalt, which is needed for the high-temperature steel used in tools and aircraft engines.

Sulfide Deposits. The sulfide deposits found on the seafloor have not been mined yet but are of interest because of their metal content. They contain not only the five metals found in the manganese nodules but also zinc, which is needed for galvanizing and for alloys such as brass; lead, which is used in batteries, scuba divers' weight belts, and as shielding for atomic reactors; silver, which is needed for coinage and in industry; and gold, with its monetary and industrial uses.

IMPACT ON INDUSTRY

Governmental agencies and numerous nongovernmental agencies play an active role in overseeing and regulating marine mining, partly because it is done in the sea.

Governmental Agencies. The International Seabed Authority (ISA), headquartered in Kingston, Jamaica, was established in 1982 under the United Nations Convention of the law of the sea. The 159 nations that ratified the law of the sea make up the membership of the ISA. The organization's mission is to supervise and control activities regarding the seabed and the subsoil below it whenever these activities lie beyond the limits of any single nation's jurisdiction. The law of the sea gives each nation full jurisdiction over its territorial waters, which extend twelve nautical miles out from the nation's shores. In addition, each nation has an exclusive economic zone (EZZ) that extends offshore for two hundred nautical miles, or even more if the edge of the continental shelf lies farther out than two hundred nautical miles. Within the boundaries of its economic zone, each nation is able to regulate fishing policies, the extraction of mineral resources, the control of pollution, and scientific research.

The ISA is developing regulations that will govern commercial prospecting for minerals in the part of the seafloor outside the various member nations' exclusive economic zones. The ISA has the power to make licensing decisions for this area, and part of the profits from mining operations in this area will be used for the benefit of the world's developing nations. Although the United States participated in the writing of the law of the sea and signed the treaty that implemented it, the U.S. Senate has never ratified the treaty. As a result, the United States is not a member nation of the ISA and therefore has only observer status at the organization's meetings.

Industry and Business. An industry-sponsored organization concerned with marine mining is the International Marine Minerals Society, headquartered in Honolulu. The society sponsors an international forum, the Underwater Mining Institute, where leaders from universities, industry, and government agencies can exchange ideas and recommend policies regarding marine mining. The society also publishes a scientific journal, *Marine Georesources and Geotechnology*, in which researchers can share findings related to seafloor resources.

Another example of an industry-sponsored organization in marine mining is the American Petroleum Institute, the leading trade association for the oil and gas industry, founded in 1919 and headquartered in Washington, D.C. The institute has nearly four hundred corporate members from all segments of the industry: producing companies, refiners, suppliers such as drillers and electric log companies, pipeline operators, and marine transporters. Its mission is to represent the oil and gas industry to the public, the U.S. Congress, the executive branch, state governments, and the media. It also negotiates with regulatory agencies regarding rules and fees and represents the industry in legal proceedings. Among the services it provides for members are sponsoring research projects, collecting and publishing statistics regarding oil and gas production, setting standards for the industry and recommending approved procedures, certifying equipment as safe and individuals as competent, and running educational programs.

CAREERS AND COURSE WORK

The only marine mining done in the United States as of 2010 is the dredging of sand and gravel for beach nourishment and oil and gas production in

---

Fascinating Facts About Marine Mining

- Salt is used to make homemade ice cream because salt lowers the freezing point of water. This is also the reason that salt is put on roads in the northern part of the United States for snow removal and why northern harbors do not freeze when temperatures dip below 32 degrees Fahrenheit.
- Offshore sand is frequently used to replace the sand eroded from a beach in a process known as beach nourishment. The cost of dredging this sand can be high—as much as several million dollars a mile—and there is always a danger that the next storm will wash the sand away.
- De Beers Marine Namibia is mining diamonds off the shores of Africa with specialized ships that serve as mines.
- According to the U.S. Geological Survey, there may be more organic carbon in methane hydrate than in all the coal, oil, and nonhydrate natural gas in the entire world.
- Scientists believe that some large sulfide deposits on land may have originally been formed in the deep sea and thrust up by land movements. The ancient Romans got their copper from the thirty large sulfide deposits on the island of Crete.
- Nautilus Minerals, a Canadian company, is planning to mine sulfide deposits on the ocean floor near Papua New Guinea for copper, gold, silver, and zinc. Its actions have been the focus of some criticism by scientists.

---

the Gulf of Mexico, along the Southern California coast, and off the north slope of Alaska. Dredging jobs are limited because the work is intermittent and seasonal and takes place largely in resort areas where beach erosion is a problem. Advanced training and professional degrees are not required. A captain is needed to run the dredge, a hydrographic surveyor has to chart the bottom and locate the deposits of sand and gravel, a mechanic has to keep the equipment running, and deckhands are required to do routine tasks. By contrast, a wide variety of jobs are available in the oil and gas industry, both on land and at sea.

A good place for the beginner to start is with an undergraduate, college-level course in geology. This could lead to graduate work in geology and a career in oil and gas exploration, or to a degree in petroleum engineering and a job designing, constructing, and maintaining offshore drilling platforms. The oil industry also offers jobs doing seismic surveying, drilling and logging wells, and producing, refining, and marketing the oil that is produced. Management positions also are available, so courses in business management are also recommended

SOCIAL CONTEXT AND FUTURE PROSPECTS

As glaciers and ice sheets melt because of global warming, the water added to the sea by this melting will result in rising sea levels. This, in turn, will increase the amount of beach erosion around the world and consequently the need to nourish beaches in resort areas. Jobs in the dredging industry should continue to be available as a result. Jobs in the oil and gas industry will continue to be available as well. More than three-quarters of world oil production goes into powering cars, trucks, ships, and planes, so the demand for oil is likely to grow as more and more people in the developing nations are able to afford to buy cars.

One problem that may lie ahead in marine mining involves the manganese nodules and sulfide crusts found on the seafloor. The developed nations are already investigating ways to mine these deposits. They are the only nations with the technological expertise to bring up these nodules and crusts. However, the law of the sea states that the profits from mining minerals in the area in which the nodules and crusts are found must be shared with the developing nations, and the developed nations may find it hard to agree to this condition.

Donald W. Lovejoy, B.S., M.S., Ph.D.

FURTHER READING

Drew, Lisa W. "The Promise and Peril of Seafloor Mining: Can Minerals Be Extracted from the Seafloor Without Environmental Impact?" *Oceanus* 47, no. 3 (2009): 8-14. Summarizes the way in which seafloor mineral deposits form, the metals they contain, how they can be mined, and the environmental damage that might result.

_____. "Who Regulates Mining on the Seafloor?" *Oceanus* 47, no. 3 (2009): 15. Describes the role played by the International Seabed Authority and the International Marine Minerals Society in overseeing seafloor mining and determining the rules that govern it.

Garrison, Tom. *Oceanography: An Invitation to Marine Science.* Belmont, Calif.: Thomson Brooks/Cole, 2007. An excellent introduction to the world's oceans; chapters cover the seafloor and its structures, the many marine resources, and the environmental concerns regarding the extraction of these resources.

McLachlan, Anton, and Alec Brown. *The Ecology of Sandy Shores.* Burlington, Mass.: Elsevier/Academic Press, 2006. Discusses beach erosion and the problems caused by mining the beach and the offshore zone; has an extensive treatment of beach nourishment and its environmental effects.

Rice, Tony. *Deep Ocean.* Washington, D.C.: Smithsonian Institution Press, 2009. An excellent introduction to the deep-ocean environment: profusely illustrated, and with sections on oil and gas as well as the metal-rich nodules and sulfide crusts.

Welland, Michael. *Sand: The Never-Ending Story.* Berkeley: University of California Press, 2009. A good summary of the ways sand is important; several examples of beach nourishment are given, as well as examples of the problems that beach nourishment may cause.

WEB SITES

American Petroleum Institute
http://www.api.org

International Marine Minerals Society
http://www.immsoc.org

International Seabed Authority
http://www.isa.org.jm

Underwater Mining Institute
http://www.underwaterming.org

See also: Gemology and Chrysology; Mineralogy; Oceanography.

MEASUREMENT AND UNITS

FIELDS OF STUDY

Metrology; agriculture; geographic information systems (GIS); geography; history; sociology; ballistics; surveying; military applications and service; navigation; construction; civil engineering; mechanical engineering; chemical engineering; physics; biochemical analysis.

SUMMARY

Measurement is the act of quantifying a physical property, an effect, or some aspect of them. Seven fundamental properties are recognized in measurements: length, mass, time, electric current, thermodynamic temperature, amount of a substance, and luminous intensity. In addition, two supplementary or abstract fundamental properties are defined: plane and solid angles. The base units for the seven fundamental properties can be manipulated to produce derived units for other quantities that are the effect of combinations of these fundamental properties. For instance, a Newton is a derived unit measuring force and weight, and a square meter is a derived unit used to measure area.

Historically, a number of units representing different amounts of the same properties have been used in various cultures around the world. For example, the United States traditionally uses the U.S. customary system (miles, cups, pints, ounces, and so on), while most other industrialized nations use the metric system (kilometer, milliliter, liters, grams, and so on). The metric system, developed in France in the late eighteenth century, represents the first true standard measurement system. The theory and physical practice of measurement is constant no matter what system of units is being used.

KEY TERMS AND CONCEPTS

- **Base Unit:** Unit of measurement that is used for one of the fundamental properties, also known as defined unit.
- **Calibration:** Process of ensuring that the measurements obtained with a specific measuring device agree with the value established by the base unit of that measured property or effect.
- **Conversion Factor:** Ratio of one unit of measure in a particular system to the unit of measure of the same property in a different system of measurements.
- **Derived Unit:** Unit of measurement that is derived from the relationship of base units.
- **Dimensional Analysis:** Consideration and manipulation of the units involved in a measurement and their relationship to one another.
- **Fundamental Property:** Property ascribed to a phenomenon, substance, or body that can be quantified: length, mass, time, electric current, thermodynamic temperature, the amount of a substance, and luminous intensity.
- **International Standard:** Unit of measurement whose consistent value is recognized as a common unit of measurement by international agreement.
- **Measurement:** Ascribing of a value to the determination of the quantity of a physical property.
- **Standard Temperature And Pressure (STP):** Value of the mean atmospheric pressure and mean annual temperature at sea level, chosen by international agreement to be designated as 1 standard atmosphere or 760 torr and 0 degrees Celsius/32 degrees Fahrenheit or 273.15 degrees Kelvin.
- **Unit:** Single definitive basic magnitude of portions used to measure various properties.

DEFINITION AND BASIC PRINCIPLES

Measurement has the purpose of associating a dimension or quantity proportionately with some fixed reference standard. Such an association is intended to facilitate the communication of information about a physical property in a manner that allows it to be reproduced in concept, or in actuality if needed. The function of an assigned unit that is associated with a definite dimension is to provide the necessary point of reference for someone to comprehend the exact dimensions that have been communicated. For example, a container may be described as having a volume of eight cubic feet. Such a description is incomplete, however, because it does not state the shape or relative proportions of the container. The description applies equally well to containers of any shape, whether they be cubic, rectangular, cylindrical, conical, or some other shape. At a more basic level, however, there is the assumption that the person who receives the description has the same understanding of what is meant by "cubic feet" as does the person who provided the description. This is the fundamental principle of any measurement system: to provide a frame of reference that is commonly understood by a large number of people and that indicates exactly the same thing to each of them.

A measurement system, no matter what its basic units, must address a limited group of fundamental properties. These are length, mass, temperature, time, electric current, amount of a substance, and luminous intensity. In addition, it must also be able to describe angles. All other properties and quantities can be described or quantified by a combination of these fundamental properties.

In practice, any of these fundamental properties can be defined relative to any randomly selected relevant object or effect. Logically, though, for a measurement system to be as effective as possible, the objects and effects that are selected as the defined units of fundamental properties must be readily available to as many people as possible and readily reproducible. If, for example, a certain king were to decree that the "foot" to be used for measurement was the human foot, a great deal of confusion would result because of the variability in the size of the human foot from person to person and from a person's left foot to right foot. Should he instead decree that the "foot" would correspond to his own right foot and no other, the unit of measurement becomes significantly more precise, but at the same time, the decree raises the problem of how to verify that measurements being taken are in fact based on the decreed length of the foot. A physical model of the length of the king's right foot must then be made available for comparison. The same logic holds true for any and all defined properties and units. All measurements made within the definitions of a specific system of measurement are therefore made by comparing the proportional size of an effect or property to the defined standard units.

BACKGROUND AND HISTORY

Historically, measurement systems were generally based on various parts of the human body, and some of these units have remained in use in the modern world. The height of horses, for example, is generally given as being so many hands high. In many

languages, the word for "thumb" and the word for "inch" are closely related, if not identical: In French, both words are *pouce*; in Hungarian, *hüvelyk* and *hüvelykujj*; in Norwegian, *tomme* and *tommelfinger*; and in Swedish, *tum* and *tumme*. Although using the thumb as a basis for measurement is convenient in that almost everyone has one, in practice, the generally accepted thumb size tended to vary from city to city, making it impossible to interchange parts made in different locations by separate artisans.

Units of measurement traditionally have been defined by a decree issued by a political leader. In ancient times, units of length often corresponded to certain parts of the human body: The foot was based on the human foot and the cubit on the length of the forearm. Other units used to weigh various goods were based on the weight of commonly available items. Examples include the stone (still used widely but not officially in Britain) and the grain. Invariably, the problem with such units lay in their variability. A stone weight may be defined as equivalent to the size of a stone that a grown man could enclose by both arms, but grown men come in different sizes and strengths, and stones come in different densities. A grain of gold may be defined as equivalent to the weight of a single grain of wheat; however, in drier years, when grains of wheat are smaller and lighter, the worth of gold is significantly different from its worth in years of good rainfall, when grains of wheat are larger and heavier.

HOW IT WORKS

Measurements and units are most useful when they are standardized so that measurements mean the same thing to everyone and are comparable. One method of solving the problem of variability in measurement is to establish a standard value for each unit and to regularly compare all measuring devices to this standard.

Historically, many societies required that measuring devices be physically compared with and calibrated against official standards. Physical representations of measuring units were made as precisely as possible and carefully stored and maintained to serve as standards for comparison. In ancient Egypt, the standard royal cubit was prepared as a black granite rod and most likely kept as one of the royal treasures by the pharaoh's chief steward. That simple stone object would have been accorded such a status because of its economic value to trade and construction and because its helped maintain the pharaoh's reputation as the keeper of his kingdom.

Later nations and empires also kept physical representations of most standard units appropriate to the economics of trade. The British made and kept definitive representations of the yard and the pound, just as the French made and kept definitive representations of the meter and the kilogram after developing the metric system in the late 1700's. Early on, countries recognized that the representative of the unit must not be subject to change or alteration. The Egyptian standard royal cubit was made of black granite; the standard meter and the kilogram, as well as the foot and the pound, were made from a platinum alloy so that they could not be altered by corrosion or oxidation.

France adopted the metric system as its official measuring system in 1795, and that system was standardized in 1960 as the International System of Units (Le Système International d'Unités; known as SI). The units of the metric system were defined based on unchanging, readily reproducible physical properties and effects rather than any physical object. For example, the standard SI unit of length, the meter, was originally defined to be one ten-millionth of the distance from the equator to the north polar axis along the meridian of longitude that passed through Paris, France. In 1960, for greater accuracy, the definition of the meter was based on a wavelengths of light emitted by the krypton-86 isotope. In 1983, it was changed to the length traveled by light in a vacuum during $1/299,792,458$ of a second. Similarly, the length of a standard second of time had been defined as a fraction of one rotation of the Earth on its axis, until it was realized that the rate of rotation was not constant but rather was slowly decreasing, so that the length of a day is increasing by 0.0013 second per hundred years. In 1967, it was formally redefined as a duration of 9,192,631,770 periods of the radiation corresponding to the transition between the two hyperfine levels of the ground state of the cesium-133 atom. As technology develops, it becomes possible to measure smaller quantities with finer precision. This capability has been the principal that permits ever more precise definitions of the basic units of measurement.

The application of measuring procedures is of fundamental importance in the economics of trade, especially in the modern global economy. In manufacturing, engineering, the sciences, and other fields, the accuracy and precision of measurement are essential to statistical process control and other quality-control techniques. All such measurement is a process of comparing the actual dimensions or properties of an item with its ideal or design dimensions or properties. The definition of a standard set of measuring units greatly facilitates that process.

APPLICATIONS AND PRODUCTS

It is quite impossible to calculate the economic effects that various systems of measurement, both good and bad, have had throughout history. Certainly, commonly understood and accepted units of weight, distance, and time have played a major role in facilitating trade between peoples for thousands of years. In many ways, all human activity can be thought of as dependent on measurement.

The study of measurement and measurement processes is known as metrology. In essence, metrology is the determination and application of more precise and effective means of measuring quantities, properties, and effects. The value in metrology derives from how the information obtained is used. This is historically and traditionally tied to the concepts of fair trade and of well-made products. Measurement serves to ensure that trade is equitable, that people get exactly what they are supposed to get in exchange for their money, services, or other trade goods, and that as little goes to waste as possible.

By far, the largest segment of metrology deals with the design, production, and calibration of the various products and devices used to perform measurements. These devices range from the simplest spring scale or pan balance to some of the most sophisticated and specialized scientific instruments ever developed. In early times, measurements were restricted to those of mass, distance, and time because these were the foremost quantities used in trade. The remaining fundamental properties of electric current, amount of a substance, temperature, and luminous intensity either remained unknown or were not of consequence.

Weight Determination. Originally, weights were determined with relative ease by the use of the pan balance. In the simplest variation of this device, two pans are suspended from opposite ends of a bar in such a way that they are at equal heights. The object to be weighed is placed in one of the pans and objects of known weights are placed in the other pan until the two pans are again at equal heights. The precision of the method depends on the ability of the bar to pivot as freely as possible about its balance point; any resistance will skew the measurement by preventing the pans from coming into proper balance with each other.

Essentially, all balances operate on the principle of comparing the weight of an unknown object against the weight of an accurately known counterweight, or some property such as electric current that can in turn be measured very accurately and precisely. The counterweight may not actually be a weight but rather an electronic pressure sensor or something that can be used to indicate the weight of an object such as the tension of a calibrated spring or a change in electrical resistance. Scales used in commercial applications, such as those in grocery stores, grain depots, and other trade locations, are inspected and calibrated on a regular basis according to law and regulations that govern their use.

Length Determination. For many practical purposes, a linear device such as a scale ruler or tape measure is all that is needed to measure an unknown length. The device used should reflect the size of the object being measured and have scale markings that reflect the precision with which the measurement must be known. For example, a relatively small dimension being measured in a machine shop would be measured by a trained machinist against a precision steel scale ruler with dimensional markings of high precision. Training in the use of graduated scale markings typically enables the user to read them accurately to within one-tenth of the smallest division on the scale. High-precision micrometers are generally used to make more precise measurements of smaller dimensions, and electronic versions of such devices provide the ability to measure dimensions to extremely precise tolerances. Smaller dimensions, for which the accuracy of the human eye is neither sufficient nor sufficiently reproducible, are measured using microscope techniques and devices. Accurate measurements of larger dimensions have always been problematic, especially when the allowable tolerance of the measurement is very small. This has been overcome in many cases by the measuring machine, a semirobotic device that uses electronic control and logic programming to determine the distance between precise points on a specific object.

The 1960 definition of the meter was achieved through the use of a precision interferometer, a device that uses the interference pattern of light waves such that the number of wavelengths of a specific frequency of light can be counted. Using this device, the meter was precisely defined as the distance equal to 1,650,763.73 wavelengths of the $2p_{10}$-$5d_5$ emission of krypton-86 atoms. In 1983, however, the meter was defined as the distance traveled by light in vacuum in 1/299,792,548 of a second.

Time Measurement. Essentially, the basic unit of time measurement in all systems is the second. This is a natural consequence of the fact that the length of a day is the same everywhere on the planet. The natural divisions of that period of time according to the observed patterns of stars and their motions almost inevitably results in twenty-four equal divisions. A natural result of the metric system is that a pendulum that is one meter in length swings with a period of one second. Pendulum clockworks have been used to measure the passage of time, coordinated to the natural divisions of the day, for thousands of years. More precise time measurements have become possible as better technology has become available. With the development of the metric system, the unit duration of one second was defined to correspond to the appropriate fraction of one rotation of the planet. Until the development of electronic methods and devices, this was a sufficient definition. However, with the realization that the rotation of the planet is not constant but is slowly decreasing, the need to redefine the second in terms that remain constant in time led to its redefinition in 1967. According to the new definition, a second is the time needed for a cesium-133 atom to perform 9,192,631,770 complete oscillations.

Temperature Determination. Of all properties, measurement of thermodynamic temperature is perhaps the most relative and arbitrary of all. The thermometer was developed before any of the commonly used temperature scales, including the Fahrenheit, Celsius, Kelvin, and Rankine scales. All are based on the freezing and boiling points of water, the most readily available and ubiquitous substance on the planet. The Fahrenheit scale arbitrarily set the freezing point of saturated salt water as 0 degrees. The physical dimensions of the scale used on Fahrenheit's thermometer resulted in the establishment of the boiling point of pure water as 212 degrees. The Celsius scale designated the freezing point of pure water to be 0 degrees and the boiling point of pure water to be 100 degrees. The relative sizes of Fahrenheit and Celsius degrees are thus different by a factor of 5 to 9. Conversion of Fahrenheit temperature to Celsius temperature is achieved by subtracting 32 degrees and multiplying the result by 5/9; to convert from Celsius to Fahrenheit, first multiply by 9/5, then add 32 degrees to the result. The temperature of -40 degrees is the same in both scales.

Each temperature scale recognizes a physical state called absolute zero, at which matter contains no thermal energy whatsoever. This must physically be the same state regardless of whether Fahrenheit or Celsius degrees are being used. Because of this, two other scales of temperature were developed. The Rankine scale, established as part of the school of British engineering, uses the Fahrenheit degree scale, beginning at 0 degrees at absolute zero. The Kelvin scale uses the Celsius degree scale, beginning at 0 degrees at absolute zero.

Temperature is accurately measured electronically by its effect on light in the infrared region of the electromagnetic spectrum, although less accurate physical thermometers remain in wide use.

Amount of a Substance. Of all the fundamental properties, amount of a substance is the least precisely known. The concept is intimately linked to the modern atomic theory and atomic weights, although it predates them by almost a hundred years. Through studies of the properties of gases in the early 1800's, Italian Amedeo Avogadro concluded that a quantity of any pure material equal to its molecular weight in grams contained exactly the same number of particles. This number of particles came to be referred to as the Avogadro constant. Thus, 2 grams of hydrogen gas (molecular weight = 2) contains exactly the same number of molecules as does 342 g of sucrose (common white table sugar, molecular weight = 342) or 18 milliliters of water (molecular weight 18, density = 1 gram per milliliter). Calculations have determined the value of the Avogadro constant to be about 6.02214×10^{23}. Because all twenty-three decimal places are not known, the absolute value has not yet been determined, and therefore amount of a substance is the least precisely known unit of measurement.

The amount of material represented by the Avogadro constant (of atoms or molecules) is termed the "mole," a contraction from "gram molecular weight." The number is constant regardless of the system of measurement that is used, as a result of the indivisibility of the atom in modern chemical theory. Thus a gram molecular weight of a substance and a pound molecular weight of a substance both contain a constant number of molecules.

Electric Current. Of course, atoms are not indivisible in fact. They consist of a nucleus of protons and neutrons, surrounded by a cloud of electrons. The electrons can move from atom to atom through matter, and such movement constitutes an electric current. More generally, any movement of electronic charge between two different points in space defines an electric current. Although some controversial evidence suggests that electricity may have been known in ancient times, any serious study of electricity did not begin until the eighteenth century, after the discovery of the electrochemical cell. At that time, electricity was thought to be a mysterious fluid that permeated matter. With the discovery of subatomic particles (electrons and protons) in 1898 and the subsequent development of the modern atomic theory, the nature of electric currents came to be better understood. The ampere, named after one of the foremost investigators of electrical phenomena, is the basic unit of electric current and corresponds to the movement of one mole of electrons for a period of one second.

The development of the transistor—and the electronic revolution that followed—made it possible to measure extremely small electric currents of as little as 10^{-9} amperes, as well as corresponding values of voltage, resistance, induction, and other electronic functions. Basically, this made it possible to precisely measure fundamental properties by electronic means rather than physical methods.

Measurement of Luminous Intensity. Until it became important and necessary to know precisely the intensity of light being emitted from a light source, particularly in the fields of astronomy and physics, there was no need for a fundamental unit of luminous intensity. Light intensities were generally compared, at least in post-Industrial Revolution Europe, to the intensity of light emitted from a candle. This sufficed for general uses such as lightbulbs, but the innate variability of candle flames made them inadequate for precision measurements. The candela was set as the standard unit of luminous intensity and corresponds to the output of energy of 0.00146 watts. Modern lightbulbs typically are rated at a certain number of watts, but this is a measure of the electric power that they consume and not of their luminous intensity.

IMPACT ON INDUSTRY

Metrology is central to industry. The ability to measure quantities accurately and precisely is an absolute necessity given the scale of modern applications from nanotechnology to celestial mechanics. The tolerances of the dimensions required for modern devices and machines makes it imperative that measurement be on the same scale. If a dimension must be accurate to within a very narrow range of values, then the

metrologist's limit of measurement of that dimension must also be within that range of values.

As analytical techniques and methods are designed to identify and quantify ever smaller amounts of materials, ever faster processes, ever finer distinctions in molecular structure, ever more distant stellar emissions, and so on, metrology must also progress in its capabilities. Specialized applications aside, the most important aspect of metrology in modern society is its applicability in the fields of medicine and manufacturing. Medical testing is carried out on a greater number of people than ever before. Routine analytical procedures must identify precisely and definitively the amount of a specific material that may be present in extremely small quantities in a sample. The analytical method used may require measurement or control of any of the properties of time, distance, luminous intensity, amount of material, electric current, weight, and temperature in the quantification of a specific material.

In industrial applications, metrology is central to the techniques of statistical process control (SPC). The basic goal of SPC is the elimination of product units that are outside of their respective design parameters. The units are closely monitored as they are produced and checked against their design standards. Interpretation of the nature and number of variations in the products is used to adjust and control the process by which they are produced. SPC and other quality-control methods are based on statistical analysis, which requires the measurement of several dimensions and aspects of product units. In a simple example, a machined part such as a fluid valve body, produced in a mass-production machining facility, may have been designed with two machined surfaces that must be parallel to each other within 0.01 millimeter and with three very precisely placed holes of a specific diameter. The quality-control program will call for measurement of those respective dimensions and comparison to the ideal design. The appearance of units with surfaces that are no longer suitably parallel or with holes that do not conform in size or in shape call for the adjustment of the machinery so that subsequent product units are again within design parameters. The value of this process is in the reduction of costs through the elimination of waste.

CAREERS AND COURSE WORK

Measurement and units, as they apply to metrology, are part of every technical and scientific field. The student planning on a career in any of these areas can expect to learn how to use the metric system and the many ways in which measurement is applied in a specific field of study. Because measurement is related to the fundamental properties of matter, a good understanding of the relationships between measurements and properties will be essential to success in any chosen field. In addition, the continuing use of multiple measuring systems, such as the metric, U.S. customary, and British Imperial systems, in the production of goods means that an understanding of those systems and how to convert between them will be necessary in many careers.

Specific courses that involve measurement include geometry and mathematics, essentially all of the physical science and technology courses, and design and technical drawing courses. Geometry, which literally means "earth measurement," is the quintessential mathematics of measurement and is essential for careers in land surveying, architecture, agriculture, civil engineering, and construction. Gaining an understanding of trigonometry and angular relationships is particularly important. The physical sciences, such as physics, chemistry, and geology, employ analytical measurement at all levels. Specializations in which measurement plays a prominent role are analytic chemistry and forensic research. Technology programs such as mechanical engineering, electronic design, and biomedical technology all rely heavily on measurement and the application of metrological techniques in the completion of projects that the engineer or technologist undertakes.

SOCIAL CONTEXT AND FUTURE PROSPECTS

Measurement and an understanding of the units of measurement are an entrenched aspect of modern society, taught informally to children from early childhood and formally throughout the course of their schooling. They are fundamental to the continued progress of technology and essential to the determination of solutions to problems as they arise, as well as to the development of new ideas and concepts. The need for individuals who are trained in metrology and who understand the relationship and use of measurements and units will be of increasing importance in ensuring the viability of both new and established industries.

Particular areas of growth and continuing development include the fields of medical research and analysis, transportation, mechanical design, and aerospace. The accuracy and precision of measurement in these fields is critical to the successful outcome of projects.

Richard M. J. Renneboog, M.Sc.

FURTHER READING

Butcher, Kenneth S., Linda D. Crown, and Elizabeth J. Gentry. *The International System of Units (SI): Conversion Factors for General Use.* Gaithersburg, Md.: National Institute of Standards and Technology, 2006. A guide to the International System of Units produced by the U.S. government.

Cardarelli, Francois. *Encyclopedia of Scientific Units, Weights and Measures: Their SI Equivalences and Origins.* New York: Springer, 2003. Systematically reviews the many incompatible systems of measurement that have been developed throughout history. It clearly relates those quantities to their modern SI equivalents and provides conversion tables for more than 19,000 units of measurement.

Kuhn, Karl F. *In Quest of the Universe.* 6th ed. Sudbury, Mass.: Jones and Bartlett, 2010. This excellent introduction to astronomy illustrates the uses and importance of measurement in that field.

Tavernor, Robert. *Smoot's Ear: The Measure of Humanity.* New Haven, Conn.: Yale University Press, 2007. An entertaining yet scholarly work that discusses measuring systems and the act of measurement in the context of the societies in which the systems were developed.

WEB SITES

Bureau International des Poids et Mesures
http://www.bipm.org

National Aeronautics and Space Administration
A Brief History of Measurement Systems
http://standards.nasa.gov/history_metric.pdf

National Institute of Standards and Technology
Weights and Measures
http://www.nist.gov/ts/wmd/index.cfm

See also: Applied Mathematics; Barometry; Electrical Measurement; Geometry; International System Of Units; Laser Interferometry; Quality Control; Temperature Measurement; Time Measurement; Weight and Mass Measurement.

MECHANICAL ENGINEERING

FIELDS OF STUDY

Acoustics; algebra; applied mathematics; calculus; chemistry; control theory; drafting; dynamics; economics; electronics; fluid dynamics; graphics; heat transfer; kinematics; materials science; mechanics; optics; physics; product design; robotics; statics; structural analysis; system design; thermodynamics.

SUMMARY

Mechanical engineering is the field of technology that deals with engines, machines, tools, and other mechanical devices and systems. This broad field of innovation, design, and production deals with machines that generate and use power, such as electric generators, motors, internal combustion engines, and turbines for power plants, as well as heating, ventilation, air-conditioning, and refrigeration systems. In many universities, mechanical engineering is integrated with nuclear, materials, aerospace, and biomedical engineering. The tools used by scientists, engineers, and technicians in other disciplines are usually designed by mechanical engineers. Robotics, microelectromechanical systems, and the development of nanotechnology and bioengineering technology constitute a major part of modern research in mechanical engineering.

KEY TERMS AND CONCEPTS

- **Computer-Aided Design (CAD):** Using computer software to design objects and systems and develop and check the graphical representation of a design.
- **Computer-Aided Manufacturing (CAM):** Use of computer software and software-guided machines to perform manufacturing operations starting with computer-aided design.
- **Dynamics:** Application of the laws of physics to determine the acceleration and velocity of bodies and systems.
- **Engineering Economics:** Science focused on determining the best course of action in designing and manufacturing a given system to reduce uncertainty and maximize the return on the investment.
- **Fluid Mechanics:** Science describing the behavior of fluids.

- **Heat Transfer:** Science relating the flow of thermal energy to temperature differences and material properties.
- **Kinematics:** Science of determining the relationship between the movement of different elements of a system, such as a machine.
- **Machine Design:** Science of designing the most suitable individual elements and their integration into machines, relating the stresses and loads that the machine must handle, its lifetime, the properties of the materials used, and the cost of manufacturing and using the machine.
- **Manufacturing:** Generation of copies of a product, based on good design, in sufficient numbers to satisfy demand at the best return on investment.
- **Materials Science:** Study of the detailed structure and properties of various materials used in engineering.
- **Mechanical System:** Grouping of elements that interact according to mechanical principles.
- **Metrology:** Science of measurement dealing with the design, calibration, sensitivity, accuracy, and precision of measuring instruments.
- **Robotics:** Science of designing machines that can replace human beings in the execution of specific tasks, such as physical activities and decision making.
- **Statics:** Science of how forces are transmitted to, from, and within a structure, determining its stability.
- **Strength Of Materials:** Science of determining the deflection of objects of different shapes and sizes under various loading conditions.
- **Thermodynamics:** Science dealing with the relationships between energy, work, and the properties of matter. Thermodynamics defines the best performance that can be achieved with power conversion, generation, and heat transfer systems.

DEFINITION AND BASIC PRINCIPLES

Mechanical engineering is the field dealing with the development and detailed design of systems to perform desired tasks. Developed from the discipline of designing the engines, power generators, tools, and mechanisms needed for mass manufacturing, it has grown into the broadest field of engineering, encompassing or touching most of the disciplines of science and engineering. Mechanical engineers take the laws of nature and apply them using rigorous mathematical principles to design mechanisms. The process of design implies innovation, implementation, and optimization to develop the most suitable solution to the specified problem, given its constraints and requirements. The field also includes studies of the various factors affecting the design and use of the mechanisms being considered.

At the root of mechanical engineering are the laws of physics and thermodynamics. Sir Isaac Newton's laws of motion and gravitation, the three laws of thermodynamics, and the laws of electromagnetism are fundamental to much of mechanical design.

Starting with the Industrial Revolution in the nineteenth century and going through the 1970's, mechanical engineering was generally focused on designing large machines and systems and automating production lines. Ever-stronger materials and larger structures were sought. In the 1990's and first part of the twenty-first century, mechanical engineering saw rapid expansion into the world of ever-smaller machines, first in the field of micro and then nano materials, probes and machines, down to manipulating individual atoms. In this regime, short-range forces assume a completely different relationship to mass. This led to a new science integrating electromagnetics and quantum physics with the laws of motion and thermodynamics. Mechanical engineering also expanded to include the field of system design, developing tools to reduce the uncertainties in designing increasingly more complex systems composed of larger numbers of interacting elements.

BACKGROUND AND HISTORY

The engineering of tools and machines has been associated with systematic processes since humans first learned to select sticks or stones to swing and throw. The associations with mathematics, scientific prediction, and optimization are clear from the many contraptions that humans developed to help them get work done. In the third century B.C.E., for example, the mathematician Archimedes of Syracuse was associated with the construction of catapults to hurl projectiles at invading armies, who must themselves have had some engineering skills, as they eventually invaded his city and murdered him. Tools and weapons designed in the Middle Ages, from Asia to

Engineers machine a turbine shaft with a machine tool. (Maximilian Stock Ltd./Photo Researchers, Inc.)

Europe and Africa, show amazing sophistication. In the thirteenth century, Mesopotamian engineer Al-Jazari invented the camshaft and the cam-slider mechanism and used them in water clocks and water-raising machines. In Italy, Leonardo da Vinci designed many devices, from portable bridges to water-powered engines.

The invention of the steam engine at the start of the Industrial Revolution is credited with the scientific development of the field that is now called mechanical engineering. In 1847, the Institution of Mechanical Engineers was founded in Birmingham, England. In North America, the American Society of Civil Engineers was founded in 1852, followed by the American Society of Mechanical Engineers in 1880. Most developments came through hard trial and error. However, the parallel efforts to develop retrospective and introspective summaries of these trials resulted in a growing body of scientific knowledge to guide further development.

Nevertheless, until the late nineteenth century, engineering was considered to be a second-rate profession and was segregated from the "pure" sciences. Innovations were published through societies such as England's Royal Society only if the author was introduced and accepted by its prominent members, who were usually from rich landed nobility. Publications came from deep intellectual thinking by amateurs who supposedly did it for the pleasure and amusement; actual hands-on work and details were left to paid professionals, who were deemed to be of a lower class. Even in America, engineering schools were called trade schools and were separate from the universities that catered to those desiring liberal arts educations focused on the classics and languages from the Eurocentric point of view.

Rigorous logical thinking based on the experience of hands-on applications, which characterizes mechanical engineering, started gaining currency with the rise of a culture that elevated the dignity of labor in North America. It gained a major boost with the urgency brought about by several wars. From the time of the American Civil War to World War I, weapons such as firearms, tanks, and armored ships saw significant advancements and were joined by airplanes and motorized vehicles that functioned as ambulances. During these conflicts, the individual heroism that had marked earlier wars was eclipsed by the technological superiority and scientific organization delivered by mechanical engineers.

Concomitantly, principles of mass production were applied intensively and generated immense wealth in Europe and America. Great universities were established by people who rose from the working classes and made money through technological enterprises. The Great Depression collapsed the established manufacturing entities and forced a sharp rise in innovation as a means of survival. New engineering products developed rapidly, showing the value of mechanical engineering. World War II and the subsequent Cold War integrated science and engineering inseparably. The space race of the 1960's through the 1980's brought large government investments in both military and civilian aerospace engineering projects. These spun off commercial revolutions in computers, computer networks, materials science, and robotics. Engineering disciplines and knowledge exploded worldwide, and as of 2011 there is little superficial difference between engineering curricula in most countries of the world.

The advent of the Internet accelerated and completed this leveling of the knowledge field, setting up sharper impetus for innovation based on science and engineering. Competition in manufacturing advanced the field of robotics, so that cars made by robots in automated plants achieve superior quality more consistently than those built by skilled master craftsmen. Manufacturing based on robotics can respond more quickly to changing specifications and demand than human workers can.

Beginning in the 1990's, micro machines began to take on growing significance. Integrated microelectromechanical systems were developed using the techniques used in computer production. One by one, technology products once considered highly glamorous and hard to obtain—from calculators to smart phones—have been turned into mass-produced commodities available to most at an affordable cost. Other products—from personal computers and cameras to cars, rifles, music and television systems, and even jet airliners—are also heading for commoditization as a result of the integration of mechanical engineering with computers, robotics, and micro electromechanics.

HOW IT WORKS

The most common idea of a mechanical engineer is one who designs machines that serve new and useful functions in an innovative manner. Often these machines appear to be incredibly complex inside or extremely simple outside. The process of accomplishing these miraculous designs is systematic, and good mechanical engineers make it look easy.

System Design. At the top level, system design starts with a rigorous analysis of the needs to be satisfied, the market for a product that satisfies those needs, the time available to do the design and manufacturing, and the resources that must be devoted. This step also includes an in-depth study of what has been done before. This leads to "requirements definition," where the actual requirements of the design are carefully specified. Experienced designers believe that this step already determines more than 80 percent of the eventual cost of the product.

Next comes an initial estimate of the eventual system characteristics, performed using simple, commonsense logic, applying the laws of nature and observations of human behavior. This step uses results from benchmarking what has been achieved before and extrapolating some technologies to the time when they must be used in the manufacturing of the design. Once these rudimentary concept parameters and their relationships are established, various analyses of more detailed implications become possible. A performance estimation then identifies basic limits and determines whether the design "closes," meeting all the needs and constraints specified at the beginning. Iterations on this process develop the best design. Innovations may be totally radical, which is relatively rare, or incremental in individual steps or aspects of the design based on new information, or on linking developments in different fields. In either case, extensive analysis is required before an innovation is built into a design. The design is then analyzed for ease and cost of manufacture. The "tooling," or specific setups and machines required for mass manufacture, are considered.

A cost evaluation includes the costs of maintenance through the life cycle of the product. The entire process is then iterated on to minimize this cost.

Fascinating Facts About Mechanical Engineering

- Robotic surgery enables surgeons to conduct very precise operations by eliminating the problems of hand vibrations and by using smaller steps than a human can take.

- The REpower 5M wind turbine in Germany, rated at 5 megawatts, is 120 meters high and has 61.5-meter radius blades, more than one and one-half times as long as each wing of an Airbus A380 jetliner.

- The General Electric H System integrates a gas turbine, steam turbine, generator, and heat-recovery steam generator to achieve 60 percent efficiency.

- The nanomotor built in 2003 by Alex Zettl, a physics professor at the University of California, Berkeley, and his research group is 500 nanometers in diameter with a carbon nanotube shaft 5 to 10 nanometers in diameter.

- The crawler-transporter used to move space shuttles on to the launch pad weighs 2,721 metric tons and has eight tracks, two on each corner. Its platform stays level when the crawler moves up a five-degree incline.

- Solar thermo-acoustic cooker-refrigerators use the heat from the Sun to drive acoustic waves in a tube, which convects heat and creates a low temperature on one side.

- Much of the world's telecommunications are carried by undersea fiber-optic cables that connect all continents except Antarctica. The first telegraph cable across the English Channel was laid in 1850.

- In 1650, Thomas Savery invented a steam engine to pump water out of coal mines.

The design is then passed on to build prototypes, thereby gaining more experience on the manufacturing techniques needed. The prototypes are tested extensively to see if they meet the performance required and predicted by the design.

When these improvements are completed and the manufacturing line is set up, the product goes into mass manufacture. The engineers must stay engaged in the actual performance of the product through its delivery to the end user, the customer, and in learning from the customer's experience in order to design improvements to the product as quickly as possible. In modern concurrent engineering practice, designers attempt to achieve as much as possible of the manufacturing process design and economic optimization during the actual product design cycle in order to shorten the time to reach market and the cost of the design cycle. The successful implementation of these processes requires both technical knowledge and experience on the part of the mechanical engineers. These come from individual rigorous fields of knowledge, some of which are listed below.

Engineering Mechanics. The field of engineering mechanics integrates knowledge of statics, dynamics, elasticity, and strength of materials. These fields rigorously link mathematics, the laws of motion and gravitation, and material property relationships to derive general relations and analysis methods. Fundamental to all of engineering, these subfields are typically covered at the beginning of any course of study.

In statics, the concept of equilibrium from Newton's first law of motion is used to develop free-body diagrams showing various forces and reactions. These establish the conditions necessary for a structure to remain stable and describe relations between the loads in various elements.

In dynamics, Newton's second law of motion is used to obtain relations for the velocity and acceleration vectors for isolated bodies and systems of bodies and to develop the notions of angular momentum and moment of inertia.

Strength of materials is a general subject that derives relationships between material properties and loads using the concepts of elasticity and plasticity and the deflections of bodies under various types of loading. These analyses help the engineer predict the yield strength and the breaking strength of various structures if the material properties are known. Metals were the preferred choice of material for engineering for many decades, and methods to analyze structures made of them were highly refined, exploiting the isotropy of metal properties. Modern mechanical engineering requires materials the properties of which are much less uniform or exotic in other ways.

Graphics and Kinematics. Engineers and architects use graphics to communicate their designs precisely and unambiguously. Initially, learning to draw on paper was a major part of learning engineering skills. As of 2011, students learn the principles of graphics using computer-aided design (CAD) software and computer graphics concepts. The drawing files can also be transferred quickly into machines that fabricate a part in computer-aided manufacturing (CAM). Rapid prototyping methods such as stereo lithography construct an object from digital data generated by computer graphics.

The other use of graphics is to visualize and perfect a mechanism. Kinematics develops a systematic method to calculate the motions of elements, including their dependence on the motion of other elements. This field is crucial to developing, for instance, gears, cams, pistons, levers, manipulator arms, and robots. Machines that achieve very complex motions are designed using the field of kinematics.

Robotics and Control. The study of robotics starts with the complex equations that describe how the different parts satisfy the equations of motion with multiple degrees of freedom. Methods of solving large sets of algebraic equations quickly are critical in robotics. Robots are distinguished from mere manipulator arms by their ability to make decisions based on the input, rather than depend on a telepresence operator for commands. For instance, telepresence is adequate to operate a machine on the surface of the Moon, which is only a few seconds of round-trip signal travel time from Earth using electromagnetic signals. However, the round-trip time for a signal to Mars is several minutes, so a rover operating there cannot wait for commands from Earth regarding how to negotiate around an obstacle. A fully robotic rover is needed that can make decisions based on what its sensors tell it, just as a human present on the scene might do.

Entire manufacturing plants are operated using robotics and telepresence supervision. Complex maneuvers such as the rendezvous between two space-craft, one of which may be spinning out of control, have been achieved in orbits in space, where the dynamics are difficult for a human to visualize. Flight control systems for aircraft have been implemented using robotics, including algorithms to land the aircraft safely and more precisely than human pilots can. These systems are developed using mathematical methods for solving differential equations rapidly, along with software to adjust parameters based on feedback.

Materials. The science of materials has advanced rapidly since the late twentieth century. Wood was once a material of choice for many engineering products, including bridges, aircraft wings, propellers, and train carriages. The fibrous nature of wood required considerable expertise from those choosing how to cut and lay sections of wood; being a natural product, its properties varied considerably from one specimen to another. Metals became much more convenient to use in design and fabrication because energy to melt and shape metals cheaply became available. Various alloys were developed to tailor machinery for strength, flexibility, elasticity, corrosion resistance, and other desirable characteristics. Detailed tables of properties for these alloys were included in mechanical engineering handbooks.

Materials used to manufacture mass-produced items have migrated to molded plastics made of hydrocarbons derived from petroleum. The molds are shaped using such techniques as rapid prototyping and computer-generated data files from design software. Composite materials are tailored with fiber bundles arrayed along directions where high-tensile strength is needed and much less strength along directions where high loads are not likely, thus achieving large savings in mass and weight.

Fluid Mechanics. The science of fluid mechanics is important to any machine or system that either contains or must move through water, air, or other gases or liquids (fluids). Fluid mechanics employs the laws of physics to derive conservation equations for specific packets of fluid (the Lagrangian approach) or for the flow through specified control volumes (Eulerian approach). These equations describe the physical laws of conservation of mass, momentum, and energy, relating forces and work to changes in flow properties. The properties of specific fluids are related through the thermal and caloric equations expressing their thermodynamic states. The speed of propagation of small disturbances, known as the speed of sound, is related to the dependence of pressure on density and hence on temperature. Various nondimensional groupings of flow and fluid properties—such as the Reynolds number, Mach number, and Froude number—are used to classify flow behavior. Increasingly, for many problems involving fluid flow through or around solid objects, calculations starting from the conservation equations are able to predict the loads and flow behavior reliably using the methods of computational fluid mechanics (CFD). However, the detailed prediction of turbulent flows remains beyond reach and is approximated through various turbulence models. Fluid-mechanic drag and the movements due to flow-induced pressure remain very difficult to calculate to the accuracy needed to improve vehicle designs.

Methods for measuring the properties of fluids and flows in their different states are important tools for mechanical engineers. Typically, measurements and experimental data are used at the design stage, well before the computational predictions become reliable for refined versions of the product.

Thermodynamics. Thermodynamics is the science behind converting heat to work and estimating the best theoretical performance that a system can achieve under given constraints. The three basic laws of temperature are the zeroth law, which defines temperature and thermal equilibrium; the first law, which describes the exchange between heat, work, and internal energy; and the second law, which defines the concept of entropy. Although these laws were empirically derived and have no closed-form proof, they give results identical to those that come from the law of conservation of energy and to notions of entropy derived from statistical mechanics of elementary particles traced to quantum theory. No one has yet been able to demonstrate a true perpetual-motion machine, and it does not appear likely that anyone will. From the first law, various heat-engine cycles have been invented to obtain better performance suited to various constraints. Engineers working on power-generating engines, propulsion systems, heating systems, and air-conditioning and refrigeration systems try to select and optimize thermodynamic cycles and then use a figure of merit—a means of evaluating the performance of a device or system against the best theoretical performance that could be achieved—as a measure of the effectiveness of their design.

Heat Transfer. Heat can be transferred through conduction, convection, or radiation, and all three modes are used in heat exchangers and insulators. Cooling towers for nuclear plants, heat exchangers for nuclear reactors, automobile and home air-conditioners, and the radiators for the International Space Station are all designed from basic principles of these modes of heat transfer. Some space vehicles are designed with heat shields that are ablative. The Thermos flask (which uses an evacuated space between two silvered glass walls) and windows with double and triple panes with coatings are examples of widely used products designed specifically to control heat transfer.

Machine Design. Machine design is at the core of mechanical engineering, bringing together the various disciplines of graphics, solid and fluid mechanics, heat transfer, kinematics, and system design in an organized approach to designing devices to perform specific functions. This field teaches engineers how to translate the requirements for a machine into a design. It includes procedures for choosing materials and processes, determining loads and deflections, failure theories, finite element analysis, and the basics of how to use various machine elements such as shafts, keys, couplings, bearings, fasteners, gears, clutches, and brakes.

Metrology. The science of metrology concerns measuring systems. Engineers deal with improving the accuracy, precision, linearity, sensitivity, signal-to-noise ratio, and frequency response of measuring systems. The precision with which dimensions are measured has a huge impact on the quality of engineering products. Large systems such as airliners are assembled from components built on different continents. For these to fit together at final assembly, each component must be manufactured to exacting tolerances, yet requiring too much accuracy sharply increases the cost of production. Metrology helps in specifying the tolerances required and ensuring that products are made to such tolerances.

Acoustics and Vibrations. These fields are similar in much of their terminology and analysis methods. They deal with wavelike motions in matter, their effects, and their control. Vibrations are rarely desirable, and their minimization is a goal of engineers in perfecting systems. Acoustics is important not only because minimizing noise is usually important, but also because engineers must be able to build machines to generate specific sounds, and because the audio signature is an important tool in diagnosing system status and behavior.

Production Engineering. Production engineering deals with improving the planning and implementation of the production process, designing efficient and precise tools to produce goods, laying out efficient assembly sequences and facilities, and setting up the flow of materials and supplies into the production line, and the control of quality and throughput rate. Production engineering is key to implementing the manufacturing step that translates engineering designs into competitive products.

APPLICATIONS AND PRODUCTS

Conventional Applications. Mechanical engineering is applied to the design, manufacture, and testing of almost every product used by humans and to the machines that help humans build those products. The products most commonly associated with mechanical engineering include all vehicles such as railway trains, buses, ships, cars, airplanes and spacecraft, cranes, engines, and electric or hydraulic motors of all kinds, heating, ventilation and air-conditioning systems, the machine tools used in mass manufacture, robots, agricultural tools, and the machinery in power plants. Several other fields of engineering such as aerospace, materials, nuclear, industrial, systems, naval architecture, computer, and biomedical developed and spun off at the interfaces of mechanical engineering with specialized applications. Although these fields have developed specialized theory and knowledge bases of their own, mechanical engineering continues to find application in the design and manufacture of their products.

Innovations in Materials. Carbon nanotubes have been heralded as a future super-material with strength hundreds of times that of steel for the same mass. As of the first decade of the twenty-first century, the longest strands of carbon nanotubes developed are still on the order of a few centimeters. This is a very impressive length-to-diameter ratio. Composite materials incorporating carbon already find wide use in various applications where high temperatures must be encountered. Metal matrix composites find use in primary structures even for commercial aircraft. Several "smart structures" have been developed, where sensors and actuators are incorporated into a material that has special properties to respond to stress and strain. These enable structures that will twist in a desired direction when bent or become stiffer or more flexible as desired, depending on

electrical signals sent through the material. Materials capable of handling very low (cryogenic) temperatures are at the leading edge of research applications. Magnetic materials with highly organized structure have been developed, promising permanent magnets with many times the attraction of natural magnets.

Sustainable Systems. One very important growth area in mechanical engineering is in designing replacements for existing heating, ventilation, and air-conditioning systems, as well as power generators, that use environmentally benign materials and yet achieve high thermodynamic efficiencies, minimizing heat emission into the atmosphere. This effort demands a great deal of innovation and is at the leading edge of research, both in new ways of generating power and in reducing the need for power.

IMPACT ON INDUSTRY

Having developed as a discipline to formalize knowledge on the design of machines for industry, mechanical engineering is at the core of most industries. The formal knowledge and skills imparted by schools of mechanical engineering have revolutionized human industry, bringing about a huge improvement in quality and effectiveness. The disciplined practice of mechanical engineering is responsible for taking innumerable innovations to market success. In the seventeenth through twentieth centuries, rampant industrialization destroyed many long-lasting community skills and occupations, replacing them with mass manufacturing concentrated and collocated with water resources, power sources, and transportation hubs. This has led to many problems as rural communities atrophied and their young people migrated to the unfamiliar and crowded environment of cities in search of well-paying jobs.

The effects on the environment and climate have also been severe. Heightened global concerns about the environment and climate change and new technological innovation may find mechanical engineers again at the head of a new revolution. This may start a drive to decentralize energy resources and production functions, permitting small communities and enterprises to flourish again.

CAREERS AND COURSE WORK

Mechanical engineers work in nearly every industry, in an innumerable variety of functions. The curriculum in engineering school accordingly focuses on giving the student a firm foundation in the basic knowledge that enables problem solving and continued learning through life. The core curriculum starts with basic mathematics, science, graphics and an introduction to design and goes on to engineering mechanics and the core subjects and specialized electives. In modern engineering schools, students have the opportunity to work on individual research and design projects that are invaluable in providing the student with perspective and integrating their problem-solving skills.

After obtaining a bachelor's degree, the mechanical engineer has a broad range of choices for a career. Traditional occupations include designing systems for energy, heating, ventilation, air-conditioning, pressure vessels and piping, automobiles, and railway equipment. Newer options include the design of bioengineering production systems, microelectromechanical systems, optical instrumentation, telecommunications equipment, and software. Many mechanical engineers also go on to management positions.

SOCIAL CONTEXT AND FUTURE PROSPECTS

Mechanical engineering attracts large numbers of students and offers a broad array of career opportunities. Students in mechanical engineering schools have the opportunity to range across numerous disciplines and create their own specialties. With nano machines and biologically inspired self-assembling robots becoming realities, mechanical engineering has transformed from a field that generally focused on big industry to one that also emphasizes tiny and efficient machines. Energy-related studies are likely to become a major thrust of mechanical engineering curricula. It is possible that the future will unfold a post-industrial age where the mass-manufacture paradigm of the Industrial Revolution that forced the overcrowding of cities and caused extensive damage to the environment is replaced by a widely distributed industrial economy that enables small communities to be self-reliant for essential services and yet be useful contributors to the global economy. This will create innumerable opportunities for innovation and design.

Narayanan M. Komerath, Ph.D.

FURTHER READING

Avallone, Eugene A., Theodore Baumeister III, and Ali M. Sadegh. *Marks' Standard Handbook for Mechanical Engineers.* 11th ed. New York: McGraw-Hill, 2006. Authoritative reference for solving mechanical engineering problems. Discusses pressure sensors and measurement techniques and their applications in various parts of mechanical engineering.

Calvert, Monte A. *The Mechanical Engineer in America, 1830-1910: Professional Cultures in Conflict.* Baltimore, Md.: Johns Hopkins University Press, 1967. Discusses the life of the mechanical engineer in nineteenth-century America. The author describes the conflict between the shop culture originating in the procedures of the machine shop and the school culture of the engineering colleges that imparted formal education.

Freitas, Robert A., Jr., and Ralph C. Merkle. *Kinematic Self-Replicating Machines.* Georgetown, Tex.: Landes Bioscience, 2004. A review of the theoretical and experimental literature on the subject of self-replicating machines. Discusses the prospects for laboratory demonstrations of such machines.

Hill, Philip G., and Carl R. Peterson. *Mechanics and Thermodynamics of Propulsion.* 2d ed. Reading, Mass.: Addison-Wesley, 1992. This textbook on propulsion covers the basic science and engineering of jet and rocket engines and their components. Also gives excellent sets of problems with answers.

Lienhard, John H., IV, and John H. Lienhard V. *A Heat Transfer Textbook.* 4th ed. Mineola, N.Y.: Dover Publications, 2011. An excellent undergraduate text on the subject.

Liepmann, H. W., and A. Roshko. *Elements of Gas Dynamics.* Reprint. Mineola, N.Y.: Dover Publications, 2001. Classic textbook on the discipline of gas dynamics as applied to high-speed flow phenomena. Contains several photographs of shocks, expansions, and boundary layer phenomena.

Pelesko, John A. *Self Assembly: The Science of Things That Put Themselves Together.* Boca Raton, Fla.: Chapman and Hall/CRC, 2007. Discusses natural self-assembling systems such as crystals and soap films and goes on to discuss viruses and self-assembly of DNA cubes and electronic circuits. Excellent introduction to a field of growing importance.

Shames, Irving H. *Engineering Mechanics: Statics and Dynamics.* 4th ed. Upper Saddle River, N.J.: Prentice Hall, 1997. A classic textbook that integrates both statics and dynamics and uses a vector approach to dynamics. Used by undergraduates and professionals all over the world since the 1970's in its various editions. Extensive work examples.

Shigley, Joseph E., Charles R. Mischke, and Richard G. Budynas. *Mechanical Engineering Design.* 7th ed. New York: McGraw-Hill, 2004. Classic undergraduate textbook showing students how to apply mathematics, physics, thermal sciences, and computer-based analysis to solve problems in mechanical engineering. Includes sections on quality control, and the computer programming sections provide an insight into the logic used with the high-level languages of the 1980's.

Siciliano, Bruno, et al. *Robotics: Modelling, Planning and Control.* London: Springer-Verlag, 2010. Rigorous textbook on the theory of manipulators and wheeled robots, based on kinematics, dynamics, motion control, and interaction with the environment. Useful for industry practitioners as well as graduate students.

WEB SITES

American Society of Heating, Refrigerating and Air-Conditioning Engineers
http://www.ashrae.org

American Society of Mechanical Engineers
http://www.asme.org

National Society of Professional Engineers
http://www.nspe.org/index.html

Society of Automotive Engineers International
http://www.sae.org

Society of Manufacturing Engineers
http://www.sme.org

See also: Acoustics; Algebra; Applied Mathematics; Calculus; Computer-Aided Design and Manufacturing; Engineering; Robotics.

METABOLIC ENGINEERING

FIELDS OF STUDY

Biology; biochemistry; molecular biology; organic chemistry; analytical chemistry; microbiology; genetic engineering; biotechnology.

SUMMARY

Metabolic engineering is a new science that appeared in the 1990's. It is associated with biology and chemistry. Metabolic engineering allows the designing of biochemical pathways that do not exist in the natural world, as well as the redesign of existing biochemical pathways often with the use of genetic engineering. Metabolic engineers often modify biochemical pathways by reducing cellular energy use or waste production, by changing the nutrient flow to the cells, or improving the productivity and yield of a particular pathway. In addition, metabolic engineers may potentially design new organisms that are tailor-made for the desired chemicals and production processes. Many novel compounds of industrial and medical interest can be produced by metabolic engineering. In the twenty-first century, the main efforts of metabolic engineers are concentrated on biofuels and pharmaceuticals.

KEY TERMS AND CONCEPTS

- **Bioreactor:** Apparatus for cell growth with practical purpose under controlled conditions.
- **DNA Sequencing:** Determining of the precise order of nucleotides (such as adenine, guanine, cytosine, and thymine) in a DNA.
- **Enzymes:** Biological catalysts made of proteins.
- **Genetic Engineering:** Modification of genetic material to achieve specific goals.
- **Metabolism:** Sum of biochemical reactions within an organism.
- **Substrate:** Substance that is acted on (as by an enzyme).

DEFINITION AND BASIC PRINCIPLES

Metabolic engineering is a relatively new field that deals with the modification and optimization of metabolic pathways, mainly in microorganisms, by altering genes, nutrient uptake, or metabolic flow to allow production of novel compounds that are of industrial and medical interest. Metabolic pathways of living organisms are not optimal for specific practical applications, but they can be modified using the tools of modern biotechnology such as genetic engineering. The redesign of existing, natural metabolic pathways for useful purposes is a main objective of metabolic engineering. Metabolic engineering usually includes two phases: careful analysis of the metabolic pathway and genes involved in the pathway (analytical phase) and its modification (synthesis phase). Pathway analysis often includes the metabolic control analysis: determining which compounds can control the productivity and yield of particular pathway. Different tasks of metabolic engineering are as follows: improvements of productivity and yield of particular pathway; expansion of substrate range; elimination of waste; improvement of process performance; improvements of cellular activities; and extension of product array. Metabolic engineering is becoming one of the principal fields of biotechnology.

Production of many chemicals and fuels uses nonrenewable resources or limited natural resources. Metabolic engineering creates many alternatives to replace dangerous chemicals and petroleum-based transportation fuels with clean, green, and renewable chemicals and biofuels.

BACKGROUND AND HISTORY

The term "metabolic engineering" first appeared in the early 1990's. Since that time, the range of products that can be generated has increased significantly, partly because of remarkable advances in other fields related to metabolic engineering, such as DNA sequencing and genetic engineering. With DNA sequencing, scientists were able to identify the majority of metabolic genes and enzymes in many organisms. In the post-sequencing era, the obtained information is used for practical construction of biochemical pathways or whole organisms with optimized functions through metabolic engineering.

In the 1990's, scientists developed new genetic tools that gave metabolic engineers more precise control over metabolic pathways. They also created analytical tools that allowed the metabolic engineer to track metabolites in a cell to identify new biochemical pathways more precisely.

Earlier in the twenty-first century, metabolic engineers joined other scientists in their quest for alternative fuels, which are in high demand because of increasing oil prices and concern about climate change.

How It Works

Metabolic engineering is based mainly on microbial metabolism. Microbes produce different kinds of substances that they use for the growth and maintenance of their cells. These substances can be useful for humans. The goal of metabolic engineering is to enhance the microbial production of useful substances. To achieve this goal, metabolic engineers must follow a particular route. They need to choose a friendly organism (host) for their metabolic manipulations. They need to find cheap and available substrates to use for modified metabolic pathways. Finally, metabolic engineers must be able to perform genetic manipulations of metabolic routes. Metabolic engineers can also alter nutrient uptake or metabolic flow. All these steps are dependent on each other. For example, genes cannot be manipulated in every organism; products or metabolic intermediates may be toxic to its host.

Host and Host Design. Generation of products by metabolic engineering has been achieved by transferring product-specific enzymes or entire metabolic pathways into so-called user-friendly microorganism hosts, which were used traditionally in industry. These industrial microorganisms grow rapidly on inexpensive culture media available in bulk quantities, are open to genetic manipulation (and genetic manipulation tools are available), and are nonpathogenic (do not cause disease). In addition, it is important that the host can survive (and thrive) under the desired process conditions (ambient versus extremes of temperature, pH). It is essential that the host is genetically stable (with the introduced pathway) and not susceptible to virus or another microbe's attack. Among the host microorganisms most widely used are *Saccharomyces cerevisiae* and *Escherichia coli*. *Saccharomyces cerevisiae*, or baker's yeast, has been used for making bread and alcohol for thousands of years. It is one of the earliest domesticated organisms. This organism has come to be used in a large number of different processes within the biotechnological and pharmaceutical industries. Comprehensive knowledge of *S. cerevisiae* has been accumulated over a long period of time. In addition, the complete genome sequence of yeast is available, and yeast is nonpathogenic. The well-established fermentation and process technology for large-scale production with *S. cerevisiae* in bioreactors makes this organism very attractive for several industrial purposes.

Escherichia coli, commonly known is *E. coli*, is a bacterium that is widely used as a research (model) organism. It is easy to grow and genetically manipulate this bacterium, and its genome sequence is available. Several important products such as interferon (flu-fighting drug), insulin, and growth hormone are manufactured by genetically modified *E. coli*.

In addition to *E. coli* and *S. cerevisiae*, several other microorganisms are widely used as hosts for metabolic engineering manipulations, including bacteria *Bacillus subtilis* and *Streptomyces coelicolor*.

Finally, in addition to redesigning particular metabolic processes, metabolic engineers may also design de novo artificial cells that will produce desired products.

Substrates. To make metabolically engineered products, chemical substrates are needed. To make these products economically viable, inexpensive sources of substrates are required. Substrates must contain different chemical components, such as carbon, nitrogen, oxygen, and hydrogen. For example, metabolic engineers are looking at sugars from cellulosic biomass as potential substrates for biofuel production. Cellulosic biomass is a very attractive biofuel feedstock because of its abundant supply. On a global scale, plants produce almost 100 billion tons of cellulose per year, making it the most abundant organic compound on Earth.

Genetic Manipulation of Metabolic Routes. Genetic manipulation of metabolic pathways by adding or deleting genes or modifying the expression of existing genes in the host can serve several useful purposes. It can extend the existing pathways or shifting metabolic route into a desired pathway or increase the rate-determined step of the particular metabolic route. Adding genes into the host consists of the following steps.

- The gene the for desired pathway is obtained from the non-host organism.
- The gene is inserted into the host cell.

- Host cells are induced to express (to cause the gene to manifest its effects) this "foreign" gene in order to produce the desired product.

One example of how gene manipulation is used in areas relevant to metabolic engineering is as follows: In the mold *Aspergillus terreus*, the producer of cholesterol-lowering drug lovastatin, genes were modified to increase their expression levels in order to change its metabolism in terms of drug production.

Another example is the introduction of bovine lactic acid pathway into *S. cerevisiae*. As a part of this, a gene responsible for speeding up removal of hydrogen, which participates in lactic acid production, was expressed in *S. cerevisiae*, and lactic acid was produced at rate of eleven grams per liter per hour.

Fascinating Facts About Metabolic Engineering

- In 2010, a team of scientists led by J. Craig Venter created the first synthetic cell. The team synthesized the artificial chromosome, which was then transplanted into the recipient cell. The artificial chromosome was able to take over the recipient cell. This research opens the door for creation of useful artificial cells to make products such as vaccines and biofuels.
- Antimalarial drug artemisinin has been produced from metabolically engineered laboratory yeast. The antimalarial comes from the *Artemisia annua* plant, which grows in Southeast Asia. Artemisinin could also possibly be used in cancer treatment.
- Scientists are able to synthesize large DNA molecules in the laboratory. Researchers have made artificial DNA containing all twenty-one genes encoding the small ribosomal subunit (cell protein factory) from *Escherichia coli* (*E. coli*).
- Using metabolic-engineering methods, researchers were able to modify *E. coli* bacterium to produce butanol. Among the types of biofuels that are on the road to commercialization, butanol has been the most promising. It is another alcohol fuel, but when compared with ethanol, it has higher energy content (roughly 80 percent of gasoline energy content). Butanol can also be stored and transported using existing infrastructure—and it does not occur naturally in *E. coli*.

Because it tolerates acid, yeast may serve as an alternative to bacteria, which is usually used in industry for lactic acid production. Lactic acid is widely used as a food preservative.

Altering Nutrient Uptake or Metabolic Flow. Alteration of nutrient uptake or metabolic flow can be done not only by genetic manipulation but also by using inhibitors—simple chemicals or physical factors such as light or temperature.

The alteration of molecular hydrogen (H_2) production in green algae using high-intensity light is an example of metabolic flow modification by physical factors. H_2 is one of the possible energy carriers of the future. Microscopic green algae produce H_2 in photosynthetic reactions from water using sunlight as an energy source, usually in anoxic (without oxygen) conditions. Oxygen (O_2) produced by photosynthesis in green algae is an inhibitor of H_2 production. Brief illumination of algal cells by high-intensity light was accompanied by rapid suppression of photosynthetic O_2 evolution. The decline in the rate of O_2 evolution was accompanied by stimulation of H_2 production in algal cells.

Production Systems. All of the above-mentioned considerations are very important in metabolic engineering, although it is also important to ensure that the production of desired compounds by modified cells can be reproduced. This can be achieved by using bioreactors, in which the important parameters such as pH, temperature, substrate supply, and other variables are controlled. It is even possible to modify cell metabolism by using bioreactors.

APPLICATIONS AND PRODUCTS

There are a wide range of metabolic engineering products and applications. Undoubtedly, a number of novel applications and products will arise in the future.

Pharmaceuticals. Metabolic engineering is most promising in the production of pharmaceuticals. These include pharmaceuticals from different classes of natural products: alkaloids, isoprenoids, and flavonoids. Biosynthesis of natural products is an emerging area of metabolic engineering that offers significant advantages over conventional chemical methods. Some pharmaceutical compounds are too complex to be chemically synthesized or extracted from biomass organisms inexpensively.

Alkaloids are mainly plant-derived compounds that have been used as drugs such as morphine. Alkaloids are produced by simple extraction from plants. Studies show that alkaloids can be synthesized from amino acids by metabolic engineering in *E. coli* and *S. cerevisiae*.

Isoprenoids, organic compounds composed of two or more hydrocarbons, have a range of functions: pigments, fragrances, and vitamins. Isoprenoids are also the precursors to sex hormones. Many isoprenoids have been produced using microorganisms, including carotenoids and various plant-derived terpenes. Metabolic engineers are using *S. cerevisiae* as a cell factory for the biosynthesis of isoprenoids. One metabolic-engineering success is the production of Taxol, which is used to treat breast cancer. It is an isoprenoid that was first isolated in the bark of the Pacific yew (*Taxus brevifolia*). The demand for Taxol greatly exceeds the supply that can be obtained from its natural source. A partial Taxol biosynthetic pathway has been engineered in *S. cerevisiae*.

Another metabolic engineering success is the production of isoprenoids-carotenoids. Carotenoids are naturally occurring yellow, orange, and red pigments commonly found in plants such as carrots as well as in bacteria, algae, and fungi and play an important role in fighting disease. Metabolic engineers have successfully introduced carotenoid biochemical pathways into nonproducing carotenoid microbes such as *E. coli* and *S. cerevisiae*.

Flavonoids are a group of secondary plant metabolites. These compounds can be used as antioxidants or antiviral, antibacterial, and anticancer drugs. Many flavonoid biosynthetic pathways are known, and a wide array of flavonoid compounds from *S. cerevisiae* are expected to be produced by metabolic engineering in the near future.

Chemicals. Numerous chemicals, such as amino acids, organic acids, vitamins, flavors, fragrances, and nutraceuticals can be manufactured by metabolic engineering.

Glycerol (or glycerin) is a chemical produced by metabolic engineering. Glycerol is used to synthesize many products, ranging from cosmetics to lubricants. It is a by-product of soap or biodiesel manufacturing and its production is 1.2 billion liters annually. It can be also used a fuel. Metabolically engineered *S. cerevisiae* strain produced more than 200 grams of glycerol per liter of liquid medium.

Another example of chemicals produced with help of metabolic engineering are sterols. The most well-known sterol is cholesterol. Sterols are important for living organisms as they are a part of the cellular membrane, participate in the synthesis of several hormones, and are also nutrient supplements. Several sterols are being produced from metabolically engineered *S. cerevisiae*.

Fuels. Metabolic engineering can be used in the production of biofuels. Several scientific laboratories have demonstrated the feasibility of manipulating microorganisms to produce molecules similar to oil-derived products, although the yield is very low. Adjusting metabolic pathways of microbes to produce fuels similar to gasoline has the potential to save an enormous amount of money. These fuels can be used in existing engines, unlike other biofuels that require modified engines or fueling stations.

Several research groups are trying to metabolically engineer microorganisms to produce ethanol fuel using cellulose as substrate. Another example of the work of metabolic engineers is biodiesel production. Biodiesel is a diesel substitute primarily obtained from vegetable oils such as soybean. However, the production of this fuel is limited by the absence of sufficient vegetable oil feedstocks. Another problem is that in order to produce biodiesel, oils should be modified by transesterification, a chemical reaction with methanol, catalyzed by acids or bases (such as sodium hydroxide). *E. coli* has been metabolically engineered to produce biodiesel directly, using low-cost materials.

IMPACT ON INDUSTRY

Metabolic engineering may one day play a major role in a number of multibillion-dollar industries, including pharmaceutical, biotechnology, and biofuel. The United States maintains a dominant position in the world of metabolic engineering. The first large-scale industrial process of the human hormone insulin by metabolic engineering was developed in the United States. Other developed countries are researching the use of metabolic engineering to produce a variety of products, such as pharmaceuticals and biofuels.

Government and University Research. Governmental agencies such as the National Science Foundation (NSF) and the U.S. Department of Energy (DOE) provide funding for research in metabolic engineering. A vast majority of metabolic engineering research is concentrated in the areas of pharmaceuticals and biofuel generation. The Joint BioEnergy Institute (JBEI) is a major player in the metabolic engineering area. It is funded by the U.S. Department of Energy, and its main goal is to develop next-generation biofuels such as cellulosic ethanol. It is working in partnership with Sandia National Laboratories and the University of California, Berkeley. Other major players in metabolic engineering are University of Chicago and Rice University.

Industry and Business. Scientists in industry traditionally carry out a significant percentage of metabolic engineering research, a good portion of which is directed to health care products, such as pharmaceuticals. Successful scientific projects include creating of lysine, riboflavin, coenzyme Q-10, the aminoshikimate pathway, and beta-carotene by metabolically engineered microorganisms.

Major Corporations. At present, just a few companies are using metabolic engineering alone to achieve their goals, including Amyris, Integrated Genomics, and LS9. Some major biotechnological companies, such as Genentech, employ a metabolic engineering approach combined with traditional biotechnological techniques such as recombinant DNA technology.

CAREERS AND COURSE WORK

Biotechnology, pharmaceutical, and biofuel companies are the biggest employers of metabolic engineers. As new biology-based products move from research into production, more metabolic engineers will be needed in industry, universities, and government laboratories.

Metabolic engineering is an interdisciplinary science. Course work includes biochemistry, molecular biology, chemistry, genetic engineering, analytical chemistry, biochemistry, biochemical and bioprocess engineering, and microbiology. Most metabolic engineers have a bachelor of science in biology, biochemistry, genetic engineering, microbiology, or biotechnology. Advanced degrees (master's and doctorate) in molecular biology, biochemistry, biotechnology, or genetics are necessary for research and teaching positions. Some governmental institutions such as the Joint BioEnergy Institute help students to develop metabolic engineering educational paths by providing opportunities for internships.

SOCIAL CONTEXT AND FUTURE PROSPECTS

Though the redesign of life forms for the benefit of mankind is definitely an exciting career, metabolic engineers are paying particular attention to ethical, legal, and political issues. To continue in this work, the field as a whole will need sustained support from the public and government.

At present, metabolic engineering is more a collection of successful experiments than an established science. In the future, metabolic engineering may play a significant role in production of chemicals and fuels from inexpensive and renewable starting materials. Continued development of the techniques of metabolic engineering will be necessary to expand the range of products. The role of metabolic engineering in science is likely to expand in the future as a result of increasing needs for pharmaceuticals and biofuels.

Sergei A. Markov, Ph.D.

FURTHER READING

Bailey, James E., and David F. Ollis. *Biochemical Engineering Fundamentals.* 2d ed. New York: McGraw-Hill, 1986. Classic textbook on biochemical engineering.

Bourgaize, David, Thomas R. Jewell, and Rodolfo G. Buiser. *Biotechnology: Demystifying the Concepts.* San Francisco: Benjamin Cummings, 2000. Excellent introduction to biotechnology.

Lewin, Benjamin. *Genes VIII.* San Francisco: Benjamin Cummings, 2003. In-depth look at genes and molecular biology.

Madigan, Michael T., et al. *Brock Biology of Microorganisms.* 12th ed. San Francisco: Benjamin Cummings, 2008. Several chapters of this popular textbook describe microbial metabolism and the application of microorganisms in industry.

Marguet, Philippe, et al. "Biology by Design: Reduction and Synthesis of Cellular Components and Behavior." *Journal of the Royal Society Interface* 4, no. 15 (2007): 607-623. Review on metabolic engineering and synthetic biology written for the general public.

Ostergaard, Simon, Lisbeth Olsson, and Jens Nielsen. "Metabolic Engineering of *Saccharomyces cerevisiae.*" *Microbiology and Molecular Biology Reviews* 64, no. 1 (2000): 34-50. Describes metabolic engineering techniques using *S. cerevisiae* as an example.

Stephanopoulos, Gregory N., Aristos A. Aristidou, and Jens Nielsen. *Metabolic Engineering: Principles and Methodologies.* San Diego: Academic Press, 1998. Classic text on metabolic engineering.

WEB SITES
Biotech Career Center
http://www.biotechcareercenter.com/biotech.html

Biotechnology Industry Organization
http://www.bio.org

Nature Technology Corporation
http://www.natx.com

See also: Biochemical Engineering; Cloning; DNA Sequencing; Genetically Modified Organisms; Genetic Engineering; Proteomics and Protein Engineering.

METALLURGY

FIELDS OF STUDY

Materials science; materials engineering; mechanical engineering; physical engineering; mining; chemical engineering; electrical engineering; environmental engineering.

SUMMARY

Starting as an art and a craft thousands of years ago, metallurgy has evolved into a science concerned with processing and converting metals into usable forms. The conversion of rocky ores into finished metal products involves a variety of activities. After the ores have been mined and the metals extracted from them, the metals need to be refined into purer forms and fashioned into usable shapes such as rolls, slabs, ingots, or tubing.

Another part of metallurgy is developing new types of alloys and adapting existing materials to new uses. The atomic and molecular structure of materials are manipulated in controlled manufacturing environments to create materials with desirable mechanical, electrical, magnetic, chemical, and heat-transfer properties that meet specific performance requirements.

KEY TERMS AND CONCEPTS

- **Bessemer Process:** Steelmaking process in which air is blown through molten pig iron contained in a furnace so that impurities can be removed by oxidation.
- **Blast Furnace:** Smelting furnace for the production of pig iron in which hot air is blown upward into the furnace as iron ore, coke, and limestone are supplied through the top, producing chemical reactions; the molten iron and slag are collected at the bottom.
- **Carburizing:** Process of adding carbon to the surface by exposing a metal to a carbon-rich atmosphere under high temperatures, allowing carbon to diffuse into the surface, making the surface more wear resistant.
- **Ductility:** Characteristic of metal that enables it to be easily molded or shaped, without fracturing.

- **Extrusion:** Process in which a softened metal is forced through a shaped metal piece or die, creating an unbroken ribbon of product.
- **Flux:** Substance added to molten metals to eliminate impurities or encourage fusing.
- **Forging:** Process of shaping metal by heating it in a forge, then beating or hammering it.
- **Galvanizing:** Process of coating steel with zinc to prevent rust.
- **Metal Alloy:** Homogeneous mixture of two or more metals.
- **Ore:** Mineral from which metal is extracted.
- **Plastic Deformation:** Permanent distortion of a metal under the action of applied stresses.
- **Recrystallization:** Process by which deformed grains in a metal or alloy are replaced by undeformed grains; reduces the strength and hardness of a material.
- **Recrystallization Temperature:** Approximate minimum temperature at which a cold-worked metal becomes completely recrystallized within a specified time.
- **Sintering:** Process of turning a metal powder into a solid by pressure and heating it to a temperature below its boiling point.
- **Slag:** By-product consisting of impurities produced during the refining of ore or melting of metal.

DEFINITION AND BASIC PRINCIPLES

Metallurgy is the science of extracting metals and intermetallic compounds from their ores and working and applying them based on their physical, chemical, and atomic properties. It is divided into two main areas: extractive metallurgy and physical metallurgy.

Extractive Metallurgy. Extractive metallurgy, also known as process or chemical metallurgy, deals with mineral dressing, the converting of metal compounds to more treatable forms and refining them. Mineral dressing involves separating valuable minerals of an ore from other raw materials. The ore is crushed to below a certain size and ground to powder. The mineral and waste rock are separated, using a method based on the mineral's properties. After that, water is removed from the metallic concentrate or compound. Because metallic compounds are often complex mixtures (carbonates, sulfides, and oxides), they need to be converted to other forms for easier processing and refining. Carbonates are converted to oxides; sulfides to oxides, sulfates, and chlorides; and oxides to sulfates, and chlorides. Depending on the type of metallic compound, either pyrometallurgy or hydrometallurgy is used for conversion. Both processes involve oxidation and reduction reactions. In oxidation, the metallic element is combined with oxygen, and in reduction, a reducing agent is used to remove the oxygen from the metallic element. The difference between pyrometallurgy and hydrometallurgy, as their names imply, is that the former uses heat while the latter uses chemicals. These two processes also include refining the metallic element in the final stage of extractive metallurgy when heat and chemicals are used. Electrometallurgy refers to the use of the electrolytic process for refining metal elements, precipitating dissolved metal values, and recovering them in solid form.

Physical Metallurgy. Physical metallurgy deals with making metal products based on knowledge of the crystal structures and properties (chemical, electrical, magnetic, and mechanical) of metals. Metals are mixed together to make alloys. Heat is used to harden metals, and their surfaces can be protected with metallic coating. Through a process called powder metallurgy, metals are turned into powders, compressed, and heat-treated to produce a desired product. Metals can be formed into their final shapes by such operations as casting, forging, or plastic deformation. Metallography is a subfield of metallurgy that studies the microstructure of metals and alloys by various methods, especially by light and electron microscopes.

BACKGROUND AND HISTORY

Metallurgy came into being in the Middle East around 5000 B.C.E. with the extraction of copper from its ore. The discovery of the first alloy, bronze, the result of melting copper and tin ores together, initiated the Bronze Age (4000-3000 B.C.E.). Melting iron ore with charcoal to obtain iron marked the beginning of the Iron Age in Anatolia (2000-1000 B.C.E.). Gold, silver, and lead were separated from lead-bearing silver in Greece about 500 B.C.E. Mercury was produced from cinnabar around 100 B.C.E., and it was later used to recover and refine various metals. Around 30 B.C.E., brass, the second alloy, was made from copper and zinc in Egypt, and another alloy, steel, was produced in India.

A vehicle removing a glowing hot ingot of titanium from a furnace. Titanium is a metal with low density but high strength, good thermal resistance and excellent corrosion resistance. This makes it an ideal metal for the aerospace industry. (RIA Novosti/Photo Researchers, Inc.)

From the sixth to the nineteenth centuries, metallurgy focused on the development and improvement of the processes involved in obtaining iron, making steel, and extracting aluminum and magnesium from their ores. The blast furnace was developed in the eighth century and spread throughout Europe. During the sixteenth century, the first two books on metallurgy were written by Vannoccio Biringuccio, an Italian metalworker, and by Georgius Agricola, a German metallurgist.

Modern metallurgy began during the eighteenth century. Abraham Darby, an English engineer, developed a new furnace fueled by coke. Another English engineer, Sir Henry Bessemer developed a steelmaking process in 1856. Great Britain became the greatest iron producer in the world, and Spain and France also produced large amounts of iron. About 1886, American chemist Charles Martin Hall and a French metallurgist Paul-Louis-Toussaint Héroult independently developed a way to extract aluminum from its ore, which became known as the Hall-Héroult process. Aluminum soon became an important metal in manufactured goods.

Metallurgy did not emerge as a modern science with two branches, extractive and physical, until the twentieth century. The development and improvement of metallurgy were made possible by the application of knowledge of the chemical and physical principles of minerals.

HOW IT WORKS

Crushing and Grinding of Ores. In the first step of mineral dressing, two kinds of mechanized crushers are used to reduce ores. Jaw crushes reduce ores to less than 150 millimeters (mm) and cone crushers to less than 10-15 mm. Different kinds of grinding mills are used to reduce crushed ores to powder: cylinder mills filled with grinding bodies (stones or metal balls), autogenous mills (coarse crushed ores grinding themselves), semiautogenous mills using some grinding bodies, and roll crushers, which combine crushing and grinding.

Separating Valuable and Waste Minerals. The process used in the next step of mineral dressing depends on the properties of the minerals. Magnetic separation is used for strongly magnetic minerals such as iron ore and iron-bearing ore. Gold, tin, and tungsten ores require gravity separation. A process called flotation separation is widely used for hydrophilic (water-attracting) intergrown ores containing copper, lead, and zinc. Electrostatic separation work best with particles of different electric charges such as mineral sands bearing zircon, rutile, and monazite.

Pyrometallurgy. Pyrometallurgy is a method of converting metallic compounds to different forms for easier processing and refining by using oxidation and reduction reactions.

The first conversion process, roasting, has two main types: One type changes sulfide compounds to oxides, and the other reduces an oxide to a metal. Other types of roasts convert sulfides to sulfates or change oxides to chlorides. These processes are carried out in different kinds of steel roasters.

The second conversion process, smelting, separates a metallic compound into two parts: an impure molten metal and a molten slag. The two types of smelting are reduction and matte, and the processes are done in many kinds of blast furnaces. Coke is used for fuel and limestone as a flux for making slag. Reduction smelting converts an oxide feed material to a metal and an oxide slag. Matte smelting converts a sulfide feed material to a mixture of nickel, copper, cobalt, and iron sulfides as well as an iron and silicon oxide slag.

Refining, a process of removing any impurities left after roasting or smelting, also can be done in a blast furnace. Iron, copper, and lead can be refined in oxidation reaction that removes impurities as an oxide slag or an oxide gas. Fire refining can separate copper from its impurities of zinc, tin, iron, lead, arsenic, and antimony. Similarly, lead can be separated from such impurities as tin, antimony, and arsenic, and zinc from impurities of cadmium and lead.

Hydrometallurgy. Another method of converting metallic compounds to different forms is hydrometallurgy. It uses several types of leach solvents: ammonium hydroxide for sulfides and carbonates; sulfuric acid, sodium carbonate, or sodium hydroxide for oxides; and sulfuric acid or water for sulfates. The dissolved metal values are then recovered from the leaching solution in solid form. Although numerous recovery processes exist, they usually involve electrolysis. By a process called precipitation, gold that has been dissolved in sodium cyanide and placed in contact with zinc is separated from the solution and gathers on zinc. In another process called electrolytic deposition, or electrowinning, an electric current is passed through the leach solution with dissolved metals, causing metal ions to deposit at the cathode. Copper, zinc, nickel, and cobalt can be obtained this way.

Electrometallurgy. Electrolysis can be used to refine metallic elements as well as to recover them after hydrometallurgical treatment. Copper, nickel, lead, gold, and silver can be refined this way. In this method, for example, impure copper is used as the anode. When the electric current passes through the solution, atoms of pure gold travel to the cathode, acquire electrons, and become neutral copper atoms. Electrolysis is also the process for recovering copper, aluminum, and magnesium in hydrometallurgy.

Alloys. Alloys are made by mixing pure metals together to obtain a substance with increased strength, increased corrosion resistance, lower cost, lower melting points, or desirable magnetic, thermal, or electrical properties. They are usually made by melting the base metals and adding alloying agents. Stainless steel, a mixture of steel, nickel, and chromium, is stronger and more chemically resistant than the base metals from which it was formed.

Powder Metallurgy. Powder metallurgy is the process of reducing metals and nonmetals to powder and shaping the powder into a solid form under great heat and pressure. Metal powders are usually produced by atomization of streams of molten metal with a spinning disk or with a jet of water, air, or inert gas. After the powders are cold pressed for initial adhesion, they are heated to temperatures about 80 percent below the melting point of the major component. Friction between powders and pressing dies is reduced by adding lubricants, and porosity in the final product is eliminated by applying pressure.

Metal Forming. Metals are usually cast into ingots in iron molds. Casting is also carried out in molds made of sand, plaster of Paris, or glue. Permanent casting uses pressure or centrifugal action. Plastic deformation is performed on metals to change their properties and dimensions. If done below the recrystalization temperature, the process is called cold working; above this temperature but below the melting or burning point, it is called hot working. Techniques involved include rolling, pressing, extrusion, stamping, forging, and drawing. Surface treatments of metals include protective coating and hardening. In metallic coating, zinc and other metals such as chromium, cadmium, lead, and silver are often used. Surface hardening of metals is usually done with heat in a gas rich in carbon or in ammonia and hydrogen.

APPLICATIONS AND PRODUCTS

The most important applications of metallurgy involve common metals and alloys and powder metallurgy technology.

Copper. Copper is ductile and malleable, and it resists corrosion and conducts heat and electricity. Copper and its alloy brass (copper plus zinc) are used to make coins, household fixtures (doorknobs, bolts), and decorative-art objects (statues, sculptures, imitation-gold jewelry). It is also used in transportation vehicles and has many electrical applications (transformers, motors, generators, wiring harnesses). Its alloy bronze (copper plus tin) is used in plumbing and heating applications (water pipes, cooking utensils). Aluminum-bronze is used to make tools and parts for aircraft and automobiles. Manganese-bronze is used to make household fixtures and ship propellers.

Iron. Iron is ductile, malleable, and one of the three magnetic elements (the others are cobalt and nickel). Cast iron is resistant to corrosion and used to make manhole covers and engine blocks for gasoline and diesel engines. Wrought iron is used to make cooking utensils and outdoor household items such as fencing and furniture. Most iron is used to make steel. Steel is used as a structural material in the construction of large, heavy projects (bridges, ships, buildings) and automobile parts (body frames, radial-ply tires).

When chromium and nickel are combined with steel, steel becomes stainless, and it is used to make flatware and surgical tools. Steel combined with cobalt is used to make jet engines and gas turbines.

Gold. Applications of gold are based on such properties as its electrical and thermal conductivity, ductility, malleability, resistance to corrosion, and infrared reflectivity. Gold serves as a medium of exchange and money. Its decorative applications include jewelry, golf leaf on the surfaces of buildings, and flourishes on ceramics or glassware. More practical applications include components for electronic devices (telephones, computers), parts for space vehicles, and dental fillings, crowns, and bridges.

Silver. Silver is ductile and very malleable, conducts heat, and has the highest electrical conductivity of all metals. It is used to make cutlery, jewelry, coins, long-life batteries, photographical films, and electronic components (circuits, contacts), and in dentistry. Its alloy, sterling silver (silver plus copper) is also used to make jewelry and tableware. German silver (silver plus nickel) is another alloy used for silverware.

Platinum. This ductile and malleable material is one of the densest and heaviest metals. It is resistant to corrosion and conducts electricity well. It is used to make jewelry, electronic components (hard disk drive coatings, fiber-optics cables), and spark plug components. It is important in making the glass for liquid crystal displays (LCDs), in the petrol industry as an additive and refining catalyst, in medicine (anticancer drugs, implants), and in dentistry. Its alloys (platinum plus cobalt or metals in the platinum groups) are mostly used to make jewelry.

Mercury. Sometimes called quicksilver, mercury is the only common metal that is liquid at ordinary temperatures. It is a fair conductor of electricity and of high density. It is used in barometers and thermometers, to recover gold from its ore, and to manufacture chlorine and sodium hydroxide. Its vapor is used in street lights, fluorescent lamps, and advertizing signs. Mercury compounds have various uses, such as insecticides, rat poisons, disinfectants, paint pigments, and detonators. Mercury easily is alloyed with silver, gold, and cadmium.

Lead. Lead is malleable, ductile, resistant to corrosion, and of high density. Its softness is compensated for by alloying it with such metals as calcium,

Fascinating Facts About Metallurgy

- Iron is the most abundant element, making up 34.6 percent of Earth, and the most used of all metals.
- The magnetic property of steel allows recyclers to reclaim millions of tons of iron and steel from garbage.
- Meteors were the source of much of the wrought iron used in early human history.
- Because gold is so malleable, 1 gram of gold can be hammered into a sheet 1 square meter in size. It can also be made so thin that it appears transparent.
- Nitric acid can be used to determine if gold is present in ore. This "acid test," proving an ore's value, has come to mean a decisive test proving an item's worth or quality.
- Half of the world's gold is held by the Republic of South Africa. The second gold-producing nation, in terms of volume, is the United States.
- A lead pencil might more accurately be called a graphite pencil, as it contains a shaft of graphite, not lead.
- An alloy of equal parts of silver and aluminum is as hard as bronze.
- The amount of platinum mined each year is 133 tons, less than one-tenth of the 1,782 tons of gold mined each year.
- Uranium is very dense. A one-gallon container filled with uranium would weigh about 150 pounds, while that same container filled with milk would weigh around 8 pounds.
- To produce about 40 million kilowatt hours of electricity, it would take 16,000 tons of coal or 1 ton of natural uranium.
- In Japan, all government-subsidized dental alloys must have a palladium content of at least 20 percent.

antimony, tin, and arsenic. Lead is a component in lead-acid batteries, television and computer screens, ammunition, cables, solders, and water drains, and is used as a coloring element in ceramic glazes.

Magnesium. Magnesium is the lightest structural metal, with low density (two-thirds that of aluminum), superior corrosion performance, and good mechanical properties. Because it is more expensive than aluminum, its applications are somewhat limited.

Magnesium and its alloys are used in the bicycle industry, racing car industry (gearbox casings, engine parts), and aerospace industry (engines, gearbox casings, generator housings, wheels).

Manganese. Manganese is a hard but very brittle, paramagnetic metal. Mostly it is used in steel alloys to increase strength, hardness, and abrasion resistance. It can be combined with aluminum and antimony to form ferromagnetic compounds. It is used to give glass an amethyst color, in fertilizers, and in water purification.

Cobalt. Cobalt has a high melting point and retains its strength at high temperatures. It is used as a pigment for glass, ceramics, and paints. When alloyed with chromium and tungsten, it is used to make high-speed cutting tools. It is also alloyed to make magnets, jet engines, and gas turbine engines.

Tungsten. Tungsten has the highest melting point and the lowest thermal expansion of all metals, high electrical conductivity, and excellent corrosion resistance. It is used to make lightbulb filaments, electric contacts, and heating elements; as an additive for strengthening steel; and in the production of tungsten carbide. Tungsten carbide is used to make dies and punches, machine tools, abrasive products, and mining equipment.

Chromium. Chromium is a hard but brittle metal of good corrosion resistance. It is mostly alloyed with other metals, especially steel, to make final products harder and more resistant to corrosion. It is also used in electroplating, leather tanning, and refractory brick making, and as glass pigments.

Cadmium. Cadmium is resistant to corrosion, malleable, ductile, and of high electrical and thermal conductivity. It is mostly used in rechargeable nickel-cadmium batteries. It is also used to make electronic components and pigments for plastics, glasses, ceramics, enamels, and artists' colors.

Nickel. Nickel is a hard, malleable, and ductile metal that is highly resistant to corrosion. Like chromium, it is used to make stainless steel. Alloyed with copper, it is used for ship propellers and chemical industry plumbing. Other uses include rechargeable batteries, coinage, foundry products, plating, burglar-proof vaults, armor plates, and crucibles.

Aluminum. Aluminum has a density about one-third that of steel, high resistance to corrosion, and excellent electrical and thermal conductivity. Moreover, this nontoxic metal reflects light and heat well.

This versatile metal can be used to replace other materials depending on the application. It is widely used in such areas as food packaging and protection (foils, beverage cans), transportation (vehicles, trains, aircraft), marine applications (ships, support structures for oil and gas rigs), and buildings and architecture (roofing, gutters, architectural hardware). Other applications of aluminum include sporting goods, furniture, road signs, ladders, machined components, and lithographic printing plates.

Special Alloys. Fusible alloys are mixtures of cadmium, bismuth, lead, tin, antimony, and indium. They are used in automatic sprinklers and in forming and stretching dies and punches. Superalloys are developed for aerospace and nuclear applications: columbium for reactors, tantalum for rocket nozzles and heat exchangers in nuclear reactors, and a nickel-based alloy for jet and rocket engines and electric heating furnaces. The alloy of tin and niobium has superconductivity. It is used in constructing superconductive magnets that generate high field strengths without consuming much power.

Powder Metallurgy Applications and Products. Powder metallurgy was developed in the late 1920's primarily to make tungsten-carbide cutting tools and self-lubricating electric motor bearings. The technique was then applied in the automobile industry, where it is used to make precision-finished machine parts, permanent metal filters, bearing materials, and self-lubricating bearings. It is useful in fabricating products that are difficult to make by other methods, such as tungsten-carbide cutting tools, super magnets of aluminum-nickel alloy, jet and missile applications of metals and ceramics, and wrought powder metallurgy tool steel.

IMPACT ON INDUSTRY

Metallurgy is used in many industries, and the challenges associated with it vary by industry. All industries, for example, seek ways to achieve fuel savings, and where metallurgy is concerned, they want high metallic yields and high-quality products.

Aluminum production requires large amounts of electricity for the electrolytic smelting process. Electricity accounts for 25 percent of the cost of producing aluminum. The techniques of ladle metallurgy (degassing, desulfurization) require high temperatures for preheating, also resulting in high energy costs. Hydrogen and oxides can cause porosity in solidified aluminum, which decreases the mechanical properties of the final product. High oxygen levels ranging from 25 percent in converters to 90 percent in flash-smelting furnaces are used for primary smelting, resulting in low productivity as well as high operating costs.

New technologies in metallurgy not only solve operational problems but also try to offer such benefits as high productivity, availability and reliability of power, safety, and environmental acceptance. For example, Alstom, a company in transport and energy infrastructure, provides air-quality control systems for the aluminum industry. To abate emissions from pot lines, anodes, and green anodes, Alstom systems take the gas from the pots and absorb hydrofluorides on alumina. Additional sulfur is removed from the gas before it is released to the atmosphere. Alstom has also developed an energy recovery system to decrease pot amperage in smelters, thereby ensuring high availability and reliability of power.

Air Liquide provides solutions to the problems caused by the need to preheat the ladle by using oxy-fuel burners, based on the evolution in refractory materials. The necessary calories needed to keep a metal at the suitable pouring temperature during the metal transfer into the ladle are stocked in the refractory. Air Liquide also provides solution to the porosity problem in aluminum. Its technique involves injecting hydrogen through porous plugs. The same company also developed an innovative scheme to lower the use of oxygen in smelting furnaces. It includes a process that allows the direct production of oxygen at 95 percent purity under medium pressure without an oxygen compressor and a balancing system for liquid and gaseous oxygen for handling fast-changing regimes and optimizing working conditions at low operating costs.

Problems in metallurgy are also solved by simply replacing the old technologies. For example, operating problems in the copper and nickel converter include tuyere blockage and refractory erosion at the tuyere line. Using an oxygen injector to replace conventional tuyeres not only permits enrichment levels of up to 60 percent of oxygen and increases converter productivity but also reduces volumes of toxic off-gases.

The metallurgy industry, like many other industries, must constantly deal with new laws and regulations and expends considerable time and money to meet environmental requirements. The industry must adapt its extraction processes and invest in emission-reducing and energy-saving technologies. For example, the steel industry has replaced all its open-hearth furnaces. The aluminum industry has used prebaked instead of Soderberg electrodes in the electrolytic cells. The copper industry has replaced the reverberatory furnace with the flash-smelting furnace. The common reason behind these changes was to reduce toxic off-gases as much as possible. Stacks have become much taller to dispose of sulfur dioxide.

A process based on the combination of biology and hydrometallurgy—also called bioleaching, biohydrometallurgy, microbial leaching, and biomining—has gained some interest from those who want to replace some of the traditional metallurgical processes. Although conventional metallurgy uses smelting of ores at high temperatures, bioleaching involves dissolving metals from ores using microorganisms. Copper, for example, can be leached by the activity of the bacterium *Acidithiobacillus ferrooxidans*. Canada used this process to extract uranium in 1970, and South Africa experimented it with gold during the 1980's. Several countries have used bioleaching for a number of metals (copper, silver, zinc, nickel). The process is helpful in the recovery of low-grade ores, which cannot be economically processed with chemical methods; however, it creates a problem because it results in excessive disposal and handling of a mixture of ferrous hydroxide and gypsum.

CAREERS AND COURSE WORK

High school students who wish to work as metallurgical technicians must take at least two years of mathematics and two years of science, including a physical science. Shop courses of any kind are also helpful. Positions in the metallurgical industry are typically in the areas of production, quality control, and research and development, which share many concerns and often require similar skills from prospective metallurgical technicians. Two years of study in metallurgy or materials science at a community college or technical college is therefore strongly recommended. Metallurgical technicians occupy a middle ground between engineers and skilled trade workers. Representative entry-level jobs include metallurgical laboratory technicians, metallographers, metallurgical observers, metallurgical research technicians,

and metallurgical sales technicians. Students who are interested in these kinds of jobs should have an interest in science and average mathematical ability. Prospective technicians must be willing to participate in a wide variety of work and must be able to communicate well. Companies employing metallurgical technicians can be found in a wide variety of industries. Working environments vary depending on the area of activities.

A number of colleges and universities offer four-year programs in metallurgy. If students wish to become metallurgical engineers, they will need a bachelor's degree in materials science or metallurgical engineering. The first two years of college focus on subjects such as chemistry, physics, mathematics, and introductory engineering. In the following years, courses will focus on metallurgy and related engineering areas. Students who wish to become metallurgical engineers should be interested in nature and enjoy problem solving. They also need to have good communication skills. There are basically three areas in which metallurgical engineers work: extractive, physical, and mechanical metallurgy. Their work environment varies depending on their area of specialty. Companies employing metallurgical engineers include metal-producing and processing companies, aircraft companies, machinery and electrical equipment manufacturers, the federal government, engineering consulting firms, research institutes, and universities.

SOCIAL CONTEXT AND FUTURE PROSPECTS

Metallurgy faces the challenges of reducing the effect of its processes on the air and land, making more efficient use of energy, and increasing the amount of recycling. To these ends, industries are increasingly using clean technologies and developing methods of oxygen combustion that drastically reduce emissions of carbon dioxide and other pollutants. For example, zinc, copper, and nickel are being recovered from their ores through a technique in which pressure leaching is performed in an acid medium, followed by electrolysis in a conventional sulfuric acid medium. The technique produces no dust, no slag, and no sulfur dioxide and therefore is environmentally acceptable. It has been applied for zinc sulfide concentrates in 1980's, for nickel sulfides in Canada, and for copper sulfide concentrates in the United States. In addition, between 1994 and 2003, the steel industry reduced its releases of chemicals to the air and water by 69 percent. However, it still releases manganese, chromium, and lead to the air, and efforts are concentrating on this problem.

Becoming more energy efficient has long been a goal of the metallurgy industry, especially as it benefits the bottom line. A wide range of approaches has been employed, including lowering generation costs (for example, by generating energy rather than purchasing it), capturing energy (such as gases) produced during various metallurgy processes, making use of energy-efficient equipment and techniques, and monitoring the production process. The steel industry in North America has reduced its consumption of energy by 60 percent since World War II. One way that the industry reduced energy consumption was by using scrap steel instead of natural resources to produce steel.

Metallurgy companies have made efforts to increase recycling. Each year, more steel is recycled in the United States than paper, aluminum, plastic, and glass combined. In 2008, more than 75 million tons of steel were either recycled or exported for recycling. All new steel produced in the United States in 2008 contained at least 25 percent steel scrap on average.

Despite these environmental challenges, metallurgy is an important area and will continue to grow and develop. In addition to developing ways to lessen metallurgy's impact on the environment, engineers are likely to develop improved processes for extraction of ores and processing materials into products. New techniques are likely to develop in response to the need to impart metals with additional qualities and to conserve natural resources by reusing metals and finding uses for by-products. Waste disposal and reduction also are likely to remain areas of research.

Anh Tran, Ph.D.

FURTHER READING

Abbaschian, Reza, Lara Abbaschian, and Robert E. Reed-Hill. *Physical Metallurgical Principles.* Stamford, Conn.: Cengage Learning, 2009. A comprehensive introduction to physical metallurgy for engineering students.

Boljanovic, Vukota. *Metal Shaping Processes: Casting and Molding, Particulate Processing, Deformation Processes, and Metal Removal.* New York: Industrial Press, 2010. Describes the fundamentals of how metal is shaped into products.

Bouvard, Didier. *Powder Metallurgy.* London: ISTE, 2008. Looks at the thermo-mechanical processes used to turn powdered metal into metal parts and also describes applications.

Brandt, Daniel A., and J. C. Warner. *Metallurgy Fundamentals.* 5th ed. Tinley Park, Ill.: Goodheart-Willcox, 2009. Examines metallurgy, focusing on iron and steel but also covering nonferrous metals.

Pease, Leander F., and William G. West. *Fundamentals of Powder Metallurgy.* Princeton, N.J.: Metal Powder Industries Federation, 2002. A primer on powder metallurgy.

Popov, K. I., Stojan D. Djokić, and Branimir N. Grgur. *Fundamental Aspects of Electrometallurgy.* New York: Plenum, 2002. Examines the theory and mechanisms of electrometallurgy.

Vignes, Alain. *Handbook of Extractive Metallurgy.* Hoboken, N.J.: John Wiley & Sons, 2010. Examines how metals are transformed from ore into liquids ready for pouring.

WEB SITES

American Institute of Mining, Metallurgical, and Petroleum Engineers
http://www.aimeny.org

The Minerals, Metals, and Materials Society
http://www.tms.org

Mining and Metallurgical Society of America
http://www.mmsa.net

See also: Chemical Engineering; Electrometallurgy; Environmental Engineering; Steelmaking Technologies.

METEOROLOGY

FIELDS OF STUDY

Climatology; hydrology; atmospheric physics; atmospheric chemistry, oceanography.

SUMMARY

Interdisciplinary study of physical phenomena occurring at various levels of the Earth's atmosphere. Practical applications of meteorological findings all relate in some way to understanding longer-term weather conditions. On the whole, however, meteorological weather forecasts are—in contrast to the research goals of climatology—mainly short term in nature. They concentrate on contributing factors, including temperature, humidity, atmospheric pressure, and winds.

KEY TERMS AND CONCEPTS

- **Cyclone:** Air mass closed in by spiraling (circular) winds that can become a moderate or violent storm, depending on factors of humidity, temperature, and the force, changing direction, and altitude of the winds.
- **Dewpoint:** Temperature at which air becomes saturated and produces dew.
- **Front:** Interface between air masses with different temperatures or densities.
- **Isotherms:** Graphically recorded lines connecting all points in a given region (large or more limited) having exactly the same temperature.
- **Jet Stream:** Narrow but very fast air current flowing around the globe at altitudes between 23,000 and 50,000 feet.
- **Ozone:** Variant form of oxygen made up of three atoms; forms a layer in the upper atmosphere that helps absorb ultraviolet rays from the Sun.
- **Relative Humidity:** Percentage value indicating how much water is in an air sample in relation to how much it can hold, or its saturation point, at a given temperature and pressure.
- **Saturation Point:** Point where the water vapor in the air is at its maximum for a given temperature and pressure; point where condensation occurs.
- **Stratosphere:** Atmospheric layer above the troposphere; temperatures in this layer increase as the altitude becomes greater.
- **Temperature Inversion:** Phenomenon that is the opposite of normal atmospheric conditions. When air close to ground level remains colder and denser than air at higher levels, the warmer air can form a sort of cover, trapping the colder air, with resultant increases in ground-level fog and smog.
- **Troposphere:** Lowest layer of the atmosphere; contains 80 percent of the total molecular mass of the atmosphere.

DEFINITION AND BASIC PRINCIPLES

Meteorology is the study of the Earth's atmosphere, particularly changes in atmospheric conditions. The three main factors affecting change in the atmosphere are humidity, temperature, and barometric pressure. Dynamic short-term interaction among these three atmospheric factors produces the various phenomena associated with changeable weather. Low barometric pressure conditions are generally associated with greater capacity for the atmosphere to absorb water vapor (resulting in various cloud formations), whereas high pressure prevents absorption of humidity.

Meteorological calculation of relative humidity, for example, reveals how much more moisture can be absorbed by the atmosphere at specific temperature levels before reaching the saturation point. Changes in temperature (either up or down) will affect this dynamic process. Rainfall occurs when colder air pushes warmer, moisture-laden air upward into higher altitudes, where the warmer air mass begins to cool. Cooler air cannot hold as much water as warmer air, so the relative humidity of the warmer air mass changes, resulting in condensation and precipitation. This phenomenon is closely associated with the presence of surface winds.

BACKGROUND AND HISTORY

Meteorology, like several other applied sciences that stem from observations of natural phenomena, has a long history. The term "meteorology" comes from the Greek word for "high in the sky." Aristotle's work on meteorology maintained that the Sun attracted two masses of air from the Earth's surface, one humid and moist (which returned as rain) and the other hot and dry (the source of wind currents). His student, Theophrastus, described distinct atmospheric signs associated with eighty types of rain, forty-five types of wind, and fifty storms.

During the Renaissance, Europeans developed instruments that could refine these ancient Greek theories. The Italian scientist Galileo, for example, used a closed glass container with a system of gauges that showed how air expands and contracts at different temperatures (the principle of the thermometer). The French philosopher Blaise Pascal developed what became the barometer, a device to measure surrounding levels of atmospheric pressure.

Although many important small-scale experiments would be carried out in the eighteenth and nineteenth centuries, a major breakthrough occurred in the first quarter of the twentieth century when the Swede Vilhelm Bjerknes and his son Jacob Bjerknes developed the theory of atmospheric fronts, involving large-scale interactions between cold and warmer air masses close to the Earth's surface. In the 1920's, a Japanese meteorologist first identified what came to be known as jet streams, or fast-moving air currents at altitudes of 23,000 and 50,000 feet.

The turning point for modern meteorology, however, came in April, 1960, when the United States launched TIROS 1, the first in a series of meteorological satellites. This revolutionary tool enabled scientists to study atmospheric phenomena such as radiation flux and balance that were known but had not been measured with high levels of accuracy.

How It Works

Meteorologists employ a variety of basic tools and methods to obtain the data needed to put together a comprehensive picture of local or regional atmospheric conditions and changes. At the most basic level, meteorologists direct their attention to three essential factors affecting the atmosphere: temperature, air pressure, and humidity.

Drawing on empirical data, meteorologists not only analyze the effects of temperature, pressure, and humidity in the area of the atmosphere they are studying but also apply their findings to ever-widening areas of the globe. From their analyses, they are able to predict the weather—for example, the

direction and strength of the wind and the nature and the probable intensity of storms heading toward the area, even if they are still thousands of miles away.

Wind Strength and Direction. Wind, like rain and snowstorms, is a common weather phenomenon, but the meteorological explanations for wind are rather complicated. All winds, whether local or global, are the product of various patterns of atmospheric pressure. The most common, or horizontal, winds arise when a low pressure area draws air from a higher pressure zone.

To build a complete picture of the likely strengths and directions of winds, however, meteorologists must gather much more than simple barometric data. They must consider, for example, the dynamics of the Coriolis effect (the influence of the Earth's rotation on moving air). Except in the specific latitude of the equator, the Coriolis effect, which is greater near the poles and less near the equator, makes winds curve in a circular pattern. Normal curving from the Coriolis effect can be altered by another force, centrifugal acceleration, which is the result of the movement of air around high and low pressure areas (in opposite directions for high and low pressure areas). Tornadoes and hurricanes (giant cyclones) occur when centrifugal acceleration reaches very high levels.

When a near balance exists between the Coriolis effect and centrifugal acceleration, the resultant (still somewhat curved) wind pattern is called cyclostophic. If no frictional drag (deceleration associated with physical obstacles at lower elevations) exists, something close to a straight wind pattern occurs, especially at altitudes of about two-thirds of a mile and greater. This straight wind, called a geostrophic wind, is characterized by a balance between the pressure gradient and the Coriolis effect.

In some parts of the Northern Hemisphere, massive pressure gradient changes produced by differences in the temperature of the land and the ocean create monsoon winds (typically reversed from season to season), which in the summer are followed by storms and heavy rains. The best-known example of a monsoon occurs in India, where the rising heat of summer creates a subcontinent-wide thermal low pressure zone, which attracts the moisture-laden air from the Indian Ocean into cyclonic wind patterns and much needed, but sometimes catastrophic, heavy rainfall.

Atmospheric Absorption and Transfer of Heat. Meteorologists worldwide use various methods to determine how much heat from the Sun (solar radiation) actually reaches the Earth's surface. Heat values for solar radiation are calculated in relation to a universal reference, the solar constant. The solar constant (1.37 kilowatts of energy per square meter) represents the density of radiation at Earth's mean distance from the Sun and at the point just before the Sun's heat (shortwave infrared waves) enters the Earth's atmosphere. Not all this heat actually reaches Earth's surface. The actual amount is determined by various factors, including latitudinal location, the degree of cloud cover, and the presence in the atmosphere of trace gases that can absorb radiation, such as argon, ozone, sulfur dioxide, and carbon dioxide.

At the same time, data must be gathered to calculate the amount of heat leaving the Earth's surface, mainly in the form of (longwave) infrared rays. Meteorologists attempt to calculate, first on a global scale and then for specific geographic locations, ecologically appropriate energy budgets.

Cyclones. Cyclones include a number of forms of severe weather, the most violent of which is known as a tornado. Cyclones are characterized by circular or turning patterns of air centered on a zone of low atmospheric pressure. The direction of cyclone rotation is counterclockwise in the Northern Hemisphere but clockwise in the Southern Hemisphere. Frontal cyclones, the most common type, usually develop in association with low pressure troughs that form along the polar front, a front that separates arctic and polar air masses from tropical air masses. Typical cyclogenesis occurs when moisture-laden air above the center of a relatively warm low pressure area begins to rotate under the influence of converging and or diverging winds at the surface or at higher levels. Simply stated, the dynamic forces operating within a cyclone can pull broad weather fronts toward them, causing increasingly strong winds and precipitation. Anticyclones occur under opposite conditions, originating in areas of high pressure where air masses are pushed down from upper areas of the atmosphere.

APPLICATIONS AND PRODUCTS

Meteorology has both practical and professional, scientific applications. Everyday weather reports are probably the most common application of meteorology. Weather reports and forecasts are available through traditional sources such as newspapers, television, and radio broadcasts and can be obtained by calling the National Oceanic and Atmospheric Administration's National Weather Service, using a smartphone application, or checking one of the many Web sites devoted to weather. Real-time weather reports, including radar, and hourly, daily, and ten-day forecasts are available for the United States and other parts of the world.

Knowledge of existing and forecasted weather conditions is invaluable to companies that provide public transportation, such as airlines. Knowledge of weather conditions and forecasts is essential to ensuring the safety of passengers on airplanes, trains, buses, boats, and ferries. Motorists, whether traveling for pleasure or commuting, need to plan for weather conditions. People who participate in outdoor recreational activities or sports, such as hiking, biking, camping, hang gliding, fishing, and boating, depend on accurate forecasts to ensure that they are adequately prepared for the conditions and to avoid getting into dangerous situations. Those participating in outdoor activities and sports often purchase various meteorological instruments—such as lightning detectors, weather alert radios, and digital weather stations—designed to keep them informed and aware.

A meteorologist tracks wind speed on a map at the National Hurricane Center in Miami, Florida. (Bloomberg via Getty Images)

Weather Equipment. Meteorology involves the measurement of temperature, air pressure, and humidity, and many companies produce equipment for this purpose, ranging from portable products for outdoor use, to products for home use, to products for industry use, to weather balloons and satellites. Some companies, such as Columbia Weather Systems in Oregon, which provides weather station systems and monitoring, focus on the professional level, and others, including Oregon Scientific, Davis Instruments, and La Crosse Technology, concentrate on the many hobbyists who enjoy monitoring the weather. Equipment ranges from simple mechanical rain gauges to digital temperature sensors to complete home weather stations. Some companies offer software that provides specific weather information in a timely manner to companies or individuals who need it.

Scientific Applications. The World Data Center in Asheville, North Carolina, distributes meteorological data that it has gathered and processed. Its facilities are open on a limited basis to visiting research scientists sponsored by recognized parent organizations or international programs. A variety of meteorological data are readily available on an exchange basis to counterparts of the center in other countries. One organization participating in the information exchange is the World Climate Programme, an international program headquartered in Switzerland devoted to understanding the climate system and using its knowledge to help countries that are dealing with climate change and variability.

IMPACT ON INDUSTRY

A wide range of industries and commercial concerns depend on meteorological data to carry out their operations. These include commercial airlines, the fishing industry, and agricultural businesses.

Aviation. Perhaps the most obvious industry that relies on the results of meteorology is the airline industry. No flight, whether private, commercial, or military, is undertaken without a clear idea of the predicted weather conditions from point of departure to point of arrival. Changing conditions and unexpected pockets of air turbulence are a constant concern of pilots and their crews, who are trained in methods of analyzing meteorological data while in flight.

Every day, the National Weather Service issues nearly 4,000 weather forecasts for aviation. Meteorologists at the Aviation Weather Center in Kansas City, Missouri; the Alaska Aviation Weather Center in Anchorage; and at twenty-one Federal Aviation Administration Air Route Traffic Control Centers across the nation use images from satellites as well as real-time information from observation units at major airports and from Doppler radar to create reliable forecasts.

Fishing Industry. Rapidly changing and violent weather on the seas and oceans can result in the loss of fishing vessels, their crews, and their catch, or it can cause severe damage to boats and ships, all of which hurt the fishing industry. In addition, when boats get caught in dangerous storms, rescuers must attempt to reach survivors, often endangering themselves. Therefore, accurate, up-to-the-minute forecasts are essential for the fishing industry, which also uses weather reports to help determine the areas where the fish are most likely to be found.

Agriculture. In agriculture, short meteorological predictions are of less importance than longer-term predictions, particularly for agricultural sectors that depend on rainfall rather than irrigation to grow their crops and that rely on predictable windows without rainfall to harvest and dry them. If the possibility of a prolonged drought menaces crops, both large- and small-scale agriculturalists turn to meteorologists to learn if changing weather patterns may bring them relief. However, short-term predictions are helpful, in that they help farmers deal with problems that could create catastrophic losses, such as too much rain in a short period, which could create flooding; extreme heat or cold, which, for example, can ruin the citrus crop; or strong winds and hail, which can flatten corn.

CAREERS AND COURSE WORK

Those seeking a career in meteorology can consider a wide range of possibilities. The American Meteorological Association requires at least twenty semester credits in the sciences, which include geophysics, earth science, physics, and chemistry, as well as computer science and mathematics. Specialized courses beyond such basic science courses (such as atmospheric dynamics, physical meteorology, synoptic meteorology, and hydrology) represent slightly over half the required courses. A number of governmental agencies and private commercial businesses employ full- or part-time meteorologists to provide technical information needed to carry out their operations.

Fascinating Facts About Meteorology

- The 53rd Weather Reconnaissance Squadron, known as the Hurricane Hunters of the Air Force Reserve, has been flying into tropical storms and hurricanes since 1944. The squadron's planes fly into the storms and send data directly to the National Hurricane Center by satellite.

- Geology departments at some colleges and universities, such as Ball State University in Muncie, Indiana, have led storm chasing groups, which train students in basic storm knowledge and lead them in pursuit of tornadoes and other violent storms.

- The popularity of storm chasing, featured in a Discovery channel television series, has led to the creation of storm chasing commercial tours. One group warns potential customers that while storm chasing, the group may not be able to stop for dinner.

- Without the continual pull of gravity toward the center of the Earth, the atmosphere would gradually disperse into space. Because air is made up of molecules in a gaseous rather than solid material state, it clings to the Earth.

- The chemical content of sedimentary layers on ocean floors and gases trapped in Arctic ice reveal data concerning the composition of the Earth's atmosphere in earlier geological times.

- The amount of oxygen contained in the air in higher mountain environments is markedly less than at sea level or mid-range altitudes. People coming to the mountains from lower elevations can experience serious shortness of breath and even altitude sickness, which is characterized by headaches, loss of appetite, fatigue, and nausea.

- At times, the atmosphere cannot keep up with the Earth it surrounds. This is particularly true at higher levels of the atmosphere, especially near the equator. At the same time, some parts of the atmosphere travel faster than others because of strong jet stream winds.

Probably the most familiar job of a meteorologist is to prepare weather reports for broadcast on television or to be published in newspapers. The task of a weather forecaster for a local television station is more complex than it may seem judging from the briefness of the televised forecast. Local predictions are created from data gathered from diverse sources, often sources that gather information for an entire region or the whole country, including pulse-doppler radar operators, who receive special training to enter the profession.

Government Agencies. Opportunities for employment in agencies that gather meteorological data are to be found within the National Oceanic and Atmospheric Administration (NOAA), part of the U.S. Department of Commerce. The NOAA contains a number of organizations that deal with meteorology, including the National Environmental Satellite, Data, and Information Service, a specialized satellite technology branch that provides global environmental data and assessments of the environment. Another NOAA organization is the National Weather Service, which provides forecasts, maps, and information on water and air quality. The Office of Oceanic and Atmospheric Research contains the Climate Program Office, which studies climate variability and predictability.

The National Centers for Environmental Prediction, part of the National Weather Service, oversees operations by several key specialized service organizations run by professional meteorologists. The most important of these are the Hydrometeorological Prediction Center in Washington, D.C., the Tropical Prediction Center (includes the National Hurricane Center in Miami, Florida), and the Storm Prediction Center in Norman, Oklahoma, which maintains a constant tornado alert system for vulnerable geographical regions. Other operations include the Ocean Prediction Center, the Aviation Weather Center, the Climate Prediction Center, Environmental Modeling Center, and the Space Weather Prediction Center.

Private Commercial Operations. A number of private companies are involved in the development and production of instruments used for meteorological data gathering. The instruments range from simple devices to sophisticated electronic equipment, complete with software. The development of new weather satellites requires instruments that can be used by and can make the most of the gathered information.

SOCIAL CONTEXT AND FUTURE PROSPECTS

The worldwide need for accurate daily and short-term meteorological forecasts, delivered in various formats, will result in continuing development of more accurate instruments and better predictive models, as well as improved and additional methods of packaging and delivering the information to users. Accurate prediction of where extreme weather, such as hurricanes and tornadoes, will occur and how intense the storms will be has the potential to save many lives and possibly minimize economic damage, and meteorologists will continue to conduct research in this area. Data gathered by meteorologists also are gaining importance in analyzing global issues such as air pollution and changes in the ozone layer. The ability to gather information by satellites allowed meteorology to make significant advances, and future research is likely to focus on increasing satellites' data-sensing ability and the speed and quality of data transmission as well as on software to interpret and analyze the information.

Satellite Technology. Development of satellite technology continues to revolutionize the science of meteorology. Various types of polar-orbiting satellites have been devised since the early 1960's. The task of the satellite bus—the computer-equipped part of the satellite without sensitive recording instruments—is to transmit data gathered by an increasingly sophisticated variety of devices designed to collect vital data.

One such scanning device is the advanced very-high-resolution radiometer (AVHRR) used to measure heat radiation rising from localized areas on the Earth's surface. An AVHRR, very much like a telescope, projects a beam that is split by a set of mirrors, lenses, and filters that distribute the work of data recording to several different sensor devices, or channels. To obtain an accurate reading of radiation rising from a given target, the AVHRR must calibrate data received from these different sensors. AVHRR technology produces images that depict the horizontal structure of the atmosphere, and another radiation-recording instrument, the high-resolution infrared radiation sounder (HIRS), produces soundings based on the vertical structure of the atmosphere.

In zones of widespread cloud cover, the microwave sounding unit (MSU) may be used. The area scanned by an MSU is about a thousand times greater than that scanned by a device using infrared wavelengths. Other specialized sounding devices used by meteorological satellites include stratospheric sounding units and solar backscatter ultraviolet radiometers (SBUV), which measure patterns of reflection of solar radiation coming back from the Earth's surface.

The functions of SBUVs in particular became more and more critical as concern about changes in the composition of the ozone layer emerged in the last decades of the twentieth century. Ozone-depleting substances such as chlorofluorocarbons (used for air-conditioning systems and as a propellant for aerosol sprays) have had a negative effect on the protective ozone layer. Ozone depletion allows increased levels of ultraviolet radiation to reach the Earth, raising the incidence of skin cancer and damaging sensitive crops. Beyond the obvious utility of using satellites to predict global weather, these devices play a major role in helping meteorologists monitor the all-important radiation budget (the balance between incoming energy from the Sun and outgoing thermal and reflected energy from the Earth) that ultimately determines the effectiveness of the atmosphere in sustaining life on Earth.

Air Pollution. Although the presence of pollutants in the air has been recognized as an undesirable phenomenon since the onset of the Industrial Revolution, by the end of the twentieth century, the question of air quality began to take on new and alarming dimensions. As countries industrialized, the combustion of fossil fuels to power industry and later automobiles released ever-increasing amounts of solar-radiation-absorbing greenhouse gases into the atmosphere. The most alarming effects stem from carbon dioxide, but meteorologists are also concerned about other serious gaseous pollutants and a wide variety of chemical particles suspended in gases (aerosols). Even if industries do not pollute by burning fossil fuels, many of them release sulfur dioxide, asbestos, and silica in quantities large enough to seriously damage air quality. Natural phenomena, including massive volcanic eruptions, also can produce atmospheric chemical imbalances that challenge analysis by meteorologists.

Byron D. Cannon, M.A., Ph.D.

FURTHER READING

Budyko, M. I., A. B. Ronov, and A. L. Yanshin. *History of the Earth's Atmosphere*. Berlin: Springer-Verlag,

1987. Deals with various ways scientists can determine the probable composition of the atmosphere in earlier geological ages.

David, Laurie, and Gordon Cambria. *The Down-to-Earth Guide to Global Warming.* New York: Orchard Books, 2007. A nontechnical discussion of practical factors, many daily and seemingly inconsequential, that contribute to rising levels of carbon dioxide in the atmosphere.

Fry, Juliane L, et al. *The Encyclopedia of Weather and Climate Change: A Complete Visual Guide.* Berkeley: University of California Press, 2010. Covers all aspects of weather, including what it is and how it is monitored, as well as the history of climate change. Many photographs, diagrams, and illustrations.

Grenci, Lee M., Jon M. Nese, and David M. Babb. 5th ed. *A World of Weather: Fundamentals of Meteorology.* Dubuque, Iowa: Kendall/Hunt, 2010. Examines the study of meteorology and how it is monitored.

Hewitt, C.N., and Andrea Jackson, eds. *Handbook of Atmospheric Science.* Oxford, England: Blackwell, 2003. Examines research approaches to and data on the chemical content of the atmosphere.

Kidder, Stanley Q., and Thomas H. Vonder Haar. *Satellite Meteorology.* 1995. Reprint. San Diego: Academic Press, 2008. Covers ways in which satellite instruments can monitor meteorological phenomena at several levels of the atmosphere.

Williams, Jack, and the American Meteorological Society. *The AMS Weather Book: The Ultimate Guide to America's Weather.* Chicago: University of Chicago Press, 2009. Examines common weather patterns in the United States and discusses the science of meteorology.

WEB SITES
Air Force Weather Observer
http://www.afweather.af.mil/index.asp

American Meteorological Society
http://www.ametsoc.org

International Association of Meteorology and Atmospheric Sciences
http://www.iamas.org

National Oceanic and Atmospheric Administration
National Weather Service
http://www.weather.gov

World Meteorological Organization
http://www.wmo.int/pages/index_en.html

See also: Aeronautics and Aviation; Air-Quality Monitoring; Atmospheric Sciences; Barometry; Climate Engineering; Climate Modeling; Climatology; Industrial Pollution Control.

MICROLITHOGRAPHY AND NANOLITHOGRAPHY

FIELDS OF STUDY

Physics; chemistry; mathematics; calculus; optics; mechanical engineering; materials science; thermodynamics; electrical engineering; microtechnology; nanotechnology; electromagnetics; semiconductor manufacturing; radiology; laser imaging.

SUMMARY

Microlithography and nanolithography are two closely related fields in the area of electronics manufacturing. Microlithography is a process used in making integrated circuits in semiconductors. The process relies on the projection of an image onto a light-sensitive plate. Circuits made by microlithographic techniques can carry features as small as 100 nanometers in width. Nanolithography refers to the making of circuits and other features even smaller and finer than those possible through traditional microlithography. New developments in nanolithography are making it possible to apply the technology in areas beyond semiconductors, such as nanoelectromechanical systems (NEMS) in highly sensitive measuring devices.

KEY TERMS AND CONCEPTS

- **Exposure:** Measured amount of light shone onto a light-sensitive surface to create an image.
- **Integrated Circuit:** Circuit manufactured as a single, rigid object; also known as a microchip.
- **Lithography:** Printing process in which an image is transferred from a flat plate to another surface.
- **Photomask:** Template used to project an image onto a light-sensitive surface.
- **Resist:** Light-sensitive material found in a thin layer on the surface of a microchip.
- **Self-Assembly:** Manufacturing method used in nanolithography in which individual chemical molecules are used to create highly precise images.
- **Substrate:** Rigid surface that carries the resist layer on a microchip.
- **Wafer:** In electronics, a sheet of material, most commonly silicon and as thin as 200 microns, carrying up to several hundred microchips.

DEFINITION AND BASIC PRINCIPLES

Some sources say that the difference between microlithography and nanolithography has to do with the size and precision of the electronic components being made. Microlithography is associated with a level of precision in manufacturing that can be measured in microns, while nanolithography is associated with nanometers.

In reality, the differences are more complicated than feature size. Microlithography is a term associated with the making of integrated circuits and semiconductors. It refers to a process of projecting images from a master pattern onto a light-sensitive surface. The process shares some basic traits with traditional lithography, a form of ink-based printing in which an image is transferred from a flat plate to a surface such as paper or cardboard.

Nanolithography refers not just to microlithography on a smaller scale. It involves a wide range of technologies, including the uses of electron beams and scanning probes, to make components in products ranging from video screens to cultures for the growing of organic tissue used in transplant surgery. While the terms are not used interchangeably, the fields are often grouped together, such as in the *Journal of Micro/Nanolithography, MEMS, and MOEMS*. (MEMS stands for microelectromechanical systems and MOEMS for micro-opto-electromechanical systems.)

BACKGROUND AND HISTORY

The roots of microlithography can be found in a program funded from 1957 to 1963 by the U.S. Army Signal Corps. The Micro-Module Program was launched shortly after the discovery of the transistor by scientists at Bell Telephone Laboratories in 1947. Before transistors, electronic components relied on vacuum tubes, which were large, broke easily, and generated heat. Transistors offered the possibility of making much smaller and more durable devices that needed less energy to run. However, transistors presented a number of manufacturing problems. They were made from numerous parts consisting of different materials that had to be assembled and soldered by hand. The Micro-Module Program encouraged the development of newer technologies in which components could be made with built-in wiring and joined together to form circuits. Texas Instruments engineer Jack Kilby discovered a way to make integrated circuits in a more cost-effective way in 1958. At nearly the same time, Robert Noyce at Fairchild Semiconductor made a related breakthrough using silicon transistors. Microlithography grew rapidly with the demand for microchips in electronics.

One of the earliest advances in the field of nanolithography was a 1983 NATO Advanced Research Workshop in Rome. Leading experts from a range of fields, including X-ray and electron-beam technology, met for a cross-disciplinary discussion of the ways in which nanolithography could be applied to technological problems. The field has since evolved beyond semiconductors to areas such as chemistry and medicine.

HOW IT WORKS

Microlithography is a process used specifically for the manufacturing of microchips. It allows a series of circuits to be placed on a hard surface with a tremendous amount of precision. Improvements in technology have made it possible for microchips to carry increasing numbers of circuits in new combinations and patterns.

Microlithography begins with the creation of a template, or pattern, known as a photomask. The photomask plays a role similar to that of a plate on a printing press. It carries an image of the circuits to be built on the surface of a chip. The image on the photomask is projected onto the chip's surface with the help of a lithographic lens. This lens is one of the most accurate and advanced pieces of equipment needed to make a microchip. An individual lens can be made out of more than thirty components and weigh almost a half ton. A source of intense light, such as a high-pressure mercury arc lamp or a laser, is shone through a series of refraction elements and eventually through the photomask itself. The light projects the photomask's image onto the surface of a silicon wafer by shining through the lithographic lens, which condenses the image down to a suitably small size for placement on the surface of the wafer. The wafer's surface has been coated with a thin layer of a light-sensitive substance. When the image is shone onto this layer, known as a photoresist or simply a resist, the intensity of the light burns the image into the resist's surface within a fraction of a second. This procedure is known as an exposure.

A single microchip goes through many exposures in the manufacturing process. At each step, the accuracy of the procedure must be absolute or the microchip will be unusable. The tolerance for errors in focusing and placing the image from the photomask is below 1 percent, or a few nanometers. Wafers are held in place by a vacuum system, and their positions are verified by lasers before each exposure occurs. For a manufacturing process to be profitable, errors must be as low as possible while wafers pass through each stage quickly. A standard system might process fifty to one hundred or more wafers in an hour.

Unlike microlithography, there are a large number of manufacturing methods used in nanolithography, each with its own steps and specialized equipment. Some of these methods rely on radiation such as X rays or ultraviolet light rather than lasers. Electron projection lithography falls into a similar category. It relies on an electronic lens and electric and magnetic fields to guide charged particles into specific shapes and patterns with the help of a mask. Many of these particles move on wavelengths short enough to pass directly through a mask. In these cases, the masks deflect the particles away from the resist rather than stopping the energy entirely. Other methods do not use photomasks at all, which are expensive and difficult to produce for small manufacturing batches, but focus a beam of light directly onto the resist surface to make the needed patterns.

These methods belong to a category of nanolithography known as top-down. In contrast, bottom-up manufacturing methods use combinations of chemical solutions to assemble images directly on a surface. These images are so precise that the assembly process takes place on the scale of individual molecules.

APPLICATIONS AND PRODUCTS

Semiconductors. Virtually all manufacturing that involves microlithography is used to make semiconductors. Like printing with lithographic plates, microlithography allows a highly complex pattern of electronic circuits to be transferred to many individual microchips at the same time. This process makes it possible to invest a great deal of time and financial resources into the creation of a sophisticated technological design, then duplicate the design on a broad scale precisely and quickly.

The process of microlithography is only one set of steps in a chain that leads to a finished microchip. Before the surface of a wafer is ready to receive a microlithographic image, the wafer itself must be formed from highly refined silicon. After refining, silicon is formed into single-crystal rods known as ingots and sliced into wafers. The refining and wafer-slicing process requires such specialized technology that this stage is often handled by companies devoted exclusively to it.

The silicon wafers are then purchased by semiconductor manufacturers, which prepare the wafers for microlithography by adding silicon dioxide (a form of sand) and elements with specific electrochemical properties such as boron, phosphorus, or arsenic. Silicon dioxide serves as an insulating base for the other additives, which are formed on the surface of the chip into transistors and circuits through repeated exposures from a photomask in the process of microlithography. The transistors are connected to each other through a metallization process that uses aluminum, tungsten, or a number of other metals and their compounds. The same metallization process is also used to make bonding pads on the surface of the chip. These pads are the points through which the chip communicates electronically with other chips and with the machine in which it will be installed.

Once the image on the chip's surface is complete, the surface itself is smoothed chemically and covered by a protective layer. The chip's transistors and circuits are tested before the chip is sold by the manufacturer. At this stage, the chip may be altered to a larger product's specifications, a process known as die cutting, before being installed inside another device.

Carbon Nanotubes. In contrast to microlithography, nanolithography as a manufacturing process has applications in a wide range of fields. Nanolithography has been found to be useful in the making of nanoelectromechanical systems (NEMS). One type of structure that shows the most promise is the carbon nanotube. Through nanolithography, carbon atoms are bonded together into molecules shaped like long, hollow tubes with closed ends. The result is no wider than a few nanometers but can be up to several inches long. Carbon nanotubes belong to a family of carbon molecules known as fullerenes, a group that also includes structures such as buckyballs. Because carbon molecules are strong conductors of electricity, nanotubes have many potential uses as components in transistors and circuits. They can also be used in electrical sensors that need a high level of precision and sensitivity, such as those used to track changes in gases.

The potential uses of carbon nanotubes are still being discovered. The presence of carbon atoms in compounds such as steel makes the resulting product lighter and stronger. A company in Finland has developed a way to add carbon nanotubes to the blades of windmills. The nanotubes make the windmill blades

---

Fascinating Facts About Microlithography and Nanolithography

- Microlithography makes up about one-third of the manufacturing cost of each microchip.
- Microchips become more complex as they go through each microlithographic procedure. The most advanced designs have fifty or more images layered on top of one other.
- Semiconductor expert Gordon Moore states in Moore's law that the number of devices that can be installed on a single microchip will double every year. Moore published his theory in an article in 1965. Though the cycle is now closer to a year and a half, the rule has held true for more than forty years.
- Silicon wafers are made into semiconductors in areas known as "clean rooms." Microlithograpic procedures are so precise that even dust and airborne contaminants can destroy the etching process on the surface of a wafer. Employees working in clean rooms wear garments over their clothes called "bunny suits."
- In 1999 an application of nanolithography known as dip-pen nanolithography was developed by Chad Mirkin at Northwestern University. The technology allows patterns at the nanoscale level, such as certain electronic circuits and biological cell material, to be copied.
- Texas Instruments engineer Jack Kilby, credited with developing one of the first designs for an integrated circuit, came up with his idea in July, 1958, while he was a new employee. He had not yet earned vacation time, so he was alone in the office while most of his colleagues were taking time off. Kilby received the Nobel Prize in Physics in 2000.

twice as light as those made from glass fibers and several times stronger, which allows for larger blades that move more efficiently. Other applications are still in the research stage. Scientists have noted, for example, that carbon nanotubes resemble tiny needles and may be used to carry antibodies and pharmaceuticals to highly targeted areas within the body, such as cancerous tumors.

Biosensors and Cell Biology. Dip-pen nanolithography (DPN) has shown great promise in the field of cell biology. DPN allows a chemical pattern of molecules, such as those found in the DNA of a cell, to be copied onto the surface of a microchip. DPN makes it possible to copy this pattern not once but thousands of times within a single manufacturing process. Unlike microlithography, which depends on light being shone through a photomask, DPN deposits a chemical agent directly onto the surface of a chip with the help of a "dip pen"—the highly precise tip of an atomic-force microscope. Direct contact between the pen and the surface means a higher possibility of problems, such as contaminating agents.

However, this challenge is outweighed by the possibilities presented by the technology in developing new kinds of cell-based therapies and other medical applications. Scientists at Northwestern University in Chicago have found that DPN may be used to replicate electrodes in DNA patterns. With the information from these patterns, customized biosensors could be developed that could, in theory, be reintroduced into living cells. These sensors could then transmit data that would be used to monitor a body's vital functions. The sensors could also track the progress of a disease or drug therapy.

Advancements beyond the technology of DPN are already being pursued. Polymer-pen lithography (PPL) uses many of the same processes as DPN, but it involves larger arrays of pens—up to 11 million, by an estimate from Northwestern—as well as the ability to push the pens with varying amounts of force against the writing surface. These features allow many details to be transferred to the writing surface during a single procedure, which makes the manufacturing process faster and more efficient. Beam-pen lithography (BPL) blends lithographic techniques with the technology behind scanning electron microscopes. These approaches are still in the early research stages, but they present possibilities for developing new treatments in fields ranging from genetics to heart disease.

IMPACT ON INDUSTRY

Microlithography's role in semiconductor manufacturing will continue to be critical in the early twenty-first century. New research in the field is primarily focused on finding ways to make manufacturing processes more efficient. Because of this, there is not a significant amount of federal or academic research funding devoted to finding new applications of the technology. However, institutes at schools such as the Georgia Institute of Technology and Rice University have built facilities devoted to microlithography research.

Nanolithography is receiving more attention and support from governments, academia and the private sector. Major efforts include:

Government Agencies and Initiatives. One of the leading providers of U.S. government agency support for new work in nanolithography is the National Nanotechnology Initiative (NNI). Launched in 2000, the NNI brings together more than twenty federal agencies and helps to coordinate their work in the field. The NNI serves as a communications hub as well as a source of financial support for research. It also tracks educational programs, particularly for college and graduate students, and is a good resource for learning more about careers in the field.

The NNI has also been instrumental in supporting nanotechnology development in other countries, primarily through the backing of cross-border efforts. Its international equivalent is the Global Issues in Nanotechnology (GIN) working group. Other programs led or supported by NNI include the Organisation for Economic Co-operation and Development (OECD) Working Party on Manufactured Nanomaterials and the OECD Working Party on Nanotechnology. Countries such as Japan, Australia, and the United Kingdom have also established government agencies or specially funded efforts to boost the development of nanotechnology. Nanolithography is too specialized a field to receive significant attention at the level of national government, but it is most commonly included under the banner of nanotechnology.

Universities and Research Institutions. Nanolithography is being explored in a variety of forms at universities and research institutions around the world. Because nanolithography has applications in more than one field, work contributing to its advancement can be carried out in departments such as electrical engineering, mechanical engineering, physics, chemistry, and others. In the United States, Northwestern and Stanford universities are leading contributors to new research in nanolithography. Institutions making significant contributions outside the United States include the University of Toronto, the University of Twente in the Netherlands, and the University of Strathclyde in Scotland.

Private Sector. The applications of nanolithography within the private sector involve many industries spread across the globe. One of the best ways to understand current developments is to look at the activities of cross-sector forums and consortiums. The memberships of these groups are made up of organizations of all kinds, but they include manufacturers that make nanolithography equipment as well as those that use the technology itself. Some of the most visible and active international groups are the Global Nanotechnology Network (GNN), the International Council on Nanotechnology (ICON), the European Nano Forum, and the Asia Pacific Nanotechnology Forum (APNF). These groups host conferences and publications supporting the sharing of information about nanotechnology.

CAREERS AND COURSE WORK

Microlithography and nanolithography are highly specialized areas when it comes to developing a career. At the same time, they involve the intersection of many different academic disciplines, such as engineering, physics, and chemistry.

Students interested in working in microlithography and nanolithography are likely to pursue bachelor's degrees with majors in fields such as electronics engineering, electrical engineering, or materials science. A student with a background in a related field such as mechanical engineering, physics, or computer science would also be well-positioned for a job in microlithography or nanolithography. Relevant course work starts with a foundation in the physical sciences and mathematics. Depending on the institution, advanced course work can be highly specialized. More than ten schools in the United States alone offer bachelor's degrees with majors in nanotechnology.

Due to the complex and specialized knowledge required to work in microlithography or nanolithography, many job candidates complete master's degrees or doctorates. Advanced degrees improve the earning

potential of graduates in the field as well as prepare one for higher-level positions such as research team leader. Among the degree programs tracked by the National Nanotechnology Initiative, none specializes exclusively in microlithography or nanolithography. However, some large research universities such as Georgia Institute of Technology host interdisciplinary centers that offer students and faculty members the opportunity to gain experience working with microlithography and nanolithography applications.

Outside of designing new systems and applying specialized knowledge, the job market for semiconductor manufacturing is very limited. The U.S. Bureau of Labor Statistics finds that the number of jobs in the field has fallen sharply in the first decade of the twenty-first century due to increased efficiencies in manufacturing processes. Most factory workers are required to wear full protective suits and to work in clean rooms to avoid introducing contaminants, which can affect the high precision of the components being made.

SOCIAL CONTEXT AND FUTURE PROSPECTS

The demand for new applications of microlithography and nanolithography is expected to continue for the foreseeable future. Microlithography remains one of the most precise and cost-effective ways to manufacture integrated circuits. An increasing number of consumer devices rely on microchips to function, ensuring that the need for inexpensive components will stay in place for many years. However, some sources say that further innovations in the field of microlithography are not expected because of the limitations of the technology. Without major changes to manufacturing processes, microchip makers are likely to continue to move their plants to markets around the world where manual labor is least expensive. A significant share of semiconductor manufacturing already takes place in regions such as Asia for cost-related reasons.

The outlook for nanolithography is more optimistic. As new applications for the technology are discovered, there is an increasing need for specialists. At present, some of the most promising areas of opportunity are in biotechnology, chemistry, and electronics outside of traditional integrated circuits. There is debate within the field about the level of precision that can be achieved in manufacturing through nanolithography. If the process is no longer cost-effective below a certain point, a next-generation technology will need to be developed.

Julia A. Rosenthal, B.A., M.S.

Further Reading

Cao, Guozhong, and Ying Wang. *Nanostructures and Nanomaterials: Synthesis, Properties, and Applications.* 2d ed. Singapore: World Scientific Publishing, 2011. Chapter 7 examines photolithography and nanolithography in detail as used in making semiconductors.

Guo, Zhen, and Li Tan. *Fundamentals and Applications of Nanomaterials.* Norwood, Mass.: Artech House, 2009. Chapter 6 places nanolithography into the context of nanotechnology as a field.

Levinson, Harry J. *Principles of Lithography.* 3d ed. Bellingham, Wash.: SPIE Press, 2011. Contains a detailed but readable explanation of the process of microlithography in making semiconductors.

Prasad, Paras N. *Nanophotonics.* Hoboken, N.J.: John Wiley & Sons, 2004. Excellent reference that covers photonics on a nano scale, written by a leader in the field.

Suzuki, Kazuaki, and Bruce W. Smith, eds. *Microlithography: Science and Technology.* 2d ed. Boca Raton, Fla.: CRC Press, 2007. Provides a thorough examination of microlithography and its applications, including technologies using electron projection and extreme ultraviolet light.

Web Sites

Global Nanotechnology Network (GNN)
http://www.globalnanotechnologynetwork.org

Journal of Micro/Nanolithography, MEMS, and MOEMS
http://spie.org/x865.xml

National Nanotechnology Initiative (NNI)
http://www.nano.gov

See also: Calculus; Computer Science; Electrical Engineering; Electromagnet Technologies; Mechanical Engineering; Nanotechnology; Radiology and Medical Imaging; Transistor Technologies.

MICROSCOPY

FIELDS OF STUDY

Electronics; electrical engineering; physics; optical physics; atomic physics; mathematics; statistics; image analysis; materials science; photomicrography; interferometry; electromagnetics; quantum electrodynamics; computer science; nanotechnology; metallography; electron microscopy; optical microscopy; scanning probe microscopy; cell biology; chemistry.

SUMMARY

Microscopy is the science of creating, observing, analyzing, and capturing visible images of objects and their components that are too small to be seen by the naked eye. It also refers to research conducted with the aid of microscopes, or instruments used for visual magnification. Microscopy is an essential tool for conducting research in a large number of scientific disciplines, including chemistry, biology, and medicine. For example, microscopy enables biologists to examine, in fine detail, the structure and function of individual components of a cell. The field also has a variety of industrial, materials science, and other practical applications. Powerful microscopes are used, for instance, to inspect the composition of the tiny silicon crystals used to manufacture semiconductors and integrated circuits and to detect minute defects in glass.

KEY TERMS AND CONCEPTS

- **Cantilever:** Flexible bracket to which the tip of an atomic force microscope is attached.
- **Compound Microscope:** Optical microscope containing two lenses, the objective and the eyepiece (or ocular lens).
- **Condenser:** Lens in a microscope that focuses, or condenses, a beam of light or electrons onto the object being studied.
- **Magnifying Power:** Ability of a lens or lenses to enlarge an object; measured in the number of times the magnified image is larger than the object appears to the naked eye.
- **Numerical Aperture (NA):** Measure of how much light an objective lens in an optical microscope is capable of gathering.
- **Objective:** Lens located closest to the object being studied; an objective is the primary magnifying lens in an optical microscope.
- **Ocular Lens:** Upper lens, or eyepiece, in a microscope, through which the viewer looks.
- **Optical Aberrations:** Errors in the image produced by an optical microscope, caused by the inability of a lens to accurately focus rays of light.
- **Photomicrography:** Photography of magnified objects through a microscope.
- **Resolving Power:** Extent to which a particular lens or microscope is able to distinguish between two points (to see them as separate from each other) at a given distance.
- **Stereomicroscope:** Microscope with two optical systems, one for each eye. Objects viewed through a stereomicroscope appear three-dimensional.
- **Working Distance:** Distance between the objective lens and the object being studied—at the point when the magnified image is in clear focus.

DEFINITION AND BASIC PRINCIPLES

Microscopes are scientific instruments whose purpose is to create enlarged visual images of objects so tiny they cannot be seen by the unaided human eye. Microscopy is an applied science concerned with ways of developing and improving microscope technology and relies heavily on knowledge gained from physics, mathematics, and engineering. Different varieties of microscopes function in various ways, but it is useful to understand two basic principles that apply to how well a microscope performs.

First, it might seem that the fundamental purpose of a microscope is to magnify an object. In reality, however, photographic enlargements can always be used to further enlarge the image any given microscope creates. Therefore, for a microscope to be truly useful to scientists and other researchers, it must not only magnify the specimen being observed but also properly separate (or "resolve") the details of individual components within the image. In effect, the more resolving power a microscope has, the crisper and clearer the magnified image it can produce and the more information it can provide about the object in question. If the resolution of a microscope is not high enough, for instance, two tiny dots next to each other might be perceived as a single element, no matter how much the image of the specimen was magnified.

Second, the image a microscope creates must possess a high enough degree of contrast to allow the viewer to clearly distinguish the object from its background and to differentiate various details within the object. A mostly translucent specimen, for example, might be impossible to make out against a bright background. Special techniques are used to increase the contrast in microscopic images. Phase contrast microscopy takes advantage of differences in the refractive indexes (the extent to which a material bends light) of various components of the specimen. Microscopes using this technology translate these variations into differences in the amplitude of the light waves reflected from each component. This results in light and dark areas that can be seen by the viewer. Interferometry is another important technique used to increase contrast. It does so by creating two images of a single specimen, superimposed on top of each other.

BACKGROUND AND HISTORY

The first ground glass lenses that had the ability to magnify objects were created in the late Middle Ages by monks who used them for reading. In the sixteenth century, the first microscopes were created in the Netherlands by inventors Hans Janssen and Zacharias Janssen, who placed two lenses into a series of tubes to create a primitive compound microscope. The device was focused by drawing one of the tubes in and out of the other, and it had a magnifying power of about 10 times (10x). The seventeenth century saw a flurry of interest in microscope technology. The Dutch amateur scientist Antoni van Leeuwenhoek designed hundreds of microscopes in which a bi-convex lens was placed between two glass plates. He was the first person to ever observe bacteria and other single-celled organisms under a microscope. In the eighteenth century, cuff-style microscopes were created, whose design prefigured the modern laboratory optical microscope. These were instruments with two brass tubes, one fixed and one sliding. By sliding the assembly up and down and turning a small thumbscrew, the object under observation could be brought into fine focus. The whole mechanism was mounted on a solid wooden base and had a magnifying power of up to 100x.

In the nineteenth century, advances in optical science and lens production pushed microscope technology forward by leaps and bounds. Two people whose work was prominent in this era were the German physicists and engineers Ernst Abbe and Carl Zeiss; together they designed sophisticated lenses that cut down drastically on spherical aberration (which causes points of lights to look like discs) and chromatic aberration (which distorts the colors of objects). The twentieth and twenty-first centuries have seen major changes in the field of microscopy, with the development of new technologies such as electron and scanning probe microscopes and huge improvements in the design of optical microscopy. Digital imaging—the transformation of an image created in an optical microscope to digital form—is making it easier than ever to analyze magnified specimens.

How It Works

Since the 1800's, hundreds of different varieties of microscopic technologies have been developed, each useful for performing certain types of observations on specific kinds of materials. Three broad categories of microscopy exist into which the majority of these technologies can be categorized.

Optical Microscopy. Optical microscopes, also known as light microscopes, create a magnified image by using a series of glass lenses to manipulate visible light, or light from the portion of the electromagnetic spectrum that can be seen by the naked eye. A condenser focuses light onto the specimen to be observed. As this light passes through or is reflected off the specimen, it is collected by one or more objectives. (Most microscopes have multiple objectives contained in a long tube with magnification powers ranging from 4x to 100x.) The objectives focus the light they have gathered into parallel rays; the result is a magnified image of the specimen. This image, however, is projected to a distance of infinity. To focus the image at a distance comfortable for the human eye, an ocular lens is required, through which the viewer looks. Most ocular lenses further magnify the image by another 10x.

In a conventional bright-field optical microscope, the source of this light is usually an incandescent or halogen lightbulb positioned directly below the specimen to be examined. Bright-field microscopy is useful for observing specimens that are either naturally dark or can be stained a dark color—such as cells or thin cross sections of biological material, usually placed on a glass slide. Images produced by a bright-field microscope appear dark against a bright white background.

Some specimens, such as living organisms, are difficult to see under bright-field microscopes. Other forms of optical microscopy, such as dark-field microscopy, have been developed to combat this problem. In dark-field microscopy, opaque material inside the condenser blocks the most central source of light, causing light to hit the specimen at oblique angles. When this angled light hits even the tiniest particle, the light scatters and makes the particle visible—like dust motes catching the angled light coming in from a window. Dark-field microscopes show specimens as bright points of light against a dark background.

Fluorescence microscopy is a special form of optical microscopy in which the specimen itself acts as the source of light. A fluorescence microscope irradiates the specimen with light of a certain wavelength, causing its atoms to become excited and emit energy as visible light. Some specimens, like chlorophyll, fluoresce naturally; others can be made to fluoresce through the use of chemicals. Other special forms of optical microscopy include phase contrast microscopy, in which small changes in the wavelength of light as it passes through transparent regions of the specimen are intensified so that they show up as areas of greater brightness, and confocal microscopy, in which light coming from out-of-focus regions of the specimen is filtered out of the final image through a pinhole aperture, eliminating blurry regions in the image.

Electron Microscopy. Electron microscopes operate using the same basic principles as optical microscopes—with one important difference. Where optical microscopy manipulates focused beams of visible light, electron microscopy manipulates focused beams of highly excited electrons, which have wavelengths much shorter than those of visible light. This technique enables objects to be magnified at far higher levels and resolved in far finer detail than optical microscopy. In addition, electron microscopes have larger depths of field, allowing a larger area of an object to be in focus at one time. The source of the electrons in an electron microscope is most often a thermionic electron gun—a device that shoots out a stream of electrons produced by heating a charged

electrode. The path these electrons take is shaped by a series of lenses, just as in an optical microscope. However, rather than being made out of glass, the lenses in an electron microscope consist of coils of wire (solenoids). An electric current passed through a solenoid creates an electromagnetic field that can direct the flow of electrons and focus it into a thin beam that can be directed toward the object under study.

In a transmission electron microscope (TEM), the beam of electrons enters the specimen and passes through it. When electrons hit dense regions of the specimen, they bounce off and are not included in the resulting image. The remaining electrons travel through the object and then pass through more electromagnetic lenses that create a final, magnified image on a fluorescent screen. In a scanning electron microscope (SEM), the beam of electrons sweeps across the surface of the specimen in a back-and-forth pattern of parallel lines known as a raster. As the beam scans over the object, it causes atoms within it to become excited and emit electrons that escape from the object. These electrons, known as deflected secondary electrons, are collected, counted, and measured. The information from this analysis is then used to create a magnified pixelized image on a

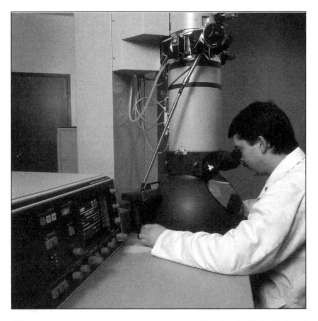

Using an electron microscope in a medical lab. (M. Gabr/ Publiphoto/Photo Researchers, Inc.)

computer screen. Because the electron beam sweeps over the entire surface of the object under study, electron microscopes are able to produce an image of the specimen's structure in three dimensions. Electron microscopes generally require samples to be placed in a vacuum in order to operate because molecules of air might disturb the movement of the electrons used to form images.

Scanning Probe Microscopy. Scanning probe microscopy abandons lenses altogether and makes use of very fine mechanical tips, or probes, attached to a cantilever. The probes delicately scan back and forth over the surface of the specimen being studied in order to inspect it. Scanning probe microscopes can deliver information about not only the topography of an object but also its internal properties. Some can even map a specimen's properties on a nanoscale.

Scanning tunneling microscopes (STM) rely on a phenomenon discovered by quantum mechanics, in which electrons—which have wavelike properties—are able to "tunnel" outside of the surface of a solid object into surrounding space. Scanning tunneling microscopes have incredibly sharp metallic probes, often made of tungsten or an alloy of platinum and iridium, with tips that are a mere one or two atoms in size. The tip of the probe does not touch the surface of the specimen but is held very close to it. An electric current of low voltage is applied to the gap between the two. In response to the current, electrons from the object tunnel across the gap. Changes in the intensity of the tunneling electrons are analyzed to produce an image of the object that can then be magnified. Scanning tunneling microscopes can be used only to examine specimens that conduct electricity.

Atomic force microscopes (AFM), whose tips are typically made of silicon or diamond, can probe surfaces made of practically any material. As the tip of an atomic force microscope is dragged across the surface of a specimen, it is either deflected by or drawn toward the object, depending on whether the atoms in the object are repelled by or attracted to the microscope's tip. By measuring these forces, a magnified representation of the physical structure of the sample can be created. By changing the modes in which these microscopes operate, different properties such as magnetism, friction, and electrical conductivity can be assessed.

Near-field scanning optical microscopes (NSOM) use a probe that emits an incredibly fine beam of

Fascinating Facts About Microscopy

- Microscopes need not be large themselves in order to enlarge other things. One of the world's smallest microscopes, the Cellvizio microscope, is less than one-tenth of an inch in diameter and can be inserted down a patient's throat to observe live cells.

- In 1986, no fewer than three recipients of the Nobel Prize in Physics were awarded their honors based on their work in improving microscopy technology. Ernst Ruska won for the invention of the electron microscope, and Gerd Binnig and Heinrich Rohrer were recognized for the invention of the scanning tunneling microscope.

- Microscopes have helped chemists at the University of California, Irvine, detect the presence of fat in strands of human hair. Their experiment seeks to determine whether fat is a natural component of hair or is deposited in hair by hair-care products.

- Atomic force microscopes, which probe the forces between atoms to produce an image of incredibly tiny particles, enable scientists to look at—and even pick up and move around—single strands of DNA or individual atoms.

- By examining either a rough or a cut-and-polished gem beneath a powerful microscope, a gemologist can easily tell whether it is an authentic natural stone or one that has been synthetically manufactured.

- Swiss inventor George de Mestral first got the idea for Velcro hook-and-loop fasteners when he used a microscope to examine the intricate hook-and-loop structure of the tiny burrs that had gotten firmly caught on his pant legs while he was walking through the forest.

- In 2008, a microscope attached to the National Aeronautics and Space Administration's Phoenix Mars Lander took a photograph of a single particle of the incredibly fine red dust that swirls around Mars and forms its soil. Dust particles on Mars are about 100 nanometers, about one-thousandth the width of a human hair—or even smaller.

laser light very close to a specimen. These microscopes use the intensity of the reflected light to produce a magnified topographical image of the object. Scanning probe acoustic microscopes (SPAM) direct a focused, ultrasonic (high-pitched) sound wave toward the specimen being observed and form a magnified image of it based on how and how much the wave is reflected by the object's surface.

APPLICATIONS AND PRODUCTS

Scientific Research. At heart, all scientific research rests on the power of observation. Microscopes make it possible for scientists from fields such as biology, chemistry, metallurgy, mineralogy, and countless other disciplines to make more accurate and more complete observations of microscopic structures. Biologists, for example, use fluorescent microscopes to analyze the structure and function of minute intracellular organelles such as ribosomes, mitochondria, and even single strands of DNA. Using the microscope, researchers have been able to watch as individual cells undergo mitosis and viruses invade healthy cells to spread their own genetic material and also to identify the precise manner in which different kinds of proteins are folded. Analytical chemists and physicists use microscopes to conduct research at the scale of the atom or even on a nanoscale. For instance, researchers use scanning tunneling microscopy and atomic force microscopy to observe how peptides—organic compounds composed of two or more amino acids—interact with carbon and graphite nanotubes.

Electron microscopes are being used by botanical researchers to examine how leaves protect themselves from insects by forming crystals inside themselves, by ornithologists to figure out how minute structures on the surface of certain bird feathers create an iridescent effect, and by geoscientists to identify the weather-induced changes to geological features such as rocks. The field is even enabling complex scientific research to take place on other planets. Robotic space vehicles such as the Mars Exploration Rover are often equipped with autonomously operated scanning probe microscopes capable of studying the properties associated with the surfaces encountered by the vehicle. In all these applications, microscopy allows investigators to transcend the limitations of the human sense of sight and expand people's scientific understanding of the world.

Microscopy has a place in many applied sciences as well. It is a useful tool in food science, where it has provided a better understanding of the chemical properties of foods and how processing them alters their natural properties and also has helped isolate food contaminants. Microscopes enable materials scientists to analyze the three-dimensional structure of the plastics and polymers they are developing. Mechanical engineers use microscopes to develop sharper and more sophisticated edges on tools used for cutting.

Microscopy is indispensable in forensic science, where it is used to examine crime scene evidence such as blood, hair, dust, fingerprints, tiny shards of glass, and threads of fiber. Criminologists use high-powered microscopes to help them study the minute hand motions that were used to construct signatures on suspicious documents, looking for frequent stops and starts or other signs of possible forgery. Counterfeit currency makers are often foiled by microscopes, which help scientists detect very subtle discrepancies in the color and texture of the paper fibers used to manufacture counterfeit bills.

Medical Applications. Virtually all biomedical and bioengineering research projects make use of microscopy at some point. Scientists in the pharmaceutical industry, for example, need to closely examine the physical structure and dispersion characteristics of the active components (chemicals) used in the development of drugs, as well as the materials used to coat medical devices such as pacemakers or other implants. Fluorescence microscopy and confocal microscopy are commonly used for these purposes.

Besides their use in preclinical biomedical research, microscopes play a role in at least two other important areas of medical practice: diagnosis and surgery. In diagnosis, microscopes help physicians and laboratory technicians detect whether cells in a patient's tissue samples show signs of disease. When a female patient undergoes a Pap smear, for example, cells are taken from her cervix and analyzed under an optical or electron microscope. If the sample is cancerous or precancerous, a microscopic examination will show changes in the cervical cells that make them look flat or scaly. Microscopes are also used to detect the presence of pathogens in tissue samples. For example, blood cells from patients who have been infected by the malaria virus may appear enlarged or stippled; the malaria parasites themselves will also be visible under magnification of about 100x.

Surgery with the use of an operating microscope (sometimes called microsurgery) has become common in nearly all surgical fields, but it is vital for performing many brain, eye, and ear surgeries.

Magnifying the sometimes minute biological structures involved in a procedure can help a surgeon perform delicate tasks that were practically impossible before the age of microscopes. One particularly significant microsurgery application, for example, is the ability to reattach limbs, fingers, or toes that have been severed from a patient's body. By magnifying the individual nerve fibers, blood vessels, and tendons both in the severed part and at the site of separation, a surgeon can connect them one by one.

The typical microsurgery setup involves a surgeon looking at an operating site through a microscope (or sometimes a television screen connected to the microscope) rather than facing the site directly. Many microsurgery procedures are minimally invasive, making use of instruments inserted through small cuts, or ports, in a patient's skin. Often, robotic instrumentation is used in conjunction with microsurgical tools to track a surgeon's hand motions and correct for tiny tremors in his or her movements. Using a combination of microscopes, remotely controlled tools, and large video screens, surgeons can conduct coronary artery bypasses without ever opening up a patient's chest.

Nanotechnology. Nanotechnology is an example of a scientific discipline whose very existence simply would not be possible without the use of extremely powerful microscopes. This emerging field takes advantage of the special ways in which molecules behave at the nanoscale to create nanoscale machinery such as tiny sensors that can detect and tally the number of specific types of molecules in a sample of chemicals or nanoparticles that systematically seek out and destroy cancerous cells within a patient's body. Optical microscopes do not have the magnifying power necessary to clearly resolve objects at the nanoscale (about 1-100 nanometers, or about one-thousandth the width of a strand of human hair), so electron microscopes and scanning probe microscopes serve as the foundational tools of nanoscientists.

Atomic force microscopes are particularly important in nanotechnology for several reasons. Like electron microscopes, they are capable of imaging structures that are incredibly small (including single atoms or molecules). Unlike electron microscopes, they do not have to operate in a vacuum, giving scientists the ability to work with a greater variety of samples, such as living biological cells. Most significantly, researchers can use the probes attached to atomic force microscopes not just for observing specimens but also for actually manipulating them.

Although most nanotechnology applications are still in the research and development stage, nanotechnology is already causing a transformation in manufacturing. With the help of microscopes that can characterize the behavior of nanoscale structures, scientists have created, for example, grease- and mildew-resistant paints, bacteria-killing storage containers, and nanoscale drug-delivery systems that introduce drugs directly into cells affected by disease.

Microscopy and Art. The applications of microscopy stretch far beyond the boundaries of science. Art historians make extensive use of microscopes to study paintings, sculptures, and other works of art in minute detail, using them to uncover insights about materials, artistic techniques, and what a piece has been through over the course of its history. For example, they often conduct microscopic analyses of the chemical and structural properties of the specific pigments used to create a painting. Museums also use microscopic inspections to authenticate artworks and accurately date them. Using an optical microscope, art historians in Belgium were able to detect minute quantities of a cobalt blue pigment mixed in with an ultramarine pigment in a painting that had been attributed to the Dutch master painter Jan Vermeer. However, Vermeer lived and worked in the seventeenth century, and cobalt blue pigment was not developed until the nineteenth century. Without the ability to scrutinize the precise morphology and crystalline structure of the pigments involved, the historians would never have been able to determine that the painting was, in fact, a forgery.

IMPACT ON INDUSTRY

Manufacturing. In many manufacturing industries, microscopes are essential for various stages of production and inspection. By examining materials and finished products on a microscopic scale, manufacturers can catch flaws, remove contaminants, and ensure the quality and safety of their products. They can also easily assemble products whose individual components are too small to be seen by the naked eye. For example, high-powered microscopes are essential for the manufacture of the silicon wafer microchips—whose circuits can be as small as 0.001 millimeter—used in computers and other electronic devices. Microscopes are particularly important in metallurgical industries because incredibly tiny discrepancies in the crystal structure of a metal alloy can cause significant differences in its physical properties, including hardness, toughness (how likely it is that small fissures in a metal will expand and cause it to break), and tensile strength (how much the metal can stretch before it is unable to return to its original shape). Both transmission and scanning electron microscopes are commonly used in the metal industry to examine the microscopic crystals, or grains, in samples of metal under production as a quality-control measure.

Product Development. Microscopy is an important tool in the research and development stages of many different products. Optical, electron, and atomic force microscopy enable manufacturers of contact lenses and intraocular implants, for example, to characterize features such as the topography, adhesive quality, hardness, elasticity, and viscosity of a lens, as well as to determine how uniformly it has been made. Cosmetics firms use microscope technology to help them evaluate the effectiveness of products ranging from face creams to skin whiteners. Electron microscopes, for example, help biochemists determine how well shampoos smooth down the rough edges of hair, making it feel softer and look glossier.

Microscope Market. The sale of electron microscopes of all kinds is responsible for generating the biggest revenues in the market for microscopy products, largely because these machines are relatively large and expensive—the most sophisticated can cost up to $1 million—in comparison with other types of microscopes. Optical microscopes tend to dominate in contexts in which high-powered microscopes are not necessarily required, such as schools and smaller research laboratories. Scanning probe microscopes are less common than both electron and optical microscopes because they tend to be used in more specialized fields. The two biggest markets for microscopy applications are the life sciences research sector and the biomedical sector, including pharmaceutical companies and medical facilities. Other major consumers of microscope technology include manufacturers of semiconductors and textiles. Among the most important global corporations involved in the development, manufacture, and sale of microscopic equipment are Carl Zeiss, Nikon, Olympus, Leica Microsystems, JEOL, and Hitachi.

CAREERS AND COURSE WORK

A great number of career options exist for individuals with training in microscopy. Among the many professional settings into which microscopists are hired are scientific and clinical laboratories, medical facilities, the research departments of pharmaceutical firms, food processing companies, nanotechnology firms, metallurgy manufacturers, consumer product development laboratories, forensics departments, archaeological digs, and museums.

Ideally, preparation for a career in microscopy should begin with a comprehensive set of advanced high-school courses in physics, chemistry, biology, and mathematics. Students should follow this early training by pursuing a bachelor's degree in science. There are many appropriate areas of concentration for the budding microscopist at the undergraduate level, including physics, biology, chemistry, nanotechnology, and geology. A few colleges and universities in the United States have offered specialized degrees in applied microscopy, though this has not become common. Whatever one's major, important topics to cover include optics, electromagnetism, and electronics. Practical laboratory experience with microscopes is also essential. An interest in photography and postprocessing of photographs would be a helpful addition to a microscopy student's list of qualifications, because these skills are relevant to photomicroscopy and the handling of microscopic images. A bachelor's degree alone (or an associate's degree with additional training or professional certification) is sufficient background for most laboratory technician positions involving microscopy, but those wishing to conduct independent research, either in microscopy itself or in a field such as cell biology or nanotechnology, will need a master's degree or a doctorate.

SOCIAL CONTEXT AND FUTURE PROSPECTS

Microscopy breaks down the barrier to knowledge created by the limitations of the human sense of sight. If knowledge is power, then microscopy represents one of science and technology's most powerful contributions to society. It enables researchers to discover more about the precise structure and behavior of healthy and diseased cells, pinpoint the mechanisms by which pathogens such as bacteria and viruses act in the body, and explore the chemical properties of potential pharmaceutical therapies. It even assists surgeons in performing difficult operations more safely and accurately, thereby saving countless lives. By providing scientists with an intimate knowledge of the way molecules, atoms, and subatomic particles interact, microscopy has propelled the formation of theories about the fundamental nature of the universe.

The growing needs of nanotechnology are inspiring further developments in microscopy. A scanning probe microscope built for the Argonne National Laboratory in Illinois allows researchers to "see" into an individual atom and observe its magnetic spin. The microscope, which cost $2 million, is itself very small—but it must be placed inside a machine 16 feet high and located in a soundproof room, so as to prevent even the tiniest vibration to throw off its focus.

On the other hand, microscopes intended for use in clinical settings in the developing world point the way toward ever smaller and cheaper instruments. A dime-sized microscope that sells for about $10 has no lenses but instead is made of a layer of metal set on top of an array of charge-coupled device (CCD) sensors arranged in a grid. The CCDs translate light into an electric signal; then, a great number of tiny channels are pierced into the metal. As a sample of blood, water, or other liquid flows over the channels, the particles in it block light from passing through to the CCDs in certain areas. The information about which channels are blocked and which remain open to light is used to create an image of the specimen.

M. Lee, B.A., M.A.

FURTHER READING

Cardell, Carolina, Isabel Guerra, and Antonio Sánchez-Navas. "SEM-EDX at the Service of Archaeology to Unravel Historical Technology." *Microscopy Today* 17, no. 14 (August, 2009): 28-33. An overview of the use of scanning electron microscopy to analyze archaeological materials. Includes diagrams and full-color photomicrographs.

Dykstra, Michael J., and Laura E. Reuss. *Biological Electron Microscopy: Theory, Techniques, and Troubleshooting.* 2d ed. New York: Kluwer Academic, 2003. A guide to using microscopic instrumentation in cytological research. Covers conventional light microscopy, transmission electron microscopy, scanning electron microscopy, and photomicroscopy.

Reitdorf, Jens, et al., eds. *Microscopy Techniques.* New York: Springer, 2005. A technical reference book designed for those with a biomedical background, including numerous tables and diagrams, plus appendixes detailing mathematical formulas.

Sluder, Greenfield, and D. E. Wolf, eds. *Digital Microscopy.* 3d ed. Boston: Elsevier Academic Press, 2007. A guide to coordinating microscopes with digital cameras to capture and analyze microscopic images. Includes detailed laboratory exercises to demonstrate principles in action.

Yao, Nan, and Zhong Lin Wang, eds. *Handbook of Microscopy for Nanotechnology.* New York: Kluwer Academic, 2005. An overview of microscopy applications in nanotechnology. Each of the twenty-two chapters contains a discussion of a specific microscopic instrument or technique by nanotechnology specialists working in different fields.

WEB SITES

Florida State University
Microscopy Primer
http://micro.magnet.fsu.edu/primer/index.html

Microscopy Society of America
http://www.microscopy.org/index.cfm

See also: Electron Microscopy; Histology; Liquid Crystal Technology; Mirrors and Lenses; Nanotechnology; Optics.

MICROWAVE TECHNOLOGY

FIELDS OF STUDY

Communications; electronics; radar technology; radio technology; radio astronomy; spectroscopy; telecommunications.

SUMMARY

A microwave is an electromagnetic wave the wavelength of which ranges from one meter to one millimeter. Microwave energy has a frequency ranging from 0.3 gigahertz (GHz) to 300 GHz. The high frequency of microwaves provides the microwave band with a very large information-carrying capacity; the band has a bandwidth thirty times that of the rest of the radio spectrum below it. Microwave signals propagate in straight lines; thus, they are limited to line-of-sight transmission. Unlike lower-frequency radio waves, they cannot pass around hills or mountains and are not refracted or reflected by atmospheric layers. In addition to the familiar microwave oven, applications include communications, radar, radio astronomy, navigation, and spectroscopy.

KEY TERMS AND CONCEPTS

- **Amplification:** Process of increasing the strength of an electronic transmission or a sound wave.
- **Amplitude:** Refers to the height of a radio wave.
- **Antenna:** Device that either converts an electric current into an electromagnetic radiation (transmitter) or converts electromagnetic radiation into an electric current (receiver).
- **Modulation:** Process of varying one or more properties of an electromagnetic wave. Three parameters can be altered via modulation: amplitude (height), its phase (timing), and its frequency (pitch). Two common forms of radio modulation are amplitude modulation (AM) and frequency modulation (FM).
- **Radio Frequency:** An oscillation of a radio wave in the range of 3 kilohertz (kHz) to 300 GHz.
- **Radio Wave:** Electromagnetic radiation that travels at the speed of light; radio waves are of a longer wavelength than infrared light.

DEFINITION AND BASIC PRINCIPLES

In contrast to sound waves, which require a medium such as air or water for propagation, microwaves can travel through a vacuum. In a vacuum such as outer space they travel at the speed of light (299,800 kilometers per second). Microwaves travel in a straight line, limiting them to line-of-sight applications. In space, microwaves conform to the inverse-square law: The power density of a microwave is proportional to the inverse of the square of the distance from a point source. All microwaves weaken as they travel a distance. At some point, depending on the strength of the signal, the microwave will no longer be discernible. Interference can weaken or destroy a radio signal. Other microwave transmitters in the same frequency range produce interference; however, their small wavelength allows small antennae to direct the electromagnetic energy in narrow beams, which can be pointed directly at the receiving antenna. This feature allows nearby microwave equipment to broadcast on the same frequencies without interfering with each other, as lower-frequency radio waves do.

Microwave ovens pass radiation, usually at a frequency of 2.45 GHz, through food. Energy from the microwaves is absorbed by fat, water, and other substances in the food through a process known as dielectric heating. Water and many other molecules have a partial positive charge at one end and a partial negative charge at the other. They rotate in an attempt to align themselves with the alternating electric field of the microwaves. This molecular movement produces heat.

BACKGROUND AND HISTORY

Electromagnetic waves were discovered in 1877 by the German physicist Heinrich Hertz, whose name is used to describe radio frequencies in cycles per second. Eight years later, American inventor Thomas Edison obtained a patent for wireless telegraphy by discontinuous (intermittent) wave. A far superior system was developed in 1894 by the Italian inventor Guglielmo Marconi. Marconi initially transmitted telegraph signals over a short distance on land. Subsequently, an improved system was capable of transmitting signals across the Atlantic Ocean.

Much of microwave technology was developed during World War II for radar applications. The technology was developed secretly; it became available for public use only after the war. In 1951, AT&T's new microwave radio-relay skyway carried a telephone call via a series of 107 microwave towers that were spaced about 30 miles apart; this was the first microwave application that could carry telephone conversations across the United States via radio (as opposed to wire or cable). The system could also carry television signals; three weeks after the first telephone call, at least 30 million people saw and heard President Harry Truman open the Japanese Peace Treaty Conference in San Francisco. Then, in 1946, Percy Spencer, an engineer at the Raytheon Corporation, developed the now-ubiquitous microwave oven.

HOW IT WORKS

Communication. Because of the short wavelength, microwave radio transmission employs small, highly directional antennae, which are smaller and therefore more practical than ones used for longer wavelengths (lower frequencies). Considerably more bandwidth is available in the microwave spectrum than in lower frequencies. This wider bandwidth is suitable for the transmission of video and audio. Since microwave transmission is line of sight, distance is limited by the curvature of the Earth. A higher antenna can transmit over a greater distance. Much greater transmission distances between the Earth's surface and an orbiting satellite are possible because the only limitation is the attenuation that occurs from the atmosphere; however, even in the vacuum of outer space, the signal degrades over a distance.

Energy Transmission. The concept of using microwaves for the transmission of power emerged following World War II; high-power microwave emitters, also known as cavity magnetrons, were developed. These emitters can transfer electrical energy from a power source to a target without interconnecting wires. Wireless transmission is reserved for cases in which interconnecting wires are inconvenient, dangerous, or impossible. To be effective as well as economical, a large part of the energy sent out by the generating plant must arrive at the receiver(s). The short wavelengths of microwave radiation can be made more directional than lower radio frequencies; thus, allowing power beaming over longer distances.

A rectenna (rectifying antenna) at the target can convert the microwave energy back into electricity. Conversion efficiencies of more than 95 percent have been achieved with rectennae.

Microwave Ovens. A microwave oven passes non-ionizing microwave radiation, usually at a frequency of 2.45 GHz, through food. Water, fat, and other substances in the food absorb energy from the microwaves in a process called dielectric heating. Many molecules (such as those of water) are electric dipoles, which means they have a partial positive charge at one end and a partial negative charge at the other. They rotate in order to align themselves with the alternating electric field of the microwaves. This movement represents heat, which is then dispersed as the rotating molecules strike each other.

Navigation. Global Positioning Systems (GPSs) operate in microwave frequencies ranging from about 1.2 GHz to 1.6 GHz. GPS is a space-based global navigation satellite system, which provides accurate location and time information at any place on Earth where there is an unobstructed line of sight to four or more GPS satellites. GPS can function under any weather conditions and anywhere on the planet. Since it cannot function underwater, a submarine must surface to use a GPS system. The technology depends upon triangulation, just as a land-based systems do, to locate a discrete point. GPS is composed of three segments: the space segment; the control segment; and the user segment. The U.S. Air Force operates and maintains both the space and control segments. The space segment is made up of satellites, which are in medium-space orbit. The satellites broadcast signals from space, and a GPS receiver (user segment) uses these signals to calculate a three-dimensional location (latitude, longitude, and altitude). The signal transmits the current time, accurate within nanoseconds.

Radar. Radar is an acronym for radio detecting and ranging. The device consists of a transmitter and receiver. The transmitter emits radio waves, which are deflected from a fixed or moving object. The receiver, which can be a dish or an antenna, receives the wave. Radar circuitry then displays an image of the object in real time. The screen displays the distance of the object from the radar. If the object is moving, consecutive readings can calculate the speed and direction of the object.

If the object is airborne, and the device is so equipped, the altitude is displayed. Radar is invaluable in foggy weather when visibility can be severely reduced.

Radio Astronomy. Radio astronomy is a subfield of astronomy that examines celestial objects, which emit radio frequencies. Much of radio astronomy is focused on the microwave band.

Spectroscopy. Spectroscopy involves the use of spectrometers and spectroscopes to analyze the distribution of atomic or subatomic particles in a system, such as a molecular beam. Microwave spectroscopy is a form of spectroscopy in which information is obtained on the structure and chemical bonding of molecules and crystals via measuring the wavelengths of microwaves emitted or absorbed by them. Microwave spectroscopy can be conducted only on gases.

APPLICATIONS AND PRODUCTS

Communication. Microwaves are commonly used by communication systems on the Earth's surface, in satellite communications, and in deep-space radio communications. Microwaves are commonly used by television news media to transmit audio and video from a specially equipped van to a television station. Mobile telephone networks operate in the lower end of the microwave band, while others operate at frequencies just beneath the microwave band. Networks of microwave relay links have largely been replaced by fiber-optic networks. Wireless transmission employing local area network (LAN) protocols such as Bluetooth operate in the 2.4 GHz microwave band. Other LAN protocols operate at higher microwave frequencies. Wireless Internet services operate in the 3.5-4.0 GHz range.

Energy Transmission. Although wireless power transmission via microwaves is well proven, many applications are still in the experimental stage. In 1964, a miniature helicopter propelled by microwave power was demonstrated. In 2008, 20 watts of power were transmitted 92 miles from a mountain on Maui to the big island of Hawaii. The U.S. military is currently using microwave energy transmission as a form of sublethal weaponry. The application, known as an Active Denial System, uses microwaves to heat a thin layer of human skin to an unbearable temperature, which forces the recipient to move away from the energy source. The skin can be heated to a temperature of 54 degrees Celsius at a depth of 0.4 mm with a two-second burst of a 95-GHz-focused beam.

Microwave Ovens. A microwave oven is a common household appliance. It rapidly heats frozen foods, pops popcorn, bakes potatoes, and boils water. Its compact size makes it beneficial when space is at a premium. It is used on commercial airlines for meal preparation and is found in many other locations, including offices. In addition, microwave heating is employed in many industrial processes for drying and curing products.

Navigation. GPS applications, which operate in the microwave band, are used for navigation over land and water, and numerous military, civilian, and commercial applications exist. GPS devices are used for navigation on military and commercial vessels, on pleasure boats, and in automobiles. The device gives a visual display of the vehicle's position on a map overlay. Useful information such as distance to the next turn and the destination are given visually and aurally, if the GPS is equipped with an audio system. Some vehicle GPS devices also provide information such as the nearest gasoline stations, rest areas, restaurants, and hospitals. GPS units for off-road use are also available. The motorist on a budget can purchase an inexpensive handheld GPS; however, these devices are popular with campers and hikers. Many small aircraft and virtually all large aircraft contain a GPS. GPS is also used for space navigation. Another application of GPS is an ankle monitor, which tracks the location of individuals under house arrest.

Radar. Radar is an essential component of any vessel that operates offshore. Navy vessels and many pleasure boats have radar. Larger vessels often have several. Aircraft are guided by land-based radar installations, known as airport surveillance radar (ASR), located at civilian and military airfields. An ASR tracks airport positions and weather conditions in the vicinity of the airport.

Radio Astronomy. Most radio astronomy applications operate in the microwave spectrum. Usually, naturally occurring microwave radiation is observed; however, radio astronomy has been used to measure distances precisely within the solar system. Radio astronomy has also been employed to map the surface of Venus, which is not visible via optical telescopes because of its dense cloud cover. The technology has expanded astronomical knowledge and has led to the discovery of new objects, including radio galaxies, pulsars, and quasars. Radio astronomy allows objects that are not detectable with an optical telescope to be seen. These objects are some of the most extreme and energetic physical processes that exist in the universe. Since microwaves penetrate dust, radio astronomy techniques can study the dust-shrouded environments where stars and planets are born. Radio astronomy is also used to trace the location, density, and motion of the hydrogen gas, which constitutes about 75 percent of the ordinary matter in the universe.

Spectroscopy. Microwave radiation is used for electron paramagnetic resonance, also known as electron spin resonance, analysis, which has been crucial in the development of the most fundamental theories in physics, including quantum mechanics, the special and general theories of relativity, and quantum electrodynamics. Microwave spectroscopy is an essential tool for the development of scientific understanding of electromagnetic and nuclear forces.

IMPACT ON INDUSTRY

Microwave technology has had a significant impact on industry. As of 2011, cell phones are widely used; in fact, some individuals rely on a cell phone rather than a landline telephone for communication. Cell phone service providers and manufacturers are currently experiencing significant growth. Wireless microwave applications such as Bluetooth are widespread and increasing in popularity as are GPS devices for military, commercial, and personal use. Radar is an essential navigation tool. Microwave ovens represent a significant segment of the appliance industry. Microwave devices also support a repair industry, and the constantly evolving technology fuels replacement of these devices with an upgraded model often before the device fails.

Regulation. Radio transmission is regulated and closely monitored by government agencies. In the United States, it is regulated by the Federal Communications Commission (FCC). The FCC is responsible for regulating interstate and international communications by radio, television, wire, satellite, and cable in all fifty states, the District of Columbia, and U.S. possessions.

Industry and Business. Although the civilian population is a significant consumer of microwave equipment, commercial and military applications represent a large market share. In the commercial sector, the civilian aircraft and shipping industries are major purchasers of microwave equipment. The National Aeronautics and Space Administration (NASA) requires extremely precise, complex, and expensive navigational and communications equipment for its

Microwave relay station, at an elevation of 5438 feet, Boucher Hill, California. (Jack Dermid/Photo Researchers, Inc.)

missions. All branches of the military—Army, Navy, Air Force, Marines, and Coast Guard—use microwave equipment extensively.

Government and University Research. One of the biggest sources of funding for microwave research in the United States is the Defense Advanced Research Projects Agency (DARPA). It is also the biggest client for certain kinds of microwave and navigation applications. The agency is primarily concerned with radio systems with a military focus, such as guidance systems. It also conducts research on satellite communication. Another source of funding for radio research is the U.S. Department of Energy (DOE). The DOE is currently focused on energy efficiency and renewable energy. The DARPA and the DOE supply funds to many universities in the United States for microwave research and development. For example, as far back as 1959, DARPA began work with Johns Hopkins Applied Physics Laboratory to develop the first satellite positioning system.

Fascinating Facts About Microwave Technology

- Science fiction author Robert Heinlein foresaw the development of the microwave oven in two of his novels. His 1948 novel *Space Cadet* mentioned the uses of "high frequency heating" to prepare food. His 1950 *Farmer in the Sky* described the preparation of a twenty-first century meal: "I grabbed two Syntho-Steaks out of the freezer and slapped them in quickthaw, added a big Idaho baked potato for Dad . . . and stepped up the gain on the quickthaw so that the spuds would be ready when the steaks were."
- In the 1926, Japanese researcher Hidetsugu Yagi and his colleague Shintaro Uda designed a directional array antenna, which they named the Yagi antenna. Although the antenna did not prove to be particularly useful for power transmission, it is used widely throughout the broadcasting and wireless telecommunications industries because of its excellent performance characteristics.
- Radio gained acceptance more rapidly in the United States among amateur radio operators than the general public. In 1913, 322 amateurs were licensed, and by 1917 there were 13,581—primarily boys and young men. Many older individuals considered radio to be a fad. They reasoned that listening to dots and dashes or the occasional experimental broadcast of music or speech over earphones was a worthless endeavor.
- The microwave oven was a by-product of radar research. In 1945, Raytheon engineer Percy Spencer was testing a new vacuum tube called a magnetron and found that a candy bar in his pocket had melted. Deducing that electromagnetic energy was responsible, he placed some popcorn kernels near the tube; the kernels sputtered and popped. His next trial was with an egg, which exploded in short order.

CAREERS AND COURSE WORK

Many technical and nontechnical careers are available in microwave technology. The technical fields require a minimum of a bachelor's degree in engineering or other scientific field; however, many also require a master's or doctorate degree. Course work should include mathematics, engineering, computer science, and robotics. Positions are available in both the government and private sector. The ability to be a team player is often of value for these positions because ongoing research is often a collaborative effort.

Technicians are needed in a variety of fields for equipment repair. Many of these positions require some training beyond high school at a community college or trade school. If a company employs a number of technicians, supervisory positions may be available. Numerous opportunities are available for sales.

SOCIAL CONTEXT AND FUTURE PROSPECTS

Microwave technology is an integral component of everyday life in all but the most primitive societies. In view of the continuous advances in radio technology, including microwave, further advances are extremely likely.

The microwave oven demonstrates the fact that electromagnetic waves are a form of energy, which is capable of heating, and thus, damaging tissue. Microwave ovens are shielded to prevent exposure; however, other microwave devices are not. For example, if one stands near or touches a microwave transmitter, severe burns can result. The heating effect of an electromagnetic wave varies depending on its power and the frequency. The Institute of Electrical and Electronics Engineers (IEEE) as well as many national governments have established safety limits for exposure to various frequencies of electromagnetic energy. Controversy still exists as to whether microwave energy can be harmful to humans and animals. Low levels of microwave radiation have no proven harmful effect. Examples of low-level microwave radiation are the small amount of leakage from a microwave oven and microwave transmission from a cell phone. Some experts express concern that prolonged low-level exposure (for example, holding a cell phone to one's ear for extended periods of time) might be harmful. Countering this concern is research that has involved exposing multiple generations of animals to microwave radiation at the cell phone intensity or higher, and no health issues have been found. High levels are definitely harmful. An example is U.S. military's Active Denial System, which can heat human skin to an intolerable level.

Robin L. Wulffson, M.D., F.A.C.O.G.

FURTHER READING

Balbi, Amedeo. *The Music of the Big Bang: Cosmic Microwave Background and the New Cosmology.* Berlin:

Springer-Verlag, 2010. Focuses on how the exploration of the cosmic background radiation has shaped the picture of the universe, leading even the non-specialized readers toward the frontier of cosmological research, helping them to understand the mechanisms behind the universe.

Hallas, Joel R. *Basic Radio: Understanding the Key Building Blocks*. Newington, Conn.: American Radio Relay League, 2005. An introduction to radio that discusses its components: receivers, transmitters, antennae, propagation, and their applications to telecommunications, radio navigation, and radio location.

Rudel, Anthony. *Hello, Everybody! The Dawn of American Radio*. Orlando, Fla.: Houghton Mifflin Harcourt, 2008. Addresses early radio, covering the entrepreneurs, evangelists, hucksters, and opportunists who exploited the new technology.

Scott, Alan W. *Understanding Microwaves*. Hoboken, N.J.: Wiley-Interscience, 2005. Covers terminology, devices, and systems, and shows how the technology works to make communications, navigation, and radar equipment. Aimed at engineers, technicians, managers, and students.

WEB SITES
Institute of Electrical and Electronics Engineers
http://www.ieee.org

National Association of Broadcasters
http://www.nab.org

Society for Applied Spectroscopy
http://www.s-a-s.org

Society of Amateur Radio Astronomers
http://www.radio-astronomy.org

See also: Computer Engineering; Computer Science; Radio; Radio Astronomy; Spectroscopy; Telecommunications.

MILITARY SCIENCES AND COMBAT ENGINEERING

FIELDS OF STUDY

Physics; chemistry; weapons-systems engineering; civil engineering; materials science; metallurgy; electrical engineering; mechanical engineering; ocean engineering; oceanography; meteorology; aeronautical engineering; avionics; electronics; cryptography; medicine; medical technology; biology; biomechanics; mathematics; artificial intelligence; computer science; computer engineering; computer programming; naval architecture; geography; environmental engineering; nuclear engineering; systems engineering; aerospace engineering.

SUMMARY

Military sciences are scientific, engineering, and technical activities undertaken by those who identify, design, and produce innovative weapons, including such items as improved rifles for individual fighters and larger strategic weapons such as laser-guided missiles; equipment, ranging from improved clothing and night-vision goggles to armored tanks; and communications devices for use in warfare. Combat engineering includes activities such as building bridges, harbors, roads, temporary shelters, or improvised airfields used to assist combat troops, or removing obstacles from the battlefield.

KEY TERMS AND CONCEPTS

- **Asymmetric Warfare:** War between two opposing countries or armed forces in which one possesses significant advantages in technology, often forcing the weaker combatant to resort to unconventional means of warfare.
- **Ballistics:** Science that studies the characteristics of projectiles, including the mechanics of flight and effects on impact.
- **Electronic Warfare:** Using electronic energy to conduct offensive and defensive operations.
- **Information Warfare:** Collecting, managing, and using information to conduct military operations, either offensively or defensively.
- **Military-Industrial Complex:** Network of relationships between government and various industries

involved in research, development, and delivery of weapons and equipment to the armed forces.
- **Network-Centric Warfare:** Activities aimed at gaining an advantage over an enemy through computer-based information systems.
- **Ordnance:** Exploding devices, or the weapons that propel explosive devices in combat.
- **Revolution in Military Affairs:** Significant shift in the conduct of warfare brought on by the emergence of new technologies.
- **Technology Transfer:** Process of sharing information about technology or manufacturing processes that allows governments or private businesses to benefit from the work done by other agencies.
- **Weapons Of Mass Destruction (WMD):** Armaments (often chemical, biological, or nuclear) that can inflict significant casualties, cause major property damage, and incite terror among military units or civilian populations.
- **Weapons Systems:** Integrated network of ordnance, delivery mechanisms, and control mechanisms (often computerized) that allows for effective employment of specific weapons in combat.

DEFINITION AND BASIC PRINCIPLES

The term "military sciences" is used to designate the broad scope of activities undertaken to develop weapons, weapons systems, and equipment for the military. This broad category includes basic research on the components used in fabrication of materials for weapons or weapons systems, including research in ballistics; design of new computers, programs, or electronic devices; engineering to fabricate armaments and ordnance, battlefield gear, or support equipment; and systems for transitioning new technologies and equipment into use by fighting forces.

"Combat engineering" is a term that describes the activities of military units that support the armed forces by performing a number of engineering functions in or near the battlefield. Many of these, such as bridge building, road building, and construction of temporary fortifications and camps for forces in the field, are handled improvisationally with materials on hand or brought into the theater of operations by military forces. By contrast, some military personnel engage in activities more akin to civil engineering.

This may involve construction of facilities at permanent bases, or in the case of the U.S. Army Corps of Engineers, projects that control the rivers and lakes within the boundaries of the United States.

It is important to distinguish "military sciences" from the closely related term, "military science." The latter refers to the broader study of war as an art and science and encompasses the study of tactics, strategy, and leadership.

BACKGROUND AND HISTORY

Technology has played an important role in the conduct of warfare since the dawn of civilization, when people first began to assemble crude weapons from natural materials. The discovery of techniques to forge metals led to improvements in weaponry, as bronze and then iron weapons made soldiers and sailors more effective fighters. The history of warfare, therefore, is inseparable from the history of technological advances that have influenced the development of weapons, equipment, fortifications, transportation, and logistics. At times, military sciences have simply been efforts to adapt existing technologies for military use; at other times, however, systematic efforts have been undertaken to design weapons of war or develop countermeasures to protect combatants and civilian populations.

It has become commonplace to describe the various epochs in the history of warfare as occurring in a series of "revolutions in military affairs," driven by advances in technology that brought about paradigmatic shifts in the conduct of warfare. Ways of waging war before the eleventh century did not change radically. Handheld weapons permitted armies to clash on the battlefield. Siege warfare was also commonplace, as invaders would surround cities and on occasion attempt to breach fortified walls. Engineers played a key role in both defensive and offensive operations, designing ever-more sophisticated fortifications from earth and stone to thwart enemy penetrations into cities (and later castles) and creating siege engines such as catapults for launching missiles over these ramparts or climbing devices for scaling them. The invention of the bow allowed armies to attack enemies from some distance. Mounted soldiers were used in limited fashion in open terrain. Shields and pikes to protect soldiers organized into tight battlefield formation were among the primary defensive weapons fashioned during these centuries. In naval battles, ships would get close with each other, grappling hooks would be tossed to secure the two vessels, and sailors would board with weapons similar to those used by infantry. Some ships carried mechanisms that allowed them to launch missiles such as fire grenades, but most naval warfare was conducted at close quarters.

The invention of gunpowder in the eleventh century revolutionized warfare around the globe. By the fourteenth century nations had learned to use the power of chemical explosions to launch projectiles (bullets, mortar, and artillery shells) that traveled greater distances and caused substantially more damage than muscle- or mechanically powered projectiles. Over the next four centuries, scientists devised more effective mixtures of chemical substances that allowed for controlled explosions, while engineers and gunsmiths designed more accurate and devastating weapons. By the eighteenth century, metallurgists had discovered ways to make artillery pieces lighter while improving their power and accuracy. Naval engineers adapted field artillery pieces for use aboard ships, providing navies new capabilities to attack other ships and provide fire support to forces onshore.

Another revolution in military sciences and engineering occurred in the late eighteenth and nineteenth centuries. Industrial-age technology led to numerous advances in military technology. The development of processes to standardize weapons production led to significant efficiencies in arming soldiers, since spare parts could be carried with troops and replaced easily. Innovations such as the invention of rifled barrels, breech-loading and repeating rifles, and the machine gun, gave armies more firepower. Though not developed specifically for military operations, rail transportation allowed troops to be deployed at greater distances in a shorter time. Advances in ship design led to the launch of ironclad warships capable of carrying cannons that could launch massive projectiles toward land-based targets from miles at sea. Clothing manufacturers and suppliers turned out gear that was more durable and better suited to soldiers' and sailors' needs.

During this time, chemists perfected a number of formulas for poisonous gases, which were deployed with devastating results during World War I. In response, protective gear was developed to counter the effects of the gas. Two additional inventions designed for peaceful purposes were quickly adapted for military use: the airplane and electronic communications devices. Planes gave field commanders opportunities for better surveillance and eventually provided platforms for delivering more sophisticated bombs over distant targets, or dropping troops behind enemy lines. Between World War I and II, naval engineers designed aircraft carriers that allowed navies to bring air power far from shore to attack enemy targets or defend friendly forces. World War II also saw the effective employment of electronic warfare, as both the Allied and Axis nations used newly created devices to intercept enemy communications, transmit messages over long distances, detect targets, or conduct countermeasures to neutralize the enemy's electronic devices.

The end of World War II saw the dawn of yet another revolution in military affairs: the introduction of atomic power into military conflict. This weapon of mass destruction allowed one combatant to inflict extensive damage on the enemy with minimal involvement of troops. The development of the atomic bomb was one of the great scientific achievements of the twentieth century, even if its deployment were morally questionable. Teams of physicists and engineers managed to harness the power of the atom to generate hitherto unseen explosive power. Atomic weapons became the signature armament of the ensuing Cold War between the Western allies and the Soviet Union. Even though no nuclear weapons were used in conflict, the threat they posed served to shape both military and political policy for four decades. Additionally, the proliferation of weapons of mass destruction (WMD) became an international concern during the second half of the twentieth century, as more stable countries grew fearful that such weapons might fall into the hands of fanatics and be used as instruments of terror. At the same time, however, nuclear-powered engines installed in surface ships and submarines gave naval vessels the ability to stay at sea for months without making port calls.

Beginning in the last decades of the twentieth century, the nature of warfare changed again with the introduction of sophisticated computer-based weapons systems, surveillance devices, and command-and-control networks. Laser-guided weapons, "smart bombs," and missiles capable of being guided to within ten meters of a target provided battlefield commanders more effective ways to hit enemy targets with minimal collateral damage. New electronic devices permitted

more sophisticated methods of gathering and processing intelligence, giving commanders better real-time data on which to base decisions. The growing presence of computers on the battlefield, networked to ones far away from the front lines, permitted commanders on both sides of a conflict to engage in network-centric war: Real-time exchange of information allowed combatants to gather intelligence and exploit weaknesses in an enemy's defenses. At the same time, weaker countries or groups engaged in asymmetric warfare often resorted to weapons using more primitive technology that could often produce casualties on combatants and civilians, often randomly, thereby creating terror among populations engaged in or living within the zone of conflict.

How It Works

With few exceptions, the application of scientific and engineering work for military purposes is carried out under the direction of a nation's defense agency. That is not to say, however, that all research and development (R&D) is performed by government employees at state-owned facilities. It is common for socialist nations or totalitarian regimes to control the entire process, while democracies tend to follow the pattern used in the United States, where R&D is carried out through a complex arrangement that involves government agencies, private industry, and academic institutions.

The United States Department of Defense has an elaborate organization to oversee research and development. The Defense Advanced Research Projects Agency (DARPA) sponsors basic scientific research (focusing on physics and chemistry) and applied research that shows promise of producing new breakthroughs in designing military weaponry and equipment. The Army, Navy, and Air Force each has its own R&D agency, employing teams of scientists, engineers, and technicians to carry out projects funded through federal appropriations. Additionally, each agency engages in partnerships with private businesses to sponsor additional research and to underwrite engineering efforts to turn basic science into usable tools for the fighting forces. These agencies also control funding that can be used to support research at academic institutions across the country. While the military services direct much of the research, fabrication of end items—weapons and weapons systems, personal gear, and high-technology

Fascinating Facts About Military Sciences and Combat Engineering

- The first recorded instance of professional weapons systems development occurred in Syracuse on the island of Sicily in the fourth century B.C.E.
- The invention of the stirrup in the fifth century C.E. allowed armies to make more effective use of cavalry and eventually resulted in the development of combined-arms operations.
- The development of firearms made the armored knight on horseback obsolete as a fighter.
- In response to the French government's need to provision troops in the field, an enterprising inventor developed the process of canning food.
- The invention of the machine gun in the nineteenth century eventually eliminated the popular bayonet charge as an offensive maneuver in battle.
- The nuclear-powered propulsion system developed for use aboard submarines allows these vessels to go without refueling for twenty years or more.
- A program initiated in 1969 to link computers for improved information sharing and research resulted in the creation of the ARPANET, the forerunner of the Internet.
- The composite materials used to construct the U.S. Air Force B-2 Spirit Stealth bomber absorb radio waves, making the plane virtually undetectable on radar.
- The military's need for precise information on the locations of friendly and enemy positions and important targets led to the development of the Global Positioning System (GPS).

equipment such as radars and surveillance devices—is more often carried out by private industry.

Combat engineers are beneficiaries of the R&D that takes place within the various organizations and activities sponsored by the federal government. While the organization of combat engineering units varies by country, the operation of such units within the United States armed forces suggests how combat engineers make use of existing technologies and equipment to support operations in the field. Army combat engineers handle tasks such as constructing or repairing roadways and temporary facilities and assembling prefabricated bridges. In the Air Force,

combat-engineering functions include constructing temporary airfields, repairing existing airstrips, constructing roads and revetments, and providing general engineering support to field commanders. Navy construction battalions (Seabees) typically build wharves and harbor facilities, airfields, field hospitals, roads, and bridges.

APPLICATIONS AND PRODUCTS

Scientists and engineers engage in work to develop thousands of products for the military. Some are large, multimillion-dollar items such as aircraft carriers or supersonic planes; others may be small but of great importance to individual fighters, such as improved lenses for night-vision goggles. A brief outline of some of the major items used to conduct warfare in the twenty-first century suggests the scope and complexity of the work military scientists, engineers, and technicians are responsible for accomplishing.

Air Warfare. The design and construction of matériel for air warfare requires thousands of individual end items built using the latest technologies. At any time, a country like the United States is deploying new aircraft, maintaining older ones, and conducting research to create new planes that will be faster, lighter, and less susceptible to detection by the enemy. The composite materials developed by chemists, metallurgists, and engineers are often key components in the body designs of new planes, and many older ones are retrofitted to accommodate new equipment that enhances overall performance.

New weapons systems—missiles, bombs, and small-arms weapons such as machine guns—are designed to be carried on these platforms. Among those in use: air-to-air and air-to-surface missiles employing complex electronic guidance systems; bunker-buster bombs guided by laser systems or from satellites capable of penetrating as much as 20 feet of concrete; and "blackout" bombs that can knock out an enemy's electrical power grid.

The United States also builds and maintains satellites that provide secure voice and data transmission capability. Equipped with antijamming devices, these form the backbone of an elaborate satellite network that affords commanders from NATO countries a reliable system for worldwide command and control. Other satellites are used to gather intelligence and serve as navigation aids to troops on the ground or at sea. Significant research is ongoing to improve the capabilities of unmanned aircraft, which can be remotely controlled and flown over enemy territory to deliver ordnance on targets or gather intelligence.

Ground Warfare. The changing nature of the battlefield and an enemy's capabilities make it necessary for armies to develop and maintain a host of new equipment to transport soldiers to the combat zone, provide them mobility once there, protect them from enemy attack, and arm them with weapons that offer sufficient firepower to subdue the enemy from close or medium range. The major rolling stock of most armies consists of tanks, armored personnel carriers, and self-propelled artillery. However, trucks used to haul supplies and transport personnel are also key components in an army's ability to remain mobile, and these are often equipped with armor and various detection devices to protect soldiers from enemy mines or other exploding devices. The weaponry designed for use by soldiers on the battlefield typically includes rifles, sidearms, and grenades. Sophisticated guidance systems and devices aid artillerists in launching ordnance accurately.

Soldiers are also equipped with a number of protective devices such as body armor (often made from composite materials), protective masks (commonly called gas masks), helmets, and special clothing that makes them less detectable by enemies. An array of products have been designed to aid in command and control, including sophisticated radios and computer devices (often handheld). Individual equipment and supplies are often subject to extensive research and design as well, so that rations, clothing, and personal gear carried for hygiene and comfort are carefully fabricated and packaged to make them usable in the difficult environment produced by combat.

Naval Warfare. Navies sail ships of various sizes and functions in conducting war at sea. Each is a floating platform for a variety of weapons systems that can project power at an enemy's navy or at targets onshore. The United States Navy employs aircraft carriers, cruisers, destroyers, submarines, and frigates as its principal fighting ships; amphibious assault ships are used to carry U.S. Marines to combat zones. The Navy also has a fleet of supply ships, hospital ships, and other support vessels. These are all equipped with guidance and navigation systems used for maneuvering and fighting. Naval ships carry missiles that can be launched from the deck, carried into

flight by naval aircraft, or launched by submarines from beneath the surface. Ships are outfitted with conventional weapons ranging from medium-size machine guns used for defense against enemy air attack to large naval guns that can fire shells for several miles at enemy ship formations or onshore targets. Ships carry substantial amounts of equipment for surveillance, supply, and maintenance, much of it specifically designed by military scientists and engineers to withstand the rigorous conditions at sea. Aircraft used aboard ship are designed for short takeoffs and landings and carry armaments similar to those used by the air force for offensive and defensive operations. As with ground forces, those aboard ships are outfitted with personal gear, much of it specially designed to protect them in battle and provide comfort and hygiene.

Combat Engineering. Combat engineers carry equipment similar to that used by civilian construction and demolition firms, much of it modified for the specific needs of working in a combat zone. These include carpenter's and other construction tools (hammers, brush cutters, vises, shovels, posthole diggers), an array of power tools, and generators built to withstand the incidental damage caused by use in rough terrain. Bulldozers, earth movers, front loaders, and similar construction equipment is standard for many engineer units. Engineers also employ breaching vehicles to remove man-made obstacles such as barbed-wire fences or mines. Some combat-engineering units are equipped with amphibious vehicles to serve as ferries and bridging vehicles that allow engineers to transport and assemble portable bridges that allow the fighting forces to cross bodies of water.

IMPACT ON INDUSTRY

In countries where the state controls both research and production of military matériel, industries see only minimal impact from increases or decreases due to changing priorities in R&D or fabrication of armaments and equipment. Such is not the case in countries where there are close ties between the government and private sector. The situation in the United States presents a good example of how decisions regarding military R&D and procurement affect industry.

Historically, businesses that engaged in manufacturing other items during peacetime (passenger cars and trucks, ball bearings, household chemicals) converted part or all of their facilities during wartime to manufacture items for the military (tanks and aircraft, ammunition, chemical weapons). Since World War II, however, many companies have specialized in doing work for the military—some exist solely for that purpose, creating what is in effect the defense industry. Certain firms have come to depend on government contracts for ships and aircraft, two items that require substantial expertise in science and engineering. These firms hire thousands of highly trained specialists to perform such work, hence employment is affected when the government increases or reduces orders for these items. Many technicians and tradespeople who work on construction of ships, planes, heavy armored equipment, and specialized items such as firearms or sophisticated surveillance equipment and guidance systems are also affected by government decisions to increase or decrease requirements for specific end items. In 2010, the United States allocated more than $82 billion for military R&D. Other countries have more modest budgets: In 2010, the United Kingdom allocated £4.6 billion (about $7 billion) for military R&D. Although some have questioned the wisdom of having this kind of permanent arrangement, the military-industrial complex has proven to be important in meeting the armed services' demands for more effective armaments that can be delivered to the fighting forces in a timely manner.

Additionally, much of the basic research undertaken in military laboratories, or by civilian researchers working on government-funded contracts, has proven to have significant benefit outside the defense industry. This is particularly true of research conducted in physics, chemistry, and basic electronics. The opportunity for technology transfer permits the adaptation or use of R&D from military agencies or defense contractors for civilian use. That transfer has been most evident to the general public in the adaptation of computer technology for the development of the Internet and Global Positioning System (GPS) technology for use by private citizens. One area where significant benefit is being realized is the medical field. The high number of incidents beginning in 2001 in Afghanistan and Iraq involving injury to extremities led to accelerated research in prosthetics. The military's willingness to share that technology has led to notable advances in the development of artificial limbs that provide patients increased mobility.

CAREERS AND COURSE WORK

Those with an interest in science and the military will find opportunities for work in government, private industry, and the academic world. A number of individuals involved in research and development for the military are members of the armed forces, often with specialized training that permits them to supplement their battlefield knowledge with classroom and laboratory preparation to carry out sophisticated scientific inquiry. Since the eighteenth century, governments have established and sponsored military academies to prepare officers for practical applications of military science and engineering, but individuals educated in other institutions can often receive commissions in the services and perform these roles.

Regardless of the institution one chooses to attend, obtaining a bachelor's degree in basic sciences (chemistry, biology, or physics), mathematics, or engineering is often adequate qualification for jobs as technicians working on an array of projects. Occasionally, those with associate's degrees in applied sciences can find work as technicians as well. By far the greatest opportunities for making significant contributions to the military sciences are available to those with advanced degrees, especially in the physical sciences, computer sciences, engineering, or medicine and medical research.

Technicians serving on active duty in military forces frequently receive on-the-job training or attend special schools established by the armed forces, although individuals wishing to pursue careers in military specialties such as avionics, ordnance disposal, medical technology, electronics equipment repair, or similar technical fields will find it helpful to have a sound foundation in mathematics and some understanding of the specific science or technology in which they plan to specialize.

Employment in military science and technology varies by country, but in almost every country opportunities for individuals to pursue their interests in these disciplines is available through commissioning or enlistment in the active service. In the United States, employment is also available with the Department of Defense at DARPA or one of the military service's laboratories and with contractors that provide products and services to the armed forces. Additionally, those with an interest in basic research that might have applications for military use can find rewarding work at a number of universities where government contracts provide funding for significant research in fields that show promise for military application.

Combat engineering is handled almost exclusively by members of the uniformed services. Those interested in working in that field must first enlist or receive a commission in one of the armed forces, and then select combat engineering as a career specialty. The academic qualifications for combat engineers are less stringent than those required of laboratory scientists or design and manufacturing specialists. Often, combat engineers are given specific instruction in the tasks they will perform as part of their military training and participate in refresher courses or advanced training to keep their skills up to date.

SOCIAL CONTEXT AND FUTURE PROSPECTS

If history is any guide, the inevitability of future conflict somewhere in the world suggests that there will be a continuing need for new technologies to wage warfare more effectively. Developments in military sciences are always carried out, however, in a social and political climate that affects both the budgets of those engaged in research and the constraints that are imposed on the kinds of weapons and equipment that may be developed. Working within those real-world parameters, military scientists, especially in countries that enjoy political stability and the financial wherewithal to support major research efforts, continue to explore new applications for existing technologies or work to create new ones that will enhance a country's ability to fight when necessary.

Many military strategists believe that the greatest prospects for advancing a country's ability to fight more effectively lie in the development of more sophisticated tools for information warfare. Devices already available, such as GPS, surveillance satellites, and radar have proven effective in combat; however, refinements to improve their accuracy and reliability, especially in the face of electronic countermeasures, will continue to be required. Electronic command and control tools—instruments such as the Internet and handheld devices that rely on satellites for broadcast capability—will also require constant updating or replacement with yet-undiscovered technologies that can give commanders improved ability to communicate with subordinates or superiors to direct activities and provide necessary support on the battlefield.

Several areas of research suggest the variety of tasks in which military scientists may be engaged. One is the construction of hypersonic aircraft. The potential to create unmanned vehicles that can travel at Mach 5 (five times the speed of sound) or more has great military value for the development of missiles that can strike with exceptional speed against high-value targets, particularly enemy soldiers and their leaders that have the potential to move about. The United States and its allies are also developing more sophisticated weapons that rely on laser technology. Directed-energy weapons, as these devices are called, are being designed using both solid-state and chemical lasers. When operational, these weapons will provide even more accurate platforms from which to engage and neutralize enemy combatants. At the same time, research in neuroscience is ongoing to produce early-warning devices that will monitor soldier's brainwave activities and alert them when their heightened subconscious senses danger.

Significant medical research continues to develop better methods of prevention, treatment, and rehabilitation for members of the armed forces engaged in combat. Of special note is work in prosthetics. Continuing research in biomechanics is leading to improvements in devices that mimic human extremities, and researchers continue to devise ways to link artificial limbs to nerve endings to provide better mobility and control. At the same time, basic research into diseases most commonly associated with battlefield conditions, as well as those that exist in potential battle zones, is under way to create more effective prophylactics that will permit military personnel to ward off disease or recover from illnesses and return to duty more quickly.

In the twenty-first century, however, a major factor influencing military R&D is the increase in incidents of asymmetric warfare worldwide. Countries with large arsenals of sophisticated weapons are finding it necessary to defeat forces with considerably less technological capability. While scientists work to create better defensive armaments and offensive weapons with greater precision to minimize collateral damage, combat engineers and explosive-ordnance-disposal specialists are facing challenges on the front lines to create effective fortifications and remove hazards from battle areas where civilians and combatants are often indistinguishable.

Laurence W. Mazzeno, B.A., M.A., Ph.D.

FURTHER READING

Amato, Ivan. *Pushing the Horizon: Seventy-Five Years of High-Stakes Science and Technology at the Naval Research Laboratory.* Washington, D.C.: Government Printing Office, 1998. Describes the activities of the Naval Research Laboratory in developing weapons systems, communications technology, and equipment during the twentieth century.

Boot, Max. *War Made New: Technology, Warfare, and the Course of History, 1500 to Today.* New York: Gotham Books, 2006. Traces the impact of new technologies on the conduct of war from the gunpowder revolution in the sixteenth century through the development of nuclear armaments and other weapons of mass destruction in the twentieth.

Evans, Nicholas D. *Military Gadgets: How Advanced Technology Is Transforming Today's Battlefield and Tomorrow's.* Upper Saddle River, N.J.: Prentice Hall, 2004. Provides an overview of existing technologies being used by armed forces and ones in development.

Langford, R. Everett. *Introduction to Weapons of Mass Destruction: Radiological, Chemical, and Biological.* Hoboken, N.J.: John Wiley & Sons, 2004. Describes elements of weapons of mass destruction, explains their capabilities, and explains technological countermeasures used to assure individual and community survival.

Levis, Alexander H., ed. *The Limitless Sky: Air Force Science and Technology Contributions to the Nation.* Washington, D.C.: Air Force History and Museums Program, 2004. Essay collection describing several significant projects in which technological developments led to significant improvements in Air Force readiness and capability.

Lewer, Nick, ed. *The Future of Non-Lethal Weapons: Technologies, Operations, Ethics and Law.* Portland, Oreg.: Frank Cass, 2002. Includes several essays on technologies used to develop and assess nonlethal weapons and evaluates claims made for their effectiveness.

O'Hanlon, Michael. *Technological Change and the Future of Warfare.* Washington, D.C.: Brookings Institution, 2000. Describes ways new technologies influence changes in military doctrine and strategy and affect budgeting and international relations.

Price, Alfred. *War in the Fourth Dimension: U.S. Electronic Warfare, from Vietnam to the Present.* Mechanicsburg, Pa.: Stackpole, 2001. Traces the history of electronic warfare and its impact on combat operations.

Richardson, Jacques. *War, Science and Terrorism: From Laboratory to Open Conflict.* Portland, Oreg.: Frank Cass, 2002. Examines the interaction of scientific research and combat, explaining the roles of scientists, engineers, and those involved in production and manufacture of new technologies.

Waltz, Edward. *Information Warfare: Principles and Operations.* Norwood, Mass.: Artech House, 1998. Explains ways advances in information technology influence offensive and defensive military operations and describes a number of technologies in some detail.

WEB SITES

Defense Advanced Research Projects Agency
http://www.darpa.mil

National Defense University
Center for Technology and National Security Policy
http://www.ndu.edu/inss/index.cfm?secID=53&pageID=4&type=section

U.S. Air Force Research Laboratory
http://www.afrl.af.mil

U.S. Army Corps of Engineers, Engineer Research and Development Center
http://www.erdc.usace.army.mil

U.S. Naval Research Laboratory
http://www.nrl.navy.mil

See also: Aeronautics and Aviation; Antiballistic Missile Defense Systems; Applied Mathematics; Applied Physics; Artificial Intelligence; Biomechanics; Bridge Design and Barodynamics; Civil Engineering; Computer Engineering; Computer Science; Cryptology and Cryptography; Electrical Engineering; Environmental Engineering; Long-Range Artillery; Mechanical Engineering; Metallurgy; Meteorology; Naval Architecture and Marine Engineering; Navigation; Robotics.

MINERALOGY

FIELDS OF STUDY

Geology; chemistry; geochemistry; physics; petrology; experimental petrology; environmental geology; forensic mineralogy; medical mineralogy; gemology; economic geology; geochronology; descriptive mineralogy; crystallography; crystal chemistry; mineral classification; geologic occurrence; optical mineralogy; mining; chemical engineering.

SUMMARY

Mineralogy is the study of the chemical composition and physical property of minerals, the arrangement of atoms in the minerals, and the use of the minerals. Minerals are naturally occurring elements or compounds. The composition and arrangement of the atoms that make up minerals is reflected in their physical characteristics. For example, gold is a naturally occurring mineral containing one element that has a definite density of 19.3 grams per milliliter and a yellow color and is chemically inactive. Sometimes mineral resources are broadened to refer to oil, natural gas, and coal, although those materials are not technically minerals.

KEY TERMS AND CONCEPTS

- **Clay Mineral:** Any of a group of tiny silicate minerals (less than 2 micrometers in size) that form by varied degrees of weathering of other silicate minerals.
- **Cleavage:** Breaking of a crystallized mineral along a plane, leaving a smooth rather than an irregular surface.
- **Hardness:** Resistance of a mineral to being scratched by another material.
- **Hydrothermal Deposit:** Mineral deposit precipitated from a hot water or gas solution.
- **Igneous Rock:** Rock formed from molten rock material.
- **Ion:** Atom or group of atoms with a positive or negative charge.
- **Lava:** Molten rock material at the Earth's surface.
- **Magma:** Molten rock material and suspended mineral crystals below the Earth's surface.

- **Major Elements:** Elements that make up the bulk of the chemical composition of a mineral or rock.
- **Petrology:** Study of the origin, composition, structure, and properties of rocks and the processes that formed the rocks.
- **Sedimentary Rock:** Rock formed from particles such as sand by the weathering of other rocks at the surface or by precipitation from water.
- **Silicate Mineral:** Mineral that contains silicon bonded with oxygen to form silicate groups bonded with positive ions; silicates make up 90 percent of the Earth's crust.
- **Trace Element:** Element present only in tiny amounts (in quantities of parts per million or less) in a mineral or rock.

DEFINITION AND BASIC PRINCIPLES

Minerals are solid elements or compounds that have a definite but often not fixed chemical composition and a definite arrangement of atoms or ions. For instance, the mineral olivine is magnesium iron silicate, with the formula of $(Mg, Fe)_2SiO_4$, meaning that it has oxygen (O) and silicon (Si) atoms in a ratio of 4:1 and a total of two ions of magnesium (Mg) and iron (Fe) in any ratio. The magnesium and iron component can vary from 100 percent iron with no magnesium to 100 percent magnesium with no iron, and all variations in between. Other minerals, however, such as gold, have nearly 100 percent gold atoms. Minerals are usually formed by inorganic means, but some organisms can form minerals such as calcite (calcium carbonate, with the formula $CaCO_3$).

Minerals make up most of the rocks in the earth, so they are studied in many fields. In geochemistry, for example, scientists determine the chemical composition of the minerals and rocks to derive hypotheses about how various rocks may have formed. Geochemists might study what rocks melt to form magmas or lavas of a certain composition. Environmental geologists attempt to solve problems regarding minerals that pollute the environment. For instance, environmental geologists might try to minimize the effects of the mineral pyrite (iron sulfide, with the formula FeS_2) in natural bodies of water. Pyrite, which is found in coal, dissolves in water to form sulphuric acid and high-iron water, which can kill some organisms. Forensic mineralogists may determine the origin of minerals left at a crime scene. Economic geologists discover and determine the distribution of minerals that can be mined, such as lead minerals (galena), salt, gypsum (for wall board), or granite (for kitchen countertops). Geophysicists may study the minerals below the Earth's surface that cannot be directly sampled. They may, for example, study how the seismic waves given off by earthquakes pass through the ground to estimate the kinds of rocks present.

BACKGROUND AND HISTORY

Archaeological evidence suggests that humans have used minerals in a number of ways for tens of thousands of years. For instance, the rich possessed jewels, red and black minerals were used in cave drawings in France, and minerals such as gold were used for barter. Metals were apparently extracted from ores for many years, but the methods of extraction were conveyed from person to person without being written down. In 1556, Georgius Agricola's *De re metallica* (English translation, 1912) described many of these mineral processing methods. From the late seventeenth century into the nineteenth century, many people studied the minerals that occur in definite crystal forms such as cubes, often measuring the angles between faces.

In the nineteenth century, the fields of chemistry and physics developed rapidly. Jöns Jacob Berzelius developed a chemical classification system for minerals. Another important development was the polarizing microscope, which was used to study the optical properties of minerals to aid in their identification.

During the late nineteenth century, scientists had theorized that the external crystal forms of minerals reflected the ordered internal arrangement of their atoms. In the early twentieth century, this theory was confirmed through the use of X rays. Also, it became possible to chemically analyze minerals and rocks so that chemical mineral classifications could be further developed. Finally, in the 1960's, the use of many instruments such as the electron microprobe allowed geologists to determine variations in chemical composition of minerals across small portions of the minerals so that models for the formation of minerals could be further developed.

HOW IT WORKS

Mineral and Rock Identification. A geologist may tentatively identify the minerals in a rock using characteristic such as crystal form, hardness, color, cleavage, luster (metallic or nonmetallic), magnetic properties, and mineral association. The rock granite, for instance, is composed of quartz (often colorless, harder than other minerals, with rounded crystals, no cleavage, and nonmetallic luster) and feldspars (often tan, softer than quartz, with well-developed crystals, two good nearly right-angle cleavages, and nonmetallic luster), with lesser amounts of black minerals such as biotite (softer than quartz, with flat crystals, one excellent cleavage direction, and shiny nonmetallic luster).

The geologist then slices the rock into a section about 0.03 millimeters thick (most minerals are transparent). The section is examined under a polarizing microscope to confirm the presence of the tentatively identified minerals and perhaps to find other minerals that could not be detected by eye because

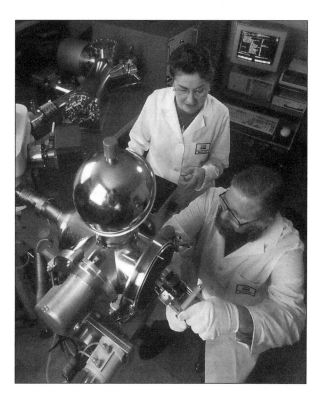

A chemist (left) and physical scientist use thermal ionization mass spectrometry to measure trackable forms of copper, called stable isotopes, in blood plasma. (Science Source)

they were too small or present in very low quantities. The geologist can determine other relationships among the minerals, such as the sequence of crystallization of the minerals within an igneous rock.

Identification Using Instruments. The minerals in a rock can be analyzed a variety of other ways, depending on the goals of a given study. For instance, X-ray diffraction may be used to identify some minerals. One of the most useful applications of X-ray diffraction is to identify tiny minerals such as clay minerals that are hard to identify using a microscope. The wavelengths of the X rays are similar to the spacing between atoms in the clay minerals, so when X rays of a single wavelength are passed through a mineral, they are diffracted from the minerals at angles that are characteristic of a particular mineral. Thus, the mixture of clay minerals in a rock or soil may be identified.

The electron microprobe has enabled the analysis of tiny portions of minerals so that changes in composition across the mineral may be determined. The instrument accelerates masses of electrons into a mineral that releases X rays with energies that are characteristic of a given element so that the elements present can be identified. The amount of energy given off is proportional to the amount of the element in the sample; therefore, the concentration of the element in the mineral can be determined when the results are compared with a standard of known concentration. The electron microprobe can also be used to analyze other materials such as alloys and ceramics.

The scanning electron microscope uses an electron beam that is scanned over a small portion of tiny minerals and essentially takes a photograph of the mineral grains in the sample. Some electron microscopes are set up to determine qualitatively what elements are present in the sample. This information is often enough to identify the mineral.

Other Analytical Techniques. Many instruments and techniques are used to analyze the major elements, trace elements, and the isotopic composition of minerals and rocks. Commonly used methods are X-ray fluorescence (XRF), inductively coupled plasma mass spectrometer (ICP-MS), and thermal ionization mass spectrometry. X-ray fluorescence is used to analyze bulk samples of minerals, rocks, and ceramics for major elements and many trace elements. A powdered sample or a glass of the sample is compressed and is bombarded by X rays so that an energy spectrum that is distinctive for each element is emitted. The amount of radiation given off by the sample is compared with a standard of known concentration to determine how much of each element is present in the sample.

The inductively coupled plasma mass spectrometer is used to analyze many elements in concentrations as low as parts per trillion by passing vaporized samples into high-temperature plasma so that all elements have positive charges. A mass spectrometer sorts out the ions by their differing sizes and charges in a magnetic field, which permits a determination of the elemental concentrations when the results are compared with a standard of known concentration. Up to seventy-five elements can be rapidly analyzed in a sample at precisions of 2 percent or better.

Thermal ionization mass spectrometers can be used to analyze the isotopic ratios of higher mass elements such as rubidium, strontium, uranium, lead, samarium, and neodymium, which may be used to interpret the geologic age of a rock. The mineral or rock is placed on a heated filament so that the isotopes are ejected into a magnetic field in which the ions are deflected by varied amounts depending on the mass and charge of the isotope. The data may then be used to calculate the amount of a certain isotope in the sample and eventually the isotopic age of the sample.

Other specialized instruments are also available. For instance, gemologists use specialized instruments to study and cut gemstones.

APPLICATIONS AND PRODUCTS

Abundant Metals and Uses. The most abundant metals are iron, aluminum, magnesium, silicon, and titanium. Iron, which is mostly obtained from several minerals composed of iron and oxygen (hematite and magnetite), accounts for 95 percent by weight of the metals used in the United States. Much of the hematite and magnetite is obtained from large sedimentary rock deposits called banded-iron formations that are up to 700 meters thick and extend for up to thousands of square kilometers. The banded-iron formations formed 1.8 billion to 2.6 billion years ago. They are abundant, for instance, around the Lake Superior region in northern Minnesota, northern Wisconsin, and northwestern Michigan. The banded-iron formations have produced billions of tons of iron ore deposits. The ores are made into pellets and are mixed with limestone and coke to burn at 1,600 degrees Celsius in a blast furnace. The iron produced in this process is molten and can be mixed with small amounts of scarce metals (called ferroalloy metals) to produce steel with useful properties. For instance, the addition of chromium gives steel strength at high temperatures, and it prevents corrosion. The addition of niobium produces strength, and the addition of copper increases the resistance of the steel to corrosion.

Much of the aluminum occurs in clay minerals in which the aluminum cannot be economically removed from the other elements. Thus, most of the aluminum used comes from bauxite, which is a mixture of several aluminum-rich minerals such as gibbsite ($Al(OH)_3$) and boehmite ($AlOOH$). Bauxite forms in the tropics to subtropics by intense chemical weathering from other aluminum-rich minerals such as the clay minerals. The production of aluminum from bauxite is very expensive because the ore is dissolved in molten material at 950 degrees Celsius and the aluminum is concentrated by an electric current. Electricity represents about 20 percent of the total cost of producing aluminum. Recycling of aluminum is economical because the energy expended in making a new can from an old aluminum can is about 5 percent of the energy required to make a new aluminum can from bauxite.

Scarce Metals and Uses. At least thirty important trace metals occur in the Earth's crust at concentrations of less than 0.1 percent. Therefore, these elements are trace elements in most minerals and rocks until special geologic conditions concentrate them enough so that they become significant portions of some minerals. For instance, gold is present in the Earth's crust at only 0.0000002 percent by weight. Gold often occurs as gold uncombined with other elements and precipitated from hydrothermal solutions as veins. Gold may also combine with the element tellurium to form several kinds of gold-tellurium minerals. Also gold does not react very well chemically during weathering, so it may concentrate in certain streams to form placer deposits. Gold has been used for thousands of years in jewelry, dental work, and coins. In the past, much world exploration has been motivated by the drive to find gold.

Gold is a precious metal. Other precious metals are silver and the platinum group elements (for

Fascinating Facts About Mineralogy

- In 1822, Friedrich Mohs developed a hardness scale for minerals, using ten minerals. From hardest to softest, they are diamond, corundum, topaz, quartz, potassium feldspar, apatite, fluorite, calcite, gypsum, and talc.
- Diamonds are the hardest natural material on Earth and are the most transparent material known. Ultra-thin diamond scalpels retain their sharp edge for long periods and can transmit laser light to cauterize wounds.
- Blood diamonds, also known as conflict diamonds, are produced or traded by rebel forces opposed to the recognized government in order to fund their activities. Most of the blood diamonds come from Sierra Leone, Angola, the Democratic Republic of Congo, Liberia, and the Ivory Coast.
- The mineral kaolin is used to create a hard clay used in rubber tires for lawn equipment, garden hoses, rubber floor mats and tiles, automotive hoses and belts, shoe soles, and wire and cable.
- In China, ownership of Imperial jade, a fine-grained jade with a rich, uniform green color, was once limited to nobility. Jade is believed to have powers to ward off evil and is associated with longevity.
- The average automobile contains about 0.9 mile of copper wiring. The total amount of copper used in an automobile ranges from 44 to 99 pounds.
- The manufacture of jewelry accounts for 78 percent of the gold used each year. However, because gold conducts electricity and does not corrode, it is used as a conductor in sophisticated electronic devices such as cell phones, calculators, television sets, personal digital assistants, and Global Positioning Systems.

example, platinum, palladium, and rhodium). Silver, like gold, has been used for thousands of years in jewelry and coins. In modern times, silver also finds uses in batteries and in photographic film and papers because some silver compounds are sensitive to light. Silver occurs in nature as silver sulfide, and it substitutes for copper in copper minerals so that much of the silver produced is a side product from copper mining. The platinum metals occur together as native elements or combined with sulfur and tellurium. They concentrate in some dark-colored igneous rocks or as placers. The platinum group metals are useful as catalysts in chemical reactions, so they are, for example, used in catalytic convertors in vehicles.

Another group of scarce metals are base metals. The base metals are of relatively low economic value compared with the precious metals. The base metals include copper, lead, zinc, tin, mercury, and cadmium. Copper minerals are native copper and various copper minerals combined with sulfate, carbonate, and oxygen. Copper minerals occur in some igneous rocks, including some hydrothermal deposits, mostly as dilute copper-sulfur minerals. These minerals may be concentrated during weathering, often forming copper minerals combined with oxygen and carbonate. Copper conducts electricity very well, and much of it is used in electric appliances and wires. Lead and zinc tend to occur together in hydrothermal deposits in combination with sulfur. Lead is used in automobile batteries, ammunition, and solder. Lead is very harmful to organisms, which restricts its potential use. Zinc is used as a coating on steel to prevent it from rusting, in brass, and in paint.

Fertilizer and Minerals for Industry. Fertilizers contain minerals that have nitrogen, phosphorus, potassium, and a few other elements necessary for the growth of plants. In Peru, deposits of guano (seabird excrement), which is rich in these elements, have been mined and used as fertilizer. Saltpeter (potassium nitrate) has also been mined in Peru. A calcium phosphate mineral, apatite, occurs in some sedimentary rocks, so some of these have been mined for the phosphorus.

Some minerals are a source for sodium and chlorine. Halite, commonly known as rock salt, is used as a flavoring and to melt ice on roads. Halite is produced in some sedimentary rocks from the slow evaporation of water in closed basins over long periods of time, a process that is occurring in the Great Salt Lake in Utah.

A variety of rocks or sediment—including granite, limestone, marble, sands, and gravels—are used for the exteriors of buildings, construction materials, or for countertops. Sands and gravels may be mixed with cement to make concrete.

IMPACT ON INDUSTRY

The materials of the Earth interact, which means that all societies depend on mineral resources. For example, crops grow in soil containing clay minerals formed from the weathering of rocks. Fertilizers composed of nitrogen, phosphorus, and potassium minerals must be added to the soil to grow enough food to support the growing population of the world. Machinery such as tractors, trains, ships, and trucks made of steel and other materials obtained from processing metal ores must be used to fertilize fields and to move food to markets. Electricity, much of which is obtained by burning coal, must be used to process food, whether in factories or in homes. Most of these mineral resources cannot be renewed quickly, so many are being rapidly used up.

Industry and Business. The largest industries directly related to mineralogy are public or private corporations engaged in the exploration, production, and development of minerals. As of 2009, the United States was the world's third largest producer of copper, after Chile and Peru, and of gold, after China and Australia. In 2009, net exports of mineral raw materials and old scrap materials was $10 billion. In 2009, Freeport-McMoRan Copper and Gold's Morenci open-pit mine in Arizona produced about 460,000 pounds of copper (about 40 percent of its total North American production), Barrick Gold Corporation's Goldstrike Complex in Nevada produced 1.36 million ounces of gold, and Teck's Red Dog Mine in Alaska produced 79 percent of the total U.S. production of zinc.

Government Regulation. Although most governments do not directly control the mineral resources in their countries, they often regulate mining to ensure the workers' safety and support mining through subsidies and funding. In the United States, government agencies regulating mining include the Mine Safety and Health Administration of the U.S. Department of Labor and the mining division of the National Institute for Occupational Safety and Health. The U.S. Geological Survey maintains statistics and information on the worldwide supply of and demand for minerals.

Academic Research. Engineering departments in colleges and universities engage in mineralogy research such as improving the safety and efficiency of mining operations and searching for better, cleaner ways to process ores. For instance, some engineers analyze ways to break up ores and separate the ore minerals or develop ways to use more of the metals in the ores.

CAREERS AND COURSE WORK

Anyone interested in pursuing a career in the minerals industry should study geology, chemistry, physics, biology, chemical engineering, and mining engineering. A bachelor's degree in one of these subjects is the minimum requirement for working in oil exploration or as a mining technician or water-quality technician. A master's degree or a Ph.D. is required for more challenging and responsible jobs in geochemistry, geophysics, environmental geology, geochronology, forensic mineralogy, and academics. Geochemists should have a background in geology and chemistry. Geophysicists use physics, mathematics, and geology to remotely tell what kinds of rocks are below the surface of the earth. Environmental geologists should have backgrounds in geology, chemistry, and biology. Geochronologists use natural radioactive isotopes to give estimates of the time when some kinds of rocks formed. They should have a background in geology, physics, and chemistry. Forensic mineralogists should have a good background in mineralogy and criminology. Economic geologists should major in geology, with a concentration in courses concerned with ore mineralization. Those interested in an academic career should have a Ph.D. in geology, mining engineering, or chemical engineering. They will be expected to teach and do research in their subject area.

SOCIAL CONTEXT AND FUTURE PROSPECTS

Mineral resources are not evenly distributed throughout the world. In the nineteenth century, the United States did not import many mineral resources, but it has increasingly become an importer of mineral resources. The need to import is driven by many factors, including an increase in the minerals used. U.S. production of final products, including cars and houses, using mineral materials accounted for about 13 percent of the 2009 gross domestic product. Other factors include the depletion of some U.S. sources of minerals and the increased use of minerals that are not naturally available in the United States. In 2009, foreign sources supplied more than 50 percent of thirty-eight mineral commodities consumed in the United States. Nineteen of those minerals were 100 percent foreign sourced. The United States must import much of the manganese, bauxite, platinum group minerals, tantalum, tungsten, cobalt, and petroleum that it uses. In contrast, the United States still has abundant resources of gold, copper, lead, iron, and salt. Japan, however, has few mineral resources, so it must import most of them.

This need to import and export mineral resources has forced countries to cooperate with one another to achieve their needs. The mineral resources traded in the largest quantities, in descending order, are iron-steel, gemstones, copper, coal, and aluminum. The total annual world trade in mineral resources is approaching $700 billion and is likely to rapidly increase as countries such as China begin to use and produce more mineral resources.

As the supply of minerals decreases worldwide and environmental and safety concerns raised by mining gain in importance, it is likely that mineralogy research may turn to improving extraction and processing methods and looking for ways to recycle materials, capture minerals remaining in wastes, and restore mining areas after the minerals have been depleted.

Robert L. Cullers, B.S., M.S., Ph.D.

FURTHER READING

Craig, James R., David J. Vaughan, and Brian Skinner. *Earth Resources and the Environment.* Upper Saddle River, N.J.: Prentice Hall, 2010. Gives an overview of mineral resources in the Earth and describes environmental problems regarding fossil fuels, metals, fertilizers, and industrial minerals. Appendix, glossary, index, and many illustrations.

Guastoni, Alessandro, and Roberto Appiani. *Minerals.* Buffalo, N.Y.: Firefly Books, 2005. This guide describes the common minerals, gives their chemical formulas, and describes where they are found. Contains illustrations.

Kearny, Philip, ed. *The Encyclopedia of the Solid Earth Sciences.* London: Blackwell Science, 1994. Contains an alphabetical list of defined geologic terms, minerals, and geologic concepts.

Klein, Cornelis, and Barbara Dutrow. *Manual of Mineral Science.* Hoboken, N.J.: John Wiley & Sons, 2008. Describes how to identify minerals using physical properties and modern analytical techniques.

Pellant, Chris. *Smithsonian Handbook: Rocks and Minerals.* London: Dorling Kindersley Books, 2002. This identification guide for minerals and rocks uses many pictures combined with text.

Sinding-Larsen, Richard, and Friedrich-Wilhelm Wellmer, eds. *Non-renewable Resource Issues: Geoscientific and*

Societal Challenges. London: Springer, 2010. Looks at minerals as nonrenewable resources and discusses the challenges that societies face.

WEB SITES
Geology and Earth Sciences
http://geology.com

Mineralogical Society of America
http://www.minsocam.org

National Institute for Occupational Safety and Health
Mining
http://www.cdc.gov/niosh/mining

U.S. Department of Labor
Mine Safety and Health Administration
http://www.msha.gov

U.S. Geological Survey
Minerals Information
http://minerals.usgs.gov/minerals

See also: Diffraction Analysis; Electronic Materials Production; Gemology and Chrysology; Land-Use Management; Marine Mining; Soil Science; Steelmaking Technologies.

MIRRORS AND LENSES

FIELDS OF STUDY

Physics; astronomy; mathematics; geometry; chemistry; mechanical engineering; optical engineering; aerospace engineering; environmental engineering; medical engineering; computer science; electronics; photography.

SUMMARY

Mirrors and lenses are tools used to manipulate the direction of light and images, mirrors by using flat or curved glass in the reversion, diversion, or formation of images, and lenses by using polished material, usually glass, in the refraction of light. Scientific applications of mirrors and lenses cover a broad spectrum, ranging from photography, astronomy, and medicine to electronics, transportation, and energy conservation. Without mirrors and lenses, it would be impossible to preserve memories in a photograph, view far-away galaxies through a telescope, diagnose diseases through a microscope, or create energy using solar panels. Miniaturized mirrors are also implemented in scanners used in numerous electronic devices, such as copying machines, bar-code readers, compact disc players, and video recorders.

KEY TERMS AND CONCEPTS

- **Achromatic Lens:** Lens that refracts light without separating its components into the various colors of the spectrum.
- **Aperture:** Opening through which light travels in a camera or optical device, determining the amount of light delivered to the lens, and affecting the quality of an image.
- **Catadioptric Imaging:** Imaging system that uses a combination of mirrors and lenses to create an optical system.
- **Catoptric Imaging:** Imaging system that uses only a combination of mirrors to create an optical system.
- **Diffraction:** Spreading out of light by passing through an aperture or around an object that bends the light.
- **Diopter:** Unit of measurement that determines the optical power of a lens by calibrating the amount of light refracted by the lens, in correlation with its focal length.
- **Fresnel Lens:** Flat lens that reduces spherical aberration and increases light intensity due to its numerous concentric circles, often creating a prism effect.
- **Negative Lens:** Concave lens with thick edges and a thinner center, used most often in eyeglasses for correcting nearsightedness.
- **Positive Lens:** Convex lens with thin edges and a thicker center, used most often in eyeglasses for correcting farsightedness.
- **Refraction:** Bending of light waves, used by a lens to magnify, reduce, or focus images.
- **Spherical Aberration:** Distortion or blurriness of image caused by the geometrical formation of a spherical lens or mirror.

DEFINITION AND BASIC PRINCIPLES

Mirrors and lenses are instruments used to reflect or refract light. Mirrors use a smooth, polished surface to revert or direct an image. Lenses use a piece of smooth, transparent material, usually glass or plastic, to converge or diverge an image. Because light beams that strike dark or mottled surfaces are absorbed, mirrors must be highly polished in order to reflect light effectively. Likewise, since rough surfaces diffuse light rays in many different directions, mirror surfaces must be exceedingly smooth to reflect light in one direction. Light that hits a mirrored surface at a particular angle will always bounce off that mirrored surface at an exactly corresponding angle, thereby allowing mirrors to be used to direct images in an extremely precise manner.

Images are reflected according to one of three main types of mirrors: plane, convex, and concave. Plane mirrors are flat surfaces and reflect a full-size upright image directly back to the viewer but left and right are inverted. Convex mirrors are curved slightly outward and reflect back a slightly smaller upright image, but in a wider angled view, and left and right are inverted. Concave mirrors are curved slightly inward and reflect back an image that may be larger or smaller, depending on the distance from the object, and the image may be right-side up or upside down.

BACKGROUND AND HISTORY

The ancient Egyptians used mirrors to reflect light into the dark tombs of the pyramids so that workmen could see. The Assyrians used the first known lens, the Nimrud lens, which was made of rock crystal, 3,000 years ago in their work, either as a magnifying glass or as a fire starter. Chinese artisans in the Han dynasty created concave sacred mirrors designed for igniting sacrificial fires, and Incan warriors wore a bracelet on their wrists containing a small mirror for the focusing of light to start fires.

It was not until the end of the thirteenth century, however, when an unknown Italian invented spectacles, that lenses subsequently became used in eyeglasses. Near the end of the sixteenth century, two spectacle makers inadvertently discovered that certain spectacle lenses placed inside a tube made objects that were nearby appear greatly enlarged. Dutch spectacle maker Zacharias Janssen and his son, Hans Janssen, invented the microscope in 1590. In 1608, Zacharias Janssen, in collaboration with Dutch spectacle makers Hans Lippershey and Jacob Metius, invented the first refractory telescope using lenses. However, it was Galileo Galilei who, in 1609, took the first rudimentary refracting telescope invented by the Dutch, drastically improved on its design, and went on to revolutionize history by using his new telescopic invention to observe that the Earth and other planets revolve around the Sun. In 1668, Sir Isaac Newton invented the first reflecting telescope, which, using mirrors, conveyed a vastly superior image than the refractory telescope, since it greatly reduced distortion and spherical aberrations often conveyed by the refractory telescope's lenses.

Almost two hundred years later, in 1839, with the invention of photography by Louis Daguerre, the convex lens became used for the first time in cameras. In 1893, Thomas Edison patented his motion-picture camera, the kinetoscope, which used lenses, and by the 1960's, mirrors were being used in satellites and to reflect lasers.

HOW IT WORKS

The law of reflection states that the angle of incidence is equal to the angle of reflection, meaning that if light strikes a smooth mirrored surface at a 45-degree angle, it will also bounce off the mirror at a corresponding 45-degree angle. Whenever light hits an object, it may be reflected, or it may be absorbed, which is what happens when light hits a dark surface. Light may also hit a rough shiny surface, in which case the light will be reflected in many different directions, or diffused. Light may also simply pass through an object altogether if the material is transparent.

When light travels from one medium to another, such as air to water, the speed of the light slows down and refracts, or bends. Lenses take advantage of the fact that light bends when changing mediums by manipulating the light to serve a variety of purposes. The angle that light is refracted by a lens depends on the lens's shape, specifically its curvature. A glass or plastic lens can be polished or ground so that it gathers light toward its edges and directs it toward the center of the lens, in which case the lens is concave, and the light rays will be diverged. Conversely, a lens which is convex, because it is thicker in the middle and has thinner edges, will cause light rays to converge.

Both concave and convex lenses rely on the light rays' focal point either to magnify, reduce, or focus an image. The focal point is simply the precise point where light rays come together in a pinpoint and focus an image. The distance from the center of the lens to the exact point where light rays focus is called the lens's focal length. Light that passes through a convex lens is refracted so that the rays join and focus out in front of the lens, creating a convergence. Light rays passing through a concave lens are refracted outward and create a divergence because the focal point for the light rays appears to be originating from behind the lens. Aberrations such as blurriness or color distortion may sometimes occur if lenses are made using glass with impurities or air bubbles or if the lens is not ground or polished properly to make it precisely curved and smooth.

APPLICATIONS AND PRODUCTS

Electronics. Lenses and mirrors are used extensively in consumer electronics, primarily as part of systems to read and write optical media such as CDs and DVDs. These optical disks encode information in microscopic grooves, which are read by a laser in much the same manner as a turntable's needle reads a record. Good-quality lenses are essential to focus the laser beam onto the disk and to capture the reflection from the disk's surface.

Cameras also make use of both mirrors and lenses. In single-lens reflex (SLR) cameras, the film or

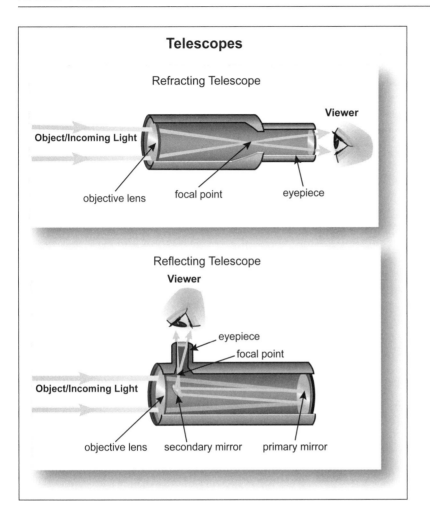

Telescopes

Refracting Telescope

Viewer

Object/Incoming Light

objective lens   focal point   eyepiece

Reflecting Telescope

Viewer

eyepiece

focal point

Object/Incoming Light

objective lens   secondary mirror   primary mirror

Lens and mirror systems are also indispensable as tools in a larger research apparatus. Microscopes, high-speed cameras, and other digital equipment are commonplace in biological, chemical, and physical science laboratories, and many experimental setups include custom-made imaging equipment all based on combinations of mirrors and lenses.

Retail. JC Penney and Macy's department stores have installed full-size digital touch-screen mirrors in their stores, revolutionizing the retail experience. The mirrors enable consumers to model clothes, makeup, and jewelry digitally, without actually having to try on the products. Because an extraordinary amount of time and energy is saved by the interactive retail mirrors, customers have more flexibility to shop creatively by sampling a broader range of merchandise. The interactive digital mirrors also offer those shopping for friends or relatives the added convenience of being able to gauge sizes correctly, greatly reducing the number of returns. The increased ease and efficiency of touch-screen digital mirrors has generated more customers and sales for both retailers.

image sensor is protected from light by a flip mirror. When a photograph is taken, the flip mirror rotates, directing light onto the film or digital sensor to expose the image to a size that will fit onto the film. Modern camera lenses are actually composed of multiple simple lenses, which helps to improve the overall image quality.

Science. Astronomy has long been the driving force behind advances in optics, and many varieties of lenses and mirrors were developed specifically to enable astronomical observation. Large modern telescopes are limited not by the optical quality of their lenses and mirrors but by distortion caused by turbulence in the atmosphere. To address this, state-of-the-art telescopes use adaptive optical systems, which can change the shape of the telescope's primary mirror in response to fluctuations in the atmosphere, which are measured with a powerful laser.

IMPACT ON INDUSTRY

Military. There are numerous military applications for lens- and mirror-based systems, and these have had an enormous influence on industry. Relatively simple applications include binoculars, range-finding systems, and weapon scopes. More complex optics are employed in guidance systems for aircraft and missiles, which often pair telescopic systems with image-recognition software. Reconnaissance satellites are also an important use for mirror and lens systems, and the general concept of a spy satellite is similar to a large space-based telescope, with multiple lenses and mirrors serving to magnify an image before projection onto an imaging sensor. Cutting-edge corporations such as Raytheon, Elbit Systems of America, and Christie were recently awarded multimillion-dollar defense contracts for military-based research involving lens and mirror applications in radar, lasers, and projectors.

Medicine. Lens technology has, of course, revolutionized the practice of optometry, and contact lenses in particular have become a multibillion-dollar industry worldwide. Because eyeglasses are essentially a single-lens optical that corrects for abnormalities in the shape of the eye, the basic geometry of the eyeglass lens has not changed for many years. However, major progress has been made in lens coatings and materials, which improve the aesthetics, durability, weight, and cost of eyeglasses. Additionally, modern manufacturing techniques have allowed greater lens diversity with more complex designs, creating progressive lenses that have a continuously variable amount of focus correction across the lens wearer's field of vision (as opposed to traditional bifocals, which provide only two levels of correction). Bausch + Lomb is at the forefront of engineering progressive lenses and adjustable-focus lenses, which allow the lens wearer to manually adjust the focus correction depending on the distance of the object being viewed.

Government and University Research. Undoubtedly, the greatest ongoing government research projects involving lenses and mirrors are undertaken by the National Aeronautics and Space Administration (NASA). The Hubble Space Telescope, launched by NASA in 1990, marks the culmination of decades of scientific research and the apex of mirror engineering and design. NASA's impact on the mirror and lens industry at large is paramount, since the majority of lens and mirror innovations have been a direct result of lens and mirror engineering for telescopes largely developed by NASA scientists. Although the Hubble Space Telescope contains the most advanced and sophisticated mirrors ever created, Hubble's successor, the James Webb Space Telescope, promises to surpass even Hubble by being able to see farther into space by monitoring infrared radiation.

Composite Mirror Applications, in Tucson, Arizona, has worked closely with NASA to build mirrors used in space, most notably building a mirror used by space shuttle Endeavor on its final mission to search for dark matter and anti-matter in the universe. Steward Observatory at the University of Arizona has also worked in close conjunction with Composite Mirror Applications in the development of special optical projects with applications for the telescopic industry.

CAREERS AND COURSE WORK

Courses in physics, astronomy, advanced mathematics, geometry, optical engineering, mechanical engineering, aerospace engineering, computer science, and chemistry provide background knowledge and training needed to pursue a future working with various applications utilizing mirrors and lenses. A bachelor's degree in any of the above fields would assist in the basic applications of mirrors and lenses, but careers involving research or development of lenses in the corporate domain or academia would almost certainly necessitate a graduate degree in one of the above disciplines—ideally a doctorate. Although employment at any university as an educator and researcher is one potential career path, with an

Fascinating Facts About Mirrors and Lenses

- In the sixteenth century, mirror makers were held prisoner on the Venetian island of Murano, and they were sentenced to death if they left the island, for fear their secret mirror-making techniques would be revealed to the outside world.

- In 2010, French artist Michel de Broin created the world's largest disco mirror ball. Measuring almost twenty-five feet across and containing more than 1,000 mirrors, the mirror ball was suspended above Paris using a skyscraper crane.

- Other than humans and a handful of other higher primates, such as chimpanzees, orangutans, and bonobos, scientists studying self-awareness have discovered only four species that are able to recognize their own reflection in a mirror: bottlenose dolphins, killer whales, Asian elephants, and European magpies.

- In 1911, at the inaugural race of the Indianapolis 500, winner Ray Harroun used the first known automobile rearview mirror in his race car, the Marmon Wasp.

- The mirrors in the Hubble Space Telescope were ground to within $1/800,000$ of an inch of a perfect curve and were the most highly polished mirrors on Earth.

- Roman emperor Nero, who was nearsighted, watched gladiator games at the Roman Colosseum through a large handheld emerald lens centuries before the invention of spectacles.

emphasis in physics and astronomy, working at an observatory is also a career possibility. Having an additional concentration of course work in aerospace engineering is also outstanding preparation for employment with NASA and government space projects.

Additional career opportunities working with lenses abound in the medical field, especially as an optical engineer. Either a master's or doctorate in physics or optics is a prerequisite for a career as an optical engineer; however, an associate's degree is all that is necessary to become trained to work as a contact lens technician. To work as an optometrist, one must graduate from a college of optometry after completing a bachelor's degree, and to work as an ophthalmologist, one must obtain a medical degree.

Additional career opportunities in the "green" industry using mirrors in solar panels requires a background in environmental engineering, and work in the photographic industry requires extensive knowledge of photography as well as lenses.

SOCIAL CONTEXT AND FUTURE PROSPECTS

Although the increased necessity of using mirrors in green technology (such as solar panels) has become universally recognized, mirrors may conceivably play a key role in another environmental technology in the future. Scientists researching the long-term impact of global warming, increasingly alarmed by the rapid escalation of carbon dioxide buildup in the Earth's atmosphere, have begun seriously investigating the potential for mirrors to help lower the temperature of the Earth's atmosphere. Scientists are studying the possibility of launching a series of satellites to orbit above the Earth, each one containing large mirrors, which could be controlled, like giant window blinds, to reflect back out into space as much of the Sun's rays as desired, thereby regulating the temperature of the Earth's atmosphere like a thermostat.

Another futuristic trend foresees the transformation of mirrors in the home, programming bathroom mirrors, for example, to monitor an individual's health by analyzing a person's physical appearance. Vital signs, such as heart rate, blood pressure, and weight can be measured by the mirror and sent to a computer. Other mirrors may project a person's possible future appearance if certain negative lifestyle habits, such as smoking, overeating, excessive drinking, and lack of exercise, are practiced. Cameras placed throughout the home may record an individual's personal habits and then relay the information to an interactive home mirror linked to a computer. The mirror, in turn, may then portray possible physical results of a negative lifestyle to someone, traits such as obesity, discolored teeth, wrinkled skin, or receding hairline, turning mirrors into modern-day oracles.

Mary E. Markland, B.A., M.A.

FURTHER READING

Andersen, Geoff. *The Telescope: Its History, Technology, and Future.* Princeton, N.J.: Princeton University Press, 2007. A comprehensive survey of telescopes since their invention more than four hundred years ago, plus provides step-by-step instructions describing how to build a homemade telescope and observatory.

Burnett, D. Graham. *Descartes and the Hyperbolic Quest: Lens Making Machines and Their Significance in the Seventeenth Century.* Philadelphia: American Philosophical Society, 2005. Fascinating account of Descartes's lifelong dream of building a machine to manufacture an aspheric hyperbolic lens to end spherical aberration in lenses. Contains extensive drawings of Descartes's designs.

Conant, Robert Alan. *Micromachined Mirrors.* Norwell, Mass.: Kluwer Academic, 2003. Using profuse illustrations, examines how the miniaturization of mirrors has transformed modern technology, especially their use in scanners and in the electronics industry.

Kingslake, Rudolph. *A History of the Photographic Lens.* San Diego: Academic Press, 1989. Documents the development of photographic lenses from the beginning of photography in 1839 to modern telescopic lenses, providing ample pictures and sketches illustrating each lens type.

Pendergrast, Mark. *Mirror Mirror: A History of the Human Love Affair with Reflection.* New York: Basic Books, 2003. An exhaustive examination of mirrors and their influence on science, astronomy, history, religion, art, literature, psychology, and advertising, with an abundance of biographical material, supported by photographs, about scientists and mirror innovators.

Zimmerman, Robert. *The Universe in a Mirror: The Saga of the Hubble Space Telescope and the Visionaries Who Built It.* Princeton, N.J.: Princeton University Press,

2008. The behind-the-scenes story of the building of the Hubble Space Telescope, from its first conception in the 1940's, to its launch, repair, and triumph in the 1990's. Contains spectacular photographs of outer space sent back to Earth by the Hubble.

WEB SITES
American Astronomic Society
http://aas.org

The International Society for Optics and Photonics
http://spie.org

National Aeronautics and Space Administration
The James Webb Space Telescope
http://www.jwst.nasa.gov

See also: Computer Science; Mechanical Engineering; Optics; Photography.

MUSIC TECHNOLOGY

FIELDS OF STUDY

Acoustic science; computer programming; music theory; analogue signal processing; digital signal processing; MIDI recording; mixing and mastering; music synchronization; composition; music history.

SUMMARY

Music technology is the application of computer software, electronic instruments, and other sound-manipulation equipment in the composition, performance, and recording of music. The boundaries of personal expression through music are expanding with computer-instrument hybrids, such as those based on the piano and the guitar. Software facilitates the composition of scores with automatic notation and immediate playback features. Technology also allows the precise reproduction and efficient storage of music as well as advanced editing, mixing, and mastering capabilities. It even makes music more accessible, such as the creation of ring tones for cellular phones and electronic music effects for video games.

KEY TERMS AND CONCEPTS

- **Acoustic Sound:** Natural sound that is not produced or enhanced electronically.
- **Analogue Signal:** Information is transmitted as a continuous signal that is responsive to change.
- **Digital Signal:** Information is transmitted as a string of binary code without distortion.
- **Electronic Sound:** Sound that is produced or enhanced with technological manipulation, such as with computer software.
- **Equalization:** Boosting or weakening bass or treble frequencies to achieve a desired sound.
- **MIDI:** Musical instrument digital interface.
- **Modulation:** Altering the frequency or amplitude of a sound wave.
- **Oscillator:** Electronic circuit that generates a regularly repeating sound wave.
- **Synthesizer:** Electronic apparatus that combines signals of different frequencies to produce music.
- **Voice:** Musical sound that is produced from the combined output of multiple oscillators.

DEFINITION AND BASIC PRINCIPLES

Music technology is the integration of personal expression through instrumental and vocal music and cutting-edge applied science that enhances rather than replaces traditional music theory, composition techniques, and philosophy. It also facilitates the invention of novel instrumental and choral sounds and styles of music. Music technology easily incorporates mathematic underpinnings to create synergy in composition and performance among electronic instruments, such as synthesizers, sequencers, and samplers, and classic instruments. Musicians now have access to technical equipment that enables them to make stronger harmonic decisions and more fully manifest their artistic visions. The social culture of music has changed with the advent of personal MP3 players and downloadable open-sharing files; more people have immediate access to new music from around the world than ever before.

Music technology is often confused with sound technology. Sound technology involves the reproduction of sounds with adjustments for volume and clarity, and is applied in the development of hearing aids and acoustic halls. It also differs from audio or recording engineering, which is concerned with the capturing, editing, and storing of digital music files. Music technology takes into account aesthetic composition and innovative performance, as well as reproduction with fidelity.

BACKGROUND AND HISTORY

In 1961, engineering physicist Robert Moog revived the theremin, the archetypal electronic instrument invented by Russian engineer Léon Thérémin in 1920 and patented in 1928. It consists of two loop antennae that control oscillators; one regulates amplitude (volume) and the other regulates frequency (pitch). The antennae sense movements of the performer's hands without actual physical contact to create changes in the electronic signal and the resulting eerie music is amplified and transmitted through speakers. Moog studied the theremin and applied some of its principles to the invention of the Moog synthesizer, introduced in 1964. His invention used electronic circuit boards to alter the sound of music produced by a keyboard. It quickly caught on with the advent of rock and funk music styles; it was used in concerts by the Beatles and the Rolling Stones in the late 1960's. Moog is acknowledged to be an American pioneer in electronic music.

The development of hyperinstruments began in 1986 in the Massachusetts Institute of Technology Media Lab under the direction of renowned composer and professor of media arts and sciences Tod Machover. The goal of the project was to find ways to use technology to increase the power and finesse of musical instruments for superior performers. Since 1992, the focus of this project has shifted toward creating innovative interactive musical instruments for use by nonprofessional musicians and students, thus making music technology more accessible to the general public.

HOW IT WORKS

Composition. Aesthetic music compositions are typically layers of patterns that can be translated into mathematical expressions. Computer programs are able to perform this translation quickly, and the results can then be applied as objective parameters for developing new musical creations. Software can assist with rapid, clear musical notation and real-time playback during the composing period as well as with subsequent complex harmonious orchestration. Songwriting is similarly facilitated and appropriate vocal keys, harmonies, and arrangements can be easily derived.

Performance. Traditional instruments are enhanced with computer parts that create novel yet related voices. Electronic instruments are also being introduced into performances. A music synthesizer is an electronic keyboard instrument that produces electrical signals of various frequencies to generate new voices, some of which imitate other traditional instruments. Synthesizers operate on programmed algorithms and these synthesis techniques may be additive, subtractive, modulate frequency, or cause phase distortion. A music sequencer is a software program or an electronic device for recording, editing, and transmitting music in a musical instrument digital instrument (MIDI) format. A music sampler is similar to a synthesizer, except that instead of generating music, it plays back preloaded music samples, such as those from a sequencer, to compose as well as perform. Like synthesizers, samplers can modify the original sound signals to generate new voices and many can generate multiple notes simultaneously.

Recording. In the recording process, music may be modified in its analogue form or converted to digital data before alteration. Sound mixing is the process of blending sounds from multiple sources into a desired end product. It often starts with finding a balance between vocal and instrumental music or dissimilar instruments so that one does not overshadow the other. It involves the creation of stereo or surround sound from the placement of sound in the sound field to simulate directionality (left, center, or right). Equalizing adjusts the bass and treble frequency ranges. Effects such as reverberation may be added to create dimension. Auditory electrical signals may then be sent to speakers, where they are converted into acoustical signals to be heard by an audience. Digital signals may be broadcast in real time over the Internet as streaming audio. Otherwise, the processed signals may be stored for future reproduction and distribution. Analogue signals may be stored on magnetic tape. Digital signals may be stored on a compact disc or subjected to MP3 encoding for storage on a computer or personal music player.

APPLICATIONS AND PRODUCTS

Hyperinstruments. Music technology products include hyperinstruments, a term coined in 1986 for electroacoustic instruments. They were originally intended for virtuosos; customized instruments have been fabricated for Yo-Yo Ma and Prince. However, their accessibility is being expanded for use by nonprofessional musicians and children. The theremin is an archetypal electronic musical instrument that is played without actual physical contact. The performer uses fine motor control to wave his hands in proximity to a pair of antennae to control the pitch and volume produced. The instrument responds to every movement, whether deliberate or unintentional. A chameleon guitar has interchangeable soundboards in the central cavity that allow the same body, neck, and frets to have a familiar feel while producing various voices that imitate other instruments. Hyperpianos yield MIDI data that are augmented as solo music or keyboard accompaniment carefully matched to other instruments and voices. With the hypercello, the bow pressure, string contact, wrist measurements, and neck fingering are measured, the data are processed mathematically, and an enhanced signal is generated. A hyperviolin makes no sound of its own but creates electronic output when it is played with a hyperbow. The speed, force, and position of the hyperbow are measured wirelessly, the resulting data are adjusted, and the sound is appropriately enhanced for consistently intentional expression. The hyperviolin and hyperbow may be seen in performances by virtuoso Joshua Bell.

Innovative Musical Productions. The technology that has emerged from the Massachusetts Institute of Technology Media Lab has been channeled by MIT professor Tod Machover into a new form of opera in the broadest sense, telling a story through music. The first project, *The Brain Opera*, debuted in 1996. Each night's event occurred in two parts. In the first half, the audience is invited to experiment with electronic instruments such as the Rhythm Tree (by punching a node on the tree, a thump comes from a nearby speaker), Gesture Walls (advanced theremins), and Harmonic Driving devices

The pioneering Moog synthesizer, introduced in 1964, was used in concerts by the Beatles and featured prominently in the music of groups such as Yes. (Getty Images)

(video games that people drive to make musical choices). In the second half, the music created in the first half is incorporated into a multimedia presentation for a unique show every night. Its story is the psychological exploration of how we think about music. The project toured worldwide, finding a permanent home in Vienna in July 2000. The second and most recent project, *Death and the Powers*, debuted in September 2010. This one-act opera tells the story of an inventor, Simon Powers, who builds an electronic system into which he downloads his memories and humanity, which is then expressed by robots, animated bookcases, and a musical chandelier. Typically inanimate objects become personified with electronic music and voices in what Machover calls "disembodied performance." Music technology is responsible for connecting audiences with the emotion of the story.

IMPACT ON INDUSTRY

International Institutes. The International Federation of the Phonographic Industry (IFPI), with headquarters in London and regional offices in Brussels, Hong Kong, Moscow, Beijing, Zurich, and Miami, unites 1,400 members in sixty-six countries with affiliated industry associations in forty-five countries to uphold its mission to "promote the value of recorded music, safeguard the rights of record producers, and expand the commercial uses of recorded music." It plays an important role in studying ways to combat the piracy of digital music worldwide.

Music technology has a relatively long history in Europe and yet it is still emerging. The Institut de Recherche et Coordination Acoustique/Musique (IRCAM) in Paris was founded in 1970 to support composers in realizing their artistic vision with assistance from technology.

Professional Organizations. The Recording Industry Association of America (RIAA), in Washington D.C., is a trade organization whose members create, manufacture, and distribute nearly 85 percent of all legally recorded music produced and sold in the United States. This organization works to protect the First Amendment rights and copyrights of composers, performers, and record labels. Like IFPI, it also seeks to curtail digital music theft, which adversely affects the economy of the industry in a significant way. Similarly, it works with other music organizations such as the Music Publishers Association to battle copyright infringement.

The MIDI Manufacturers Association is composed of companies from various aspects of the music industry who develop and promote recommended practices and specifications for MIDI systems. This is to promote and expand the use of MIDI technology and keep systems compatible with other audio technological advancements.

Corporate Support. Advancements in music technology are often put into practice by corporations such as computer hardware and software producers, instrument manufacturers, and commercial music promoters and distributors. Significant companies

---

Fascinating Facts About Music Technology

- The Institut de Recherche et Coordination Acoustique/Musique (IRCAM) in Paris was founded to study the intersection of music and technology. Its mission is to assist composers in manifesting their artistic vision through technology.

- Music technology has been used to create a virtual castrato, reproducing the distinctive high-pitch quality of castrated boy singers in eighteenth century Europe. After hearing a careful mix of a soprano and a countertenor, even the singers could not discern their own voices.

- Music technology can serve corporations in producing subtly effective advertising. The automobile manufacturer Renault once sought expertise from a music technology institute to fine-tune the musicality of car doors closing in a commercial to support its marketing image.

- Barcodas, an iPhone app created by Dutch media artist Leo van der Veen, allows the user to scan any standard Universal Product Code bar code and translate it into a musical phrase. While entertaining, it is primarily intended as an electronic muse for composing.

- The audience-interactive musical production *The Brain Opera*, created by MIT Media Lab professor Tod Machover, was the technological forerunner of the *Guitar Hero* and *Rock Band* video games.

- The first commercially available digital music was the ABBA CD *The Visitors* in 1982.

- Record albums, even in a digital format, may be becoming obsolete. In 2009, digital single downloads outsold physical and digital albums by 250 percent.

---

in computer hardware and software include: Avid Technology, the developer and manufacturer of the Pro Tools digital audio workstation platform; Propellerhead, the Swedish developer and manufacturer of Record and Reason music software applications; and Apple, with products such as iPod and iTunes. Leading instrument manufacturers include Yamaha and Korg. The major music labels are Sony Music Entertainment, Warner Music Group, EMI Music, and Universal Music Group.

CAREERS AND COURSE WORK

In the United States, music technology degrees are offered at such prestigious schools as New York University and the Juilliard School in New York City and Massachusetts Institute of Technology in Cambridge. These degrees are also offered at universities across the country, many located in metropolitan areas, including Wayne State University in Detroit and Northwestern University in Evanston, Illinois.

The Cork Institute of Technology, Cork School of Music, in Cork, Ireland, offers a master's degree in music and technology as of 2009.

Students of music technology may pursue various levels of education. Although some associate's and doctoral degree programs are available, most are bachelor's and master's degree programs. Admission into these programs often requires the ability to read music, the ability to play a traditional instrument (piano is preferred and voice may be considered), some experience with computer hardware and software, and some experience with music recording. Nearly all programs offer courses in music history, traditional music theory, contemporary music theory, music appreciation and critical evaluation, analogue and digital signal processing, recording, mixing, mastering, synchronization, songwriting, orchestration and ensemble performance, and music business. Many colleges and universities arrange internships with local recording studios and radio stations to provide practical application.

Graduates may pursue many career paths in music technology. One is studio production, engineering, and recording for film, television, radio, video, Internet, and record labels. Another is composing, arranging, and orchestration for film, television, theater, church, advertising, and video games. Some choose to perform in live music venues. Some work in education and research, studying music design based on how the mind interprets pitch, rhythm, melody, and harmony. Similarly, some become music critics who must stay abreast of the current trends in both popular music and advancing technology. Another option is multimedia collaboration, integrating audio and video. This collaboration extends to Web design, digital video editing, and interactive media. A similar option is equipment design, creating the next generations of synthesizers, sequencers, and samplers. Another popular career path is music business and administration involving sales, marketing, and management for music retailers, production companies, and record labels.

SOCIAL CONTEXT AND FUTURE PROSPECTS

The Recording Industry Association of America estimates that recorded music is a $10.4 billion industry in the United States. In 2008, the United States Bureau of Labor Statistics reported that there were 53,600 music directors, arrangers, and composers (with a projected 10 percent increase in the subsequent decade), 114,600 broadcasting and sound engineering professionals (with a projected 8 percent increase in the subsequent decade), and 186,400 commercial musicians and singers (with a projected 8 percent increase in the subsequent decade).

Music technology is going in several exciting and novel directions beyond those of audio engineering. While audio engineering is developing music information retrieval systems, advanced digital signal processing, and mobile recording studios, music technology is pursuing more intimate avenues that will strengthen the bond between the artists and the audience.

One such area is music cognition. Technology is developing the capability to interpret and map expressive gestures made by professional musicians to personify electronic music. Similarly, researchers are mapping the brain's responses to different pieces of music to predict the effect of new compositions on an audience. Conversely, bioengineers are working to make brain waves audible so that happy, sad, fearful, or angry thoughts can be heard as musical patterns and subsequently evaluated by mental health professionals.

Another emerging area is interactive music systems, which take the audience from listening passively to thinking actively about how they react to music as they create it by making choices. This ranges from live musical performances, in which the audience is deliberately invited to participate, to blind research in which musical stairs are placed in proximity to escalators, and observers measure the rate at which pedestrians progressively choose to take the stairs, thus getting more exercise for the reward of music. Interactive music systems overall make music an inclusive experience.

Bethany Thivierge, B.S., M.P.H.

FURTHER READING

Ballora, Mark. *Essentials of Music Technology*. Upper Saddle River, N.J.: Prentice Hall, 2003. This book discusses five broad areas: the sound of music, computer software and hardware, MIDI, digital audio signals, and additional equipment.

Brown, Andrew R. *Computers in Music Education: Amplifying Musicality*. New York: Routledge, 2007. This book presents an easily understood overview of music technology to teachers and students, including sections on notation software, MIDI files, and downloading music.

Hosken, Dan. *An Introduction to Music Technology*. New York: Routledge, 2011. This covers the basics for music students to enhance their composition and performance with computer hardware and software.

Katz, Mark. *Capturing Sound: How Technology Has Changed Music*. Rev. ed. Berkeley: University of California Press, 2010. Describes how technology "has changed the way we listen to, perform, and compose music."

Middleton, Paul, and Steven Gurevitz. *Music Technology Workbook: Key Concepts and Practical Projects*. Burlington, Mass.: Focal Press, 2008. This workbook offers commonly encountered problems and their practical solutions, while discussing a variety of relevant software programs.

Roads, Curtis. *The Computer Music Tutorial*. Cambridge, Mass.: MIT Press, 1996. A comprehensive reference written by an acknowledged expert and respected professor of music technology.

Williams, David Brian, and Peter Richard Webster. *Experiencing Music Technology*. 3d ed. Belmont, Calif.: Cengage Learning, 2008. This paperback comes with a DVD and presents a complete overview of this rapidly changing field.

WEB SITES

Institut de Recherche et Coordination Acoustique/Musique
http://www.ircam.fr/?L=1

International Federation of the Phonographic Industry (IFPI)
http://www.ifpi.org

MIDI Manufacturers Association
http://www.midi.org

Recording Industry Association of America
http://www.riaa.com

See also: Acoustics; Audio Engineering; Computer
Engineering.

N

NANOTECHNOLOGY

FIELDS OF STUDY

Agriculture; artificial intelligence; bioinformatics; biomedical nanotechnology; business; chemical engineering; chemistry; computational nanotechnology; electronics; engineering; environmental studies; mathematics; mechanical engineering; medicine; molecular biology; microelectromechanical systems; molecular scale manufacturing; nanobiotechnology; nanoelectronics; nanofabrication; molecular nanoscience; molecular nanotechnology; nanomedicines; pharmacy; physics; toxicology.

SUMMARY

Nanotechnology is dedicated to the study and manipulation of structures at the extremely small nano level. The technology focuses on how particles of a substance at a nanoscale behave differently than particles at a larger scale. Nanotechnology explores how those differences can benefit applications in a variety of fields. In medicine, nanomaterials can be used to deliver drugs to targeted areas of the body needing treatment. Environmental scientists can use nanoparticles to target and eliminate pollutants in the water and air. Microprocessors and consumer products will also benefit from the use of nanotechnology, as components, and associated products, become exponentially smaller.

KEY TERMS AND CONCEPTS

- **Bottom-Up Nanofabrication:** The creation of nanoparticles that will combine to create nanostructures that meet the requirements of the application.
- **Buckyballs (Fullerenes):** Molecules made up of carbon atoms that make a soccer ball shape of hexagons and pentagons on the surface. Named for the creator of the geodesic dome, Buckminster Fuller.

- **Carbon Nanotubes:** Molecules made up of carbon atoms arranged in hexagonal patterns on the surface of cylinders.
- **Mechanosynthesis:** Placing atoms or molecules in specific locations to build structures with covalent bonds.
- **Molecular Electronics:** Electronics that depend upon or use the molecular organization of space.
- **Moore's Law:** Transistor density on integrated circuits doubles about every two years. American chemist Gordon Moore attributed this effect to shrinking chip size.
- **Nano:** Prefix in the metric system that means "10^{-9}" or "1 billionth." The word comes from the Greek *nanos*, meaning "dwarf." The symbol is *n*.
- **Nanoelectronics:** Electronic devices that include components that are less than 100 nanometers (nm) in size.
- **Nanoionics:** Materials and devices that depend upon ion transport and chemical changes at the nanoscale.
- **Nanolithography:** Printing nanoscale patterns on a surface.
- **Nanometer:** 10^{-9} or 1 billionth of a meter. The symbol is *nm*.
- **Nanorobotics:** Theoretical construction and use of nano-scaled robots.
- **Nanosensors:** Sensors that use nanoscale materials to detect biological or chemical molecules.
- **Scanning Probe Microscopy:** The use of a fine probe to scan a surface and create an image at the nanoscale.
- **Scanning Tunneling Microscopy (STM):** Device that creates images of molecules and atoms on conductive surfaces.
- **Self-Assembly:** Related to bottom-up nanofabrication. A technique that causes functionalized nanoparticles to chemically bond with or repel other atoms or molecules to assemble in specified patterns without external manipulation.

- **Top-Down Nanofabrication:** Manufacturing technique that removes portions of larger materials to create nano-sized materials.

DEFINITION AND BASIC PRINCIPLES

Nanotechnology is the science that deals with the study and manipulation of structures at the nano level. At the nano level, things are measured in nanometers, or 1 billionth of a meter (10^9). The symbol for nanometers is *nm*. Nanoparticles can be produced using a process known as top-down nanofabrication, which starts with a larger quantity of material and removes portions to create the nano-sized material. Another method being developed is bottom-up nanofabrication, in which nanoparticles will create themselves when the necessary materials are placed in contact with one another.

Nanotechnology is based upon the discovery that materials behave differently at the nano scale, less than 100 nm in size, than they do at slightly larger scale. For instance, gold is classified as an inert material because it neither corrodes nor tarnishes. However, at the nano level, gold will oxidize in carbon monoxide. It will also appear as colors other than the yellow for which it is known.

Nanotechnology is not simply about working with materials like gold at the nano level. It is about taking advantage of these differences at the nanoscale to create markers and other new structures that are of use in a wide variety of medical and other applications.

BACKGROUND AND HISTORY

In 1931, German scientists Ernst Ruska and Max Knoll built the first transmission electron microscope (TEM). Capable of magnifying objects by a factor of up to one million, the TEM made it possible to see things at the molecular level. The TEM was used to study the proteins that make up the human body. It was also used to study metals. The TEM made it possible to view these particles smaller than 200 nm by focusing a beam of electrons to pass through an object, rather than focusing light on an object, as was the case with traditional microscopes.

In 1959 the noted American theoretical physicist Richard Feynman brought nanoscale possibilities to the forefront with his talk "There's Plenty of Room at the Bottom," presented at the California Institute of Technology in 1959. In this talk, he asked the audience to consider what would happen if they could arrange individual atoms, and he included a discussion of the scaling issues that would arise. It is generally considered that Feynman's reputation and influence brought increased attention to the possible uses of structures at the atomic level.

In the 1970s scientists worked with nanoscale materials to create technology for space colonies. In 1974 Tokyo Science University professor Norio Taniguchi coined the term "nano-technology." As he defined it, nanotechnology would be a manufacturing process whose materials were built by atoms or molecules.

In the 1980s the invention of the scanning tunneling microscope (STM) led to the discovery of

Super water repellent coating is demonstrated at a facility in Palo Alto, California. Drops of water literally bounce off the surface of the material, which has a coating built from nanostructures. (AP Photo)

fullerenes in 1986. The carbon nanotube was discovered a few years later. In 1986, Eric Drexler's seminal work on nanotechnology, *Engines of Creation*, was published. In this work, Drexler used the term "nanotechnology" to describe a process that is now understood to be molecular nanotechnology. Drexler's book explores the positive and negative consequences of being able to manipulate the structure of matter. Included in his book are ruminations on a time when all the works in the Library of Congress can fit on a sugar cube and when nanoscale robots and scrubbers can clear capillaries or whisk pollutants from the air. Controversy continues as to whether the view of a world with nanotechnology envisioned by Drexler is even attainable.

In 2000 the U.S. National Nanotechnology Initiative was founded. Its mandate is to coordinate federal nanotech research and development. Great growth in the creation of improved products using nanoparticles has taken place since that time. The creation of smaller and smaller components—which reduces all aspects of manufacture, from the amount of materials needed to the cost of shipping the finished product—is driving the use of nanoscale materials in the manufacturing sector. Furthermore, the ability to target delivery of treatments to areas of the body needing those treatments is spurring research in the medical field.

The true promise of nanotechnology is not yet known, but this multidisciplinary science is widely viewed as one that will alter the landscape of fields from manufacturing to medicine.

HOW IT WORKS

Basic Tools. Nanoscale materials can be created for specific purposes, but there exists also natural nanoscale material, like smoke from fire. To create nanoscale material and to be able to work with it requires specialized tools and technology. One essential piece of equipment is an electron microscope. Electron microscopy makes use of electrons, rather than light, to view objects. Because these microscopes have to get the electrons moving, and because they need several thousand volts of electricity, they are often quite large.

One type of electron microscope, the scanning electron microscope (SEM), requires a metallic sample. If the sample is not metallic, it is coated with gold. The SEM can give an accurate image with good resolution at sizes as small as a few nanometers.

For smaller objects or closer viewing, a transmission electron microscope (TEM) is more appropriate. With a TEM, the electrons pass through the object. To accomplish this, the sample has to be very thin, and preparing the sample is time-consuming. The TEM also has greater power needs than the SEM, so that in most cases the SEM is used, reserving the TEM for times when a resolution of a few tenths of a nanometer is absolutely necessary.

The atomic force microscope (AFM) is a third type of electron microscope. Designed to give a clear image of the surface of a sample, this microscope uses a laser to scan across the surface. The result is an image that shows the surface of the object, making visible the object's "peaks and valleys."

Moving the actual atoms around is an important part of creating nanoscale materials for specific purposes. Another type of electron microscope, the scanning tunneling microscope (STM), not only images the surface of a material in the same way as the AFM; the tip of the probe, which is typically made up of a single atom, also can be used to pass an electrical current to the sample. This charge lessens the space between the probe and the sample. As the probe moves across the sample, the atoms nearest the charged atom move with the charged atom. In this way, individual atoms can be moved to a desired location in a process known as quantum mechanical tunneling.

Molecular assemblers and nanorobots are two other potential tools. The assemblers would use specialized tips to form bonds with materials that would make specific types of materials easier to move. Nanorobots might someday move through a person's blood stream or through the atmosphere, equipped with nanoscale processors and other materials that enable them to perform specific functions.

Bottom-Up Nanofabrication. Bottom-up nanofabrication is one approach to nanomanufacturing. This process builds a specific nanostructure or material by combining components of atomic and molecular scale. Creating a structure this way is time-consuming, so scientists are working to create nanoscale materials that will spontaneously join to assemble a desired structure without physical manipulation.

Top-Down Nanofabrication. Top-down nanofabrication is a process in which a larger amount of material is used at the start. The desired nanomaterial is created by removing, or carving away, the material that is not needed. This is less time-consuming than bottom-up nanofabrication, but it produces considerable waste.

Specialized Processes. To facilitate the manufacture of nanoscale materials, a number of specialized processes are used. They include nanoimprint lithography, in which nanoscale features are stamped or printed onto a surface; atomic layer epitaxy, in which a layer that is only one atom thick is deposited on a surface; and dip pen lithography, in which the tip of an atomic force microscope writes on a surface after being dipped into a chemical.

APPLICATIONS AND PRODUCTS

Smart Materials. Smart materials are materials that react in ways appropriate to the stimulus or situation they encounter. Combining smart materials with nanoscale materials would, for example, enable scientists to create drugs that would respond when encountering specific viruses or diseases. They could also be used to signal problems with other systems, like nuclear power generators or pollution levels.

Sensors. The difference between a smart material and a sensor is that the smart material will generate a response to the situation encountered and the sensor will generate an alarm or signal that there is something that requires attention. The capacity to incorporate sensors at a nanoscale will greatly enhance the ability of engineers and manufacturers to create structures and products with a feedback loop that is not cumbersome. Nanoscale materials can easily be incorporated into the product.

Medical Uses. The potential use for nanoscale materials in the field of medicine is one that is of particular interest to researchers. Theoretically, nanorobots could be programmed to perform functions that would eliminate the possibility of infection at a wound site. They could also speed healing. Smart materials could be designed to dispense medication in appropriate doses when a virus or bacteria was encountered. Sensors could be used to alert physicians to the first stages of malignancy. There is great potential for nanomaterials to meet the needs of aging populations without intrusive surgeries requiring lengthy recovery and rehabilitation.

Energy. Nanomaterials also hold promise for energy applications. With nanostructures, components of heating and cooling systems can be tailored to control temperatures with greater efficiency. This can be accomplished by engineering the materials so that

---

Fascinating Facts About Nanotechnology

- Approximately 80,000 nanos equal the width of one strand of human hair.
- Gold is considered an inert material because it does not corrode or tarnish. At the nano level, this is not the case, as gold will oxidize in carbon monoxide.
- The transmission electron microscope (TEM) allowed the first look at nanoparticles in 1931. The TEM was built by German scientists Ernst Ruska and Max Knoll.
- German physicist Gerd Binning and Swiss physicist Heinrich Rohrer, colleagues at IBM Zurich Research Laboratory, invented the scanning tunneling microscope in 1981. The two scientists were awarded the Nobel Prize in Physics in 1986 for their invention.
- Nanoparticles can be enclosed in an outer covering or coat.

---

some types of atoms, such as oxygen, can pass through, but others, such as mold or moisture, cannot. With this level of control, living conditions can be designed to meet the specific needs of different categories of residents.

Extending the life of batteries and prolonging their charge has been the subject of decades of research. With nanoparticles, researchers at Rutgers University and Bell Labs have been able to better separate the chemical components of batteries, resulting in longer battery life. With further nanoscale research, it may be possible to alter the internal composition of batteries to achieve even greater performance.

Light-emitting diode (LED) technology uses 90 percent less energy than conventional, non-LED, lighting. It also generates less heat than traditional metal filament light bulbs. Nanomanufacture would make it possible to create a new generation of efficient LED lighting products.

Electronics. Moore's law states that transistor density on integrated circuits doubles about every two years. With the advent of nanotechnology, the rate of miniaturization has the potential to double at a much greater rate. This miniaturization will profoundly affect the computer industry. Computers will become lighter and smaller as nanoparticles are used to increase everything from screen resolution and battery life while reducing the size of essential internal components, such as capacitors.

IMPACT ON INDUSTRY

Nanotechnology will have a profound impact on industry. Reducing the size of components to nanoscale will mean that the internal circuits of a computer will be etched on a surface too small to be seen by the naked eye. With circuit boards and other components reduced in size, finished products will be similarly reduced in scale.

Industries such as the automotive industry, which must have products of a certain size to function, also will take advantage of nanoparticles and materials. Their use will result in components that weigh less and are stronger because of the manufacturing use of structures such as the carbon nanotube.

The use of lighter coatings of materials on the components of vehicles destined for space will result in significantly lighter loads and reduced fuel requirements. Even space suits will be improved by the use of nanoparticles in their manufacture, as layers of bulky material are replaced with trimmer fabrics providing the same or better protection.

In the medical field, experimentation is already underway to deliver medication to tumors in places that are not readily accessible through conventional means. Researchers are also experimenting with coatings to protect tissue as medication moves through the human body and with markers to guide medication to the proper treatment site.

Governments and universities around the world have shown a strong interest in nanotechnology. Alliances are being formed to create solutions to problems such as air and water pollution. One such application under development is the use of nanostructures to create products that will clean oil spills before widespread damage is done to wildlife and their habitats. The National Nanotechnology Initiative is involved in education and research. To support its project, the organization's Web site clearly explains nanotechnology. The site also includes discussion of the expected benefits of nanotechnology. The initiative is also involved in setting standards for nanotechnology and is legislating for policies and regulations for this new branch of science.

The private sector is also involved in nanotechnology-related research, as individuals and private companies work to develop ideas and products they can patent and produce. For example, major corporations are investigating the use of nanoparticles such as carbon nanotubes to strengthen the materials they use in the components that make up cars and airplanes.

Nanotechnology is already a technology with broad applications for the business community. Commercialization of research findings is a priority. The funding from the commercialization will pay for future research as consumers take advantage of the positive effects of nanotechnology. Because nanotechnology is rising to prominence during the information age, which includes the Internet and web, research findings and writings are readily available. Another factor advancing progress in this field is the global reach of research and information dissemination.

CAREERS AND COURSEWORK

Engineering and architecture, manufacturing, and the health care industry are just a few of the career fields that will be touched by advances in nanotechnology and its applications.

For those pursuing a career in engineering, nanoscale materials will result in stronger, lighter materials that can be created specifically for their function. This will result in lighter, more durable airplanes, ships, spacecraft, and other vehicles. Nanotechnology will also be a factor in the selection of materials used by civil engineers for roads, bridges, and dams. Architects will have a new range of materials available as they design homes and commercial buildings.

In manufacturing, advances in the use of nanotechnology will result in stronger materials of lighter weight that are better suited for their purpose. Products that now must be of a certain size to accommodate internal components will not have this constraint in the future. In a move analogous to the switch from televisions with cathode ray tubes to those using liquid crystal or plasma displays, advances in the production of nanocomponents will alter design constraints on a variety of products. Those pursuing manufacturing careers will need to stay abreast of those changes to produce products at the greatest level of efficiency and at the lowest cost.

Professionals in the health industry, from personal trainers to surgeons, will feel the effect of advances in nanotechnology. From the use of nanomaterials to speed patient recovery to the use of nanostructures in prosthetics and related aids, nanotechnology will change current practices at a rapid pace.

Because nanotechnology is multidisciplinary, coursework applying to nanotechnology can be found in a variety of disciplines. Because nanotechnology is such a new field, much research is reported in journals and on reputable Web sites dedicated to the field. Keeping informed of developments and advances will benefit those seeking careers that touch on the application of nanotechnology.

SOCIAL CONTEXT AND FUTURE PROSPECTS

Whether nanotechnology will be good or bad for the human race remains to be seen. There is tremendous potential associated with the ability to manipulate individual atoms and molecules, to deliver medications to a disease site, and to build products such as cars that are lighter yet stronger than ever. There also exists the persistent worry that humans will lose control of this technology and face what one writer called "gray goo," a scenario in which self-replicating nanorobots take over and ultimately destroy the world.

Gina Hagler, M.B.A.

FURTHER READING

Drexler, K. E. *Engines of Creation.* London: Fourth Estate, 1990.

Ratner, D., and M. A. Ratner. *Nanotechnology and Homeland Security: New Weapons for New Wars.* Upper Saddle River, N.J.: Prentice Hall/PTR, 2004.

Ratner, M. A., and D. Ratner. *Nanotechnology: A Gentle Introduction to the Next Big Idea.* Upper Saddle River, N.J.: Prentice Hall, 2003.

WEB SITES

Arizona State University, Arizona Initiative for Nano-Electronics
http://www.asu.edu/aine/app_nanoionics.htm

Carnegie Mellon NanoRobotics Lab
http://nanolab.me.cmu.edu/

NanoScience Instruments
http://www.nanoscience.com/index.html

Nanotechnology Research Foundation
http://www.nanotechnologyresearchfoundation.org/

National Nanotechnology Initiative
http://www.nano.gov/

University of Illinois at Urbana-Champaign Imaging Technology Group
http://virtual.itg.uiuc.edu/training/AFM_tutorial/

See also: Applied Physics; Bionics and Biomedical Engineering; Computer Engineering; Computer Science; Electronic Materials Production; Electronics and Electronic Engineering; Engineering; Environmental Biotechnology; Industrial Pollution Control; Integrated-Circuit Design; Liquid Crystal Technology; Microlithography and Nanolithography; Surface and Interface Science; Surgery; Transistor Technologies.

NAVAL ARCHITECTURE AND MARINE ENGINEERING

FIELDS OF STUDY

Thermodynamics; fluid mechanics; mechanics of materials; machine design; engineering; hydrodynamics; ship structures; propulsion technologies.

SUMMARY

The process of designing a ship involves two complementary disciplines: naval architecture and marine engineering. Naval architecture has two subdivisions: hydrodynamics and ship structures. Hydrodynamics is concerned with the interaction between the moving ship hull and the water in which it floats. Ship structures is concerned with building a hull that has the strength needed to withstand the forces to which it is subject. Marine engineering is concerned with all the machinery that goes into the ship. The machinery must perform the following tasks: propulsion and steering, electric power generation and distribution, and cargo handling.

KEY TERMS AND CONCEPTS

- **Bridge:** Place from which the ship's movement and navigation is controlled.
- **Deck:** Horizontal surface in a ship, like the floors in a building.
- **Deckhouse:** Upper part of the ship that sits on the deck; also known as a ship's superstructure.
- **Hull:** Lower part of the ship. Part of it is submerged in the water.
- **Knot:** Unit of measure of ship speed. It represents one nautical mile per hour or about 1.15 land miles per hour.
- **Long Ton:** Unit of weight used for ships and their cargoes. It amounts to 2,240 pounds.
- **Rudder:** Movable device that projects below the ship's hull and is used to steer the ship.
- **Waterline:** Highest point on the hull that is submerged in the water when the ship is stationary.

DEFINITION AND BASIC PRINCIPLES

A ship is a very complex object. Some ships must be able to load and unload themselves, while others depend on facilities in ports of call. A ship must propel itself from place to place and control its direction of motion. It must produce its own electricity and freshwater. In addition to the space devoted to cargo, a ship must provide space for fuel, freshwater, living accommodations for crew and perhaps passengers, propulsion, and related machinery. A ship must float, and it must remain upright in all sorts of sea conditions. There are many different kinds of ships: aircraft carriers, submarines, containerships, tankers, passenger ships, ferries, and ships that transport liquefied natural gas, to name a few. The size of a ship is chosen based on cargo-carrying capacity, but it is limited by channel depth, pier length, and other characteristics of the ports it must enter. Most ships are part of a transportation business. To maximize profit, the costs of construction and operation must be minimized, while the income from transporting cargo must be maximized. Working together, naval architects and marine engineers design each ship to satisfy the requirements mentioned above. A design team is assembled that represents the needed areas of expertise. The team may work for a year or more to produce the complete design. When the design work is finished, one or more shipyards may be invited to submit bids for construction.

BACKGROUND AND HISTORY

One of the earliest well-documented ships is the funeral ship of Pharaoh Cheops that was built in about 2600 B.C.E. The Greeks and the Romans built large oar-powered ships called biremes and triremes. Bartholomew Diaz sailed three caravels around the tip of Africa in 1488, and Christopher Columbus reached the New World in 1492. All these ships were built by artisans who knew no theory of ship design. Scientific ship design began in France and Spain in the late 1600's. Steam was first used to propel watercraft in the late 1700's, and the first successful steam powered vessel, *Charlotte Dundas*, made its first voyage on the River Clyde in Scotland in 1802. By 1816, steam-powered passenger ships sailed regularly between Brighton, England, and Le Havre, France. William H. Webb designed and built clipper ships along the East River in New York City in the years before the Civil War. Webb and his contemporaries and predecessors called themselves shipbuilders, but they designed their ships too. During the twentieth century, the functions of designing ships and building them became separate.

HOW IT WORKS

The two principal issues in designing the hull of a ship are its structural strength and the force required to push it through the water at the desired speed. Other issues include stability and the way the ship moves up and down and side to side in waves.

Designing the Hull. Forces on a ship's hull may be divided into static (constant) forces and dynamic (variable) forces. There are two static forces: the force of gravity pulling down on the ship and its contents and the upward force exerted by the water on the hull.

Wave action is responsible for the dynamic forces, but some of these forces are direct results of the waves and some are indirect. When the ship's bow meets a wave, the wave exerts an upward force, lifting the bow. As the ship passes through the wave, the upward force moves along the ship from bow to stern. When the bow reaches the low point between two wave crests, the bow drops and the stern rises. Wave action causes the ship to flex, like a board that is supported at its ends does when someone jumps up and down at the middle.

Wave action also causes the ship to roll from side to side. As the ship rolls, liquids in its tanks slosh from side to side. This causes dynamic forces on the walls of the tanks. Waves may slap against the sides of the ship, and sometimes waves put water onto the decks.

Resistance and Stability. As a ship moves through the water, it must push aside the water ahead of it. This water moves away from the ship as waves. Water flowing along the sides of the ship exerts a friction force on the hull. The combination of these forces is called the resistance. The propulsion force provided by a ship's propeller must overcome the resistance. Naval architects can accurately predict the friction resistance, but the resistance associated with pushing the water aside is usually determined by testing a scale model.

A ship must be stable. This means it must float upright in still water, and it must return to an upright position after waves cause it to tilt. A naval architect must perform detailed calculations to ensure that the ship being designed meets requirements for stability.

Propulsion and Auxiliary Machinery. It is the responsibility of the marine engineer to select the machinery that will provide the required power. There are ships powered by steam turbines, gas turbines, and diesel engines. The propeller may be driven directly by the engine, it may be driven through a speed-reducing gear, or the engine may drive an electric generator that provides power to a motor that drives the propeller.

The marine engineer must also select pumps, piping, oil purifiers, speed-reducing gears, heat exchangers, and many other pieces of auxiliary machinery. The marine engineer designs the systems that connect all of the components and allow them to work together. A ship must generate its own electricity and produce freshwater from seawater. Machinery must be provided to control the direction of motion of the ship.

APPLICATIONS AND PRODUCTS

Naval architects and marine engineers are called on to design many different types of ships. Each ship type has its own unique design requirements. The designers must consider many factors, including how deep the channels are in the intended ports of call, what type of cargo will be carried, and how time sensitive it is. Several examples are discussed below.

Passenger Ships. Passenger ships may be liners or cruise ships. Liners are used to transport people from one place to another. Before air travel became common, this is how people traveled across oceans. In the early twenty-first century, most passenger ships are cruise ships. They embark passengers at a port, take them to visit interesting places, and return them to the same port where they embarked. Cruise ships have extensive passenger-entertainment facilities, which may range from rock-climbing walls to casinos. Most cruise ships are propelled by diesel engines. These engines may drive the propellers directly, or they may drive electric generators. In the latter case, the propellers are driven by electric motors.

Containerships. Shipping of cargo in standard rectangular containers has revolutionized the shipping business. At specialized containership ports, a ship can be loaded or unloaded in twelve hours or less by large container cranes mounted on the dock. The largest containerships can carry more than 10,000 twenty-foot long containers. Containers can hold all manner of cargo from cameras and flat-screen TVs to food and cut flowers. Containerships are typically the fastest category of merchant ships. Speeds between

20 and 25 knots are common because these ships carry time-sensitive cargo.

Tankers. Tankers range from ultra-large crude-oil carriers that carry more than 300,000 long tons of crude to small coastal tankers that carry 10,000 long tons or less. Ultra-large crude-oil ships transport large amounts of crude oil from sources in the Persian Gulf to refineries in Europe and North America. There are many other categories of tankers. The most specialized are chemical tankers, which are equipped to carry corrosive cargo such as sulfuric acid and highly flammable cargo such as gasoline.

LNG Ships. These tankers are designed to carry liquefied natural gas (LNG). Natural gas, which is mainly methane, changes from gas to liquid at –160 degrees Celsius. Although the tanks are heavily insulated, some heat does leak into the LNG. This causes a small amount of LNG to vaporize. One of the major decisions in designing an LNG ship is how to handle this gas. Many LNG ships are propelled by steam turbines, and the gas that boils off from the tanks is burned in the boilers. Other LNG ships are propelled by diesel engines. In some cases, the boil-off gas (BOG) is burned in the engines. On other diesel-powered ships the BOG is condensed and returned to the tanks. Many LNG ships operate at about 19 knots. Depending on their size, they may require as much as 50,000 horsepower for propulsion.

Naval Surface Ships. Naval surface ships range from huge aircraft carriers, which may weigh 100,000 long tons or more, to much smaller destroyers and frigates, which may weigh 4,000 long tons. Modern aircraft carriers and cruisers are often nuclear powered. Destroyers and frigates are often powered by gas turbines. On an aircraft carrier the main weapons are the aircraft. Other warships are armed with guns, missiles, and torpedoes. The ships must have the capability to locate and track enemy targets and to launch weapons at them. They must also be able to defend themselves against weapons launched by enemy ships and aircraft.

Submarines. Submarines range from one-person research vessels powered by batteries to the U.S. Navy's large, nuclear-powered missile submarines. Two things make submarines different from other ships: First, they must be able to operate without access to the atmosphere, and second, their hulls must withstand the pressure of the sea at the depths where they operate. Navy submarines are powered by nuclear

---

Fascinating Facts About Naval Architecture and Marine Engineering

- Modern containerships can carry more than 10,000 containers of twenty feet in length. Container shipping has drastically reduced the cost of moving cargo by sea.
- Containerships can be loaded and unloaded in twenty-four hours or less. Containers are quickly loaded onto trucks or railcars for land transportation.
- Nuclear submarines can remain underwater for several months. They produce their own oxygen by splitting water molecules, and nuclear reactors require no air to produce power.
- A crude oil tanker can carry 300,000 long tons of crude halfway around the world.
- An aircraft carrier may have more than 5,000 people aboard. It is like a small city.
- LNG ships carry liquefied natural gas at a temperature of –160 degrees Celsius. As small amounts boil off from the tanks, it may be re-liquefied or burned in the ship's engines.
- Until the middle of the twentieth century most ships were steam powered, but in the early twenty-first century most are powered by diesel engines.

---

reactors so that they do not need air for the combustion of fuel. The oxygen required to support human life aboard a submarine is produced by using electricity to split water molecules into hydrogen and oxygen, a process called electrolysis.

There are many other types of ships: ferries, roll-on/roll-off ships, heavy lift ships, fishing vessels, dredges, tugboats, and yachts of all shapes and sizes. Two very interesting specialized ships are hovercraft and hydrofoils. Hovercraft ride on a cushion of air that is trapped between the hull and the surface of the sea. Hydrofoils are supported by "wings" that are submerged in the water. Both hovercraft and hydrofoils are capable of much higher speeds than conventional ships.

IMPACT ON INDUSTRY

Ship-design companies range in size from a single individual to companies employing hundreds of naval architects, marine engineers, and support personnel. A team of experts may spend a year or more designing a large, complex ship. Detailed drawings and specifications are the final product of a ship-design company. Major ship-design companies in the United States include Gibbs & Cox, and the naval architecture and marine engineering division of Alion Science and Technology. Smaller companies include Herbert Engineering in California and Robert Allan in Canada.

There is a modest amount of ship building in the United States, much of it devoted to the U.S. Navy and Coast Guard. Northrop Grumman Shipbuilding operates shipyards on the Gulf Coast and in Newport News, Virginia. At Newport News, it builds nuclear-powered aircraft carriers and submarines as well as surface ships for the Navy and Coast Guard. General Dynamics has shipyards in Connecticut and San Diego, California. General Dynamics Electric Boat specializes in building nuclear submarines for the Navy, while its San Diego facility builds surface ships for the Navy and for commercial service. By law, ships that carry cargo from one American port to another must be registered in the United States. With some exceptions, such ships must be built in the United States as well. These are called cabotage laws, and the specific U.S. law is called the Jones Act.

In Asia, there is an extensive shipbuilding industry in Japan, Korea, and China. These three countries are the top shipbuilding nations in the world. China ranks third, but its industry is growing very rapidly. Europe has large shipyards in France, Italy, and Denmark. Drydocks World in Dubai provides ship repair and new construction, mainly in support of the oil business.

Standards for ship construction are issued by governments and by the insurance industry. In the United States, the Coast Guard issues such regulations. Companies that write insurance on ships manage their risk through organizations called classification societies. The American Bureau of Shipping is one such society. International organizations such as the International Maritime Organization, part of the United Nations, are concerned with safe operation of ships and the prevention of pollution.

Technical societies such as the Society of Naval Architects and Marine Engineers and the American Society of Naval Engineers provide a forum for the exchange of information among professionals in this field. These organizations hold annual meetings at the national level, and they have local groups, called sections, that meet more frequently in cities around the country. Such organizations also publish books and technical journals that cover the latest developments in the field. The Royal Institution of Naval Architects and the Institute of Marine Engineering, Science and Technology are organizations with a similar purpose based in the United Kingdom.

CAREERS AND COURSE WORK

A small number of colleges and universities in the United States offer degrees in naval architecture, marine engineering, or a combination of the two. Similar programs are offered in Canada, Europe, and Asia. Students of naval architecture take courses in strength of materials, ship structures, hydrodynamics, ship resistance, and propeller design. Marine engineering programs include thermodynamics, heat transfer, and machine design. Advanced mathematics is a part of both programs. A bachelor of science degree is the minimum requirement, and many working professionals in these fields have master of science degrees.

Most ship design is performed by companies that produce only the plans and specifications for a ship. The would-be shipowner takes these documents to one or more shipyards and invites bids for the actual construction. Shipyards also employ naval architects and marine engineers who deal with design issues that arise during the construction of a ship.

SOCIAL CONTEXT AND FUTURE PROSPECTS

When cargo must be transported across large bodies of water, such as the Atlantic and Pacific oceans, there are only two possible ways to do it—ships and airplanes. Lightweight, high-value, time-sensitive cargo goes by air, but more mundane cargo goes by ship. Ships are slow, but in many cases, time is not of the essence. Crude oil and natural gas are abundant in a part of the world where demand for these materials is low. A steady stream of large tankers carries crude oil nearly halfway around the world. LNG ships do the same. It is hard to imagine an alternative. Although it is possible to transport cargo from France to Algeria by road, it is not practical to do so.

Transportation by ship is far less expensive than by air. Among other factors, the fuel consumed to move a given amount of cargo a given distance is much less. Shipping products in standard rectangular containers has drastically reduced the cost and time required for handling cargo at both ends of its travel by sea.

It appears that cargo transportation by ship will remain an important business for the foreseeable future.

Edwin G. Wiggins, B.S., M.S., Ph.D.

FURTHER READING

Benford, Harry. *Naval Architecture for Non-Naval Architects.* Jersey City, N.J.: Society of Naval Architects and Marine Engineers, 1991. The author states that one of his intended audiences is "high school students contemplating a career in the marine field." Includes many good black-and-white photos and line drawings.

Ferreiro, Larrie D. *Ships and Science: The Birth of Naval Architecture in the Scientific Revolution, 1600-1800.* Cambridge, Mass.: MIT Press, 2007. Although technical in some chapters, most of the book is readable for the layperson. This is a very thorough history of naval architecture.

Gardiner, Robert, ed. *The Earliest Ships: The Evolution of Boats Into Ships.* London: Conway Maritime Press, 2004. Easy to read with many excellent black-and-white illustrations. Traces shipbuilding back to the Greeks, the Vikings, and the Celts.

Kemp, John F., and Peter Young. *Ship Construction Sketches and Notes.* 2d ed. Boston: Butterworth-Heinemann, 1997. Contains line drawings on half the pages with explanations on the facing pages.

Rowen, Alan L., et al. *Introduction to Practical Marine Engineering.* New York: Society of Naval Architects and Marine Engineers, 2005. Comprehensive, easily readable coverage of marine engineering with extensive illustrations.

Van Dokkum, Klass. *Ship Knowledge: Ship Design, Construction and Operation.* 3d ed. Enkhuizen, the Netherlands: Dokmar Maritime, 2008. Profusely illustrated in color. Very comprehensive nonmathematical coverage.

Zubaly, Robert B. *Applied Naval Architecture.* Centreville, Md.: Cornell Maritime Press, 1996. Written as a textbook for students who aim to become deck officers in the merchant marine. It contains some high-school-level math, and it is easy to read.

WEB SITES

American Society of Naval Engineers
http://www.navalengineers.org/Pages/default.aspx

Royal Institution of Naval Architects
http://www.rina.org.uk

Society of Naval Architects and Marine Engineers
http://www.sname.org

See also: Engineering; Engineering Mathematics; Submarine Engineering.

NAVIGATION

FIELDS OF STUDY

Aeronautics; aviation; cartography; celestial navigation; electronic navigation systems; geodesy; geoinformatics; space exploration; ship piloting.

SUMMARY

Navigation is the process of plotting an object from one point to another. This is accomplished by determining one's exact position on Earth by visual or electronic means and determining the course to the exact position of the intended destination. Specific points are often defined as measurements of longitude and latitude.

KEY TERMS AND CONCEPTS

- **Celestial Navigation:** Method of locating one's position by the angular measurement between celestial bodies (the Sun, Moon, stars, or planets).
- **Global Positioning System (GPS):** Space-based global navigation satellite system that provides accurate location and time information for any point on Earth where there is an unobstructed line of sight to four or more GPS satellites.
- **Gyrocompass:** Compass that points true north (rather than magnetic north) by using a fast-spinning gyroscope. Unlike a magnetic compass, a gyrocompass is unaffected by magnetic fields.
- **Latitude:** Angular measurement (degrees, minutes, and seconds) of a location on the Earth north or south of the equator. The North Pole has a latitude of 90 degree north, the South Pole has a latitude of 90 degrees south, and the equator has a latitude of 0 degree.
- **Longitude:** Angular measurement (degrees, minutes, and seconds) of a location relative to the prime meridian, which runs from the North to the South Pole through the original site of the Royal Observatory in Greenwich, England. Measurements range from 180 degrees east to 180 degrees west.
- **Magnetic Compass:** Device that contains a magnetized needle that points to magnetic north.
- **Radar:** Acronym for radio detecting and ranging; a system that emits a narrow beam of extremely high-frequency pulses that are reflected back from objects (buoys, ships, and landmasses). A radar screen updates in real time, yielding the speed and direction of moving objects.
- **Triangulation:** Determining the position of an object by calculation from two known quantities; the distance between two fixed points and the angles formed between the line described by those points and the line between a third point.

DEFINITION AND BASIC PRINCIPLES

Location of one's position is based on trigonometry and the process of triangulation. Triangulation involves determining the location of a point by measuring angles to it from two or more known points. The intersection of the lines from the known points represents one's position.

Prior to the development of electronic instruments such as radio, radar, and Global Positioning System (GPS), the determination of one's position was accomplished using a sextant. The sextant was able to pinpoint one's location accurately by measuring the angular distance between the horizon and a celestial object. Repeated sightings were plotted on a nautical chart. This method was suitable for ships and propeller aircraft.

After the development of the radio, a position could be determined by triangulating radio stations with known positions. With the development of radar, images of known landmasses or buoys could be used for triangulation.

GPS also involves triangulation. The device electronically determines one's position by triangulation of orbiting satellites and marks the position on a map overlay.

Navigation also involves correction for ocean currents and obstacles such as landmasses and other vessels. Experienced seamen past and present, aided by experience, can predict current with reasonable accuracy. Wind affects currents, which can be estimated by the sailor or measured with an anemometer (wind gauge). Airplanes are affected by currents known as jet streams. Spacecraft are affected by gravitational forces, which increase in power as the craft approaches. That force can be used to change the direction of the craft substantially.

Background and History

Human migration and discovering new lands by navigation of the oceans was accomplished by many ancient cultures, including the Phoenicians, ancient Greeks, the Norse, the Persians, and the Polynesians. Primitive navigation depended on knowledge of ocean currents and the position of celestial objects.

The magnetic compass was first used in China around 200 B.C.E. The ancient compasses used a naturally magnetic lodestone, which pointed to the magnetic North Pole. In eighth-century China, the lodestone was replaced by a magnetized needle.

Navigation charts first appeared in Italy at the close of the thirteenth century. Called portolan charts, they were rough maps based on accounts of sailors plying the coastlines of the Mediterranean and Black seas. The octant, precursor to the sextant and developed around 1730, made it possible to determine latitude accurately but not longitude. The sextant added the ability to determine longitude. At the close of the nineteenth century, radios that transmitted and received Morse code began to appear on oceangoing vessels.

The prototype of the modern radar was installed on the USS *Leary* in 1937. In 1942, the first long-range navigation system (loran) was installed at various points along the Atlantic coast of the United States. Loran is based on using the intersection of two radio waves to determine one's position. The subsequent proliferation of satellites, after the Soviet Union launched Sputnik, the first Earth-orbiting satellite, in 1957, led to the development of satellite technology and the highly accurate GPS.

How It Works

Compasses. The magnetic compass needle points to magnetic north and floats over a 360-degree compass face, which displays the cardinal points: north (0 degree), east (90 degrees), south (180 degrees), and west (270 degrees). The deviation (declination) of magnetic north is not a constant and is updated annually. Nautical charts display the amount of needed correction from the baseline, which is the declination value of the year the chart was published. The magnetic compass is a simple, reliable instrument; however, its readings can be affected by any magnetic field in the vicinity. Marine electronics located near the compass generate magnetic fields; ferrous (iron) material often has a magnetic field. To improve

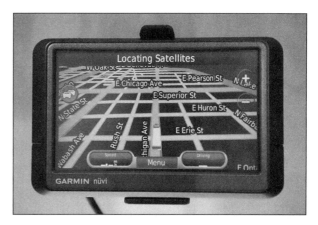

A Garmin Ltd. Nuvi 465T GPS navigation system sits on display during the 2011 International Consumer Electronics Show (CES) in Las Vegas, Nevada. (Bloomberg via Getty Images)

accuracy, the compass can be calibrated by small magnets or other devices to correct for extraneous magnetic fields. The compass must be recalibrated with the addition of any new electronics. A gyrocompass points to true north and is unaffected by stray magnetic fields. The gyrocompass employs a rapidly spinning gyroscope, which is motor driven. A disadvantage of the gyrocompass is that electrical or mechanical failure can render it useless.

Loran. Loran is an acronym for long-range navigation. The system relies on land-based low-frequency radio transmitters. The device calculates a ship's position by the time difference between the receipt of signals from two radio transmitters. The device can display a line of position, which can be plotted on a nautical chart. Most lorans convert the data into longitude and latitude. Since GPS became available, the use of loran has markedly declined.

Radar. Radar consists of a transmitter and receiver. The transmitter emits radio waves, which are deflected from a fixed or moving object. The receiver, which can be a dish or an antenna, receives the wave. Radar circuitry then displays an image of the object in real time. The screen displays the distance of the object from the radar. If the object is moving, consecutive readings can calculate the speed and direction of the object. If the object is airborne, and the radio is so equipped, the altitude is displayed. Radar is invaluable in foggy weather when visibility can be severely reduced.

Sonar. Sonar, which is an acronym for sound navigation ranging, transmits and receives sound waves for underwater navigation by submarines. Passive sonar is a related technology in which the equipment merely listens for underwater sound made by vessels. Active sonar emits a pulse of sound (a "ping") then listens for a reflection or echo of the pulse. The distance of the object is determined by the time difference from transmission to reception of the ping (the speed of sound in water is a constant). To measure the bearing, two or more separated transmitter/receivers are used for triangulation of the object. Sonar can also be used by both submarines and surface vessels to determine the depth and contour of the ocean (or other body of water). By consulting a nautical chart, which is marked with depth gradients, the distance from the shoreline can be calculated.

Global Positioning System (GPS). GPS is a space-based global navigation satellite system that provides accurate location and time information for any point on Earth where there is an unobstructed line of sight to four or more GPS satellites. GPS can function under any weather condition anywhere on the planet. It cannot function underwater—a submarine must surface to use a GPS system. The technology depends on triangulation, just as a land-based system such as loran employs. GPS is composed of three segments: the space segment, the control segment, and the user segment. The U.S. Air Force operates and maintains both the space and control segments. The space segment is made up of satellites, which are in medium-space orbit. The satellites broadcast signals from space, and a GPS receiver (user segment) uses these signals to calculate a three-dimensional location (latitude, longitude, and altitude). The signal transmits the time, accurate within nanoseconds.

Navigational Aids. According to the U.S. Coast Guard, a navigational aid is a device external to a vessel or aircraft specifically intended to assist navigators in determining their position or safe course, or to warn them of dangers or obstructions to navigation. Navigational aids include buoys, lighthouses, fog signals, and day beacons. Buoys are used worldwide to mark nautical channels. They are color coded: Red buoys mark the right (starboard) side of the channel for a vessel returning to a port; green buoys mark the port (left) side of the channel; and red-and-green-striped buoys mark the junction of two channels. Buoys often contain gongs, which are activated by wave motion and solar-powered lights. Some buoys do not mark safe channels and are referred to as "nonlateral markers." They contain shapes identifying their purpose. Squares depict information, such as food, fuel, and repairs. Diamonds warn of dangers such as rocks. Circles mark areas of reduced speed. Crossed diamonds indicate off-limits areas such as dams and places where people may be swimming.

APPLICATIONS AND PRODUCTS

Navigational aids have many applications, which are tailored to underwater, water surface, terrestrial, atmospheric, and space use.

Water-Surface Navigation. Any vessel, which is capable of venturing offshore, must contain some basic navigation equipment. A magnetic compass is mandatory, as is a marine radio, which can be used to call for help. A knot meter, which consists of an underwater paddle wheel connected to a dial, can give the relative speed over water. The vessel should contain nautical charts of the area, which are prepared by the National Oceanic and Atmospheric Administration (NOAA). Another essential is an accurate timepiece (chronometer). By combining the readings of speed registered on the knot meter, the compass heading, and chronometer, one can plot the vessel's approximate position on the nautical chart. All inboard marine engines are equipped (or can be equipped) with a tachometer, which registers engine speed in revolutions per minute (rpm). A prudent skipper can run his boat over a known distance (between measured mile markers) and construct a chart of the vessel's speed at various rpm. If the vessel does not have a knot meter, or it fails, the approximate boat speed can be determined via the tachometer. Another relatively inexpensive piece of equipment is a simple depth sounder, which displays depth (more expensive models display a graphic representation of the seabed). The depth sounder can aid in determining the vessel's position, and more importantly, alert the skipper to shallow water, which could damage the hull. Binoculars, which can help identify shoreline features, are another inexpensive necessity. The sextant, which can pinpoint one's position, is another inexpensive navigation aid; however, most recreational boaters are unfamiliar with its use.

Many recreational vessels of modest size (thirty-two to forty-two feet long) have an automatic pilot, radar, and GPS in addition to the aforementioned items. Some utilize a loran receiver, which may be less expensive than a GPS, and it can accurately display the ship's position. An automatic pilot adjusts the rudder(s) to maintain a desired heading. Radar can be adjusted for the distance it displays; ranges of less than one mile and more than thirty miles are commonplace. In many situations, the shortest range is the most valuable because it can clearly depict an approaching vessel or a small landmass (an islet or an offshore oil rig). Even in clear weather, radar can provide invaluable navigation information. Ship-mounted GPS devices can be purchased

---

Fascinating Facts About Navigation

- About 1,500 years ago, Tahitians began navigating back and forth between Tahiti and Hawaii in large, double-hulled canoes without the aid of a compass. Their techniques are not completely understood because they did not have a written language; however, their skills were based primarily on celestial navigation augmented by wave patterns, bird sightings, and even the light from active volcanoes on the Hawaiian Islands.
- At the time of his 1492 sailing, Christopher Columbus's maps showed that Japan was 2,700 miles from Europe on the other side of the Atlantic Ocean. Japan was actually 12,200 miles to the west, and North and South America are in the way.
- Native Americans were referred to as "Indians" because when Columbus first landed on the North American continent he thought that he had come ashore in India.
- In early 1942, long before laser guidance systems appeared, behavioral psychologist B. F. Skinner at the University of Minnesota investigated weapons guidance by trained pigeons. The birds were trained with slides of aerial photographs of the target, placed in a harness inside the guidance system, and kept the target in the crosshairs by pecking. Skinner never overcame official skepticism, and the project was eventually abandoned.
- The Mars Climate Orbiter, launched in December 1998, crashed on Mars nine months later because of a computational error. The orbiter's approach altitude to the planet was set at 160 kilometers, when, in fact, it was just under 60 kilometers. This is an example of human, not computer, error.

---

and installed for as little as $1,000 and are becoming commonplace on recreational vessels.

Large commercial and naval vessels (and many luxury yachts) contain sophisticated navigation equipment. Large vessels also boast onboard computers, which interpret data from radar, GPS, and the tachometers to display detailed navigation information. These computers also replace the automatic pilot found on smaller vessels. The computer can adjust the rudders and throttles to maintain the ship's heading and time of arrival. Depth sounders with graphic displays and sonar are components of most large vessels as well as fishing boats, which use them for locating schools of fish.

Underwater Navigation. Unlike surface vessels, submarines must navigate in three dimensions. Active sonar is employed by distance to a target and depth. Transmission of sound from two separate sources can give the submarine's heading. Submarines are equipped with GPS, which can be used to determine position only when the vessel is on the surface. When below the surface, an inertial navigation system (INS) is employed. The INS is composed of precise accelerometers and gyroscopes to record every change in the submarine's speed and direction. The data from these instruments is fed to a computer, which determines the vessel's position. Over time, small errors accrue; these errors are corrected when the submarine surfaces and the GPS is activated.

Terrestrial Navigation. When traveling on a road, a motorist can navigate using street and highway signs as well as a map of the area. Many motor vehicles are equipped with a GPS. The GPS gives a visual display of the vehicle's position on a map overlay. Useful information such as distance to the next turn and the destination are given. The GPS in most automobiles has an audio system, which informs the motorist of upcoming turns. The GPS of some vehicles also gives additional information such as gasoline stations, rest areas, local restaurants, and hospitals. GPS units for off-road use are also available. The motorist on a budget can purchase an inexpensive handheld GPS, devices that are also popular with campers and hikers. Foot travelers and off-roaders also rely on compasses for navigation. Celestial navigation with a sextant can also be utilized. Being able to locate Polaris (the North Star), which is the brightest star in the constellation Ursa Minor (the Little Dipper or Little Bear), is a necessary skill for all off-road travelers. The rising and setting sun or other visual aids such as the glow from a city at night can also aid in navigation.

Atmospheric Navigation. Aircraft navigate in three-dimensional space, so latitude, longitude, and elevation must be known. All aircraft contain basic instrumentation, which includes: a magnetic compass; an altimeter, which displays the altitude; an airspeed indicator; a heading indicator, which is similar to a gyrocompass; a turn indicator, which displays direction and rate of turn; a vertical-speed indicator, which displays the rate of climb or descent; and an attitude indicator. The attitude indicator, which is also known as an artificial horizon, is an invaluable navigation instrument. It contains a gyroscope that reflects both the horizontal and vertical alignment of the aircraft. When visibility is reduced, the pilot is unable to visualize the horizon. Many small aircraft and virtually all large aircraft contain a GPS. Other navigation equipment commonly found on larger aircraft include a course deviation indicator (CDI) and a radio magnetic indicator (RMI). However, a GPS is much superior to either a CDI or RMI. The CDI displays an aircraft's lateral position in relation to a track, which is provided by a very high frequency (VHF) omnidirectional range (VOR) or an instrument landing system. A VOR is a ground-based station that transmits a magnetic bearing of the ship from the station. An instrument landing system is a ground-based system that can provide precision guidance of the aircraft to a runway. It includes radio signals, and often, high-intensity lighting arrays. An RMI is usually coupled to an automatic direction finder (ADF); the RMI displays a bearing to a nondirectional beacon (NDB), which is a ground-based radio transmitter.

Unmanned aircraft (drones) and guided missiles must also be navigated. In the case of an unmanned drone, a pilot sits at a console on the ground and directs the flight via radio. Guided missiles are either internally or externally guided, sometimes both. Some missiles contain inertial navigation systems similar to those found on submarines. The system is programmed to strike a specified target. External control can be via radio waves or laser. In some cases of laser guidance, the missile contains a laser homing device in which the projectile "rides" the laser beam to the target. More commonly, a technique referred to as semi-active laser homing (SALH) is employed. With SALH, the laser beam is kept pointed at the target after the projectile is launched. Laser energy is scattered from the target, and as the missile approaches the target, heat sensors home in on this energy.

Space Navigation. Space navigation is a complex and highly technical science. In addition to involving three dimensions it also requires plotting a course between two moving objects (the Earth and a space station or an orbiting space shuttle and the Earth). Space navigation also entails calculating the gravitational force of celestial objects such as the Moon and planets. Space navigation requires the collaboration between computers (both ground and ship based) and complex instrumentation. It also requires collaborative effort between highly skilled personnel and either astronauts or navigation equipment within unmanned spacecraft.

IMPACT ON INDUSTRY

Navigation has a significant industrial impact. Navigation is required for mundane activities such as driving to the local market as well as more complex ones such as directing a space shuttle. It is a component of daily living. Although a market still exists for simple navigational aids such as magnetic compasses and nautical charts, the thrust of the market is focused on high-end, sophisticated electronics such as GPS. Since GPS was first developed for military use in 1978, it has spread into many civilian applications for navigation on the Earth's surface and the air. Electronics evolves rapidly, and new, improved products regularly reach the marketplace. Marine electronics is particularly vulnerable because it functions in an environment where corrosion from salt is an ongoing threat.

Industry and Business Sector. Although navigation equipment for owners of recreational boats and aircraft generates significant business, commercial and military applications represent the lion's share of the market. In the commercial sector, the civilian aircraft and shipping industries are major purchasers of navigation equipment. The National Aeronautics and Space Administration (NASA) requires extremely precise, complex, and expensive navigational equipment for its missions. All branches of the military—Army, Navy, Air Force, Marines, and Coast Guard—are major purchasers of navigation equipment. Many automobiles are equipped with a GPS or a compass. Although a compass is rudimentary when compared to a GPS, it can be helpful for finding one's way.

Handheld GPS devices are popular and inexpensive. They are used not only by pedestrians but also can be taken along in an automobile or boat.

Government and University Research. One of the biggest sources of funding for navigation research in the United States is the Defense Advanced Research Projects Agency (DARPA). It is also the biggest client for certain kinds of navigation applications. The agency is primarily concerned with navigation systems with a military focus, such as inertial navigation systems. It also conducts research on satellite navigation. Another source of funding for navigation research is the U.S. Department of Energy (DOE). The DOE is focused on energy efficiency and renewable energy. DARPA and the DOE supply funds to many universities in the United States for navigation research and development. As far back as 1959, DARPA began working with Johns Hopkins Applied Physics Laboratory to develop the first satellite positioning system.

CAREERS AND COURSE WORK

Any navigation course requires knowledge of trigonometry. Whether navigation is based on visual or electronic information, triangulation to fix a position is required. The military offers many careers and course work involving navigation. Following military service, this training can be applied to many civilian job opportunities, up to and including piloting an aircraft or navigating a ship. The U.S. Coast Guard Auxiliary and the U.S. Power Squadrons offer free boating-safety courses, which include navigational skills. Courses are available throughout the United States as well as online. For individuals interested in more advanced courses, such as ones for preparation for the U.S. Coast Guard captain's license, reasonably priced courses are available throughout the U.S. and online. For those not interested in a military career, navigation training is available at civilian institutions. These include California Maritime Academy in Vallejo; Great Lakes Maritime Academy in Traverse City, Michigan; Maine Maritime Academy in Castine; Massachusetts Maritime Academy in Buzzards Bay; State University of New York Maritime College in the Bronx; Texas Maritime Academy, which is part of Texas A&M University at Galveston; and the United States Merchant Marine Academy in Kings Point, New York.

Although the basic principles of navigation can be learned and applied by any high school graduate,

more advanced topics require a college degree and often a postgraduate degree such as a master of arts (M.A.) or doctorate (Ph.D.). An M.A. requires one year of study after four years of college; however, an academically aggressive student can earn an M.A. concurrently with a bachelor's degree. A Ph.D. requires two to three additional years of study followed by submission of a thesis or dissertation. Course work should include mathematics, engineering, computer science, and robotics. Positions are available for individuals with a degree in laser engineering in both the government and private sector. The ability to be a team player is often of value for these positions because ongoing research is often a collaborative effort.

SOCIAL CONTEXT AND FUTURE PROSPECTS

Although the basic concepts of navigation have remained unchanged for centuries, modern navigation technology is highly advanced and continues to evolve. The frontier of navigation research lies in military and extraterrestrial applications. The military is focused on guidance systems for missiles and drones. Extraterrestrial applications range from the navigation of Earth-orbiting shuttles to interplanetary (and beyond) navigation. Navigation is an essential component of daily life—both civilian and military.

Robin L. Wulffson, M.D., F.A.C.O.G.

FURTHER READING

Burch, David. *Emergency Navigation: Find Your Position and Shape Your Course at Sea Even if Your Instruments Fail.* 2d ed. New York: McGraw-Hill, 2008. Burch, a veteran sailor and the founder of the Starpath School of Navigation, provides all manner of how to navigate without electronics: celestial, wind, swells, and even using airliners' contrails.

Burns, Bob, and Mike Burns. *Wilderness Navigation: Finding Your Way Using Map, Compass, Altimeter and GPS.* 2d ed. Seattle: Mountaineers Books, 2004. An invaluable aid for the backpacker, hiker, and camper.

Cutler, Thomas J. *Dutton's Nautical Navigation.* 15th ed. Annapolis, Md.: Naval Institute Press, 2004. An essential textbook and reference for anyone venturing offshore in any size vessel.

Launer, Donald. *Navigation Through the Ages.* Dobbs Ferry, N.Y.: Sheridan House, 2009. Covers navigation history from antiquity to the present.

Lele, Ajey. *Strategic Technologies for the Military: Breaking New Frontiers.* Thousand Oaks, Calif.: Sage Publications, 2009. Describes the nuances of technological development in a purely scientific manner and provides a social perspective to the relevance of future warfare and issues such as disarmament and arms control and their impact on the environment.

WEB SITES

Boating Safety Resource Center
http://www.uscgboating.org/safety/boating_safety_courses_.aspx

National Aeronautics and Space Administration
http://www.nasa.gov

U.S. Department of Energy
http://www.energy.gov

U.S. Power Squadrons
http://www.usps.org

See also: Aeronautics and Aviation; Computer Science; Engineering; Geoinformatics; Maps and Mapping; Robotics; Sonar Technologies; Trigonometry.

NEPHROLOGY

Dialysis; transplantation; pediatric nephrology; proteomics; genetics; electrolyte physiology; hypertension; plasmapheresis; mineral metabolism; pharmacology; internal medicine; nephrolithiasis.

SUMMARY

Nephrology is a division of medical science associated with internal medicine, concentrated on the kidneys and the associated anatomy, physiology, diseases, and disorders. The word "nephrology" is of Greek origin: *nephros*, meaning kidney, and *logos*, meaning word, reason, thought, or discourse. Kidney diseases may include electrolyte-balance disorders and hypertension. Therapies associated with kidney disease manage limitations or failure of the renal system and may include dialysis or kidney transplant. Disorders of the kidney are often a result of systemic or congenital disorders, affecting more than one system or organ in the body, which can make treatment complicated and often challenging. A medical doctor who specializes in the diagnoses and treatment of the kidneys is a nephrologist.

KEY TERMS AND CONCEPTS

- **Calyx:** Cuplike urine-collection cavity located at the tip of each pyramid within the kidney.
- **Cortex:** Outer layer of the kidney that contains millions of microscopic nephrons.
- **Dialysis:** Process of removing toxic materials from the blood and maintaining fluid, electrolyte, and acid balance often using automated equipment when kidneys have become damaged or have been removed.
- **Kidney:** One of a pair of reddish brown, bean-shaped organs located on either side of vertebral column in the retroperitoneal area. Their function is to form urine from blood plasma by regulating water, electrolyte, and acid balance of the blood.
- **Nephrolithiasis:** Presence of calculi in the kidney, commonly called a kidney stone.
- **Nephron:** Functional unit of the kidney, consisting of a glomerulus, Bowman's capsule, renal tubule,

and peritubular capillaries. Urine is formed by nephrons by a process of filtration, reabsorption, and secretion.
- **Pyramid:** Triangular tissue in the medulla (the inner region) of the kidney, which contains the loops and collecting tubules of the nephrons.
- **Renal Artery:** Blood vessel that carries blood from the aorta into the kidney. Inside the kidney, the renal artery branches into kidney tissue until the smallest artery (arteriole) leads to a glomerulus, where filtration can begin.
- **Renal Vein:** Blood vessel that carries blood from the kidney, back to the heart.
- **Ureter:** Muscular tube with a mucosal lining that leads from each kidney to the urinary bladder.
- **Urine:** Fluid created by the kidney that consists of 95 percent water and 5 percent dissolved solids (salts and nitrogen-containing wastes), which is eliminated by the body.

DEFINITION AND BASIC PRINCIPLES

Nephrology is a branch of medical science, a specialty of internal medicine, concerned with the structure and function of the kidneys. Any pathology within this system and the management of many systemic diseases affecting the kidneys are key responsibilities of a nephrologist. A nephrologist will determine the stage or degree of kidney disease, treat any associated complications (hypertension, bone disorders, vitamin imbalances), manage anemia, educate the patient about nutrition, risk factors, treatments, and transplantation. The nephrologist also commonly provides the vascular access placement for dialysis and coordinates treatment.

The purpose of the kidneys is to regulate water, electrolytes, and acid-base content of the blood and, indirectly, all other body fluids. Filtration by the kidneys is a continuous process and the rate is affected by blood flow through the kidneys and daily fluid intake. Blood enters the kidney and passes through the glomerulus, where water and dissolved substances are filtered through capillary membranes, and the inner layer of Bowman's capsule, where it becomes glomerular filtrate. The blood cells and larger protein cells stay in the capillaries during this time. The filtrate travels through a series of renal tubules, where useful substances such as water, glucose, amino acid, minerals, and vitamins are reabsorbed into the capillaries to be used in the body. The amount of water that is reabsorbed is regulated by antidiuretic hormone and indirectly by aldosterone. The products that are not reabsorbed, such as metabolic products of medications, are considered waste and remain in the filtrate and become part of urine. The collecting tubules join together to form papillary ducts that empty urine into the calyxes and eventually into the ureter and bladder. The process from the ingestion of a large amount of liquid to the production of urine takes about forty-five minutes for most well-hydrated people. The average volume of urine produced each day is about 1,500 milliliters, but it depends on many factors—age, climate, activity, diet, and blood pressure. Patient pathology or disease may affect the urine production; however, people who have had part of a kidney removed or have only one kidney can have normal renal function.

There are five recognized stages of kidney disease.

- **Stage 1:** Slightly diminished function. Kidney damage with normal or relatively high glomerular filtration rate: >90 milliliters per minute per 1.73 m^2.
- **Stage 2:** Mild reduction in kidney function. Glomerular filtration rate: 60-89 milliliters per minute per 1.73 m^2.
- **Stage 3:** Moderate reduction in kidney function. Glomerular filtration rate: 30-59 milliliters per minute per 1.73 m^2.
- **Stage 4:** Severe reduction in kidney function. Glomerular filtration rate: 15-29 milliliters per minute per 1.73 m^2.
- **Stage 5:** Established kidney failure. Glomerular filtration rate: <15 milliliters per minute per 1.73 m^2.

BACKGROUND AND HISTORY

English physician Richard Bright first established the relationship between the symptoms and pathology of renal failure in 1827. It was not until 1854 when Scottish chemist Thomas Graham described osmotic force. In 1861, Graham went on to explain the process of dialysis using a hoop form dialyzer. In the late 1800's and early 1900's, several scientists began performing dialysis and kidney transplants on animals. Austrian surgeon Emerich Ullmann performed a kidney autotransplant on a dog, and in 1914, pharmacologist John Jacob Abel and his colleagues

Leonard Rowntree and Benjamin Turner discovered that salicylic acid could be removed from the blood of rabbits using dialysis. In 1924, German physician George Haas performed the very first dialysis procedure on a human. Dutch physician Willem Kolff treated sixteen patients with acute kidney failure between 1943 and 1944 but with limited success. The first success came in 1945 with the seventeenth patient, a sixty-seven-year-old woman in uremic coma due to acute renal failure from gram-negative sepsis. After eleven hours of hemodialysis, the patient regained consciousness and began to produce urine. She went on to live seven more years.

The Scribner shunt, a U-shaped Teflon tube, is the creation of Chicago-born Belding Scribner. The shunt, inserted between an artery and vein in a patient's forearm, could be opened and connected to the artificial kidney machine during dialysis. Teflon was relatively new to the biomedical community at the time, and its nonstick properties made it less likely to clot. Before Scribner's shunt, a patient could receive only a few dialysis treatments before doctors would run out of places to connect the machine to the patient. The shunt was first used on March 9, 1960, on Clyde Shields, who was dialyzed repeatedly for eleven years. Another patient was dialyzed for thirty-six years, undergoing 5,700 cycles of hemodialysis, before his death. In 1962, Scribner and American physician James Haviland developed the first free-standing dialysis center in the world, the three-bed Seattle Artificial Kidney Center.

HOW IT WORKS

Nephrology is the science that concentrates on the kidneys and the associated anatomy, physiology, diseases, and disorders. The improper functioning of the kidney may disrupt electrolyte balance or lead to hypertension (high blood pressure) and is often related to other systemic or congenital conditions.

Diagnosing Kidney Disease. Treating the underlying condition may be complicated by reduced kidney function, and treatment of the kidney may be a challenge because of the underlying condition. Deteriorating kidney function has very unspecific symptoms, which makes kidney disease difficult to diagnose. Patients may feel generally unwell or have a reduced appetite. Quite often, kidney disease is not recognized until a major complication such as anemia, pericarditis, or cardiovascular disease has

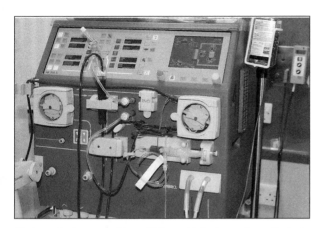

Kidney dialysis machine. Kidney dialysis, or hemodialysis, is required when the kidneys are unable to filter waste products from the blood. (Life in View/Photo Researchers, Inc.)

been detected. People diagnosed with high blood pressure or diabetes often have their kidney function assessed as part of normal screening procedure.

Chronic Kidney Disease Detection. Chronic kidney disease is detected by a blood analysis for levels of creatinine. Higher levels of creatinine indicate a decreased glomerular filtration rate resulting in a decline in normal kidney function. A glomerular filtration rate of less than 60 milliliters per minute per 1.73 m², for a period of three months, is classified as having chronic kidney disease. Red blood cells or excess protein detected in urinalysis may cause a physician to investigate more thoroughly.

Kidney Stone Detection. Patients may develop nephrolithiasis or renal calculus, also known as a kidney stone, which is a solid mass normally composed of mineral salts. In the kidney, a calculus can block the ureter and urine flow. Symptoms may include a severe, sudden pain, chills, fever, and appearance of blood in the urine. If the blockage cannot be passed on its own by relaxation of surrounding smooth muscle, it should be removed surgically or disintegrated ultrasonically.

APPLICATIONS AND PRODUCTS

Dialysis, the common name for hemodialysis, is the procedure used to treat end-stage kidney failure, transient kidney failure, and some poisoning or drug-overdose situations. Other indications for dialysis include: hyperkalemia, uremia, uremic pericarditis,

acidosis, and fluid overload. It is used to act as an artificial kidney, outside the body when a disorder is causing fluid, acids, electrolytes, and some drugs from being effectively eliminated.

Dialysis. Dialysis involves a series of five steps:

- Establishment of access to the patient's circulatory system via an arteriovenous fistula, graft, or catheter.
- Anticoagulating the patient's blood in order to prevent clotting during its circulation outside the body.
- Pumping of the patient's blood to a dialysis membrane.
- Adjusting the diffusion of solutes from the blood into a buffered dialysis solution.
- Returning the cleaned and buffered blood to the patient's circulation.

Typically, the entire process takes between three and four hours to complete and must occur several times per week. Even with regular dialysis, mortality rates remain quite high for end-stage renal disease patients. Most deaths are attributed to stroke, heart disease, or complications from diabetes. Complications of dialysis include hypotension (low blood pressure), infection at access site, sepsis, air embolism, bleeding, anemia, and muscle cramping.

Dialysis is the only treatment option for most renal disease patients, and the procedure that has not changed much since it was first performed. Unfortunately, during the days off of treatment, patients experience toxin buildup leaving them feeling bloated, tired, and uncomfortable.

There are two common types of dialysis, the most common being hemodialysis. Hemodialysis removes wastes and water by circulating blood outside the body through an external dialyzer that acts as a filter and contains a semipermeable membrane. Peritoneal dialysis uses a peritoneal membrane inside the body to filter wastes and water from the blood.

Home Dialysis Machines. In an effort to meet the growing need for dialysis treatment and provide some convenience to patients, home-treatment machines were created. The first home-treatment units were very large, very expensive, and cleaning and maintenance were difficult, but things are improving. Modern home dialysis machines are portable, and after rigorous training sessions, patients can treat themselves at home in six shorter sessions per week, rather than in several three- or four-hour sessions at a facility. Only about 1 percent of dialysis patients are being treated at home. Machines cost almost $25,000, are the size of a microwave, and perform the sophisticated filtration using a disposable cartridge.

Anemia Treatments. Anemia is an independent predictor of mortality in chronic kidney disease patients and is also associated with worsening of cardiovascular morbidity and accelerated rate of kidney damage. The administration of recombinant human erythropoietin (rHuEpo) has greatly reduced anemia in patients with chronic kidney disease.

Unfortunately, almost 15 percent of patients show limited or no response to rHuEpo.

Genomic and Proteomic Research. Genomics (the study of genomes or DNA sequence of organisms) and proteomics (the study of structure and function of proteins) in nephrology are still in the early phases of research. The application of these approaches has recently produced promising new urinary biomarkers for kidney injury and chronic kidney disease, which may provide better understanding of renal physiology and assist in the development of new therapeutic strategies.

IMPACT ON INDUSTRY

Industry has stepped in to help manage the growing number of patients requiring dialysis treatment for end-stage renal disease. More than 70 percent of patients receiving dialysis in the United States are doing so at corporate-owned facilities. Many companies are public, and concern has been raised that decisions regarding patient health are managed based on corporate policy, rather than the physician's advice. Some physicians act as health care providers in these facilities as well as medical director, which may lead to complex conflicts of interest regarding patient care and corporate responsibility. United States Renal Data System database requires that physicians provide information regarding their role in a facility and patient outcomes. Research is being done to investigate the trends of corporate involvement in dialysis treatment when compared with government or hospital facilities with regard to patient outcomes, costs, and transplantation referral rates.

A recent study concluded that kidney patients can improve their health by undergoing twice as many dialysis sessions. Industry has developed options for patients and health care providers to explore different ways to provide additional treatment while not overwhelming the health care system or the time commitment of the patient. Home-dialysis equipment makers are developing and seeking Food and Drug Administration (FDA) approval for home-based devices and programs to keep up with the rising demand for treatment.

CAREERS AND COURSE WORK

The most well-known career choice in nephrology is a nephrologist, which is a medical doctor (M.D.) who goes on to specialize in diseases and disorders of

---

Fascinating Facts About Nephrology

- The World Health Organization (WHO) reports that more than 68,300 kidney transplants are performed every year worldwide.
- Kidney disease affects more than 600 million people globally.
- The 1990 Nobel Prize in Physiology or Medicine was awarded to Joseph E. Murray, the surgeon who performed the first-ever kidney transplant between identical twins, which demonstrated that previous failures were due to immunologic incompatibilities rather than surgical methods.
- The average kidney is about 4.5 inches long and weighs between four and six ounces.
- The kidneys have a higher blood flow than the heart, liver, or brain.
- World Kidney Day started in 2006 and is now an annual event held every March in more than one hundred countries.
- Most people are born with two kidneys but can survive with a single healthy kidney.
- The Voluntary Health Association of India estimates that about 2,000 Indians sell a kidney every year.
- The largest reported kidney stone weighed just more than thirty-one pounds.
- The United States has one of the highest incidences of end-stage renal disease in the world, 363 per million people, compared with Iceland, which has less than 60 per million.
- Chinese medicine believes that kidney stones may be caused by blockage or imbalance of chi (vital energy) in the kidney and acupuncture can restore positive energy flow.

---

the kidney as a subspecialty of internal medicine. Becoming a nephrologist in the United States requires the completion of a bachelor's degree, four years of medical school, a three-year residency in internal medicine followed by an additional two-year fellowship in nephrology.

Nephrology social workers provide support and maximize the psychosocial functioning and adjustment of patients who are experiencing end-stage renal disease. They assist patients with the management of social and emotional stresses, which may include shortened life expectancy; altered lifestyle with changes in social, financial, vocational, and sexual functioning; and the complex, rigorous and time-consuming treatment required.

Dialysis technicians, also referred to as renal dialysis technicians, hemodialysis technicians, or nephrology technicians, operate the dialysis machines in hospitals and clinics, under the supervision of physicians. They may also travel to a patient's home to provide home treatment. Dialysis technicians must have a high school diploma or GED and have completed an additional training program at a technical school, community college, or hospital.

Renal scientists normally hold a master's or doctoral degree and have conducted years of research, as well as written and defended a thesis. Much of their work is funded by research grants from government or foundation sources or by academic institutions. Renal scientists normally find work in academic institutions, biotechnology laboratories, or government agencies. Renal scientists often collaborate with nephrologists to bring new findings and technology to the clinical phase of development.

SOCIAL CONTEXT AND FUTURE PROSPECTS

Patients requiring nephrology treatment are increasing at an alarming rate. In 2008, there were more than 1.64 million people on dialysis. This growth in end-stage renal disease patients is five times the world population growth (1.3 percent). Diabetes alone is expected to grow by 165 percent by 2050; therefore it is estimated that one out of every eight United States residents will have some level of kidney disease. Improving the facilitation of dialysis treatment to patients is a key concern as well as managing the growing number of people requiring treatment. The average cost to treat each patient with end-stage renal disease is about $85,000 annually.

Research has found that daily treatment provides better outcomes for patients. These findings are not accommodated by Medicare coverage, which, on average, pays for three treatments per week. Home-dialysis programs are being established to meet the growing need for treatment and allow patients the flexibility to accommodate treatment into their lives. Patients receive training on how to operate the machines as well as basic dialysis knowledge.

About one-third of patients who have a friend or family member willing to become a living kidney donor will not be able to receive the donor kidney because of an incompatible blood type or cross match. Kidney exchange programs are becoming more popular and are being facilitated by major medical centers and health providers. A kidney exchange increases the pool of available donors through an exchange between incompatible donor-recipient pairs, or a donor chain, managed by a specialized computer program that matches donors and recipients.

April D. Ingram, B.Sc.

FURTHER READING

Goldsmith, David, Satish Jayawardene, and Penny Ackland, eds. *ABC of Kidney Disease.* Malden, Mass.: Blackwell Publishing, 2007. This book contains excellent illustrations and clearly written information to provide a greater understanding of renal disease.

Greenberg, Arthur, ed. *Primer on Kidney Diseases.* Philadelphia: Saunders Elsevier, 2009. This book is endorsed by the National Kidney Foundation and provides very good background and basics of renal anatomy and physiology. Defines and classifies common kidney disorders and outlines treatment protocols and modalities.

Jörres, Achim, Claudio Ronco, and John A. Kellum, eds. *Management of Acute Kidney Problems.* Berlin: Springer-Verlag, 2010. This excellent reference contains information regarding the definition, epidemiology, pathophysiology, and clinical causes of acute kidney failure.

Lai, Kar Neng, ed. *Practical Manual of Renal Medicine: Nephrology, Dialysis and Transplantation.* Hackensack, N.J.: World Scientific, 2009. This manual provides practical information about dialysis, transplantation, and general nephrology in straightforward language. It describes treatment rationale and kidney disease management.

Schrier, Robert W. *Manual of Nephrology*. 7th ed. Philadelphia: Lippincott, Williams & Wilkins, 2009. An excellent clinical reference guide for the advanced student.

WEB SITES
American Board of Internal Medicine
http://www.abim.org

American Society of Nephrology
http://www.asn-online.org

International Society of Nephrology
http://www.isn-online.org

National Institute of Diabetes and Digestive and Kidney Diseases
http://www2.niddk.nih.gov

See also: Artificial Organs; Pharmacology; Proteomics and Protein Engineering; Surgery.